Lecture Notes in Computer Science 3796

Commenced Publication in 1973
Founding and Former Series Editors:
Gerhard Goos, Juris Hartmanis, and Jan van Leeuwen

Nigel P. Smart (Ed.)

Cryptography and Coding

10th IMA International Conference
Cirencester, UK, December 19-21, 2005
Proceedings

 Springer

Volume Editor

Nigel P. Smart
University of Bristol
Department of Computer Science
Woodland Road, Bristol, BS8 1UB, UK
E-mail: nigel@cs.bris.ac.uk

Library of Congress Control Number: 2005936359

CR Subject Classification (1998): E.3-4, G.2.1, C.2, J.1

ISSN   0302-9743
ISBN-10   3-540-30276-X Springer Berlin Heidelberg New York
ISBN-13   978-3-540-30276-6 Springer Berlin Heidelberg New York

Springer is a part of Springer Science+Business Media

springeronline.com

© Springer-Verlag Berlin Heidelberg 2005
Printed in Germany

Typesetting: Camera-ready by author, data conversion by Scientific Publishing Services, Chennai, India
Printed on acid-free paper   SPIN: 11586821   06/3142   5 4 3 2 1 0

Preface

The 10th in the series of IMA Conferences on Cryptography and Coding was held at the Royal Agricultural College, Cirencester, during 19–21 December 2005. As usual, the venue provided a relaxed and informal atmosphere for attendees to discuss work and listen to the collection of talks.

The program consisted of four invited talks and 26 contributed talks. The invited talks where given by Tuvi Etzion, Ueli Maurer, Alfred Menezes and Amin Shokrollahi, and three of these invited talks appear as papers in this volume. Special thanks must go to these four speakers as they helped to set the tone, by covering all the areas the meeting aimed to cover, from cryptography through to coding. In addition the best speakers are often the hardest to persuade to come to a meeting, as they are usually the most busy. We therefore feel privileged to have had a meeting with four such distinguished speakers.

The contributed talks were selected from 94 submissions. This is nearly twice the number of submissions for the previous meeting in 2003. This is an indication of the strength of the subject and the interest in the IMA series of meetings as a venue to present new work. The contributed talks ranged over a wide number of areas, including information theory, coding theory, number theory and asymmetric and symmetric cryptography. Subtopics included a number of current "hot topics," such as algebraic cryptanalysis and cryptographic systems based on bilinear pairings.

Assembling the conference program and these proceedings required the help of a large number of individuals. I would like to thank them all here.

Firstly, thanks must go to the Program Committee and their subreferees who aided in evaluating the contributed papers and coming to a difficult decision as to what to include and what to exclude. We had to reject a number of papers simply due to lack of space.

For the first time the IMA Program Committee used the WebReview software from K.U. Leuven. This is an excellent program, which greatly helped myself as Program Chair in collating and mediating the referee reports. Thanks must go to the authors and maintainers of this program. Particular thanks must also go to Richard Noad, one of my Research Assistants at Bristol, who dealt with all the technical issues with mounting the servers and generally acted as a right-hand-man throughout the process.

Thanks must also go to the authors of the submitted papers, and in particular to those whose papers were accepted. The authors of the accepted papers co-operated in compiling this volume, often meeting very tight deadlines imposed by the publication schedule. Thanks must also go to the staff of Springer, in particular Alfred Hofmann who helped in a large number of ways.

Valuable sponsorship of the meeting was provided by Hewlett-Packard Laboratories and Vodafone. We thank them for their contributions.

Finally, thanks must go to the staff of the IMA, in particular Pamela Bye and Lucy Nye, who dealt with all the day-to-day issues and allowed the Program Committee to concentrate on the science. A conference such as this could not take place without their help.

September 2005   Nigel Smart
                                                                    Program Chair

Cryptography and Coding 2005

December 19–21, 2005, Cirencester, United Kingdom

Sponsored by the
The Institute of Mathematics and its Applications (IMA)
in cooperation with
Hewlett-Packard Laboratories and *Vodafone Ltd.*

Program Chair

Nigel Smart .... University of Bristol

Program Committee

Steve Babbage.................................. Vodafone Group Services Ltd.
Bahram Honary .... University of Lancaster
Steven Galbraith .... Royal Holloway, University of London
Chris Mitchell .... Royal Holloway, University of London
David Naccache .... École Normale Supérieure
Matthew Parker .... University of Bergen
Kenny Paterson .... Royal Holloway, University of London
Ana Salagean .... Loughborough University
Frederik Vercauteren .... K.U. Leuven
Mike Walker .... Vodafone Group Services Ltd.
Gilles Zemor .... ENST, Paris

External Referees

Claude Barral
Eric Brier
Alister Burr
Julien Cathalo
Benoit Chevallier-Mames
Carlos Cid
Mathieu Ciet
Gerard Cohen
Nicolas Courtois
Alex Dent
Jean-François Dhem
Eran Edirisinghe

Nicolas Gresset
Helena Handschuh
Marc Joye
Caroline Kudla
Arjen Lenstra
John Malone-Lee
Keith Martin
Philippe Martins
Jean Monnerat
David M'Raihi
Gregory Neven
Katie O'Brien

Dan Page
Havard Raadum
Vincent Rijmen
Jasper Scholten
Martijn Stam
Emmanuel Thomé
Claire Whelan
Andreas Winter
Christopher Wolf

Table of Contents

Invited Papers

Coding Theory

Signatures and Signcryption

Symmetric Cryptography

Side Channels

Algebraic Cryptanalysis

Information Theoretic Applications

Number Theoretic Foundations

Public Key and ID-Based Encryption Schemes

Abstract Models of Computation in Cryptography

Ueli Maurer[*]

Department of Computer Science,
ETH Zurich, CH-8092 Zurich, Switzerland
`maurer@inf.ethz.ch`

Abstract. Computational security proofs in cryptography, without unproven intractability assumptions, exist today only if one restricts the
computational model. For example, one can prove a lower bound on the
complexity of computing discrete logarithms in a cyclic group if one considers only generic algorithms which can not exploit the properties of the
representation of the group elements.

We propose an abstract model of computation which allows to capture
such reasonable restrictions on the power of algorithms. The algorithm
interacts with a black-box with hidden internal state variables which allows to perform a certain set of operations on the internal state variables,
and which provides output only by allowing to check whether some state
variables satisfy certain relations. For example, generic algorithms correspond to the special case where only the equality relation, and possibly
also an abstract total order relation, can be tested.

We consider several instantiation of the model and different types of
computational problems and prove a few known and new lower bounds
for computational problems of interest in cryptography, for example that
computing discrete logarithms is generically hard even if an oracle for
the decisional Diffie-Hellman problem and/or other low degree relations
were available.

1 Introduction and Motivation

1.1 Restricted Models of Computation

Proving the security of a certain cryptographic system means to prove a lower
bound on the hardness of a certain computational problem. Unfortunately, for
general models of computation no useful lower bound proofs are known, and it is
therefore interesting to investigate reasonably restricted models of computation
if one can prove relevant lower bounds for them.

In a restricted model one assumes that only certain types of operations are
allowed. For example, in the monotone circuit model one assumes that the circuit
performing the computation consists only of AND-gates and OR-gates, excluding
NOT-gates. Such a restriction is uninteresting from a cryptographic viewpoint
since it is obvious that an adversary can of course perform NOT-operations.

---

[*] Supported in part by the Swiss National Science Foundation.

N.P. Smart (Ed.): Cryptography and Coding 2005, LNCS 3796, pp. 1–12, 2005.

Nevertheless, some restricted models are indeed meaningful in cryptography, for example the generic model which assumes that the properties of the representation of the elements of the algebraic structure (e.g. a group) under consideration can not be exploited. In view of the fact that for some problems, for example the discrete logarithm problem on general elliptic curves, exploiting the representation is not known to be of any help and hence generic algorithms are the best known, such an assumption is reasonable from a practical viewpoint.[1]

The purpose of this paper is to provide a simple framework for such restricted models of computation and to prove some lower bounds. Generic algorithms are the simplest case. Some of the presented results are interpretations and generalizations of previous results, for instance of [10] and [4].

1.2   Generic Algorithms and Computing Discrete Logarithms

In order to compute with the elements of a set S (e.g. a group), one must represent the elements as bitstrings (without loss of generality). A representation is a bijective mapping from S to the set of bitstrings. A generic algorithm works independently of the representation. The term generic means that one can not exploit non-trivial properties of the representation of the elements, except for two generic properties that any representation has. First, one can test equality of elements, and second one can impose a total order relation $\preceq$ on any representation, for example the usual lexicographic order relation on the set of bitstrings. However, one can generally not assume that the representation is dense or satisfies any regularity or randomness condition.

In order to motivate the model to be introduced, we briefly discuss generic algorithms for computing discrete algorithms in a cyclic group G. A cyclic group G of order n, generated by a generator g, is isomorphic to the additive group $\mathbf{Z}_n$. A generic algorithm for computing the discrete logarithm (DL) x of an element $b = g^x$ to the base g in G can be modeled as follows. The algorithm is given a black-box which contains x. It can also input constants into the box[2] and add values in the box. The only information reported back from the box is when an equality (collision) between two computed elements occurs. The algorithm's task is to extract x by provoking collisions and computing x from the collision pattern. The order relation allows to establish ordered tables of the generated values and thus reduces the number of equality tests required, but it does not allow to reduce the number of computed values and is ignored in most of the following.

If one is interested in proving a lower bound on the number of operations for any generic algorithm, then one can consider the simpler objective of only provoking a *single* collision and that all equalities of elements are reported for free. Since only additions and the insertion of constants are allowed, every value computed in the box is of the form $ax + b$ (modulo n) for known values a and

---

[1] In contrast, for computing discrete logarithms in Z_p^* for a prime p, quite sophisticated algorithms are known (e.g. index calculus) which exploit that the elements are integers that can be factored into primes.

[2] One can also assume that the box contains only 1 and x initially and constants must be computed explicitly from 1 by an addition-and-doubling algorithm.

b. For uniform x the probability that two such values $ax + b$ and $a'x + b'$ collide is easily seen to be at most $1/q$, where q is the largest prime factor of n. Hence the total probability of provoking a collision is upper bounded by $\binom{k}{2}/q$ and therefore the running time of any algorithm with constant success probability is at least $O(\sqrt{q})$.

The simplest non-trivial generic DL algorithm is the so-called baby-step giant-step algorithm with complexity $O(\sqrt{n}\log n)$. It need not know the group order n, an upper bound on n suffices, and it is the best known algorithm when the group order is unknown. The Pohlig-Hellman algorithm [7] is also generic and a bit more sophisticated. It makes use of the prime factorization of n and has complexity $O(\sqrt{q}\log q)$, which is essentially optimal.

1.3 Discussion and Generalization of the Model

This view of a generic algorithm appears to be simpler than the model usually considered in the literature, introduced by Shoup [10], where one assumes that access to group elements is via a randomly selected representation. This complicates the random experiment in which the algorithm's success probability is to be analyzed. Also, in a realistic setting one has no guarantee that the representation corresponds in any way to a random mapping.

As a generalization of the described approach, one can also model that one can exploit certain additional information from the representation of the elements, for instance that one can test certain relations efficiently. As an example, one can imagine that one can efficiently test for any three elements x, y and z whether $xy = z$, which corresponds to assuming the availability of a decisional Diffie-Hellman (DDH) oracle. For this setting one can still prove an $O(\sqrt[3]{q})$ lower bound for the discrete logarithm problem.

2 An Abstract Model of Computation

2.1 The Model

We consider an abstract model of computation characterized by a black-box **B** which can store values from a certain set S (e.g. a group) in internal state variables $V_1, V_2, \ldots, V_m$. The storage capacity m can be finite or unbounded.

The initial state consists of the values of $V^d := [V_1, \ldots, V_d]$ (for some $d < m$, usually d is 1, 2, or 3), which are set according to some probability distribution P_{V^d} (e.g. the uniform distribution).

The black-box **B** allows two types of operations, computation operations on internal state variables and queries about the internal state. No other interaction with **B** is possible.[3] We give a more formal description of these operations:

---

[3] This model captures two aspects of a restricted model of computation. The computation operations describe the types of computations the black-box can perform, and the state queries allow to model precisely how limited information about the representation of elements in S can be used. A quantum computer is another type of device where only partial information about the state can be obtained, but it could not be captured in our model.

- **Computation operations.** For a set Π of operations on S of some arities (nullary, unary, binary, or higher arity), a computation operations consist of selecting an operation $f \in \Pi$ (say t-ary) as well as the indices $i_1, \ldots, i_{t+1} \le m$ of $t+1$ state variables.[4] **B** computes $f(V_{i_1}, \ldots, V_{i_t})$ and stores the result in $V_{i_{t+1}}$.[5]
- **Queries.** For a set Σ of relations (of some arities) on S, a query consist of selecting a relation $\rho \in \Sigma$ (say t-ary) as well as the indices $i_1, \ldots, i_t \le m$ of t state variables. The query is replied by $\rho(V_{i_1}, \ldots, V_{i_t})$.

A black-box **B** is thus characterized by S, Π, Σ, m, and d. As mentioned above, one can include an abstract total order relation $\preceq$.

2.2   Three Types of Problems

We consider three types of problems for such black-boxes, where the problem instance is encoded into the initial state of the device.

- **Extraction:** Extract the initial value x of V_1 (where $d = 1$).[6]
- **Computation:** Compute a function $f : S^d \to S$ of the initial state within **B**, i.e., one must achieve $V_i = f(x_1, \ldots, x_d)$ for some (known) i, where $x_1, \ldots, x_d$ are the initial values of the state variables $V_1, \ldots, V_d$.
- **Distinction:** Distinguish two black-boxes **B** and **B'** of the same type with different distributions of the initial state V^d.

An algorithm for solving one of these problems is typically assumed to be computationally unbounded, but it is restricted in terms of the number k of interactions with the black-box it can perform. The memory capacity m can also be seen as a parameter of the algorithm.

One is often only interested in the computation queries, especially when proving lower bounds, and can then assume that, for every (say t-ary) relation $\rho \in \Sigma$, **B** provides all lists $(i_1, \ldots, i_t)$ such that $\rho(u_{i_1}, \ldots, u_{i_t})$ for free. We prove lower bounds in this model.

The success probability of an algorithm is taken over the choice of the initial state $V_1, \ldots, V_d$ and the (possible) randomness of the algorithm. The advantage of a distinguisher is defined as usual.

3   Concrete Settings

In this section we consider a few concrete instantiations of the model which are of interest in cryptography.

---

[4] This information is the input to **B**.

[5] A special case are constant functions, i.e., the operation of setting an internal state variable V_i to a particular value $c \in S$. If m is unbounded, then one can assume without loss of generality that each new result is stored in the next free state variable.

[6] More generally, one could consider the problem of extracting more general information about the initial state. This can be formalized by a function $g : S^d \to \mathcal{Q}$ for some $\mathcal{Q}$, where the task is to guess $g(V_1, \ldots, V_d)$.

3.1   Notation

We introduce some notation. Let $\mathcal{C}$ denote the set of constant (nullary) operations, which correspond to inserting a constant into the black-box. For a ring
S, let $\mathcal{L}$ denote the set of linear functions (of the form $a_1 V_1 + \cdots + a_d V_d$) on
the initial state V^d. For a multiplicatively written operation (e.g. of a ring) S,
let *square* denote the binary relation $\{(x, y) : y = x^2\}$, let *power*(e) denote
$\{(x, y) : y = x^e\}$, and let *prod* denote the ternary relation $\{(x, y, z) : z = xy\}$.

For a given set Π of operations, let $\overline{\Pi}$ be the set of functions on the initial
state that can be computed using operations in Π.

3.2   Extraction Problems with Constant and Unary Operations

The simplest case of an extraction problem to consider is when $\Pi = \mathcal{C}$ and
$\Sigma = \{=\}$, i.e., one can only input constants and check equality.[7] It is trivial
that the best strategy for the extraction problem is to randomly guess, and the
success probability of any k-step algorithm is bounded by $k/|S|$, i.e., the com-
plexity for achieving a constant success probability is $O(|S|)$. This bound holds
independently of whether one counts equality checks or whether one assumes a
total order $\preceq$ on S. This bound is trivially achievable with constant memory m.

If one would also allow to check a more general relation than equality (i.e.,
$\Sigma = \{=, \rho\}$ for some ρ), then better algorithms may exist. But the above upper
bound generalizes easily to $kd/|S|$, where

$$d = \max_{u \in S} |\{v \in S : u\rho v \vee v\rho u\}|$$

is the maximal vertex degree of the relation graph. Note that $d = 1$ for the
equality relation. If d is large, there can exist efficient algorithms. For example,
if $\Sigma = \{=, \leq\}$ and S is totally ordered by the relation $\leq$, then one can use the
binary search algorithm with running time $O(\log |S|)$, which is optimal.[8] It may
be interesting to consider other relations.

We return to the case $\Sigma = \{=\}$ but now allow some unary operations.

Theorem 1. *Let $\star$ be a group operation on S, let $\Pi = \mathcal{C} \cup \{x \mapsto x \star a \mid a \in S\}$
consist of all constant functions and multiplications by constants, and let $\Sigma = \{=\}$. The success probability of every k-step algorithm for extraction is upper
bounded by $\frac{1}{4}k^2/|S|$, and by $km/|S|$ if m is bounded.*

Proof. We use three simple general arguments which will be reused implicitly
later. First, we assume that as soon as some collision occurs (more generally, some
relation in Σ is satisfied for some state variables) in the black-box, the algorithm

---

[7] This corresponds to a card game where one has to find a particular card among n
cards and the only allowed operation is to lift a card, one at a time.

[8] Note that the previously discussed order relation $\preceq$ can not be used to perform
a binary search because it is not known explicitly, but only accessible through an
oracle.

is successful.[9] One can therefore concentrate on algorithms for provoking some collision by computing an appropriate set of values in the black-box.

Second, we observe, as a consequence of Lemma 2 in Appendix B, that if the only goal is to provoke a deviation of a system from a fixed behavior (namely that it reports no collisions), then adaptive strategies are not more powerful than non-adaptive ones.

Third, for lower-bound proofs we can assume that an algorithm can not only perform operations in Π but can, in every step, compute a function in $\overline{\Pi}$ (of the initial state V^d). This can only improve the algorithm's power. Without loss of generality we can assume that only distinct functions are chosen by the algorithm.

In the setting under consideration, the composition of two operations in Π is again in Π, i.e., $\overline{\Pi} = \Pi$. For all $x \in S$ and distinct a and b we have $x \star a \neq x \star b$. Thus collisions can occur only between operations of the form $x \mapsto x \star a$ and constant operations. Let u and v be the corresponding number of operations the algorithm performs, respectively. Then the probability of a collision is upper bounded by $uv/|S|$. The optimal choice is $u = v \approx k/2$, which proves the first claim.

If m is finite, then in each of the k steps the number of potential collisions is at most $m - 1$. The total number of x for which any of these collisions can occur is at most $k(m - 1)$.   △

The implied lower bound $k = O(\sqrt{n})$ for constant success probability can essentially be achieved even by only allowing a certain single unary operation, for example increments by 1 when $S = \mathbf{Z}_n$, i.e., $\Pi = \mathcal{C} \cup \{x \mapsto x + 1\}$. This is the abstraction of the baby-step giant-step (BSGS) algorithm: One inserts equidistant constants with gap $t \approx \sqrt{n}$ and increments the secret value x until a collision with one of these values occurs. If one considers a total order relation $\preceq$ one can generate a sorted table of stored values.[10]

3.3   The Group $\{0, 1\}^\ell$

We consider the group $\{0, 1\}^\ell$ with bit-wise XOR (denoted $\oplus$) as the group operation. As an application of Theorem 1 we have:

Corollary 1. *For $S = \{0, 1\}^\ell$, $\Pi = \mathcal{C} \cup \{\oplus\}$ and $\Sigma = \{=\}$ the success probability of every k-step extraction algorithm is upper bounded by $\frac{1}{4}k^2 2^{-\ell}$.*

Proof. Any sequence of operations is equivalent wither to a constant function or the addition of a constant, i.e., the set $\overline{\Pi}$ of computable functions is $\overline{\Pi} = \mathcal{C} \cup \{x \oplus a \mid a \in \{0, 1\}^\ell\}$. Hence we can apply Theorem 1.   △

---

[9] Phrased pictorially, we assume a genie who provides x for free when any collision occurs.

[10] Note that the BSGS algorithm can also be stated as an algorithm for a group with group operation $\star$, where $\Pi = \{1, \star\}$, $\Sigma = \{=, \preceq\}$, and the addition operation is needed to compute other constants from the constant 1.

It is easy to give an algorithm essentially matching the lower bound of $O(2^{\ell/2})$ implied by the above corollary.

3.4   Discrete Logarithms in Cyclic Groups

We now consider the additive group $\mathbf{Z}_n$. The extraction problem corresponds to the discrete logarithm (DL) problem for a cyclic group of order n.[11]

In the sequel, let p and q denote the smallest and largest prime factor of n, respectively.

Theorem 2. *For $S = \mathbf{Z}_n$, $\Pi = \mathcal{C} \cup \{+\}$ and $\Sigma = \{=\}$ the success probability of every k-step extraction algorithm is upper bounded $\frac{1}{2}k^2/q$ and by km/q if the memory m is bounded.*

Proof. We have $\overline{\Pi} = \mathcal{L} = \{ax + b \mid a, b \in \mathbf{Z}_n\}$. As argued above, we need to consider only non-adaptive algorithms for provoking a collision. Consider a fixed algorithm computing in each step (say the ith) a new value $a_i x + b_i$, keeping $m-1$ of the previously generated values in the state. A collision occurs if $a_i x + b_i \equiv_n a_j x + b_j$ for some distinct i and j, i.e., if $(a_i - a_j)x + (b_i - b_j) \equiv_n 0$. Considered modulo q, this congruence has one solution for x (according to Lemma 1). The total number of x for which any collision modulo q (which is necessary for a collision modulo n) can occur is bounded by $k(m-1)$. If m is unbounded (actually $O(\sqrt{q})$ is sufficient), then the number of such x is bounded by $\binom{k}{2}$.[12] $\triangle$

The case of unbounded m corresponds to the results of Nechaev [6] and Shoup [10], but the proof in [10] is somewhat more involved because a random permutation of the group representation is explicitly considered and makes the random experiment more complex. The Pohlig-Hellman algorithm requires $k = O(\sqrt{q} \log q)$ operations and essentially matches this bound. If the equality checks are also counted in k and no order relation is available, then $k = O(n)$ is required.

It is worthwhile to discuss the bounded-memory case. The theorem implies that the complexity of every algorithm achieving a constant success probability is $O(n/m)$, which is linear in n for constant m. Since memory is bounded in reality and $m = O(\sqrt{q})$ is typically infeasible, it appears that this result is a significant improvement of the lower bound over the unbounded memory case. However, this is in conflict with the fact that the Pollard-ρ algorithm [8] requires constant memory and also has (heuristic) complexity $O(\sqrt{q})$. The reason is that when a representation for S is explicitly available, then one can explicitly define a function on S, for example to partition the set S in a heuristically random manner into several subsets (three subsets in case of the Pollard-ρ algorithm). It is interesting to model this capability abstractly in the spirit of this paper.

---

[11] For other groups, such as $\{0,1\}^{\ell}$ discussed in the previous section, the extraction problem can be seen as a generalization of the DL problem.

[12] If no collision has occurred, one could allow the algorithm one more guess among the values still compatible with the observation of no collision, but this can be neglected.

3.5   The DL-Problem with a DDH-Oracle or Other Side Information

Let us consider the following natural question: Does a DDH-oracle help in computing discrete logarithms? Or, stated differently, can one show that even if the DDH-problem for a given group is easy, the DL-problem is still hard for generic algorithms. It turns out that the DDH oracle can indeed be potentially helpful, but not very much so.

Theorem 3. *For $S = \mathbf{Z}_n$, $\Pi = \mathcal{C} \cup \{+\}$ and $\Sigma = \{=, prod_n\}$ the success probability of every k-step extraction algorithm is upper bounded by $2k^3 + \frac{1}{2}k^2$. Every algorithm with constant success probability has complexity at least $O(\sqrt[3]{q})$.*

Proof. Each computed value is of the form $a_i x + b_i$ for some a_i and b_i. The product relation is satisfied for three computed values if

$$(a_i x + b_i)(a_j x + b_j) = a_k x + b_k$$

for some i, j, k, which is equivalent to

$$a_i a_j x^2 + (a_i b_j + a_j b_i - a_k)x + b_i b_j - b_k = 0,$$

a quadratic equation for x which has two solutions modulo q. There are k^3 such triples i, j, k. When also counting the potential collisions for the equality relation, the number of x modulo q for which one of the relations holds is bounded by $2k^3 + \binom{k}{2}$. △

A similar argument shows that when one considers a relation involving more than three variables, then the complexity lower bound decreases. For example, if we consider an oracle for the triple-product relation $\{(w, x, y, z) : z = wxy\}$, then we get a lower bound of $O(\sqrt[4]{q})$. It would be interesting to show that these bounds can be (or can not be) achieved.

A similar argument as those used above shows that when an oracle for the e-th power relation (i.e., $x_j = x_i^e$) is available, then every generic algorithm has complexity $O(\sqrt{q/e})$.

3.6   Product Computation in $\mathbf{Z}_n$ and the CDH Problem

We now consider the computation problem for the product function $(x, y) \mapsto xy$ in $\mathbf{Z}_n$. This corresponds to the generic computational Diffie-Hellman (CDH) problem in a cyclic group of order n analyzed already in [10]. Essentially the same bounds can be obtained for the squaring function $x \mapsto x^2$ in $\mathbf{Z}_n$. This theorem shows that for generic algorithms, the DL and the CDH problems are essentially equally hard.

Theorem 4. *For $S = \mathbf{Z}_n$, $\Pi = \mathcal{C} \cup \{+\}$ and $\Sigma = \{=\}$ the success probability of every k-step algorithm for computing the product function is upper bounded by $\frac{1}{2}(k^2 + 3k)/q$.*

Proof. Again, to be on the safe side, we can assume that as soon as a collision occurs among the values $a_i x + b_i$, the algorithm is successful. In addition, we need to consider the events $a_i x + b_i \equiv_n xy$ (for some i). For every i there are two solutions modulo q (according to Lemma 1). Hence the total number of x (modulo q) for which one of the collision events occurs is bounded by $\binom{k}{2} + 2k = \frac{1}{2}(k^2 + 3k)$.   △

One can also show a $O(\sqrt[3]{n})$ generic lower bound for the CDH-problem when given a DDH-oracle.

3.7   Decision Problems for Cyclic Groups

We consider the decision problem for the squaring and product relations in $\mathbf{Z}_n$.

Theorem 5. *For $S = \mathbf{Z}_n$, $\Pi = \mathcal{C} \cup \{+\}$ and $\Sigma = \{=\}$ the advantage of every k-step algorithm for distinguishing a random pair (x, y) from a pair (x, x^2) is upper bounded by k^2/p.*

Proof. Again we can assume that as soon as a collision occurs among the values $a_i x + b_i$, the algorithm is declared successful. Hence it suffices to compute the probabilities, for the two settings, that a collision can be provoked, and take the larger value as an upper bound for the distinguishing advantage. For the pair (x, x^2) the set of computable functions is $\{ax^2 + bx + c \mid a, b, c \in \mathbf{Z}_n\}$, i.e., the ith computed value is of the form

$$a_i x^2 + b_i x + c_i$$

(in $\mathbf{Z}_n$) for some a_i, b_i, c_i. For any choice of $(a_i, b_i, c_i) \neq (a_j, b_j, c_j)$ we must bound the probability that

$$a_i x^2 + b_i x + c_i \equiv_n a_j x^2 + b_j x + c_j$$

for a uniformly random value x. This is equivalent to

$$(a_i - a_j)x^2 + (b_i - b_j)x + (c_i - c_j) \equiv_n 0.$$

There must be at least one prime factor p of n (possibly the smallest one) such that (a_i, b_i, c_i) and (a_j, b_j, c_j) are distinct modulo p. The number of solutions x of the equation modulo p is at most 2 (according to Lemma 1). Hence the total probability of provoking a collision modulo p (and hence also modulo n) is upper bounded by $\binom{k}{2}2/p < k^2/p$.

This should be compared to the case where the pair (x, y) consists of two independent random values. The number of solutions (x, y) of

$$(a_i - a_j)y + (b_i - b_j)x + (c_i - c_j) \equiv_q 0$$

for any choice of $(a_i, b_i, c_i) \neq (a_j, b_j, c_j)$ is at most p. Hence the collision probability is, for all generic algorithms, upper bounded by $\binom{k}{2}/p < \frac{1}{2}k^2/p$. This concludes the proof.   △

A very similar argument can be used to prove the same bound for the decision problem for the product relation, which corresponds to the generic decisional Diffie-Hellman (DDH) problem in a cyclic group of order n (see also [10]). To illustrate our approach we prove a lower bound for the DDH problem, even when assuming an oracle for the squaring relation.

Theorem 6. *For $S = \mathbf{Z}_n$, $\Pi = \mathcal{C} \cup \{+\}$ and $\Sigma = \{=, square_n\}$ the advantage of every k-step algorithm for distinguishing a random triple (x, y, z) from a triple (x, y, xy) is upper bounded by $\frac{5}{2}k^2/p$.*

Proof. We only analyze the case where the initial state is (x, y, xy). The set $\overline{\Pi}$ of computable functions is $\{ax + by + cxy + d \mid a, b, c, d \in \mathbf{Z}_n\}$, i.e., the ith computed value is of the form

$$a_i x + b_i y + c_i xy + d_i$$

for some a_i, b_i, c_i, d_i. For any choice of $(a_i, b_i, c_i, d_i) \neq (a_j, b_j, c_j, d_j)$ we must bound the probability that

$$a_i x + b_i y + c_i xy + d_i \equiv_n a_j x + b_j y + c_j xy + d_j$$

or that

$$(a_i x + b_i y + c_i xy + d_i)^2 \equiv_n a_j x + b_j y + c_j xy + d_j$$

The latter is a polynomial relation of degree 4 that is non-zero if $(a_i, b_i, c_i, d_i) \neq (a_j, b_j, c_j, d_j)$, except when $a_i = b_i = c_i = a_j = b_j = c_j = 0$ and $d_i^2 \equiv_n d_j$. However, we need not consider this case since it is known *a priori* that such a relation holds for all x and y.[13] The fraction of pairs (x, y) for which one of these relations can be satisfied modulo p is at most $5\binom{k}{2}/p$.   △

3.8   Reducing the DL-Problem to the CDH-Problem

If one includes multiplication modulo n in the set Π of allowed operations for the generic extraction problem, i.e., one considers the extraction problem for the ring $\mathbf{Z}_n$, then this corresponds to the generic reduction of the discrete logarithm problem in a group of order n to the computational Diffie-Hellman problem for this group. The Diffie-Hellman oracle assumed to be available for the reduction implements multiplication modulo n. There exist an efficient generic algorithm for the extraction problem for the ring $\mathbf{Z}_n$ [3] (see also [5]) for most cases. For prime n the problem was called the black-box field problem in [1].

Acknowledgments

I would like to thank Dominic Raub for interesting discussions and helpful comments.

---

[13] More formally, this can be taken into account when defining the system output sequence to be deviated from according to Lemma 2.

References

1. D. Boneh and R. J. Lipton, Algorithms for black-box fields and their application to cryptography, *Advances in Cryptology - CRYPTO '96*, Lecture Notes in Computer Science, vol. 1109, pp. 283–297, Springer-Verlag, 1996.
2. W. Diffie and M. E. Hellman, New directions in cryptography, *IEEE Transactions on Information Theory*, vol. 22, no. 6, pp. 644–654, 1976.
3. U. Maurer, Towards the equivalence of breaking the Diffie-Hellman protocol and computing discrete logarithms, *Advances in Cryptology - CRYPTO '94*, Lecture Notes in Computer Science, vol. 839, pp. 271–281, Springer-Verlag, 1994.
4. U. Maurer and S. Wolf, Lower bounds on generic algorithms in groups, *Advances in Cryptology - EUROCRYPT 98*, Lecture Notes in Computer Science, vol. 1403, pp. 72–84, Springer-Verlag, 1998.
5. U. Maurer and S. Wolf, On the complexity of breaking the Diffie-Hellman protocol, *SIAM Journal on Computing*, vol. 28, pp. 1689–1721, 1999.
6. V. I. Nechaev, Complexity of a deterministic algorithm for the discrete logarithm, *Mathematical Notes*, vol. 55, no. 2, pp. 91–101, 1994.
7. S. C. Pohlig and M. E. Hellman, An improved algorithm for computing logarithms over $GF(p)$ and its cryptographic significance, *IEEE Transactions on Information Theory*, vol. 24, no. 1, pp. 106–110, 1978.
8. J. M. Pollard, Monte Carlo methods for index computation mod p, *Mathematics of Computation*, vol. 32, pp 918–924, 1978.
9. J. T. Schwartz, Fast probabilistic algorithms for verification of polynomial identities, *Journal of the ACM*, vol 27, no. 3, pp. 701–717, 1980.
10. V. Shoup, Lower bounds for discrete logarithms and related problems, *Advances in Cryptology - EUROCRYPT '97*, Lecture Notes in Computer Science, vol. 1233, pp. 256–266, Springer-Verlag, 1997.

A   Polynomial Equations Modulo n

We make use of a lemma due to Schwartz [9] and Shoup [10] for which we give a simple proof.

Lemma 1. *The fraction of solutions $(x_1, \ldots, x_k) \in \mathbf{Z}_n$ of the multivariate polynomial equation $p(x_1, \ldots, x_k) \equiv_n 0$ of degree d is at most d/q, where q is the largest prime factor of n.*[14]

Proof. A solution of a multivariate polynomial equation $p(x_1, \ldots, x_k) \equiv_n 0$ over $\mathbf{Z}_n$ is satisfied only if it is satisfied modulo every prime factor of n, in particular modulo the largest prime q dividing n, i.e., $p(x_1, \ldots, x_k) \equiv_q 0$. It follows from the Chinese remainder theorem that the fraction of solutions $(x_1, \ldots, x_k)$ in $\mathbf{Z}_n^k$ is upper bounded by the fraction of solutions $(x_1, \ldots, x_k)$ in $\mathbf{Z}_q^k$.

Note that $\mathbf{Z}_q$ is a field. It is well-known that a univariate polynomial (i.e., $k = 1$) of degree $\leq d$ over a field F has at most d roots, unless it is the 0-polynomial for which all field elements are roots. The proof for multivariate

---

[14] The degree of a multivariate polynomial $p(x_1, \ldots, x_k)$ is the maximal degree of an additive term, where the degree of a term is the sum of the powers of the variables in the term.

polynomials is by induction on k. Let e be the maximal degree of x_k in any term in $p(x_1, \ldots, x_k)$. The polynomial $p(x_1, \ldots, x_k)$ over $\mathbf{Z}_n$ can be considered as a univariate polynomial in x_k of degree e with coefficients of degree at most $d - e$ in the ring $\mathbf{Z}_n[x_1, \ldots, x_{k-1}]$. By the induction hypothesis, for any of these coefficients the number of $(x_1, \ldots, x_{k-1})$ for which it is 0 is at most $(d - e)q^{k-2}$, which is hence also an upper bound on the number of tuples $(x_1, \ldots, x_{k-1})$ for which *all* coefficients are 0, in which case all values for x_k are admissible. If one of the coefficients is non-zero, then the fraction of solutions for x_k is at most e/q. Thus the total number of solutions $(x_1, \ldots, x_k)$ in $\mathbf{Z}_q$ is upper bounded by

$$(d - e)q^{k-2} \cdot q + (q - d + e)q^{k-2} \cdot e < dq^{k-1}. \qquad \triangle$$

B   A Simple Lemma on Random Systems

Consider a general system which takes a sequence $X_1, X_2, \ldots$ of inputs from some input alphabet $\mathcal{X}$ and produces, for every input X_i, an output Y_i from some output alphabet $\mathcal{Y}$. The system may be probabilistic and it may have state.

Lemma 2. *Consider the task of provoking, by an appropriate choice of the inputs $X_1, \ldots, X_k$, that a particular output sequence $y^k := [y_1, \ldots, y_k]$ does not occur. The success probability of the best non-adaptive strategy (without access to $Y_1, Y_2, \ldots$) is the same as that of the best adaptive strategy (with access to $Y_1, Y_2, \ldots$).*

Proof. Any adaptive strategy with access to $Y_1, Y_2, \ldots$ can be converted into an equally good non-adaptive strategy by feeding it, instead of $Y_1, Y_2, \ldots$, the (fixed) values $y_1, \ldots, y_k$. As long as the algorithm is not successful, these constant inputs $y_1, y_2, \ldots$ correspond to what happens in the adaptive case. $\qquad \triangle$

Pairing-Based Cryptography at High Security Levels

Neal Koblitz[1] and Alfred Menezes[2]

[1] Department of Mathematics, University of Washington
koblitz@math.washington.edu
[2] Department of Combinatorics & Optimization,
University of Waterloo
ajmeneze@uwaterloo.ca

Abstract. In recent years cryptographic protocols based on the Weil and Tate pairings on elliptic curves have attracted much attention. A notable success in this area was the elegant solution by Boneh and Franklin [8] of the problem of efficient identity-based encryption. At the same time, the security standards for public key cryptosystems are expected to increase, so that in the future they will be capable of providing security equivalent to 128-, 192-, or 256-bit AES keys. In this paper we examine the implications of heightened security needs for pairing-based cryptosystems. We first describe three different reasons why high-security users might have concerns about the long-term viability of these systems. However, in our view none of the risks inherent in pairing-based systems are sufficiently serious to warrant pulling them from the shelves.

We next discuss two families of elliptic curves E for use in pairing-based cryptosystems. The first has the property that the pairing takes values in the prime field $\mathbb{F}_p$ over which the curve is defined; the second family consists of supersingular curves with embedding degree $k = 2$. Finally, we examine the efficiency of the Weil pairing as opposed to the Tate pairing and compare a range of choices of embedding degree k, including $k = 1$ and $k = 24$.

1 Introduction

Let E be the elliptic curve $y^2 = x^3 + ax + b$ defined over a finite field $\mathbb{F}_q$, and let P be a basepoint having prime order n dividing $\#E(\mathbb{F}_q)$, where we assume that n does not divide q. Let k be the multiplicative order of q modulo n; in other words, it is the smallest positive k such that $n \mid q^k - 1$. The number k, which is called the *embedding degree*, has been of interest to cryptographers ever since it was shown in [37] how to use the Weil pairing to transfer the discrete log problem in the group $\langle P \rangle \subset E(\mathbb{F}_q)$ to the discrete log problem in the finite field $\mathbb{F}_{q^k}$.

In recent years, the Tate pairing (introduced to cryptographers by Frey-Rück [17]) and the Weil pairing have been used to construct a number of different cryptosystems. These systems were the first elliptic curve cryptosystems not

N.P. Smart (Ed.): Cryptography and Coding 2005, LNCS 3796, pp. 13–36, 2005.

constructed by analogy with earlier versions that used the multiplicative group of a finite field. Rather, pairing-based cryptosystems use properties of elliptic curves in an essential way, and so they cannot be constructed in simpler settings (such as finite fields or the integers modulo N). In the next section we shall describe a particularly elegant example of such a cryptosystem, namely, the solution of Boneh and Franklin [8] of the problem of efficient identity-based encryption.

Meanwhile, it is becoming increasingly apparent that we are approaching a transitional moment in the deployment of cryptography. Calls are going out for heightened security standards for public key cryptosystems, so that in the future they will be capable of providing security equivalent to 128-, 192-, or 256-bit AES keys. In this paper we examine the implications for pairing-based elliptic curve cryptography of a move to higher levels of security.

Our first purpose is to describe three general questions about efficiency and security that arise. These concerns are not new to people working in the area, but they are rarely mentioned explicitly in print. By calling the reader's attention to these issues we have no intention of sounding alarmist or of discouraging deployment of these systems. On the contrary, in our view none of the considerations discussed below are sufficiently worrisome to justify abandoning pairing-based cryptography.

Our second purpose is to describe two very simple families of elliptic curves defined over a prime field $\mathbb{F}_p$ with embedding degrees $k = 1$ and $k = 2$, respectively, that could be used in pairing-based cryptosystems. The main advantage of these families is the flexibility one has in choosing the two most important parameters of the system — the field size p and the prime order n of the basepoint $P \in E(\mathbb{F}_p)$. One can easily get n and p both to have optimal bitlengths and at the same time to be Solinas primes [51] (that is, the sum or difference of a small number of powers of 2). In earlier papers on parameter selection for pairing-based systems we have not found any discussion of the advantages and disadvantages of low-Hamming-weight p.

On the negative side, when $k = 1$ one does not have any of the speedups that come from working in a subfield at various places in the pairing computations. When $k = 2$ our curves are supersingular, and so one must anticipate some resistance to their use because of the traditional stigma attached to the word "supersingular" by implementers of elliptic curve cryptography. Moreover, in both cases the use of a Solinas prime p could possibly enable an attacker to use a special form of the number field sieve. It remains to be seen whether the increased field sizes which would then be necessary offset the efficiency advantage provided by the use of such a prime.

Our third purpose is to compare different choices of k, ranging from 1 to 24, for different security levels. Our comparisons are simple but realistic, and incorporate most of the important speedups that have been discovered so far. Although much depends on the implementation details, it appears that for non-supersingular curves the choice $k = 2$ that is recommended by some authors [48] is probably less efficient than higher values of k. We also find that for very high security levels, such as 192 or 256 bits, the Weil pairing computation is sometimes faster than the Tate pairing.

Earlier work in this area has focused on providing 80 bits of security, which is sufficient for most current applications. In contrast, we are particularly interested in how the choice of parameters will be affected by the move to the higher AES standard of 128, 192, or 256 bits of security that is anticipated in the coming years.

2   Identity-Based Encryption

One of the most important applications of the Weil (or Tate) pairing is to identity-based encryption [8]. Let's recall how the basic version of the Boneh–Franklin scheme works. Suppose that E over $\mathbb{F}_q$ is an elliptic curve on which (a) the Diffie–Hellman problem is intractable and (b) the Weil pairing $\hat{e}(P,Q) \in \mathbb{F}_{q^k}$ can be efficiently computed. (For an excellent treatment of the Weil and Tate pairings, see [18].) Here P and Q are $\mathbb{F}_{q^k}$-points of prime order n, where $n \mid \#E(\mathbb{F}_q)$, and the embedding degree k (for which $E(\mathbb{F}_{q^k})$ contains all n^2 points of order n) must be small.

Bob wants to send Alice a message m, which we suppose is an element of $\mathbb{F}_{q^k}$, and he wants to do this using nothing other than her identity, which we suppose is hashed and then embedded in some way as a point I_A of order n in $E(\mathbb{F}_q)$. In addition to the field $\mathbb{F}_q$ and the curve E, the system-wide parameters include a basepoint P of order n in $E(\mathbb{F}_{q^k})$ and another point $K \in \langle P \rangle$ that is the public key of the Trusted Authority. The TA's secret key is the integer s that it used to generate the key $K = sP$.

To send the message m, Bob first chooses a random r and computes the point rP and the pairing $\hat{e}(K, I_A)^r = \hat{e}(rK, I_A)$. He sends Alice both the point rP and the field element $u = m + \hat{e}(rK, I_A)$. In order to decrypt the message, Alice must get her decryption key D_A from the Trusted Authority; this is the point $D_A = sI_A \in E(\mathbb{F}_q)$ that the TA computes using its secret key s. Finally, Alice can now decrypt by subtracting $\hat{e}(rP, D_A)$ from u (note that, by bilinearity, we have $\hat{e}(rP, D_A) = \hat{e}(rK, I_A)$).

3   Clouds on the Horizon?

The first reservation that a high-security user might have about pairing-based systems relates to efficiency. A necessary condition for security of any pairing-based protocol is that discrete logarithms cannot be feasibly found in the finite field $\mathbb{F}_{q^k}$. In practice, q is either a prime or a power of 2 or 3, in which case the number field sieve [20,45] or function field sieve [15,1,46] will find a discrete log in time of order $L(1/3)$; this means that the bitlength of q^k must be comparable to that of an RSA modulus offering the same security.

In both cases the bitlength should be, for example, at least 15360 to provide security equivalent to a 256-bit AES key [31,42].[1]

As in the case of RSA, the loss of efficiency compared to non-pairing-based elliptic curve cryptography (ECC) increases steeply as the security level grows. Unlike RSA, pairing-based systems can achieve certain cryptographic objectives — notably, identity-based encryption — that no one has been able to achieve using ordinary ECC. So one has to ask how badly one wants the features that only pairing-based methods can provide. As the security requirements increase, the price one has to pay for the extra functionality will increase sharply.

It should be noted that in certain applications bandwidth can be a reason for using pairing-based systems (see, for example, [6,7,9]). We shall not consider bandwidth in this paper, except briefly in §4.1 for Boneh–Lynn–Shacham signatures.

The other two concerns about pairing-based systems are more theoretical, and both relate to security. In the first place, in most pairing-based protocols security depends upon the assumed intractability of the following problem, which Boneh and Franklin [8] called the Bilinear Diffie–Hellman Problem (BDHP): Given $P, rP, sP, Q \in E(\mathbb{F}_{q^k})$ such that $\zeta = \hat{e}(P,Q) \neq 1$, compute ζ^{rs}.

The BDHP is a new problem that has not been widely studied. It is closely related to the Diffie–Hellman Problem (DHP) in the elliptic curve group $E(\mathbb{F}_{q^k})$, which is the problem, given P, rP, and sP, of computing rsP. Since $\zeta^{rs} = \hat{e}(rsP, Q)$, it follows that if one has an algorithm for the DHP on the curve, one can immediately solve the BDHP as well. But the converse is not known, and it is possible that the BDHP is an easier problem than the DHP on the curve.

In the early discussions of discrete-log-based cryptosystems it was a source of concern that security depended on the presumed intractability of the Diffie–Hellman Problem rather than the more natural and more extensively studied Discrete Log Problem (DLP). That is why cryptographers were very pleased when a series of papers by den Boer, Maurer, Wolf, Boneh, Lipton and others (see [35] for a survey) developed strong evidence for the equivalence of the Diffie–Hellman and Discrete Log Problems on elliptic curves. But unfortunately, no such evidence has been found for hardness of the Bilinear Diffie–Hellman Problem. Of course, no one knows of any way to solve the BDHP except by finding discrete logs, so perhaps it is reasonable to proceed as if the BDHP is equivalent to the DHP and the DLP on elliptic curves — despite the absence of theoretical results supporting such a supposition.

The BDHP is also closely related to the Diffie–Hellman Problem in the finite field $\mathbb{F}_{q^k}$, and any algorithm for the DHP in the field will immediately enable us to solve the BDHP too. But it is possible that the BDHP is strictly easier than the DHP in the field. In the DHP we are given only the values ζ, ζ^r, and ζ^s, whereas in the BDHP the input also includes the inverse images of these

---

[1] For fields of small characteristic, q^k should be significantly larger than for q a prime. In [31] the bitlengths 4700, 12300, and 24800 are suggested for security levels 128, 192, and 256 bits, respectively.

n-th roots of unity under the Menezes-Okamoto-Vanstone [37] embedding from $\langle P \rangle \subset E(\mathbb{F}_{q^k})$ to the finite field given by $X \mapsto \hat{e}(X, Q)$ for $X \in \langle P \rangle$.

This brings us to the third major concern with pairing-based cryptosystems, namely, Verheul's theorem [53].

Even if one is willing to suppose that the Bilinear Diffie–Hellman Problem on a low-embedding-degree curve is equivalent to the DHP and the DLP on the curve, in practice one really considers the DHP and DLP in the multiplicative group of a finite field, because it is there that the problem has been extensively studied and index-calculus algorithms with carefully analyzed running times have been developed. Using the MOV embedding, the DHP and DLP on the low-embedding-degree curve reduce to the corresponding problems in the finite field. At first it seems that it would be nice to have reductions in the other direction as well. That is, a homomorphism in the opposite direction to the MOV embedding would show that the problems on the curve and in the field are provably equivalent. Indeed, in special cases construction of such a homomorphism was posed as an open problem in [30] and [38]. However, in [53] Verheul dashed anyone's hopes of ever strengthening one's confidence in the security of pairing-based systems by constructing such a reduction.

Verheul proved the following striking result. Let μ_n denote the n-th roots of unity in $\mathbb{F}_{p^6}$, where $n \mid (p^2 - p + 1)$, and hence μ_n is not contained in a proper subfield; this is called an XTR group [32]. Suppose that an efficiently computable nontrivial homomorphism is found from μ_n to $\langle P \rangle \subset E(\mathbb{F}_{p^2})$, where E is an elliptic curve defined over $\mathbb{F}_{p^2}$ with $\#E(\mathbb{F}_{p^2}) = p^2 - p + 1$. Here we are assuming, as before, that P is a point of prime order n. Then Verheul's theorem states that the DHP is efficiently solvable in both μ_n and $\langle P \rangle$.

A generalization of Verheul's theorem, which was conjectured but not proved in [53], would give the same result whenever a group $\mu_n \subset \mathbb{F}_{q^k}$ can be efficiently mapped to a supersingular curve $E(\mathbb{F}_q)$. (Note that $q = p^2$ and $k = 3$ for the XTR group.) It is this generalized version that prompted Verheul to suggest that his results "provide evidence that the multiplicative group of a finite field provides essentially more...security than the group of points of a supersingular elliptic curve of comparable size."

The following observation, which was not made in [53], seems to give further support for Verheul's point of view. Given an arbitrary finite field $\mathbb{F}_q$, suppose that one can efficiently construct a trace-zero elliptic curve E over $\mathbb{F}_q$, that is, a curve for which $\#E(\mathbb{F}_q) = q + 1$. (If $q \equiv -1 \pmod 4$ or $q \equiv -1 \pmod 6$, then the curve (3) or (4) in §7 has this property; more generally, see §7.6 and Exercise 2 in Chapter 7 of [14] for the prime field case.) We then have the following theorem about the so-called class-VI supersingular curves, which can be viewed as curves of embedding degree $k = 1/2$.

Theorem 1. *Let $\mathbb{F}_q$ be an arbitrary finite field, and let E be a trace-zero elliptic curve over $\mathbb{F}_q$. Suppose that E has equation $y^2 = f(x)$ for odd q and $y^2 + y = f(x)$ for q a power of 2. Let $\beta \in \mathbb{F}_{q^2}$ be a nonsquare in $\mathbb{F}_{q^2}$ for odd q and an element of absolute trace 1 for q a power of 2 (that is, $\mathrm{Tr}_{\mathbb{F}_{q^2}/\mathbb{F}_2}(\beta) = 1$). Let $\widetilde{E}$ be the "twisted" curve over $\mathbb{F}_{q^2}$ with equation $\beta y^2 = f(x)$ for odd q and $y^2 + y + \beta = f(x)$*

for q a power of 2. Then $\widetilde{E}(\mathbb{F}_{q^2})$ is a product of two cyclic groups of order $q-1$, each of which is isomorphic to the multiplicative group of $\mathbb{F}_q$ under the MOV embedding.

This theorem is an immediate consequence of the classification of supersingular elliptic curves (see Table 5.2 in [36]). Notice that for a trace-zero curve E we have $\#E(\mathbb{F}_q) = q + 1 = q + 1 - \alpha - \overline{\alpha}$ with $\alpha^2 = -q$, and hence $\#E(\mathbb{F}_{q^2}) = q^2 + 1 - \alpha^2 - \overline{\alpha}^2 = q^2 + 1 + 2q$. Thus, for the twist we have $\#\widetilde{E}(\mathbb{F}_{q^2}) = q^2 + 1 - 2q$.

It is reasonable to think that Verheul's theorem can be generalized to the curves in the above theorem. One would need to describe algorithms for obtaining a trace-0 curve for arbitrary q and a "distortion" map in the twisted curve over $\mathbb{F}_{q^2}$. In that case the construction of a Verheul homomorphism would make the DHP easy in *all* finite fields.

Thus, there are two possible interpretations of Verheul's theorem in its (conjectured) general form. The "optimistic" interpretation is that a Verheul homomorphism will never be constructed, because to do so would be tantamount to making the Diffie–Hellman problem easy in all finite fields. Under this interpretation we are forced to conclude that the DHP that arises in pairing-based cryptography is not likely to be provably equivalent to the DHP in finite fields. The "pessimistic" interpretation is that a Verheul homomorphism might some day be constructed. Even if it were constructed just for the class-VI supersingular elliptic curves, that would be enough to render all pairing-based cryptosystems (and also many XTR protocols) completely insecure.

Remark 1. This issue does not arise in the usual non-pairing-based elliptic curve cryptography (ECC). In ECC protocols one uses nonsupersingular curves having large embedding degree k. In fact, k is generally of size comparable to n itself (see [2]), in which case even the input to the Verheul inversion function would have exponential size. Thus, the danger posed by such a map — if it could be efficiently computed — applies only to small k.

Remark 2. The third concern with pairing-based systems — that the problem that their security relies on is not likely to be provably equivalent to a standard problem that is thought to be hard unless both problems are easy — is analogous to a similar concern with RSA. In [11] Boneh and Venkatesan proved that an "algebraic" reduction from factoring to the RSA problem with small encryption exponent is not possible unless both problems are easy.

Remark 3. In [21] Granger and Vercauteren study improved index-calculus algorithms for finding discrete logarithms in subgroups $\mu_n \subset \mathbb{F}_{p^k}$ where $k = 2m$ or $k = 6m$ and $n|p^m + 1$ or $n|p^{2m} - p^m + 1$, respectively. Although their results do not give a faster algorithm for any of the parameters discussed below, they indicate that the type of group used in pairing-based protocols at high security levels may become vulnerable to "algebraic tori attacks" of the type in [21]. Like Verheul's work, the paper by Granger and Vercauteren highlights the special nature of the groups μ_n that correspond to elliptic curves with low embedding degree.

4 Parameter Sizes

For the remainder of this paper, unless stated otherwise, we shall suppose that $\mathbb{F}_q$ is a prime field, and we set $q = p$. As mentioned in the last section, in order for a pairing-based cryptosystem to be secure, the field $\mathbb{F}_{p^k}$ must be large enough so that discrete logs cannot feasibly be found using the best available algorithms (the number field and function field sieves). It is also necessary for the prime order n of the basepoint P to be large enough to withstand the Pollard-ρ attack on discrete logs in the group $\langle P \rangle$. Table 1 (see [31,42]) shows the minimum bitlengths of n and p^k as a function of the desired security level.

Table 1. Minimum bitlengths of n and p^k

security level (in bits)	80	128	192	256
b_n (min. bits of prime subgroup)	160	256	384	512
b_{p^k} (min. bits of big field)	1024	3072	8192	15360
$\gamma = $ the ratio b_{p^k}/b_n	6.4	12	$21\frac{1}{3}$	30

4.1 Short Signatures

One of the best known uses of pairings is to produce short signatures [10]. Without using pairing methods, the shortest signatures available are the ECDSA, where the length is roughly $2b_n$ bits, and the Pintsov–Vanstone [43] and Naccache–Stern [41] schemes, where the length is roughly $1.5b_n$. The pairing-based Boneh–Lynn–Shacham signatures have length approximately equal to the bitlength of p, which is ρb_n, where $\rho = \log p / \log n$.

Thus, in order to have short Boneh–Lynn–Shacham signatures, one must choose the parameters so that $\rho = \log p / \log n$ is close to 1 and hence $k = \gamma/\rho$ is nearly equal to $\gamma = b_{p^k}/b_n$ (see Table 1). Starting with [40], techniques have been developed to do this with nonsupersingular curves when k can be taken equal to 2, 3, 4, 6, or 12. For $k = 2$, 3, 4, or 6 the k-th cyclotomic polynomial is linear or quadratic, and the resulting Diophantine equations are computationally tractable. When $k = 12$, the cyclotomic polynomial is quartic, and recently Barreto and Naehrig [5] were able to achieve $\rho = 1$ using a combination of quadratic polynomials. For larger k — notably, for $k = 24$ — the best results are due to Brezing–Weng [12], who obtain $\rho = 1.25$. For example, at the 256-bit security level with 512-bit n they can produce 640-bit signatures, compared to 768 bits for Pintsov–Vanstone and Naccache–Stern and 1024 bits for ECDSA.

It should also be noted that at very high security levels the Boneh–Lynn–Shacham public keys are much larger than in the Pintsov–Vanstone, Naccache–Stern and ECDSA schemes. For instance, at the 256-bit level the latter public keys are roughly 512 bits long, whereas in the pairing-based short signature scheme the public key is a point of $E(\mathbb{F}_{p^k})$, where p^k has about 15360 bits. It suffices to give the x-coordinate of the public-key point, and for even k we may assume that this coordinate is in the smaller field $\mathbb{F}_{p^{k/2}}$ (see the end of §8.2). But even then the public key is about 7680 bits.

4.2   Changes as Security Requirements Increase

As our security needs increase, the gap between the desired sizes of n and of p^k increases (see Table 1). At the same time, the optimal choices of algorithms in implementations — and hence the decisions about what families of curves provide greatest efficiency — are likely to be affected. That is, certain tricks that were useful at lower security levels may become less important than other considerations, such as the ability to choose parameters of a special form.

Our first observation is that as b_n and b_{p^k} increase for greater security, the parameter selection methods proposed with nonsupersingular curves of embedding degree $k \geq 2$ do not seem to yield values of n (the prime order of the basepoint) and p (the size of the prime field) that are both Solinas primes. In the literature we have found one construction, due to Scott and Barreto [50], that comes close to solving this problem at the 128-bit security level. Namely, one can apply their construction for $k = 6$ in §5 of [50] with $x = 12Dz^2 + 1$, where D is a small power of 2 and z is a roughly 80- to 90-bit power of 2 or sum or difference of two powers of 2 that is chosen so that both $n = x^2 - x + 1$ and $p = (x^3 - 2x^2 + 14x - 1)/12$ are primes. Then the bitlengths of n and p^6 are roughly equal to the optimal values in Table 1; moreover, n and p are each equal to a sum or difference of a relatively small number of powers of 2. However, for higher security levels and $k \geq 2$ we do not know of any similar method to achieve nearly-optimal characteristics.

Example 1. Set $D = 1$, $z = 2^{81} + 2^{55}$. Then n is a 332-bit prime with Hamming weight 19, and p is a 494-bit prime with Hamming weight 44.

Our second observation is that for $k > 2$ at the higher security levels it is probably not possible to find suitable supersingular elliptic curves with n having the optimal bitlength that one uses for nonsupersingular curves. The greatest value of k that one can get is $k = 6$, and there are only two supersingular elliptic curves E, both defined over $\mathbb{F}_3$, that have embedding degree 6. Because of the efficiency of the function field sieve in finding discrete logs in characteristic-3 fields, it would be advisable to choose fields $\mathbb{F}_{3^m}$ such that the bitlength of 3^{6m} is larger than the value of b_{p^k} in Table 1. But even using the values in Table 1, there is a serious question of whether one can find a field extension degree $m \approx b_{p^k}/(6 \log_2 3)$ such that $\#E(\mathbb{F}_{3^m})$ has a prime factor n of the appropriate size. There are a relatively small number of possible choices of extension degree, so an elliptic curve group whose order has such a factor n might simply not exist. Moreover, even if it does exist, to find it one needs to factor $\#E(\mathbb{F}_q)$, $q = 3^m$, which cannot feasibly be done unless one is lucky and this number is fairly smooth. For example, at the 256-bit security level we would want the 2560-bit integer $\#E(\mathbb{F}_q)$ to be the product of a roughly 512-bit prime and a 2048-bit cofactor made up of primes that are small enough to be factored out of

$\#E(\mathbb{F}_q)$ by the Lenstra elliptic curve factorization method [33]. This is not very likely; in fact, standard estimates from analytic number theory imply that the probability of a random 2048-bit integer being 2^{150}-smooth is less than 2^{-50}.

Very recently, however, techniques have been developed to speed up the pairing computations in the low-characteristic supersingular case to make them virtually independent of the bitlength of n (see [3]). A detailed analysis has not yet been done, so it is still unclear how these supersingular implementations compare with the nonsupersingular ones as the security level increases. In particular, the field operations in the former case are in characteristic 2 or 3, and in the latter case they are in a large prime field.

Our third observation is that as n and p^k increase, one should look more closely at the possibility of switching back to use the Weil pairing rather than the Tate pairing. We shall examine this question when we study efficiency comparisons in §8.

5   Pairing-Friendly Fields

Suppose that we have an elliptic curve E defined over $\mathbb{F}_p$ with even embedding degree k. We shall say that the field $\mathbb{F}_{p^k}$ is *pairing-friendly* if $p \equiv 1 \pmod{12}$ and k is of the form $2^i 3^j$.[2] The following theorem is a special case of Theorem 3.75 of [34]:

Theorem 2. *Let $\mathbb{F}_{p^k}$ be a pairing-friendly field, and let β be an element of $\mathbb{F}_p$ that is neither a square nor a cube in $\mathbb{F}_p$.[3] Then the polynomial $X^k - \beta$ is irreducible over $\mathbb{F}_p$.*

The field $\mathbb{F}_{p^k}$ can thus be constructed from $\mathbb{F}_p$ as a tower of quadratic and cubic extensions by successively adjoining the squareroot or cuberoot of β, then the squareroot or cuberoot of that, and so on (see Figure 1). It is easy to see that, if an element of $\mathbb{F}_{p^k} = \mathbb{F}_p[X]/(X^k - \beta)$ is written as a polynomial $\sum_{\ell < k} a_\ell X^\ell$, then it belongs to a subfield $\mathbb{F}_{p^{k'}}$, where $k' = 2^{i'} 3^{j'}$, if and only if ℓ is a multiple of $k/k' = 2^{i-i'} 3^{j-j'}$ in all of the nonzero terms. Namely, if we set $\mathbb{F}_{p^{k'}} = \mathbb{F}_p[Y]/(Y^{k'} - \beta)$, then the map $Y \mapsto X^{k/k'}$ gives an embedding of the elements of $\mathbb{F}_{p^{k'}}$ (regarded as polynomials in Y) into $\mathbb{F}_{p^k}$. Thus, when we do arithmetic in the field $\mathbb{F}_{p^k}$, we can easily work with the tower of quadratic and cubic field extensions used to construct it.

In practice, it is easy to find a small value of β that satisfies the conditions of the theorem. In that case multiplication by β in $\mathbb{F}_p$ is much faster than a general multiplication in that field, and so can be neglected in our count of field multiplications. Then the Karatsuba method reduces a multiplication in a quadratic extension to 3 (rather than 4) multiplications in the smaller field; and the Toom–Cook method reduces a multiplication in a cubic extension to 5 (rather than 9) small field multiplications (see §4.3.3 of [27]). This means that we

---

[2] If $j = 0$, we only need $p \equiv 1 \pmod 4$.

[3] If $j = 0$, it is enough for β to be a nonsquare.

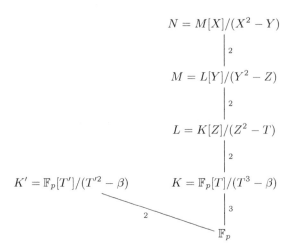

Fig. 1. Tower of pairing-friendly fields

can expect to perform a field operation in $\mathbb{F}_{p^k}$ in time $\nu(k)m$, where $\nu(k) = 3^i 5^j$ for $k = 2^i 3^j$, and m denotes the time to perform a multiplication in $\mathbb{F}_p$.

In what follows we shall occasionally perform multiplications in a quadratic subfield $\mathbb{F}_{p^{k/2}} \subset \mathbb{F}_{p^k}$. Because of the Karatsuba technique, we suppose that an $\mathbb{F}_{p^{k/2}}$-operation is equivalent to $1/3$ of an $\mathbb{F}_{p^k}$-operation.

Another nice feature of $k = 2^i 3^j$ is that many of the best examples of families of curves for pairing-based cryptography have embedding degree 2, 6, 12, or 24. For instance, we noted in §4.1 that examples with $\rho = \log p / \log n = 1.25$ were constructed in [12] with $k = 24$.

6   Curves with Embedding Degree 1

Let $p > 2$ be a prime of the form $A^2 + 1$. If $4 \mid A$, let E be the elliptic curve defined over $\mathbb{F}_p$ with equation

$$y^2 = x^3 - x. \tag{1}$$

If, on the other hand, $A \equiv 2 \pmod 4$, then let E be the curve

$$y^2 = x^3 - 4x. \tag{2}$$

Theorem 3. *The elliptic curve group* $E(\mathbb{F}_p)$ *is isomorphic to* $\mathbb{Z}/A\mathbb{Z} \oplus \mathbb{Z}/A\mathbb{Z}$. *In addition, the map* $(x, y) \mapsto (-x, Ay)$ *is a "distortion map" on this group in the sense of §4.2 of [53].*

Proof. The curve E is the reduction modulo p of a rational elliptic curve of the form $y^2 = x^3 - N^2 x$, where $N = 1$ in (1) and $N = 2$ in (2). This curve has endomorphism ring $\mathbb{Z}[i]$, where i corresponds to the map $(x, y) \mapsto (-x, iy)$;

modulo p the endomorphism i corresponds to the map $(x, y) \mapsto (-x, Ay)$ (note that A is a squareroot of -1 in $\mathbb{F}_p$). According to the theorem in §2 of [29], the Frobenius endomorphism of E is the (unique up to complex conjugation) element α of $\mathbb{Z}[i]$ having norm p and satisfying the congruence $\alpha \equiv \left(\frac{N}{p}\right) \pmod{2 + 2i}$, where $\left(\frac{N}{p}\right)$ denotes the Legendre symbol. When 4 divides A, we see that $\alpha = 1 + Ai \equiv 1 \pmod{2 + 2i}$; when $A \equiv 2 \pmod 4$, we note that $p \equiv 5 \pmod 8$ and hence $\left(\frac{2}{p}\right) = -1$, and so again $\alpha = 1 + Ai \equiv -1 \pmod{2 + 2i}$. Thus, in both cases the number of $\mathbb{F}_p$-points on E is $|\alpha - 1|^2 = A^2$. Moreover, all $\mathbb{F}_p$-points on E are in the kernel of the endomorphism $\alpha - 1 = Ai$, and $E(\mathbb{F}_p)$ is isomorphic as a $\mathbb{Z}[i]$-module to $\mathbb{Z}[i]/Ai\mathbb{Z}[i] \simeq \mathbb{Z}/A\mathbb{Z} \oplus \mathbb{Z}/A\mathbb{Z}$. In the first place, this implies that $E(\mathbb{F}_p)$ is isomorphic as an abelian group to $\mathbb{Z}/A\mathbb{Z} \oplus \mathbb{Z}/A\mathbb{Z}$. In the second place, if $P = (x, y)$ is a point of prime order $n \mid A$, the whole n-torsion group is generated over $\mathbb{Z}$ by P and $iP = (-x, Ay)$; in other words, the endomorphism i is a distortion map.

Remark 4. As noted in [2] (see Remark 2 of §2), if n is a prime dividing A, then most curves E over $\mathbb{F}_p$ with the property that $n^2 \mid \#E(\mathbb{F}_p)$ have cyclic n-part, that is, they do *not* have n^2 $\mathbb{F}_p$-points of order n, and one has to go to the degree-n extension of $\mathbb{F}_p$ to get all the points of order n. Thus, the property $E(\mathbb{F}_p) \simeq \mathbb{Z}/A\mathbb{Z} \oplus \mathbb{Z}/A\mathbb{Z}$ of the curves (1) and (2) is very unusual, statistically speaking. On the other hand, our elliptic curves are much easier to construct than ones with $n^2 \mid \#E(\mathbb{F}_p)$ and only n points of order n.

Remark 5. The coefficient of x in (1) and (2) can be multiplied by any fourth power N_0^4 in $\mathbb{F}_p$ without changing anything, as one sees by making the substitution $x \mapsto x/N_0^2$, $y \mapsto y/N_0^3$.

Remark 6. In general, a distortion map exists only for supersingular curves; it can exist for a nonsupersingular curve only when $k = 1$ (see Theorem 6 of [53]).

6.1   History of Embedding Degree 1

Although many papers have proposed different families of elliptic curves for use in pairing-based systems, until now no one has seriously considered families with embedding degree $k = 1$. Most authors stipulate from the beginning that $k \geq 2$. We know of only three papers ([23,25,53]) that briefly discuss curves E over $\mathbb{F}_p$ with $\#E(\mathbb{F}_p) = p - 1$. In [23], Joux points out that no efficient way is known to generate such curves with $p - 1$ divisible by n but not by n^2, a condition that he wants to have in order to ensure that the Tate pairing value $\langle P, P \rangle$ must always be nontrivial. In [25], Joux and Nguyen repeat this observation. Even though they then show that $\langle P, P \rangle$ is nontrivial for most P even when there are n^2 points of order n, they leave the impression that such curves are less desirable than the supersingular ones that they use in their examples.

In [53], Verheul discusses the nonsupersingular $k = 1$ curves. However, he erroneously states that the discrete logarithm problem in the subgroup $\langle P \rangle$ of prime order n reduces to the discrete log in the field $\mathbb{F}_n$, in which case one needs

$b_n \geq 1024$ to achieve 80 bits of security. This mistake leads him also to over-estimate the required bitlength of p, and apparently accounts for his negative view of the practicality of such curves. Thus, the few papers that include the $k = 1$ case quickly dismiss it from serious consideration. No valid reason has been given, however, for excluding such curves.

6.2   Choice of Parameters

We must choose $A = nh$ such that n and $p = A^2 + 1$ are prime; and, to maximize efficiency, we want

(a) n and p to have respective bitlengths approximately b_n and b_{p^k} corresponding to the desired security level (see Table 1);
(b) n to be a Solinas prime, that is, equal to a sum or difference of a small number of powers of 2;
(c) p also to be a Solinas prime.

The bitlengths of n and p in the following examples are equal to or just slightly more than the minimum values given in Table 1 for the corresponding security level.

Example 2. For 128 bits of security let n be the prime $2^{256} - 2^{174} + 1$ and let $h = 2^{1345}$. Then $p = (nh)^2 + 1 = 2^{3202} - 2^{3121} + 2^{3038} + 2^{2947} - 2^{2865} + 2^{2690} + 1$ is prime.

Example 3. For 192 bits of security let n be the prime $2^{386} - 2^{342} - 1$ and let $h = 2^{3802}$. Then $p = (nh)^2 + 1 = 2^{8376} - 2^{8333} + 2^{8288} - 2^{7991} + 2^{7947} + 2^{7604} + 1$ is prime.

Example 4. For 256 bits of security let n be the Mersenne prime $n = 2^{521} - 1$ and let $h = 2^{7216}$. Then $p = (nh)^2 + 1 = 2^{15474} - 2^{14954} + 2^{14432} + 1$ is prime.

Remark 7. If p is of a certain special form, then discrete logarithms can be found using a special version of the number field sieve (see, for example, [16,24]). Then the running time for $2b$-bit primes is roughly comparable to the running time of the general number field sieve for b-bit primes. For this reason it is important to avoid the special number field sieve when choosing p. It is clear that certain Solinas primes that one might want to use are of a form that permits the use of a modification of the special number field sieve with running time somewhere between that of the general and the special number field sieves. An analysis by Schirokauer [47] has shown that the primes in Examples 3 and 4 provide significantly less security than general primes of the same bitlength. In Example 2, however, his modified number field sieve did not yield a faster algorithm. More work remains to be done in order to determine which Solinas primes are vulnerable to faster versions of the number field sieve.

Example 5. For the prime $p = 2^{1007} + 2^{1006} + 2^{1005} + 2^{1004} - 1 = 240 \cdot 2^{1000} - 1$ discrete logs in $\mathbb{F}_p$ can be found using the special sieve. The reason is that 2^{200} is a root mod p of the polynomial $f(X) = 240X^5 - 1$, which has small degree and small coefficients.

7   Supersingular Curves with $k = 2$

Suppose that n is a prime and $p = nh - 1$ is also a prime, where $4 \mid h$. If h is not divisible by 3, we let E be the elliptic curve defined over $\mathbb{F}_p$ with equation

$$y^2 = x^3 - 3x; \tag{3}$$

if $12 \mid h$, then we let E be either the curve (3) or else the curve

$$y^2 = x^3 - 1. \tag{4}$$

It is an easy exercise to show that in these cases $\#E(\mathbb{F}_p) = p + 1 = nh$, and so E is a supersingular elliptic curve with embedding degree $k = 2$. Note also that $\beta = -1$ is a nonsquare in $\mathbb{F}_p$, and so $\mathbb{F}_{p^2} = \mathbb{F}_p[X]/(X^2 + 1)$. In addition, the map $(x, y) \mapsto (\zeta x, \varepsilon y)$ is a distortion map in the sense of [53], where $\zeta = -1$ and ε is a squareroot of -1 in $\mathbb{F}_{p^2}$ for the curve (3), and $\varepsilon = 1$ and ζ is a nontrivial cuberoot of 1 in $\mathbb{F}_{p^2}$ for the curve (4).

7.1   History of Embedding Degree 2 (Supersingular Case)

In the early days of elliptic curve cryptography, before the publication of [37] caused people to turn away from such elliptic curves, the supersingular curves (3) with $p \equiv -1 \pmod 4$ and (4) with $p \equiv -1 \pmod 6$ were the most popular examples, because of the simple form of the equation, the trivial determination of the group order, and the easy deterministic coding of integers as points (see Exercise 2 of Chapter 6 of [28]). For similar reasons, Boneh and Franklin used these curves as examples in [8]. On the other hand, authors who study implementation issues tend to shun supersingular curves, perhaps because of the subconscious association of the word "supersingular" with "insecure."

Despite the customary preference for nonsupersingular elliptic curves, there is no known reason why a nonsupersingular curve with small embedding degree k would have any security advantage over a supersingular curve with the same embedding degree. Of course, it is not inconceivable that some day someone might find a way to use the special properties of supersingular elliptic curves to attack the security of the system, perhaps by constructing a Verheul homomorphism from μ_n to the curve (see §3). However, one consequence of a generalized version of Verheul's theorem (see the end of §3) is that if supersingular curves were broken in this way, then the Diffie–Hellman problem in *any* finite field would be easy, and hence nonsupersingular curves of low embedding degree would be insecure as well. This means that the only way that supersingular curves could fall without bringing down all low-embedding-degree curves with them is through some special attack unrelated to a Verheul homomorphism.

Thus, on the one hand one has the remote possibility of a vulnerability of supersingular elliptic curves that is not shared by other curves of low embedding degree. On the other hand, one has the very real efficiency advantages of supersingular curves with $k = 2$. Namely, they provide the benefits of both the $k = 1$ case (flexibility in the choice of n and p) and also the $k \geq 2$ case (speedups coming from subfields).

7.2 Choice of Parameters

It is a simple matter to select n and h so that both n and p are Solinas primes.

Example 6. At the 80-bit security level let n be the prime $2^{160} + 2^3 - 1$, and let $h = 2^{360}$; then $p = nh - 1 = 2^{520} + 2^{363} - 2^{360} - 1$ is prime.

Example 7. At the 128-bit level let $n = 2^{256} + 2^{225} - 1$, $h = 2^{1326}$, $p = nh - 1 = 2^{1582} + 2^{1551} - 2^{1326} - 1$.

Example 8. At the 192-bit level let $n = 2^{384} - 2^{60} + 1$, $h = 2^{3847}$, $p = nh - 1 = 2^{4231} - 2^{3907} + 2^{3847} - 1$.

Example 9. At the 256-bit level let $n = 2^{521} - 1$, $h = 2^{6704}(2^{521} + 1)$, $p = nh - 1 = 2^{7746} - 2^{6704} - 1$. Note that here $12 \mid h$, so we can use either curve (3) or (4).

   As in the $k = 1$ case (see Remark 7), certain Solinas primes can be handled by a special version of the number field sieve for $\mathbb{F}_{p^2}$ which is faster than the general algorithm. In particular, Schirokauer [47] has shown that the four choices of p in Examples 6–9 provide less security than non-Solinas primes of the same bitlength. It is not yet clear whether primes p can be found that are sparse enough to provide enhanced efficiency of field arithmetic and are also resistant to speedups of the number field sieve.

8 Efficiency Comparisons

Let's briefly recall the ingredients in pairing computations. According to Proposition 8 of [39] (see also [13]), the Weil pairing $\widehat{e}(P, Q)$ is given by the formula

$$(-1)^n \frac{F_P(Q)}{F_Q(P)}, \qquad P \neq Q,$$

in which F_P, F_Q are functions whose divisors are $n(P) - n(\infty)$ and $n(Q) - n(\infty)$, respectively. Here F_P and F_Q must be normalized so that $F_P(\infty)/F_Q(\infty) = 1$.

   In recent years most authors have preferred to use the Tate pairing rather than the Weil pairing. To evaluate the Tate pairing at points P and Q, one first chooses an auxiliary point R (which must not be equal to P, $-Q$, $P - Q$, or ∞). One then evaluates the ratio

$$\frac{F_P(Q + R)}{F_P(R)},$$

with F_P as above. This preliminary value is an element of $\mathbb{F}_{p^k}^*$ that must be raised to the $((p^k - 1)/n)$-th power to convert it to an n-th root of unity.

   In pairing computations the procedure to compute $F_P(Q)$, $F_Q(P)$, or $F_P(Q + R)/F_P(R)$ resembles the double-and-add method for finding a point multiple. If n is a Solinas prime, then the number of adds/subtracts is negligible compared to the number of doublings. For each bit of n we have to perform a point doubling which leads to two functions $\ell = \ell_1/\ell_2$ and $v = v_1/v_2$ with constant denominators, and then we have a function-evaluation step of the form

$$\frac{f_1}{f_2} \leftarrow \frac{f_1^2}{f_2^2} \cdot \frac{v_2 \ell_1(Q)}{\ell_2 v_1(Q)}.$$

for the numerator (or denominator) in the Weil pairing and

$$\frac{f_1}{f_2} \leftarrow \frac{f_1^2}{f_2^2} \cdot \frac{\ell_1(Q+R)v_1(R)}{\ell_1(R)v_1(Q+R)}.$$

for the ratio in the Tate pairing (note that the denominators ℓ_2 and v_2 cancel in the Tate pairing). Such a procedure is called a "Miller operation" [39].

For this type of computation it is usually most efficient to use Jacobian coordinates (see [22], §3.2.2). A point (X, Y, Z) in Jacobian coordinates corresponds to the point (x, y) in affine coordinates with $x = X/Z^2$, $y = X/Z^3$. In Jacobian coordinates the formula for doubling a point $T = (X, Y, Z)$ takes the form $2T = (X_3, Y_3, Z_3)$ with $X_3 = (3X^2 + aZ^4)^2 - 8XY^2$, $Y_3 = (3X^2 + aZ^4)(4XY^2 - X_3) - 8Y^4$, $Z_3 = 2YZ$. The functions ℓ and v correspond, respectively, to the tangent line to the curve at T and the vertical line through the point $2T$:

$$v(x) = v_1(x)/v_2 = (Z_3^2 x - X_3)/Z_3^2;$$
$$\ell(x, y) = \ell_1(x, y)/\ell_2 = (Z_3 Z^2 y - 2Y^2 - (3X^2 + aZ^4)(xZ^2 - X))/(Z_3 Z^2).$$

8.1   The Case $k = 1$

We first examine the case $k = 1$, where E has equation (1) or (2). Using the above formulas, we count the number S of squarings and the number M of multiplications in $\mathbb{F}_p$ that must be performed for each bit of n. In the case of the Weil pairing, after initially setting $T = P$, $f_1 = f_2 = 1$, for each bit of n we do

$$T \leftarrow 2T, \qquad f_1 \leftarrow f_1^2 v_2 \ell_1(Q), \qquad f_2 \leftarrow f_2^2 \ell_2 v_1(Q). \tag{5}$$

Our field operation count is $9S + 12M$.[4] Since we must go through essentially the same procedure twice — once for $F_P(Q)$ and once for $F_Q(P)$ — the total number of operations per bit of n required to evaluate the Weil pairing is $18S + 24M$.

The Tate pairing has the advantage that the procedure is needed only once. Namely, we choose R to be the point $(0,0)$ for $k = 1$, and after initially setting $T = P$, $f_1 = f_2 = 1$, for each bit of n we do

$$T \leftarrow 2T, \qquad f_1 \leftarrow f_1^2 \ell_1(Q+R)v_1(R), \qquad f_2 \leftarrow f_2^2 \ell_1(R)v_1(Q+R). \tag{6}$$

---

[4] In the case $k \geq 2$, without loss of generality we may assume that the coefficient a in the elliptic curve equation $y^2 = x^3 + ax + b$ is equal to -3, in which case in the doubling one saves two squarings. (This is because $3X^2 + aZ^4 = 3(X + Z^2)(X - Z^2)$ when $a = -3$.) When $k = 1$, we suppose that the curve is given by (1) or (2), and so we still have the extra squarings but save one multiplication (by a). If we want to use the equation $y^2 = x^3 - 3x$ instead of (1) or (2), we may do so, provided that $3 \mid h = A/n$ (so that 3 is a quadratic residue in $\mathbb{F}_p$) and 3 is a fourth power in $\mathbb{F}_p$ when $4 \mid A$ but not when $A \equiv 2 \pmod 4$.

When $k = 1$, we have $9S + 13M$ rather than $18S + 24M$ for each bit of n, and in the case $k \geq 2$ one can gain further savings by working in subfields, as we'll see later. On the other hand, in the Tate pairing computation the preliminary result is an element of $\mathbb{F}_{p^k}^*$ that must be raised to the $((p^k - 1)/n)$-th power to convert it to an n-th root of unity. For high security levels the bitlength of p^k is large compared to that of n (the ratio is what we denoted γ in Table 1), and so the time required for this exponentiation is not negligible. If $k = 1$ and $(p - 1)/n$ has sparse binary representation, or if we use window methods, then the exponentiation is essentially $b_{p^k} - b_n = (\gamma - 1)b_n$ squarings in the field. This adds $(\gamma - 1)S$ to our operation count for each bit of n. If we suppose that $S \approx M$, then we see that the Tate method retains its advantage as long as $(\gamma - 1)S < 9S + 11M \approx 20S$. But when $\gamma > 21$ the Weil computation is faster in the case $k = 1$. According to Table 1, the cross-over point when we should switch to the Weil pairing for $k = 1$ occurs just around the 192-bit security level.

8.2   The Case $k \geq 2$

Now suppose that $k \geq 2$, and k is even. In that case one distinguishes between full field multiplications in $\mathbb{F}_{p^k}$, multiplications in the quadratic subfield $\mathbb{F}_{p^{k/2}}$ (each of which takes one third as long as a multiplication in the full field, see §5), and multiplications where one or both elements are in $\mathbb{F}_p$. We let S and M, as before, denote squaring and multiplication in the large field $\mathbb{F}_{p^k}$, and we let s and m denote squaring and multiplication in $\mathbb{F}_p$; we suppose that a multiplication of an element in $\mathbb{F}_{p^k}$ by an element in $\mathbb{F}_p$ takes time km. When we make efficiency comparisons, we shall further assume that $S \approx M$, $s \approx m$, and $M \approx \nu(k)m$, where $k = 2^i 3^j$ and $\nu(k) = 3^i 5^j$ (see §5).

In most cryptographic protocols there is some flexibility in the choice of order-n subgroups generated by P and by Q. In particular, one of the two — say, P — can be chosen in $E(\mathbb{F}_p)$. Then $\langle P \rangle$ is the unique subgroup of order n in $E(\mathbb{F}_p)$. In this case the Miller operation for computing $F_P(Q)$ in the Weil pairing is quicker than that for $F_Q(P)$, and so has been dubbed "Miller lite" by Solinas [52].

In addition, in [4] it was pointed out that when the embedding degree k is even, the subgroup $\langle Q \rangle \subset E(\mathbb{F}_{p^k})$ can be chosen so that the x-coordinates of all of its points lie in the quadratic subextension $\mathbb{F}_{p^{k/2}}$ and the y-coordinates are products of elements of $\mathbb{F}_{p^{k/2}}$ with $\sqrt{\beta}$, where β is a fixed nonsquare in $\mathbb{F}_{p^{k/2}}$ and $\sqrt{\beta}$ denotes a fixed squareroot in $\mathbb{F}_{p^k}$. We shall call such values of x and y "real" and "imaginary," respectively, by analogy with the familiar complex plane.

To see that Q can be chosen in this way, we consider the "twisted" elliptic curve $\widetilde{E}$ with equation $\beta y^2 = x^3 + ax + b$. It is easy to show that if E has $p^{k/2} + 1 - t$ points over the field $\mathbb{F}_{p^{k/2}}$, then $\widetilde{E}$ has $p^{k/2} + 1 + t$ points over $\mathbb{F}_{p^{k/2}}$. Over the big field $\mathbb{F}_{p^k}$ the number of points on E is equal to the product of the orders of E and its twist $\widetilde{E}$ over $\mathbb{F}_{p^{k/2}}$. Since n^2 divides $\#E(\mathbb{F}_{p^k})$ and only

n (but not n^2) divides $\#E(\mathbb{F}_{p^{k/2}})$, it follows that $n \mid \#\widetilde{E}(\mathbb{F}_{p^{k/2}})$.[5] Thus, there is a point $\widetilde{Q} \in \widetilde{E}(\mathbb{F}_{p^{k/2}})$ of order n. The map $(x, y) \mapsto (x, y\sqrt{\beta})$ maps $\widetilde{Q}$ and its multiples to $\mathbb{F}_{p^k}$-points of E (because $(y\sqrt{\beta})^2 = x^3 + ax + b$) that have "real" x and "imaginary" y.[6]

8.3 Operation Count for $k \geq 2$

When computing the Tate pairing, major savings can be obtained by ignoring terms that are contained in a proper subfield of $\mathbb{F}_{p^k}$ (see [18,4,48]). The reason such terms can be ignored is that when raised to the $((p^k - 1)/n)$-th power at the end of the Tate pairing computation, they become 1; this is because k is the multiplicative order of p modulo n, and so $(p^k - 1)/n$ is a multiple of $p^{k'} - 1$ for any proper divisor k' of k. In addition, in Theorem 1 of [4] it is shown that (again because of the exponentiation to the $((p^k - 1)/n)$-th power in the Tate pairing) the auxiliary point R can be ignored; that is, the Tate pairing value is $F_P(Q)^{(p^k-1)/n}$. Since the x-coordinate of Q — and hence $v_1(Q)$ — is in $\mathbb{F}_{p^{k/2}}$, it follows that we can drop the entire denominator in (6), and the function-evaluation step becomes simply

$$f_1 \leftarrow f_1^2 \ell_1(Q). \tag{7}$$

An operation count for a Miller lite point doubling and function evaluation gives $4s + 8m + S + M$ for $k = 2$ and $4s + (k + 7)m + S + M$ for $k \geq 4$ even.

The final stage of the Tate pairing computation is the exponentiation. This can be expedited if we use the fact that $n \mid \Phi_k(p)$, where Φ_k is the k-th cyclotomic polynomial; once again, this is a consequence of the assumption that k is the multiplicative order of p modulo n. We then write

$$y^{(p^k-1)/n} = \left(y^{(p^k-1)/\Phi_k(p)}\right)^{\Phi_k(p)/n}.$$

Now raising to the power $(p^k - 1)/\Phi_k(p)$ takes very little time (since the p-th power map takes negligible time in extensions $\mathbb{F}_p[X]/(X^k - \beta)$ once $X^{pi} \bmod (X^k - \beta)$ has been precomputed for $i = 1, 2, \ldots, k - 1$). Thus, our estimate for the number of field operations (squarings in $\mathbb{F}_{p^k}$) is the bitlength of $\Phi_k(p)/n$, which is $\frac{\varphi(k)}{k}b_{p^k} - b_n = (\tau_k\gamma - 1)b_n$, where we define

$$\tau_k = \frac{\varphi(k)}{k} = \begin{cases} 1/2 \text{ if } k = 2^i, \ i \geq 1; \\ 1/3 \text{ if } k = 2^i 3^j, \ i, j \geq 1. \end{cases}$$

Thus, the operation count for the exponentiation in the Tate pairing is $(\tau_k\gamma - 1)S$ for each bit of n.

---

[5] Another way to see this is to note that $n \mid (p^{k/2} + 1)$ and also $n \mid (p^{k/2} + 1 - t)$, from which it follows that $n \mid (p^{k/2} + 1 + t)$.

[6] In [5] it is shown that for certain curves with $k = 12$ a further speedup can be achieved by using a sextic rather than quadratic twist. However, we won't be considering this speedup in our operation counts.

However, a further speedup is possible because the element that is raised to the $(\Phi_k(p)/n)$-th power has norm 1 over any proper subfield of $\mathbb{F}_{p^k}$. In particular, this element is "unitary" over the quadratic subextension $\mathbb{F}_{p^{k/2}}$. As explained in [49], this means that one need only keep track of the "real" part of powers of the element and can use Lucas sequences to process each bit of the exponent using only one squaring and one multiplication in $\mathbb{F}_{p^{k/2}}$.[7] When $k = 2$, this allows us to replace $(\frac{\gamma}{2} - 1)S$ by $(\frac{\gamma}{2} - 1)(s + m)$ for the exponentiation in the Tate pairing; when $k \geq 4$ is even, the operation count is $(\tau_k\gamma - 1)(\widetilde{S} + \widetilde{M})$, where $\widetilde{S}$ denotes a squaring and $\widetilde{M}$ denotes a multiplication in the subfield $\mathbb{F}_{p^{k/2}}$.

If $k > 2$ is a multiple of 6 (as it will be for us), then instead of Lucas sequences one could use the trace representations in $\mathbb{F}_{p^{k/3}}$ that Lenstra and Verheul [32] developed in order to make their XTR cryptosystem more efficient (see [49]). This would not necessarily give a better speedup than the Lucas sequences; it is an open question whether the use of the quadratic or cubic subfield is best.

The results so far are summarized in the first two columns of Table 2.

Table 2. Operation counts for each bit of n

	Exponentiation at end of Tate pairing computation	Miller lite	Full Miller
$k = 1$	$(\gamma - 1)S$	not applicable	$9S+12M$ (Weil) $9S+13M$ (Tate)
$k = 2$	$(\frac{\gamma}{2} - 1)(s + m)$	$4s + 8m + S + M$	$4s + 8m + S + M$
$k \geq 4$ even	$(\tau_k\gamma - 1)(\widetilde{S} + \widetilde{M})$	$4s + (k + 7)m + S + M$	$km + 4\widetilde{S} + 6\widetilde{M} + S + M$

8.4   Weil or Tate? The Case $k \geq 2$

If we want to compute the Weil pairing rather than the Tate pairing, we need to go through two Miller procedures, one to find $F_P(Q)$ and the other to find $F_Q(P)$. In the case $k \geq 2$, we suppose that $P \in E(\mathbb{F}_p)$, in which case the former is the "Miller lite" part and the latter is the full Miller computation. At first glance it appears that even the Miller lite part is more time-consuming than in the case of the Tate pairing, because we can no longer neglect terms whose $((p^k - 1)/n)$-th power equals 1. However, we make the following observation. In any cryptographic application of the pairing it makes no difference if the pairing is replaced by its m-th power, where m is a fixed integer not divisible by n. In particular, for k even we can replace $\widehat{e}$ by its $(1 - p^{k/2})$-th power.[8] That means

---

[7] The use of Lucas sequences is closely analogous to computing the n-th power of a complex number on the unit circle; one can use the formula for $\cos(n\theta)$ and work only with the real part. See [49] for details of the Lucas method.

[8] In [26] it was noted that the $(1 - p^{k/2})$-th power of $\widehat{e}$ is the same as $\widehat{e}^2$; this is because $n \mid (p^{k/2} + 1)$, and so the $(1 - p^{k/2})$-th power of an n-th root of unity is the same as the $(1 - p^{k/2} + p^{k/2} + 1)$-th power.

that, just as in the case of the Tate pairing, terms in the Miller computations that lie in $\mathbb{F}_p$ or $\mathbb{F}_{p^{k/2}}$ can be ignored.

In the Miller lite computation the point Q has "real" x-coordinate and "imaginary" y-coordinate; and in the full Miller computation (where we stay with the notation in (5) but with Q now an $\mathbb{F}_p$-point) the point Q has coordinates in $\mathbb{F}_p$. In both the Miller lite and full Miller computations all of the ℓ- and v-terms in (5) except for $\ell_1(Q)$ lie in $\mathbb{F}_{p^{k/2}}$ (or are "purely imaginary"), and so the process (5) again simplifies to (7).

If we make a careful count of the number of operations required for each bit of n, we find that the operation count for the full Miller step is $km + 4\widetilde{S} + 6\widetilde{M} + S + M$, where, as before, $\widetilde{S}$ is a squaring and $\widetilde{M}$ is a multiplication in $\mathbb{F}_{p^{k/2}}$.[9]

We can decide between the Tate and Weil pairings by comparing the exponentiation column in Table 2 with the full Miller column. As before, we assume that $S \approx M$, $s \approx m$, and $M \approx \nu(k)m$; we also suppose that $\widetilde{S} \approx \frac{2}{3}M$ and $\widetilde{M} \approx \frac{1}{3}M$ (see §5). We find that when $k = 2$ the Tate pairing is quicker as long as $\gamma < 20$; but for higher values of γ — that is, starting at the 192-bit security level — we should switch to the Weil pairing. When $k \geq 4$ is even, the value of γ after which the advantage shifts to the Weil pairing is 28.8 for $k = 6$, 28.2 for $k = 12$, and 27.8 for $k = 24$. Thus, for those values of k we should switch to the Weil pairing at the 256-bit security level.

Remark 8. These conclusions about the relative speed of the Tate and Weil pairing computations are not definitive. Indeed, not nearly as much effort has been put into finding ways to speed up the full Miller operation in the Weil pairing as has been put into speeding up the exponentiation stage of the Tate pairing. So it is possible that further study of the matter will result in an earlier switch to the Weil pairing, which asymptotically at least is the faster method.

8.5 Time Comparison When $k = 1, 2, 6, 12, 24$

Let $T(b)$ denote the time required for a multiplication in $\mathbb{F}_p$ for general b-bit p, and let $\widetilde{T}(b)$ denote the time required when p is a b-bit Solinas prime. As before, we assume that $s \approx m$, $S \approx M$, $\widetilde{S} \approx \frac{1}{3}M$, $\widetilde{M} \approx \frac{1}{3}M$, $M \approx \nu(k)m$.

For $k = 1$ the operation count is $9S + 13M + \min((\gamma - 1)S, 9S + 11M)$, where the latter minimum determines the choice of Tate versus Weil pairing. For $k = 2$ the operation count is

$$4s + 8m + S + M + \min((\frac{\gamma}{2} - 1)(s + m), 4s + 8m + S + M),$$

or approximately $(16 + \min(\gamma, 20))m$. For $k = 6, 12, 24$ the operation count is

$$\approx \left(k + 11 + \frac{4}{3}\nu(k) + \min(\frac{2}{9}\gamma\nu(k), k + 6\nu(k)) \right)m.$$

---

[9] In the supersingular case (4) with $k = 2$, where $a = 0$ rather than -3, a multiplication can be replaced by a squaring in the point-duplication part of both the Miller lite and full Miller computations. Of course, this has no effect on Table 3.

These formulas give us the time estimates in Table 3. Notice that for non-supersingular curves Table 3 suggests that even at the 80-bit security level the choice $k = 2$ is less efficient than higher k, and that, more generally, for $k \geq 2$ large k has an advantage. The comparison between $k = 1$ and $k \geq 2$ is harder to make, because it depends on how much of a saving we are able to achieve when multiplying modulo a Solinas prime rather than an arbitrary prime. It is not clear, for example, whether $42\widetilde{T}(15360)$ is greater or less than $36T(7680)$ or $1049T(640)$. The limited experiments we have conducted with integer multiplication packages were inconclusive.

We estimate that $T(512)$ is at least twice $\widetilde{T}(512)$, and so for $k = 2$ supersingular curves are at least twice as fast as nonsupersingular curves at the 80-bit security level.

Finally, we emphasize that the above analysis is imprecise, and definitive conclusions will be possible only after extensive experimentation. In addition, the relative merits of $k = 1$ and $k \geq 2$ depend on the protocol being used and the types of optimization that are desirable in the particular application.

Table 3. Pairing evaluation time for each bit of n (ss="supersingular," ns="nonsupersingular")

Security (bits)	80	128	192	256
bitlength of p^k	1024	3072	8192	15360
$k = 1$	$27\widetilde{T}(1024)$	$33\widetilde{T}(3072)$	$42\widetilde{T}(8192)$	$42\widetilde{T}(15360)$
$k = 2$ (ss)	$22\widetilde{T}(512)$	$28\widetilde{T}(1536)$	$36\widetilde{T}(4096)$	$36\widetilde{T}(7680)$
$k = 2$ (ns)	$22T(512)$	$28T(1536)$	$36T(4096)$	$36T(7680)$
$k = 6$	$58T(171)$	$77T(512)$	$108T(1365)$	$133T(2560)$
$k = 12$		$203T(256)$	$296T(683)$	$365T(1280)$
$k = 24$				$1049T(640)$

For example, in identity-based encryption suppose that we are very concerned about the time it takes to convert Alice's identity to a public key, which in the Boneh–Franklin system is a point $I_A \in E(\mathbb{F}_p)$. One is then at a disadvantage when $k = 1$. The reason is that after Alice's identity is hashed into the curve, the resulting point must be multiplied by $h = \sqrt{(p-1)/n}$ to get a point I_A of order n. The bitlength of h is $\frac{1}{2}(\gamma - 1)b_n$. In contrast, when $k \geq 2$ the cofactor $h \approx p/n$ is usually small; its bitlength is $(\rho - 1)b_n$, where $\rho = \log p / \log n$ is generally between 1 and 2. In the $k = 1$ case, to avoid the point multiplication by h one might want to use a different identity-based encryption scheme, such as the one in [44] or [54], where Alice's public key is an integer rather than a point.

9   Open Problems

(1) Prove Verheul's theorem for class-VI supersingular elliptic curves, which, as we saw at the end of §3, contain subgroups isomorphic to the multiplicative groups of all finite fields.

(2) To what extent can the special number field sieve be applied to $\mathbb{F}_p$ for Solinas primes p? For what Solinas primes can we be confident that only the general number field sieve and not the special one can be used to find discrete logarithms?

(3) What Solinas primes can be used with embedding degree $k = 2$ without allowing an attacker to use the special number field sieve for $\mathbb{F}_{p^2}$?

(4) At the 80-bit security level with nonsupersingular elliptic curves, is embedding degree 6 faster than embedding degree 2, as suggested by the preliminary results in §8.5?

(5) For higher security levels such as 192 and 256 bits, is it possible to construct nonsupersingular examples with $k \geq 2$ where n and p^k have roughly b_n and b_{p^k} bits and both n and p are Solinas primes?

(6) Try to find ways to speed up the full Miller operation, and then reexamine the relative speed of the Tate and Weil pairing computations.

(7) Determine more precisely the relative efficiency of curves with embedding degree 1.

(8) When k is a multiple of 6, investigate the use of trace methods similar to the one in [32] to speed up the exponentiation stage of the Tate pairing computation.

(9) Compare implementations in large characteristic p with supersingular implementations in characteristic 2 and 3 [3].

(10) More generally, analyze the efficiency of pairing-based protocols at the AES security levels.

10   Conclusions

It is still hard to say whether pairing-based cryptosystems will be able to provide satisfactory security and efficiency as the desired level of security rises. None of the concerns raised in §3 give sufficient cause to avoid these systems, but they certainly point to the need to proceed with caution.

Despite the spate of recent papers on curve selection for pairing-based cryptosystems, the simplest cases — that of embedding degree 1 and that of supersingular curves with embedding degree 2 — have been largely neglected. To be sure, the $k = 1$ case has some drawbacks, since all of the arithmetic must be done in the large field (there being no subfield) and certain simplifications of the pairing computations when $k \geq 2$ are unavailable. On the other hand, the greater flexibility in choosing the pair (n, p) is a compensating advantage. Thus, the embedding degree 1 case should be seriously considered by implementers of pairing-based cryptography.

Similarly, unless someone finds a way to exploit some special properties of supersingular curves to attack the Bilinear Diffie–Hellman Problem — and we see no reason to believe that this will happen — implementers should pay special attention to supersingular curves with $k = 2$. Those curves have the efficiency advantages of both $k = 1$ (flexibility in the choice of n and p) and also $k \geq 2$ (speedups coming from subfields).

When $k = 1$ the Weil pairing rather than the Tate pairing should be used at security levels significantly above 192 bits, such as the 256-bit level. For $k = 2$ the Weil pairing should be used at the 192-bit level and above, and for $k \geq 4$ even the Weil pairing should be used at the 256-bit level.

For nonsupersingular curves with $k \geq 2$ our preliminary results do not seem to support the viewpoint expressed in [48] that $k = 2$ is the embedding degree that leads to the fastest implementation. Rather, at all security levels considered it appears that among the possible values of $k \geq 2$ one should choose $k = 2^i 3^j$ as large as possible.

There is a need for further study of the relative merits of different values of k as our security requirements increase from the present 80 bits to 128, 192, 256 bits and beyond.

Acknowledgments

We would like to thank Darrel Hankerson and Arjen Lenstra for answering our questions about efficient finite field arithmetic, Oliver Schirokauer for answering questions about the number field sieve, Paulo Barreto and Michael Scott for calling our attention to the papers [49] and [3], and Steven Galbraith for commenting extensively on an earlier version of the paper.

References

1. L. Adleman and M. Huang, Function field sieve methods for discrete logarithms over finite fields, *Information and Computation*, **151** (1999), 5-16.
2. R. Balasubramanian and N. Koblitz, The improbability that an elliptic curve has subexponential discrete log problem under the Menezes–Okamoto–Vanstone algorithm, *J. Cryptology*, **11** (1998), 141-145.
3. P. Barreto, S. Galbraith, C. Ó hÉigeartaigh, and M. Scott, Efficient pairing computation on supersingular abelian varieties, http://eprint.iacr.org/2004/375/
4. P. Barreto, B. Lynn, and M. Scott, On the selection of pairing-friendly groups, *Selected Areas in Cryptography – SAC 2003*, LNCS 3006, 2004, 17-25.
5. P. Barreto and M. Naehrig, Pairing-friendly elliptic curves of prime order, *Selected Areas in Cryptography – SAC 2005*, to appear; http://eprint.iacr.org/2005/133/
6. D. Boneh, X. Boyen, and E.–J. Goh, Hierarchical identity based encryption with constant size ciphertext, *Advances in Cryptology – EUROCRYPT 2005*, LNCS 3494, 2005, 440-456.
7. D. Boneh, X. Boyen, and H. Shacham, Short group signatures, *Advances in Cryptology – CRYPTO 2004*, LNCS 3152, 2004, 41-55.
8. D. Boneh and M. Franklin, Identity-based encryption from the Weil pairing, *Advances in Cryptology – CRYPTO 2001*, LNCS 2139, 2001, 213-229.

9. D. Boneh, C. Gentry, and B. Waters, Collusion resistant broadcast encryption with short ciphertexts and private keys, *Advances in Cryptology – CRYPTO 2005*, LNCS 3621, 2005, 258-275.

10. D. Boneh, B. Lynn, and H. Shacham, Short signatures from the Weil pairing, *Advances in Cryptology – ASIACRYPT 2001*, LNCS 2248, 2001, 514-532.

11. D. Boneh and R. Venkatesan, Breaking RSA may not be equivalent to factoring, *Advances in Cryptology – EUROCRYPT '98*, LNCS 1233, 1998, 59-71.

12. F. Brezing and A. Weng, Elliptic curves suitable for pairing based cryptography, *Designs, Codes and Cryptography*, **37** (2005), 133-141.

13. L. Charlap and R. Coley, An Elementary Introduction to Elliptic Curves II, CCR Expository Report 34, 1990, available from `http://www.idaccr.org/reports/reports.html`

14. H. Cohen, *A Course in Computational Algebraic Number Theory*, Springer-Verlag, 1993.

15. D. Coppersmith, Fast evaluation of logarithms in fields of characteristic two, *IEEE Transactions on Information Theory*, **30** (1984), 587-594.

16. T. Denny, O. Schirokauer, and D. Weber, Discrete logarithms: the effectiveness of the index calculus method, *Algorithmic Number Theory Symp. II*, LNCS 1122, 1996, 337-361.

17. G. Frey and H. Rück, A remark concerning m-divisibility and the discrete logarithm in the divisor class group of curves, *Math. Comp.*, **62** (1994), 865-874.

18. S. Galbraith, Pairings, Ch. IX of I. F. Blake, G. Seroussi, and N. P. Smart, eds., *Advances in Elliptic Curve Cryptography*, Vol. 2, Cambridge University Press, 2005.

19. S. Galbraith, J. McKee and P. Valença, Ordinary abelian varieties having small embedding degree, http://eprint.iacr.org/2004/365/

20. D. Gordon, Discrete logarithms in $GF(p)$ using the number field sieve, *SIAM J. Discrete Math.*, **6** (1993), 124-138.

21. R. Granger and F. Vercauteren, On the discrete logarithm problem on algebraic tori, *Advances in Cryptology – CRYPTO 2005*, LNCS 3621, 2005, 66-85.

22. D. Hankerson, A. Menezes, and S. Vanstone, *Guide to Elliptic Curve Cryptography*, Springer-Verlag, 2004.

23. A. Joux, A one round protocol for tripartite Diffie–Hellman, *J. Cryptology*, **17** (2004), 263-276.

24. A. Joux and R. Lercier, Improvements to the general number field sieve for discrete logarithms in prime fields, *Math. Comp.*, **72** (2003), 953-967.

25. A. Joux and K. Nguyen, Separating Decision Diffie–Hellman from Computational Diffie–Hellman in cryptographic groups, *J. Cryptology*, **16** (2003), 239-247.

26. B. Kang and J. Park, On the relationship between squared pairings and plain pairings, http://eprint.iacr.org/2005/112/

27. D. Knuth, *The Art of Computer Programming*, 3rd ed., Vol. 2, Addison-Wesley, 1997.

28. N. Koblitz, *A Course in Number Theory and Cryptography*, Springer-Verlag, 1987.

29. N. Koblitz, *Introduction to Elliptic Curves and Modular Forms*, 2nd ed., Springer-Verlag, 1993.

30. N. Koblitz, An elliptic curve implementation of the finite field digital signature algorithm, *Advances in Cryptology – CRYPTO '98*, LNCS 1462, 1998, 327-337.

31. A. Lenstra, Unbelievable security: matching AES security using public key systems, *Advances in Cryptology – ASIACRYPT 2001*, LNCS 2248, 2001, 67-86.

32. A. Lenstra and E. Verheul, The XTR public key system, *Advances in Cryptology – CRYPTO 2000*, LNCS 1880, 2000, 1-19.

33. H. W. Lenstra, Jr., Factoring integers with elliptic curves, *Annals Math.*, **126** (1987), 649-673.
34. R. Lidl and H. Niederreiter, *Finite Fields*, 2nd ed., Cambridge University Press, 1997.
35. U. Maurer and S. Wolf, The Diffie–Hellman protocol, *Designs, Codes and Cryptography*, **19** (2000), 147-171.
36. A. Menezes, *Elliptic Curve Public Key Cryptosystems*, Kluwer Academic Publishers, 1993.
37. A. Menezes, T. Okamoto, and S. Vanstone, Reducing elliptic curve logarithms to logarithms in a finite field, *IEEE Trans. Inform. Theory, IT-39*, 1993, 1639-1646.
38. A. Menezes and S. Vanstone, ECSTR (XTR): Elliptic Curve Singular Trace Representation, Rump Session of Crypto 2000.
39. V. Miller, The Weil pairing and its efficient calculation, *J. Cryptology*, **17** (2004), 235-261.
40. A. Miyaji, M. Nakabayashi, and S. Takano, New explicit conditions of elliptic curve traces for FR-reduction, *IEICE Trans. Fundamentals, E84-A (5)*, 2001.
41. D. Naccache and J. Stern, Signing on a postcard, *Financial Cryptography – FC 2000*, LNCS 1962, 2001, 121-135.
42. National Institute of Standards and Technology, Special Publication 800-56: Recommendation for pair-wise key establishment schemes using discrete logarithm cryptography, Draft, 2005.
43. L. Pintsov and S. Vanstone, Postal revenue collection in the digital age, *Financial Cryptography – FC 2000*, LNCS 1962, 2001, 105-120.
44. R. Sakai and M. Kasahara, ID based cryptosystems with pairing on elliptic curve, http://eprint.iacr.org/2003/054/
45. O. Schirokauer, Discrete logarithms and local units, *Phil. Trans. Royal Soc. London A*, **345** (1993), 409-423.
46. O. Schirokauer, The special function field sieve, *SIAM J. Discrete Math.*, **16** (2002), 81-98.
47. O. Schirokauer, The number field sieve for integers of low weight, preprint, 2005.
48. M. Scott, Computing the Tate pairing, *Topics in Cryptology — CT-RSA 2005*, LNCS 3376, 2005, 300-312.
49. M. Scott and P. Barreto, Compressed pairings, *Advances in Cryptology – CRYPTO 2004*, LNCS 3152, 2004, 140-156.
50. M. Scott and P. Barreto, Generating more MNT elliptic curves, *Designs, Codes and Cryptography*, to appear; http://eprint.iacr.org/2004/058/
51. J. Solinas, Generalized Mersenne numbers, Technical Report CORR 99-39, University of Waterloo, 1999, `http://www.cacr.math.uwaterloo.ca/techreports/1999/corr99-39.pdf`
52. J. Solinas, ID-based digital signature algorithms, 2003, `http://www.cacr.math.uwaterloo.ca/conferences/2003/ecc2003/solinas.pdf`
53. E. Verheul, Evidence that XTR is more secure than supersingular elliptic curve cryptosystems, *J. Cryptology*, **17** (2004), 277-296.
54. B. Waters, Efficient identity-based encryption without random oracles, *Advances in Cryptology – EUROCRYPT 2005*, LNCS 3494, 2005, 114-127.

Improved Decoding of Interleaved AG Codes

Andrew Brown, Lorenz Minder, and Amin Shokrollahi

Laboratoire des mathematiques algorithmiques (LMA),
Ecole Polytechnique Federale de Lausanne (EPFL), 1015 Lausanne
{andrew.brown, lorenz.minder, amin.shokrollahi}@epfl.ch

Abstract. We analyze a generalization of a recent algorithm of Bleichenbacher et al. for decoding interleaved codes on the Q-ary symmetric channel for large Q. We will show that for any m and any ϵ the new algorithms can decode up to a fraction of at least $\frac{m}{m+1}(1 - R - 2Q^{-1/2m}) - \epsilon$ errors, and that the error probability of the decoder is bounded by $O(1/q^{\epsilon n})$, where n is the block-length. The codes we construct do not have a-priori any bound on their length.

1   Introduction

The general Q-ary symmetric channel of communication has not been as prominently featured in the literature as the binary symmetric channel. While the case of small Q has been investigated by some authors in connection with belief-propagation algorithms, the case of large Q has been largely untouched.

Perhaps one reason for this omission is the complexity of belief-propagation type algorithms which increases with the alphabet size Q, rendering the design of efficient decoding algorithms impossible for large Q. Another possible reason is the observation that for large Q the code design problem can be reduced to the code design problem for the binary *erasure* channel, albeit at the expense of some loss in the rate of the transmission. This reduction is for example employed in the Internet: in this case the symbols are packets; each packet is equipped with a checksum, or more generally, a hash value. After the transmission, the hash value of each symbol is checked, and a symbol is declared as erased if the hash value does not match. If h bits are used for the hash value, and if $Q = 2^{mh} = q^m$, then, each symbol's effective information rate is reduced by a factor of $(m - 1)/m$. If the error rate of the Q-ary symmetric channel is p, and if the erasure code operates at a rate of $1 - p - \varepsilon$ for some ε, then the effective rate of the transmission is about $1 - (p + \varepsilon + 1/m)$, and the error probability is upper bounded by $n/2^h = n/q$, where n is the block-length of the erasure code, when an erasure code with a linear time decoding algorithm, such as a Tornado code [9] is used.

A linear decoding algorithm for the Q-ary symmetric channel using LDPC codes was recently proposed by Luby and Mitzenmacher [8]. They did not exhibit codes that come arbitrarily close to the capacity of the Q-ary symmetric channel, but it is possible to extend their methods to find such codes [11]. In their construction, the error probability of the decoder is at most $O(n/Q)$, which can be much smaller than the error probability obtained using the hashing method.

Recently, Bleichenbacher et al. [1] invented a new decoding algorithm for Interleaved Reed-Solomon Codes over the Q-ary symmetric channel. As the name suggests,

N.P. Smart (Ed.): Cryptography and Coding 2005, LNCS 3796, pp. 37–46, 2005.

the codes are constructed with an interleaving technique from m Reed-Solomon codes defined over $\mathbb{F}_q$, if $Q = q^m$. These codes are similar to well-known product code constructions with Reed-Solomon codes as inner codes, but there is an important improvement: interleaved codes model the Q-ary channel more closely than a standard decoder for the product code would. It follows that interleaved codes achieve much better rates: interleaved Reed-Solomon Codes can asymptotically have rates as large as $1 - p(1 + 1/m)$, which is much more than the rate $1 - 2p$ achieved with a standard product code decoder. Bleichenbacher et al. prove that the error probability of their decoder is upper bounded by $O(n/q)$, where n is the block length of the code. Compared to the hashing method, this decoder has about the same error probability, but the rate of the code is closer to the capacity of the channel.

A general method for decoding of interleaved codes has been discussed in [3]. The gist of the algorithm is to find a polynomial in $m + 1$ variables that passes through the points given by the interpolation points of the code and the coordinate positions of the received words. The polynomial can then be used to scan the received word, and probabilistically identify the incorrect positions. The method can decode up to a fraction of $1 - R - R^{m/(m+1)}$ errors, with an error probability of $1 - O(n^{O(m)}/q)$, where R is the rate of the code. Note that the error probability of this algorithm depends on n. Note also that this algorithm is superior to that of Bleichenbacher et al. for small rates. The interleaved decoding algorithm has also been used in conjunction with concatenated coding [6].

Another class of algorithms to which the interleaved decoding algorithm can be compared is that of list-decoding algorithms [13,12,5]. However, this comparison is not fair, since these decoding algorithms work under adversarial conditions, i.e., recover a list of closest codewords without any restriction on the noise (except the number of corrupted positions). The best known codes to-date (in terms of error-correction capability) with a polynomial time decoding algorithm are those of of Parvaresh and Vardy [10]. For these codes the authors provide a decoding algorithm which can correct up to a fraction of $1 - \epsilon$ errors with a code of length n and rate $\Omega(\epsilon/\log(1/\epsilon))$ over an alphabet of size $n^{O(\log(1/\epsilon))}$. The codes provided in this paper improve upon this bound considerably, when the rate is not too small.

We have recently shown in [2] that the error probability of the decoder in [1] is in fact $O(1/q)$, independent of n. In this paper, we present a slightly different algorithm than that of Bleichenbacher et al. for the class of algebraic-geometric codes (AG-codes). We will show that the algorithm can successfully decode e errors with an error probability that is proportional to

$$\left(\frac{1}{q-1}\right)^{\frac{m}{m+1}n(1-R)-\frac{2m-1}{m+1}g-\frac{m-1}{m+1}-e},$$

where g is the genus of the curve underlying the AG-code, R is the rate, and $Q = q^m$. In particular, when $e = \frac{m}{m+1}n(1 - R) - \frac{2m-1}{m+1}g$, the error probability is roughly proportional to $1/(q - 1)$.

Since the error probability of our algorithm does not depend on n, it is possible to consider long codes over the alphabet $\mathbb{F}_q$. In particular, using codes from asymptotically optimal curves over $\mathbb{F}_{q^2}$ [7,4], and assuming that m is large enough, our codes will be able to reliably decode over a Q-ary symmetric channel with error probability p, and maintain a rate close to $1 - p - \frac{2}{q-1}$.

Despite the proximity to channel capacity we can gain with this algorithm, the construction of codes on the Q-ary channel with both rate close to the capacity and polynomial-time decoding complexity (where we measure complexity relative to the size of the received input, i.e. as a function of $n \log(Q)$), is still an open challenge.

In the next two sections of this paper we will introduce interleaved codes and the main decoding algorithm, and analyse this algorithm. The last section gives a detailed comparison of our method with various hashing methods. For the rest of the paper we will assume familiarity with the basic theory of AG-codes.

2 Interleaved AG-Codes and Their Decoding

Let $\mathcal{X}$ be an absolutely irreducible curve over $\mathbb{F}_q$, and let $D, P_1, \ldots, P_n$ denote $n + 1$ distinct $\mathbb{F}_q$-rational points of $\mathcal{X}$. Let g denote the genus of $\mathcal{X}$. For a divisor A of $\mathcal{X}$ we denote by $\mathcal{L}(A)$ the associated linear space. The theorem of Riemann states that the dimension of this space, denoted $\dim(A)$, is at least $\deg(A) - g + 1$.

Fix a parameter α with $2g - 2 < \alpha < n$. A (one-point) AG-code associated to $D, P_1, \ldots, P_n$ and α is the image of the evaluation map $\mathrm{Ev} \colon \mathcal{L}(\alpha D) \to \mathbb{F}_q^n$, $\mathrm{Ev}(f) = (f(P_1), \ldots, f(P_n))$.

Suppose that $Q = q^m$, and let $\beta_1, \ldots, \beta_m$ denote a basis of $\mathbb{F}_Q$ over $\mathbb{F}_q$. We define a code over $\mathbb{F}_Q$ of length m in the following way: the codewords are

$$\left(\sum_{j=1}^{m} f_j(P_1)\beta_j, \ldots, \sum_{j=1}^{m} f_j(P_n)\beta_m \right),$$

where $(f_1, \ldots, f_m) \in \mathcal{L}(\alpha D)^m$. (This algebraic interpretation of interleaved coding was communicated to us by A. Vardy [14].) Note that this code does not necessarily form an $\mathbb{F}_Q$-vector space, but it does form an $\mathbb{F}_q$-vector space.

Suppose that such a codeword is sent over a Q-ary symmetric channel, and that e errors occur during the transmission. Denote by E the set of these error positions. Because of the properties of the Q-ary symmetric channel, each of the m codewords of the constituent code is independently subjected to a q-ary symmetric channel, and for each of them an error has occured in the set E of error positions.

Our task is to decode the codeword. We proceed in a way similar to [1]: let t be a parameter to be determined later, and let

$$W := \begin{pmatrix} \phi_1(P_1) & \phi_2(P_1) & \cdots & \phi_d(P_1) \\ \phi_1(P_2) & \phi_2(P_2) & \cdots & \phi_d(P_2) \\ \vdots & \vdots & \ddots & \vdots \\ \phi_1(P_n) & \phi_2(P_n) & \cdots & \phi_d(P_n) \end{pmatrix},$$

$$V := \begin{pmatrix} \psi_1(P_1) & \psi_2(P_1) & \cdots & \psi_s(P_1) \\ \psi_1(P_2) & \psi_2(P_2) & \cdots & \psi_s(P_2) \\ \vdots & \vdots & \ddots & \vdots \\ \psi_1(P_n) & \psi_2(P_n) & \cdots & \psi_s(P_n) \end{pmatrix},$$

where $\phi_1, \ldots, \phi_d$ form a basis of $\mathcal{L}((t+g)D)$, and $\psi_1, \ldots, \psi_s$ form a basis of $\mathcal{L}((t + g + \alpha)D)$. Let $\left(\sum_{j=1}^{m} y_{1j}\beta_j, \ldots, \sum_{j=1}^{m} y_{nj}\beta_j\right)$ be the received word, and let

$$
A := \begin{pmatrix} V & & & \vline & -D_1W \\ & V & & \vline & -D_2W \\ & & \ddots & \vline & \vdots \\ & & & V \vline & -D_mW \end{pmatrix}, \tag{1}
$$

where D_j is the diagonal matrix with diagonal entries $y_{1j}, \ldots, y_{nj}$. The decoding process is now as follows:

- Find a nonzero element $v = (v_1 \mid \cdots \mid v_m \mid w)$ in the right kernel of A, where $v_1, \ldots, v_m \in \mathbb{F}_q^s$ and $w \in \mathbb{F}_q^d$. If v does not exist, output a decoding error.
- Identify $v_1, \ldots, v_m$ with functions in $\mathcal{L}((t + g + \alpha)D)$, and w with a function in $\mathcal{L}((t+g)D)$. If w divides v_j for each $j = 1, \cdots, m$, then set $f_1 = v_1/w, \cdots, f_m = v_m/w$, and output $f_1, \cdots, f_m$. Otherwise, output a decoding error.

Note that we did not explicitly require that w be nonzero: one can show that for a nonzero solution the vector w has to be nonzero.

Apart from some minor side conditions, there is no a-priori condition on the value of t. Some suggestions on how to choose this value are given in the next section.

The main theorem of the paper is as follows:

Theorem 1. *Suppose that t is such that $m(n - \alpha - t) - (m - 1)(g + 1) - 1 \geq t$, let e denote the number of errors incurred during transmission, and suppose that $e \leq t$. Then we have:*

(1) *If $e + t < n - \alpha - g$, then the error probability of the above decoder is zero.*
(2) *For general $e \leq t$ the error probability of the above decoder is at most*

$$
\left(\frac{1}{q-1}\right)^{m(n-\alpha-t)-(m-1)(g+1)-e}.
$$

This theorem will be proved in the next section. Note that the condition $m(n - \alpha - t) - (m - 1)(g + 1) - 1 \geq t$ implies that

$$
t \leq \left\lfloor \frac{m}{m+1}(n - \alpha) - \frac{m-1}{m+1}g - \frac{m-2}{m+1} \right\rfloor.
$$

3   Analysis of the Decoder

To analyze the decoder of the last section, we make the following simplifying assumptions:

(a) The error positions are $1, 2, \ldots, e$.
(b) The functions $f_1, \ldots, f_m$ sent over the channel are all zero.

It is easily seen that we can assume (a) and (b) without loss of generality. This goes without saying for (a); as for (b), note that since the code is invariant under addition, the behavior of the matrix A in (1) with respect to the algorithm is the same no matter which codeword is sent.

The assumptions imply the following:

(1) For each j, the first e diagonal entries of D_j are uniform independent random variables over $\mathbb{F}_q$.
(2) For each j, the last $n - e$ diagonal entries of D_j are zero.
(3) The probability of a decoding error is upper bounded by the probability that there exists a vector $(v_1 \mid \cdots \mid v_m \mid w)$ in the right kernel of A for which at least one of the v_i is nonzero, plus the probability that the right kernel of A is trivial.

Note that because both the number of errors and the error positions have been fixed, the only randomness in the matrix A comes from the values y_{ij} for $i = 1, \ldots, e$ and $j = 1, \ldots, m$.

We will show that if $e \leq t$, then the right kernel of A is nontrivial. Hence, we only need to bound the probability that there exists a vector $(v_1 \mid \cdots \mid v_m \mid w)$ in the right kernel of A for which at least one of the v_i is nonzero. Let us call such a vector *erroneous*. Note that if the right kernel of A does not contain any erroneous vectors, then the algorithm is successful with probability one.

We bound the probability of the existence of an erroneous vector in the following way: for each nonzero v_1, we calculate the expected number of $(v_2 \mid \cdots \mid v_m)$ such that $(v_1 \mid \cdots \mid v_m \mid w)$ is in the right kernel of A, for some w. An upper bound on the desired probability is then obtained from this easily using Markov's inequality.

PROOF OF THEOREM 1. First we will show that if $e \leq t$, then the right kernel of A is nontrivial. To this end, note that by the Theorem of Riemann $\dim \left((t + g)D - \sum_{i=1}^{e} P_i \right) \geq t - e + 1 > 0$, hence $\mathcal{L} \left((t + g)D - \sum_{i=1}^{e} P_i \right)$ is nontrivial. Let w be a nonzero function in this space. Setting $v_j := w f_j$, we see that the vector $(v_1 \mid \cdots \mid v_m \mid w)$ is in the right kernel of A, and is nontrivial, as required.

It follows that the error probability of the decoder is upper bounded by the probability that the right kernel of A contains erroneous vectors. Assume that $(v_1 \mid \cdots \mid v_m \mid w)$ is in the right kernel of A, and identify $v_1, \ldots, v_m$ with functions in $\mathcal{L} \left((t + g + \alpha)D \right)$, and w with a function in $\mathcal{L} \left((t + g)D \right)$ in the usual way. Because $(v_1 \mid \cdots \mid v_m \mid w)$ is in the right kernel of A, we have:

$$\forall i = 1, \ldots, n, \ \forall j = 1, \ldots, m: \quad v_j(P_i) = y_{ij} w(P_i). \tag{2}$$

Furthermore, since we are assuming that the zero codeword was transmitted, we have $y_{ij} = 0$ for $i > e$ (since i is not an error position). From this and (2) we can deduce that

$$\forall i = e + 1, \ldots, n, \ \forall j = 1, \ldots, m, : \quad v_j(P_i) = 0. \tag{3}$$

This implies that

$$\forall j = 1, \ldots, m: \quad v_j \in \mathcal{L} \left((t + \alpha + g)D - \sum_{i=e+1}^{n} P_i \right). \tag{4}$$

In particular, this proves part (1) of the theorem: if $t+\alpha+g-n+e < 0$, or equivalently, if $t+e < n-\alpha-g$, then this linear space is trivial, and hence any element in the right kernel of A is non-erroneous (since it has the property that $v_j = 0$ for all $j = 1,\ldots,m$).

Since any $i < e$ is an error position, we have:

$$\forall i = 1,\ldots,e, \quad \forall j = 1,\ldots,m: \quad y_{ij} \neq 0. \tag{5}$$

Moreover, we have:

$$\forall i = 1,\ldots,e: \quad \frac{v_1(P_i)}{y_{i1}} = \frac{v_2(P_i)}{y_{i2}} = \cdots = \frac{v_m(P_i)}{y_{im}} = w(P_i). \tag{6}$$

Because of (4), we know that $v_j \in \mathcal{L}\big((t+g+\alpha)D - \sum_{i=e+1}^n P_i\big)$ for $j = 1,\ldots,m$. Hence v_j is uniquely determined by its values on $\ell_1 := t + g + \alpha - n + e + 1$ of the points $P_1,\ldots,P_e$, and for at most ℓ_1 of these points does v_j evaluate to zero, if v_j is nonzero. By the same token, if v_j is nonzero, then $v_j(P_i)$ is nonzero for at least $\ell_2 := e - \ell_1 = e - (t + g + \alpha - n + e + 1) = n - t - g - \alpha - 1$ of these points.

Now fix some nonzero $v_1 \in \mathcal{L}\big((t + g + \alpha)D - \sum_{i=e+1}^n P_i\big)$. Let us call N (for nonzero) the first ℓ_2 points in $\{P_1,\ldots,P_e\}$ at which v_1 is nonzero. Let $\overline{N}$ be the complement of N in $\{P_1,\ldots,P_e\}$ (so $|\overline{N}| = e - \ell_2 = \ell_1$).

The values of v_1 on the points in $\overline{N}$ uniquely determine the values of the functions $v_2,\ldots,v_m$ on the points in $\overline{N}$ (because of (6)), which in turn determine the entire functions $v_2,\ldots,v_m$ (because $|\overline{N}| = \ell_1$ and for each j the value of v_j is uniquely determined by it values on ℓ_1 of the points in $\{P_1,\ldots,P_e\}$, as stated above).

To satisfy the equalities in (6) for the points in N, we need to have $\frac{y_{ij}}{y_{i1}} = \frac{v_j(P_i)}{v_1(P_i)}$ for each one of these points. Since $\frac{y_{ij}}{y_{i1}}$ is uniformly distributed in $\mathbb{F}_q^\times$, the probability that it is equal to the given value $\frac{v_j(P_i)}{v_1(P_i)}$ is $\frac{1}{q-1}$. Since this has to be true for all i such that $P_i \in N$, and for all $j = 2,\ldots,m$, and since the y_{ij} are independent random variables on $\mathbb{F}_q^\times$, we see that the expected number of erroneous vectors in the right kernel of A for which the first coordinate is v_1 is at most

$$\left(\frac{1}{q-1}\right)^{(m-1)\ell_2} = \left(\frac{1}{q-1}\right)^{(m-1)(n-t-\alpha-g-1)}.$$

Hence, the expected number of erroneous vectors in the right kernel of A is at most

$$\frac{q^{t+\alpha+e-n+1} - 1}{(q-1)^{(m-1)(n-t-\alpha-g-1)}} \leq \left(\frac{1}{q-1}\right)^{m(n-\alpha-t)-(m-1)(g+1)-1-e}.$$

If a vector is erroneous, then any nonzero $\mathbb{F}_q$-scalar multiple of that vector is also erroneous. Thus, the probability that the number of erroneous vectors is larger than 0 equals the probability that the number of erroneous vectors is at least $q - 1$. By Markov's inequality, this probability is at most the expected number of erroneous vectors divided by $q - 1$. This implies that

$$\Pr[\text{exists erroneous vector in } \ker(A)] \leq \left(\frac{1}{q-1}\right)^{m(n-\alpha-t)-(m-1)(g+1)-e}.$$

t		zero error prob.	error prob. exponent
$\frac{m}{m+1}(n-\alpha) - \frac{m-1}{m+1}g - \frac{m-2}{m+1}$		$\frac{n-\alpha-2g-1}{m+1}$	$\frac{m}{m+1}(n-\alpha) - \frac{m-1}{m+1}g - \frac{2m-1}{m+1} - e$
$\frac{m}{m+1}(n-\alpha-g-1)$		$\frac{n-\alpha-g-1}{m+1}$	$\frac{m}{m+1}(n-\alpha) + \frac{g+1}{m+1} - e$

Fig. 1. Error probabilities for some values of t

The error probability of the algorithm is upper bounded by the probability of existence of erroneous vectors in the right kernel of A. This proves the theorem. □

Figure 1 shows for two values of t the bound for reliable decoding, and the exponent of the error probability in case of probabilisitic decoding. The first value for t is the largest possible. For $m = 1$ this bound translates to $(n - \alpha - 1)/2$, and if e is equal to this number, then the error probability is at most $1/(q-1)$, while the bound for reliable decoding is $(n - \alpha - 1)/2 - g$. Since $n - \alpha$ is the designed distance of the underlying AG-code, we see that in this case probabilistic decoding up to the designed distance is possible using our algorithm, and reliable decoding is only possible up to $(d-1)/2 - g$, where d is the designed distance.

The second choice for t leads on the one hand to a smaller decoding radius for probabilistic decoding, but on the other hand to a better decoding radius for reliable decoding. For $m = 1$ the latter bound is $(n - \alpha - 1 - g)/2$, which is $(d-1)/2 - g/2$, where d is the designed distance. The bound for probabilistic decoding is the same if $m = 1$. For $m > 1$ the bounds for probabilistic and reliable decoding differ.

We remark that in the analysis of the decoding we are assuming that we will pick an erroneous vector in the right kernel of A in the first step of the algorithm, if such a vector exists. In reality, we can pick any element in the right kernel of A uniformly at random. If e is smaller than t, the right kernel of A contains $q^{t-e} - 1$ non-erroneous vectors. Hence, depending on how many erroneous vectors are actually contained in the right kernel, the error probability of the algorithm can be significantly lower than given in Theorem 1.

We conclude the section with the following observation: Choosing for t the first choice in Figure 1, and observing that $Q^{\alpha-g+1}$ is a lower bound for the number of codewords in the interleaved code, we see that the upper bound for the number of errors is

$$\frac{m}{m+1}n(1-R) - \frac{2m-1}{m+1}g + \frac{2}{m+1}.$$

If m is very large, if q is a square, and if a sequence of very good algebraic curves are used to construct the underlying AG-code, then on a Q-ary symmetric channel with error probability p the maximum achievable rate for vanishing error probability of the decoder is roughly

$$1 - p - \frac{2}{\sqrt{q}-1}.$$

(This follows from the fact that for a very good sequence of AG-codes the ratio g/n tends to $1/(\sqrt{q}-1)$.) This shows that these codes and these decoding algorithms can get very close to the capacity of the Q-ary symmetric channel.

4 Comparison to the Hashing Method

In this final chapter of this paper we give an extensive comparison of our method to other hashing methods. These methods effectively reduce the number of errors, albeit at the expense of reducing the rate of transmission.

The classical method for coding over large alphabets is to dedicate a part of each symbol as check positions. These positions can be used to store a hash value of the symbol. The advantage is that the hash can be used at the receiver side to detect corrupted symbols: If it does not match the symbol, the symbol is corrupted and can be discarded. This way, the decoding problem is effectively reduced to an erasure decoding problem. There is a downside however: each erroneous symbol has a small probability of having a matching hash. The decoder will fail if such a symbol is used, and therefore such decoders have an error term which is linear in the blocklength n.

If we use an $[n, k, n - k + 1 - g]$ AG-code over $\mathbb{F}_Q$, and ℓ bits are used in each symbol for the hashing value, then only the remaining $\log(Q) - \ell$ bits can be used per symbol, so the effective rate of the code is

$$r = \frac{\log(Q) - \ell}{\log Q} \cdot \frac{k}{n}.$$

There are two possible failure modes for this decoder. First, if too many symbols have to be discarded, then decoding will fail. A Chernoff-bound argument can be used to show that this happens with exponentially small probability if the symbol error probability bounded away from

$$\frac{n - k - 1 + g}{n} = \left(1 - \frac{\log Q}{\log(Q) - \ell} r\right) + \frac{g - 1}{n}.$$

The second failure mode is when an incorrect symbol passes the hashing test and is therefore used in the decoding process. This happens with probability

$$\frac{np}{2^\ell},$$

where p is the symbol error probability. Note that this error probability is linear in n, unlike the bounds we get for interleaved codes.

However, it is possible to do better also with the hashing method by adding a second stage to the decoder. After removing symbols with mismatching hash, a few erroneous symbols remain; if there are not too many such symbols, those can be corrected in a second step with the decoder for AG codes. The reasoning is as follows. Let X_1 be the number of received erroneous symbols which have mismatching hash values, and let X_2 be the number of received erroneous symbols for which the hash matches. Then after removing the mismatching X_1 symbols, we are left with an $[n - X_1, k, n - X_1 - k + 1 - g]$ AG-code. Such a code is correctable for errors up to half the minimum distance, hence the correctability condition is

$$n - k + 1 - g > X_1 + 2X_2.$$

If p is the symbol error probability, then we have

$$E[X_1 + 2X_2] = (1 - 2^{-\ell})p + 2p2^{-\ell}$$

A Chernov-bound can then be used to show that if the symbol error probability p is bounded away from

$$\frac{(1 - R)n + 1 - g}{(1 + 2^{-\ell})n},$$

the resulting failure probability will be exponentially small ($R = k/n$). To summarize, such codes are decodable, if the overall rate (including loss via hashing) is chosen such that

$$r < \frac{\log(Q) - \ell}{\log(Q)}\left(1 - (1 + 2^{-\ell})p + \frac{g}{n} - \frac{1}{n}\right).$$

To compare this to interleaved AG codes, note that the factor

$$\frac{\log(Q) - \ell}{\log(Q)}$$

corresponds to the $(m - 1)/m$ term we have for interleaved codes. So, hashing is away by the factor $(1 + 2^{-\ell})$. On the other hand, the advantage of hashing is that g/n can be made much smaller than in the interleaved case, since we are working in a much larger field.

Unfortunately, this fact has another downside in itself: Working on the larger field increases the complexity of the decoder considerably. For interleaved code, it is $O(n^{1+\varepsilon}\log(q)^2)$ where for the hashing method, it is $O(n^{1+\varepsilon}\log(Q)^2)$.

Hashing can also be combined with an interleaved code to produce a much faster decoder which is also extremely simple. The idea is as follows: We dedicate the first of the m interleaved words just for error detection. That is, the first codeword will always be a transmitted zero. On the receiver side, symbols which have a nonzero value in this first interleaved word are again considered erasures. The other interleaved words can then all be decoded seperately, using the standard decoder. That way, it is possible to get a decoder which operates on the small field only, and which thus has decoding complexity similar to the interleaved decoder. The error analysis is the same as for the hashing code over the large field; the downside is that we are back to the case where g/n tends to $1/(\sqrt{q} - 1)$. Hence, these codes have slightly worse rates than interleaved AG codes.

References

1. D. Bleichenbcher, A. Kiyayias, and M. Yung. Decoding of interleaved Reed-Solomon codes over noisy data. In *Proceedings of ICALP 2003*, pages 97–108, 2003.
2. A. Brown, L. Minder, and A. Shokrollahi. Probabilistic decoding of interleaved Reed-Solomon-codes on the Q-ary symmetric channel. In *Proceedings of the IEEE International Symposium on Information Theory*, page 327, 2004.
3. D. Coppersmith and M. Sudan. Reconstructing curves in three (and higher dimensional) space from noisy data. In *Proceedings of the 35th Annual ACM Symposium on Theory of Computing (STOC)*, 2003.

4. A. Garcia and H. Stichtenoth. A tower of Artin-Schreier extensions of function fields attaining the Drinfeld-Vladut bound. *Invent. Math.*, 121:211–222, 1995.

5. V. Guruswami and M. Sudan. Improved decoding of Reed-Solomon and algebraic-geometric codes. In *Proceedings of the 39th IEEE Symposium on Foundations of Computer Science*, pages 28–37, 1998.

6. J. Justesen, Ch. Thommesen, and T. Høholdt. Decoding of concatenated codes with interleaved outer codes. In *Proc. International Symposium on Information Theory*, page 328, 2004.

7. G.L. Katsman, M.A. Tsfasman, and S.G. Vladut. Modular curves and codes with a polynomial construction. *IEEE Trans. Inform. Theory*, 30:353–355, 1984.

8. M. Luby and M. Mitzenmacher. Verification codes. In *Proceedings of the 40th Annual Allerton Conference on Communication, Control, and Computing*, 2002.

9. M. Luby, M. Mitzenmacher, A. Shokrollahi, and D. Spielman. Efficient erasure correcting codes. *IEEE Trans. Inform. Theory*, 47:569–584, 2001.

10. F. Parvaresh and A. Vardy. Correcting errors beyond the guruswami-sudan radius in polynomial time. In *Proceedings of the 46th Annual IEEE Symposium on the Foundations of Computer Science (FOCS)*, 2005. To appear.

11. A. Shokrollahi and W. Wang. LDPC codes with rates very close to the capacity of the q-ary symmetric channel for large q. In *Proceedings of the International Symposium on Information Theory, Chicago*, 2004.

12. A. Shokrollahi and H. Wasserman. List decoding of algebraic-geometric codes. *IEEE Trans. Inform. Theory*, 45:432–437, 1999.

13. M. Sudan. Decoding of Reed-Solomon codes beyond the error-correction bound. *J. Compl.*, 13:180–193, 1997.

14. A. Vardy. Private communication. 2005.

Performance Improvement of Turbo Code Based on the Extrinsic Information Transition Characteristics

Woo Tae Kim[1], Se Hoon Kang[2], and Eon Kyeong Joo[1]

[1] School of Electrical Engineering and Computer Science,
Kyungpook National University,
Daegu 702-701, Korea
{state, ekjoo}@ee.knu.ac.kr
http://dcl.knu.ac.kr
[2] Application Engineer Organization,
Agilent Technologies Korea Ltd.,
Seoul 150-711, Korea
sehoonkang@agilent.com

Abstract. Good performance of turbo code can be obtained by updating extrinsic information in the iterative decoding process. At first, transition patterns are categorized by the characteristics of extrinsic information in this paper. The distribution of these patterns is surveyed according to signal-to-noise ratio. The dominant error pattern is determined based on the results. And error performance improvement is expected by correcting it. Thus, a scheme to correct the bit with the lowest extrinsic information is proposed and analyzed. The performance is improved as expected from simulation results as compared to the conventional scheme especially in error floor region.

1   Introduction

It is required to use a powerful error correcting code to provide high-quality as well as high-speed multimedia services in future communication systems. Turbo code is known to be one of the most powerful error correcting codes due to its solid performance by updating extrinsic information [1-3]. But the iterative decoding induces delay problem. In addition, turbo code shows error floor phenomenon [4], that is, the performance is not improved even if signal-to-noise ratio (SNR) is increased. The researches on the extrinsic information of the turbo code have been conducted to reduce the delay in decoding [4-10]. Among them, the scheme to use the cyclic redundancy check (CRC) code is known to be the most efficient one [11].

The transition patterns of extrinsic information are investigated and classified in order to improve error performance especially at error floor region in this paper. Also, the distribution of transition patterns is found according to various SNR values and the dominant pattern in error floor region is analyzed by observing the distribution of these patterns. Finally, a new scheme to improve

N.P. Smart (Ed.): Cryptography and Coding 2005, LNCS 3796, pp. 47–58, 2005.
© Springer-Verlag Berlin Heidelberg 2005

error performance especially in error floor region is proposed and analyzed in this paper.

2 Transition Characteristics of Extrinsic Information

Computer simulation is performed in the additive white Gaussian noise (AWGN) channel to investigate the transition characteristics of extrinsic information. The code rate is 1/3 and the frame length is 1024. An eight-state, recursive systematic convolutional (RSC) code is used for the constituent encoder.

The transition patterns are categorized by the characteristics of extrinsic information according to the number of iterations. The patterns are classified into two groups such as oscillation and convergence as shown from Fig. 1 to 6. The oscillation patterns are again classified into three patterns. The extrinsic information of the first pattern oscillates in wide range and that of the second one oscillates in relatively narrow range. The extrinsic information of the last pattern oscillates in wider range as the number of iterations is increased. The oscillation patterns have two modes. If the absolute value of average extrinsic information is large enough, the corresponding bit is definitely either a correct or an error bit. Therefore, there is no performance improvement even if iteration is continued. On the other hand, it is fluctuated between the correct and error region if it is small. So error performance is determined by the stopping instance of iteration.

The convergence patterns are also classified into three patterns. The extrinsic information of the first pattern is converged into a large value and that of the second one is increased almost linearly. The extrinsic information of the last pattern is changed very little regardless of the number of iterations. These six transition patterns are observed at error as well as correct bits. So it is difficult to determine the existence of an error bit by observing these transition patterns only.

The dominant error pattern is different according to SNR in general. The oscillation patterns are usually found at low SNR. On the other hand, the convergence patterns are dominant at high SNR. In addition, the first and second pattern of convergence are observed dominantly in this range where performance is not improved due to the error floor phenomenon. The distribution of the first and second pattern of convergence in error is shown in Table 1. It is obvious that they are the dominant error patterns at high SNR.

These convergence patterns in error are usually different from the others that correspond to the correct bits in the same frame. The example frames with the correct and error bits are shown in Fig. 7 and 8. The dotted lines of Fig. 7 show the error bits which are the second pattern of convergence. The solid lines which correspond to the correct bits show the first pattern of convergence. The correct bits are converged after about 5 iterations. But the error bits show slowly increasing extrinsic information as compared to the correct bits. So the difference between error and correct bits is obvious. The second frame pattern is shown in Fig. 8. All bits show the first pattern of convergence. The error bits show

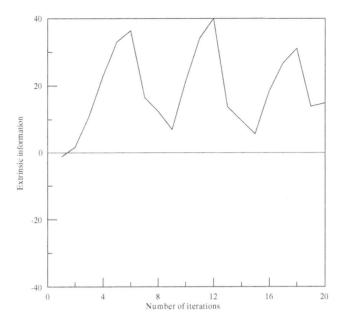

Fig. 1. The first oscillation pattern

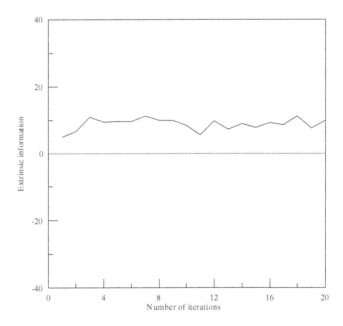

Fig. 2. The second oscillation pattern

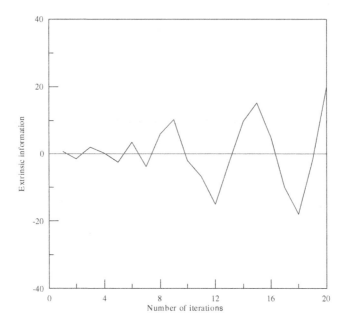

Fig. 3. The third oscillation pattern

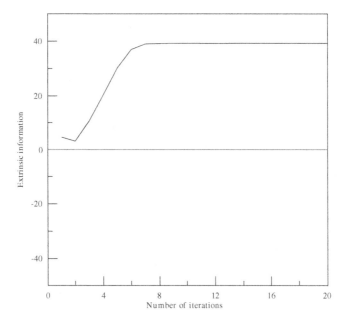

Fig. 4. The first convergence pattern

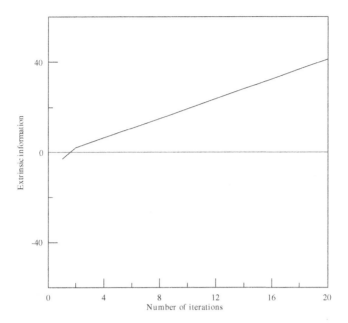

Fig. 5. The second convergence pattern

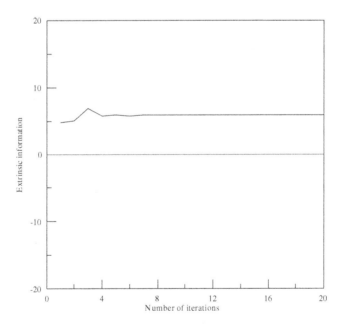

Fig. 6. The third convergence pattern

Table 1. Distribution of error patterns according to SNR

SNR (dB)	The number of error bits with the first and second pattern of convergence	The number of other error bits
0.8	1186(4.60%)	24635(95.4%)
1.0	579(19.8%)	2348(80.2%)
1.2	165(47.8%)	180(52.2%)
1.4	104(66.7%)	52(33.3%)
1.6	60(77.0%)	18(23.0%)
1.8	29(82.9%)	6(17.1%)
2.0	26(86.7%)	4(13.3%)

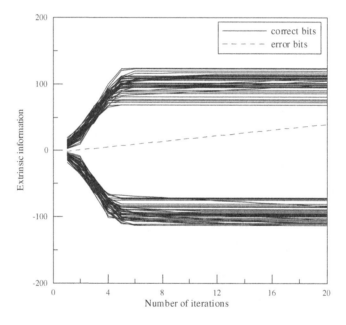

Fig. 7. The first frame pattern with correct and error bits

slowly increasing and smaller absolute extrinsic information values as compared to the correct bits even if both bits have similar values after 12 iterations. The difference is more obvious within the first 10 iterations. Therefore, it is possible to detect and correct the error bits in these frames by observing the difference in the transition patterns of extrinsic information between the correct and error bits.

The number of error bits of the first and second pattern of convergence is generally less than 10 at the error floor region. These error bits have lower extrinsic information than the correct bits during the first several iterations. So almost all bits which have lower extrinsic information than the other bits are

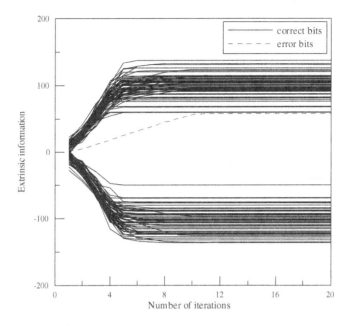

Fig. 8. The second frame pattern with correct and error bits

considered to be in error. Thus, the performance is expected to be improved if inverting the bit with the lowest extrinsic information after checking error existence at each iteration.

3   The Proposed Scheme and Simulation Results

A turbo decoder which contains the proposed scheme is shown in Fig. 9. Iterative decoding is performed at first as in the conventional turbo decoder. Then the existence of errors is checked by the CRC code with the hard decision values of the extrinsic information. If there is no error, then the whole decoding process is completed. Otherwise, the existence of errors is checked again after inverting the bit with the lowest extrinsic information. If no error is found at any step, the whole decoding process is finished. If there are still errors, then the bit with the next lowest extrinsic information is inverted until the predetermined maximum number of bits to be inverted. After the procedure is finished, iterative decoding is continued after reinverting the inverted bits up to the predetermined maximum number of iterations.

Computer simulation is performed in the AWGN channel in order to investigate the error performance of the proposed system. The code rate is 1/3 and the frame length is 1024. The maximum number of iterations is set to 20. The ethernet-32 CRC code [12] is used to check the existence of errors and the maximum number of bits to be inverted is set to 4.

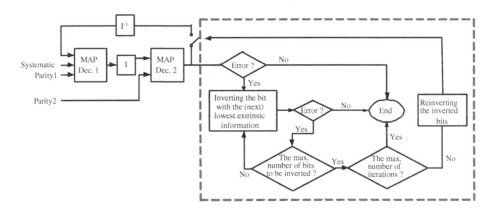

Fig. 9. Turbo decoder with the proposed scheme

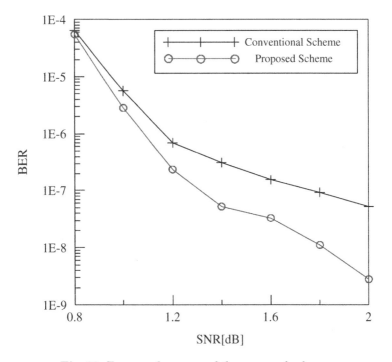

Fig. 10. Error performance of the proposed scheme

The results of the error performance of the conventional scheme where the CRC code is used only to stop the iterative decoding and the proposed system which inverts the bit with the lowest extrinsic information after checking the error existence are shown in Fig. 10. It can be easily verified that the performance of the proposed scheme has improved as compared to the conventional

Table 2. Distribution of error bits (number of frames)

Number of error bits	1.8dB	2.0dB
1	2	1
2	128	76
3	5	4
4	4	1
5	5	3
6	1	1
others	2	1

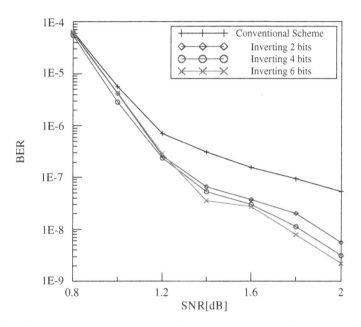

Fig. 11. Error performance according to the number of inverting bits

one especially in the error floor region. For example, about 0.1dB and 0.5dB gain can be obtained at BER of 10^{-6} and 10^{-7}. Also, it is expected that the performance difference is increased further if SNR is increased.

It is considered that the number of bits to be inverted may affect the performance of the proposed scheme. In other words, error performance may be improved if it is increased. On the other hand, the decoding complexity may be increased also. So the distribution of error bits which is classified as the first and second pattern of convergence is investigated to get the appropriate number of bits to be inverted. The distributions at 1.8 and 2.0dB where the error floor phenomenon occurs are shown in Table 2 according to the number of error bits. The number of frames used in this simulation is 3×10^6. The frames which have two

Table 3. Average number of iterations

SNR (dB)	Without inverting	Inverting 2 bits	Inverting 4 bits	Inverting 6 bits
0.8	3.4846	3.4212	3.3808	3.3734
1.0	2.9766	2.9112	2.8709	2.8638
1.2	2.5050	2.5333	2.4919	2.4840
1.4	2.3092	2.2561	2.2207	2.2130
1.6	2.1302	2.0938	2.0731	2.0692
1.8	2.0374	2.0148	2.0035	2.0016
2.0	1.9726	1.9532	1.9401	1.9370

Table 4. Processing time(msec) of the whole decoding procedure and the proposed inverting algorithm only

SNR (dB)	Whole decoding procedure (Inverting 2 bits)	Proposed inverting algorithm only (Inverting 2 bits)	Whole decoding procedure (Inverting 6 bits)	Proposed inverting algorithm only (Inverting 6 bits)
0.8	7.80E+07	7.50E+04	7.74E+07	7.84E+04
1.0	3.49E+08	1.02E+05	3.47E+08	1.08E+05
1.2	2.94E+08	6.15E+04	2.93E+08	6.44E+04
1.4	3.53E+08	1.47E+05	3.52E+08	1.56E+05
1.6	3.53E+08	9.82E+03	3.53E+08	1.06E+04
1.8	1.40E+08	2.34E+04	1.15E+08	2.41E+04
2.0	1.03E+08	2.08E+04	9.55E+07	2.09E+04

error bits are observed the most frequently and those are the dominant number of error bits. In addition, the error bits do not always have the lowest extrinsic information if there are more than 7 error bits.

The simulation is executed using 2 to 6 bits in order to investigate the error performance of the system. Better error performance can be obtained if the number of bits to be inverted is increased in the high SNR region as shown in Fig. 11. On the other hand, the first and second convergence pattern of the error bits are relatively sparse in the low SNR region below 1.2dB. So the effect of increase in the number of bits to be inverted cannot be obtained. Accordingly, it is considered that maximum 6 bits are sufficient to be inverted within considerable increase in terms of the decoding complexity.

The average number of iterations of the conventional and proposed scheme is shown in Table 3 according to various SNR values. The average number of iterations of the proposed scheme is decreased as compared to that of the conventional one. In addition, it is also decreased as the number of bits to be inverted is increased. This is due to the fact that the frames which are still in error after finishing the maximum number of iterations are reduced by the proposed scheme.

It is expected that the decoding time is also increased as the number of bits to be inverted is increased. So the processing time of whole decoding procedure and that of the proposed algorithm only shown in the dotted box in Fig. 9 are compared in Table 4 with Intel Pentium IV processor. The processing time of the proposed inverting scheme only is considerably short as compared to the total processing time which is measured by msec. The time needed to invert 6 bits is longer than that for 2 bits. But the total processing time for 6 bits is shorter than 2 bits. This is because the required time for inverting algorithm is relatively small portion of the whole decoding process and the average number of iterations is decreased as shown in Table 3. Therefore, the scheme of inverting maximum 6 bits can reduce the whole decoding time as well as improve error performance as compared to the other ones.

4   Conclusions

High-speed and high-quality multimedia services are required in future communication systems. So powerful error correcting code such as turbo code is recommended. The turbo code shows excellent performance by updating extrinsic information through iterative decoding. But it shows a drawback of error floor in which the performance is not improved even if SNR is increased. Thus, the transition characteristics of extrinsic information in high SNR where the error floor is occurred are analyzed in this paper.

As a result, the transition patterns are categorized into three convergence and oscillation patterns. The first and second pattern of convergence are dominant at error floor region among these patterns. And the error bits show different extrinsic information transition pattern or shape as compared to the correct bits in the same frame. That is, they have low extrinsic information values as compared with the correct bits. Therefore, the error bits in a frame can be detected by observing this characteristic and then the performance is expected to be improved by correcting those detected error bits.

So the scheme which inverts the bit with the lowest extrinsic information is proposed and analyzed. The error performance is improved as compared to the conventional scheme especially in the error floor region from the simulation results. In addition, the average number of iterations is also reduced and the total processing time is decreased although the time needed to detect and correct error patterns is increased. Accordingly, the proposed scheme is expected to be applied easily to future high-speed and high-quality multimedia communication systems.

Acknowledgments

This work was supported by grant no. B1220-0501-0153 from the Basic Research Program of the Institute of Information Technology Assessment.

References

1. C. Berrou and A. Glavieux, "Near optimum error correcting coding and decoding: Turbo-codes," *IEEE Trans. Commun.*, vol. 44, no. 10, pp. 1261-1271, Oct. 1996.
2. S. Benedetto and G. Montorsi, "Unveiling turbo codes-Some results on parallel concatenated coding schemes," *IEEE Trans. Inform. Theory*, vol. 42, no. 2, pp. 409-428, Feb. 1996.
3. S. Benedetto and G. Montorsi, "Average performance of parallel concatenated block codes," *Electron. Lett.*, vol. 31, no. 3, pp. 156-158, Feb. 1995.
4. J. D. Andersen, "The TURBO coding scheme," Rep. IT-146, Tech. Univ. of Denmark, June 1994.
5. A. C. Reid, T. A. Gulliver, and D. P. Taylor, "Convergence and Errors in Turbo-Decoding," *IEEE Trans. Commun.*, vol. 49, no. 12, pp. 2045-2051, Dec. 2001.
6. R. Shao, S. Lin, and M. Fossorier, "Two simple stopping criteria for turbo decoding," *IEEE Trans. Commun.*, vol. 47, no. 8, pp. 1117-1120, Aug. 1999.
7. B. Kim and H. Lee, "Reduction of the number of iterations in turbo decoding using extrinsic information," *Proc IEEE TENCON'*99, Cheju, Korea, vol. 1, pp. 494-497, Sep. 1999.
8. J. Hagenauer, E. Offer, and L. Papke, "Iterative decoding of binary block and convolutional codes," *IEEE Trans. Inform. Theory*, vol. 42, no. 2, pp. 429-445, Mar. 1996.
9. P. Robertson, "Illuminating the structure of parallel concatenated recursive systematic (TURBO) codes," *Proc. IEEE GLOBECOM'*94, San Fransisco, CA, vol. 3, pp. 1298-1303, Nov. 1994.
10. A. Shibutani, H. Suda, and F. Adachi, "Reducing average number of turbo decoding iterations," *Electron. Lett.*, vol. 35, no. 9, pp. 701-702, Apr. 1999.
11. A. Shibutani, H. Suda, and F. Adachi, "Complexity reduction of turbo decoding," *Proc. IEEE VTC'*99, Houston, TX, vol. 3, pp. 1570-1574, Sep. 1999.
12. G. Castagnoli, S. Brauer, and M. Hemmann, "Optimization of cyclic redundancy check codes with 24 and 32 parity bits," *IEEE Trans. Commun.*, vol. 41, no. 6, pp. 883-892, June 1993.

A Trellis-Based Bound on $(2, 1)$-Separating Codes

Hans Georg Schaathun[1],[*] and Gérard D. Cohen[2]

[1] Dept. Informatics, University of Bergen,
Pb. 7800, N-5020 Bergen, Norway
`georg@ii.uib.no`
[2] Dept. Informatique et Reseaux,
Ecole Nationale Supérieure des Télécommunications,
46, rue Barrault, F-75634 Paris Cedex 13, France
`cohen@enst.fr`

Abstract. We explore some links between higher weights of binary codes based on entropy/length profiles and the asymptotic rate of $(2, 1)$-separating codes. These codes find applications in digital fingerprinting and broadcast encryption for example. We conjecture some bounds on the higher weights, whose proof would considerably strengthen the upper bound on the rate of $(2, 1)$-separating codes.

Keywords: Trellis, coding bounds, separating codes.

1 The Problem

The concept of (t, u)-separating codes has been studied for about 35 years in the literature, with applications including fault-tolerant systems, automata synthesis, and construction of hash functions. For a survey one may read [15]. The concept has been revived by the study of digital fingerprinting [3]; a $(t, 1)$-separating code is the same as a t-frameproof code, on which we elaborate now.

In broadcast encryption, a company distributes a unique decoder to each user. Users may collude and combine their decoders to forge a new one. The company wants to limit this or trace back illegal decoders to the offending users. Among the forbidden moves: framing an innocent user. This goal can be achieved with frameproof codes. One can consult e.g. [2,16] for more.

In this paper we study binary $(2, 1)$-separating codes $((2, 1)$-SS$)$. An (n, M) code is a subset of size M from the set of binary vectors of length n. The code or a set of codewords will often be regarded as matrices, with the codewords forming the rows.

Definition 1. *Let* $\mathbf{a}, \mathbf{b}, \mathbf{c}$ *be three vectors. We say that* $\mathbf{a}$ *is separated from* $(\mathbf{b}, \mathbf{c})$ *if there is at least one postion* i *such that* $a_i \neq b_i$ *and* $a_i \neq c_i$.

An (n, M) *code is* $(2, 1)$-separating *if for every ordered triplet* $(\mathbf{a}, \mathbf{b}, \mathbf{c})$, $\mathbf{a}$ *is separated from* $(\mathbf{b}, \mathbf{c})$.

---

[*] Research was supported by the Norwegian Research Council under Grant Number 146874/420 and the AURORA program.

N.P. Smart (Ed.): Cryptography and Coding 2005, LNCS 3796, pp. 59–67, 2005.

Table 1. Rate bounds for $(2,1)$-SS

	Linear		Nonlinear	
	Rate	Ref.	Rate	Ref
Known construction	0.156	[6]	0.1845	[8]
New construction			0.2033	Theorem 2
Existence	0.2075	Well-known e.g. [6]	0.2075	Well-known e.g. [10]
Upper bound	0.28	Well-known e.g. [6]	0.5	[10]

Many interesting mathematical problems are equivalent to that of $(2,1)$-SS. An overview of this is found in [10]. A linear $(2,1)$-separating code is equivalent to an intersecting code [6], and some results have been proved independently for intersecting and for separating codes. For further details on intersecting codes, see [5] and references therein.

The rate of the code is

$$R = \frac{\log M}{n}.$$

For an asymptotic family of codes (n_i, M_i) codes $(C_1, C_2, \ldots)$ where $M_i > M_{i-1}$ for all i, the rate is defined as

$$R = \limsup_{i \to \infty} \frac{\log M_i}{n_i}.$$

The known bounds on asymptotic families of $(2,1)$-SS are shown in Table 1. By abuse of language, an asymptotic family of codes will also be called an asymptotic code.

We observe a huge gap between the upper and lower bounds for non-linear codes. Our goal is to reduce this gap. The references in the table are given primarily for easy access and are not necessarily the first occurrences of the results, which are sometimes folklore.

Section 2 gives a minor result, namely a new construction slightly improving the lower bound. In Section 3, we make some observations about the trellis of a $(2,1)$-SS. In Section 4, we discuss higher weights of arbitrary codes and we introduce the 'tistance' of a code and make some conjecture. Section 5, we prove bounds for $(2,1)$-SS in terms of the 'tistance' and show how the conjectures would give major improvements of the upper bounds if proved.

2   A New Asymptotic Construction

Theorem 2. *The codes obtained by concatenating an arbitrary subcode of 121 words from the $(15, 2^7)$ shortened Kerdock code $K'(4)$ with codes as described by Xing [18] $(t = 2)$ over $\mathsf{GF}(11^2)$, is a family of $(2,1)$-SS of asymptotic rate $R = 0.2033$.*

Proof. It is well known that the concatenation of two $(2,1)$-SS is a $(2,1)$-SS. The shortened Kerdock code $K'(4)$ was proved to be a $(2,1)$-SS in [11]. The

Xing codes were proved to be $(2,1)$-SS in [18]. Let us recall for convenience their parameters.

Suppose that $q = p^{2r}$ with p prime, and that t is an integer such that $2 \leq t \leq \sqrt{q} - 1$. Then there is an assymptotic family of $(t,1)$-separating codes with rate

$$R = \frac{1}{t} - \frac{1}{\sqrt{q}-1} + \frac{1 - 2\log_q t}{t(\sqrt{q}-1)}.$$

We take $K'(4)$ which is a $(15, 2^7)$ $(2,1)$-SS, wherefrom we pick 11^2 arbitrary codewords. This code can be concatenated with a Xing code over $\mathsf{GF}(11^2)$, for which a rate of approximately 0.4355 and minimum distance more than 0.5 is obtainable. This gives a concatenated code which is $(2,1)$-separating with the stated rate.

It is a bit unclear how easily we can construct the sequences of curves on which the Xing codes are based; we have found no explicit construction in the literature, but it is hinted that the construction should be feasible. The alternative is to use the random construction of [6,10] for a rate of 0.2075, but that is certainly computationally intractable even for moderate code sizes.

3 Trellises for $(2,1)$-SS

We know that a code can always be described as a trellis, and trellises have been studied a lot as devices for decoding. We will not rule out the possibility that someone will want to use trellis decoding of separating codes at some point, but that *is not* our concern. We want to derive bounds on the rate of $(2,1)$-separating codes, and it appears that such bounds may be derived by studying the trellis.

A trellis is a graph where the vertices are divided into $(n+1)$ classes called *times*. Every edge goes from a vertex at time i to a vertex at time $i+1$, and is labeled with an element from some alphabet Q. The vertices of a trellis are called *states*. A trellis also have the property that time 0 and time n each has exactly one vertex, called respectively the *initial* and the *final* states. There is at least one path from the initial state to each vertex of the graph, and at least one path to the final state from each vertex of the graph.

A binary (n, M) code corresponds to a trellis with label alphabet $\{0,1\}$. Every path from time 0 to time n defines a codeword by the labels of the edges. Every codeword is defined by at least one such path.

A trellis is most often considered as an undirected graph, but we will nevertheless say that an edge between a state v at time $i-1$ to some state w at time i goes from v to w.

If each codeword corresponds to exactly one trellis path, then we say that the trellis is *one-to-one*. A *proper* trellis is one where two edges from the same vertex never have the same label. If, in addition, no two edges into the same vertex have the same label, then we say that the trellis is *biproper*. It is known that every block code corresponds to a unique minimal proper trellis, i.e. the proper trellis with the minimal number of edges.

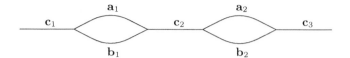

Fig. 1. The impossible subtrellis in Proposition 3

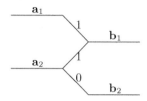

Fig. 2. The impossible subtrellis in Proposition 5

Proposition 3. *In a trellis corresponding to a $(2,1)$-separating code, if two distinct paths join in some vertex v, the joint path cannot rebranch at any later time.*

Proof. If the trellis were to contain two paths which first join and later rebranch, it would mean a sub-trellis as given in Figure 1. If so, we can consider the three vectors

$$\mathbf{v}_1 = \mathbf{c}_1||\mathbf{a}_1||\mathbf{c}_2||\mathbf{a}_2||\mathbf{c}_3$$
$$\mathbf{v}_2 = \mathbf{c}_1||\mathbf{a}_1||\mathbf{c}_2||\mathbf{b}_2||\mathbf{c}_3$$
$$\mathbf{v}_3 = \mathbf{c}_1||\mathbf{b}_1||\mathbf{c}_2||\mathbf{a}_2||\mathbf{c}_3$$

Now $\mathbf{v}_1$ is not separated from $(\mathbf{v}_2, \mathbf{v}_3)$, so the trellis cannot be $(2,1)$-separating.

Corrollary 4. *The trellis of a $(2,1)$-separating code cannot contain a vertex with both two incoming and two outgoing edges (often called a butterfly vertex).*

Proposition 5. *Every $(2,1)$-separating code has a biproper trellis.*

Proof. We consider the minimal proper trellis of a $(2,1)$-separating code. Let v be a vertex with two incoming edges with the same label, say 1. This must mean that we have a subtrellis like the one drawn in Figure 2, where $\mathbf{a}_1||0||\mathbf{b}_2$ is not a codeword. Observe the three codewords

$$\mathbf{v}_1 = \mathbf{a}_1||1||\mathbf{b}_1$$
$$\mathbf{v}_2 = \mathbf{a}_2||1||\mathbf{b}_1$$
$$\mathbf{v}_3 = \mathbf{a}_2||0||\mathbf{b}_2.$$

Here, $\mathbf{v}_2$ is not separated from $(\mathbf{v}_1, \mathbf{v}_3)$.

4   Entropy/Length Profiles and Higher Weights

This section deals with general codes, not necessarily separating. Higher weights, or generalised Hamming weights, have been studied for linear codes since 1977 [9] and have received considerable interest with the definition of the weight hierarchy in 1991 [17]. For non-linear codes, different definitions have been suggested [4,1,14]. We will primarily use the entropy/length profiles (ELP) from [13]. The ELP was used to define the weight hierarchy in [14].

Let X be a stochastic variable, representing a codeword drawn uniformly at random from some code C. Write $[n] = \{1, 2, \ldots, n\}$. For any subset $I \subseteq [n]$, let X_I be the vector $(X_i : i \in I)$, where $X = (X_i : i \in [n])$. Clearly X_I is also a stochastic variable, but not necessarily uniformly distributed.

Definition 6 (Entropy). *The (binary) entropy of a discrete stochastic variable X drawn from a set $\mathcal{X}$ is defined as*

$$H(X) = - \sum_{x \in \mathcal{X}} P(X = x) \log P(X = x).$$

The conditional entropy of X with respect to another discrete stochastic variable Y from $\mathcal{Y}$ is

$$H(X|Y) = - \sum_{y \in \mathcal{Y}} P(Y = y) \sum_{x \in \mathcal{X}} P(X = x|Y = y) \log P(X = x|Y = y).$$

We define the (unordered) conditional ELP to be the sequence $(h_i : i \in [n])$ where

$$h_i = \max_{\#I = i} H(X_I | X_{[n] \setminus I}),$$

we also have the ordered conditional ELP $(g_i : i \in [n])$ where

$$g_i = H(X_{[i]} | X_{\{i+1, \ldots, n\}}).$$

Evidently, g_i depends on the coordinate ordering and may thus be different for two equivalent codes. On the other hand, h_i is the maximum of g_i for all equivalent codes, and thus invariant throughout an equivalence class.

The weight hierarchy as defined in [14] is $\{i | h_i > h_{i-1}\}$. It is an interesting point that the weight hierarchy of a linear code always has k elements, while there is no way to predict the number of elements in the weight hierarchy of a non-linear code, no matter which definition is used.

In this paper we use the parameters

$$t_j := \min\{i : g_i \geq j\}, \quad \text{where } j = 0, \ldots, \lfloor \log M \rfloor.$$

We have particular interest in the first parameter, $t := t_1$, which we are going to call the 'tistance' of the code. For all codes $t \geq d$, and for linear codes we have $t = d$. We also define a normalised measure $\tau := t/n$.

Lemma 7 [14, Lemma 2] *For any $(n, M)_q$ code C, we have $h_l(C) \geq r$ where $r = l + log_q M - n$.*

Proposition 8 (Tistance Singleton bound). *For any $(n, M)_q$ code with tistance t, we have $t < n - log_q M + 2$.*

The proposition follows directly from Lemma 7, by setting $l = t$ and noting that $h_t < 2$. Note that if $M = q^k$ for some integer k, then $t \leq n - k + 1$, which is the more common Singleton form of the bound.

Corrollary 9. *For an asymptotic class of codes, we have $\tau \leq 1 - R$.*

Conjecture 10 (Plotkin-type bound). *For all asymptotic codes, we have $R \leq R_P(\tau) := 1 - 2\tau$.*

The regular asymptotic Plotkin bound states that $R \leq 1 - 2\delta$. Since $\tau \geq \delta$, the conjecture is stronger than this.

Conjecture 11. *Let $R_{LP}(\delta)$ be the MRRW bound [12]. For an asymptotical non-linear code, it holds that $R \leq R_{LP}(\tau)$.*

Obviously, Conjecture 11 implies Conjecture 10, because the MRRW bound is stronger than the Plotkin bound. We state the conjectures separately to encourage work on a Plotkin-type bound in terms of t. The usual Plotkin bound has a cleaner expression and a simpler proof than the MRRW bound, and thus Conjecture 11 may well be considerably harder to prove.

5   Trellis Bounds on $(2, 1)$-SS

At time i, let $\Sigma_i = \{\sigma_1, \ldots, \sigma_a\}$ be the set of states with more than one incoming path. For any state σ, let $P(\sigma)$ be the number of distinct incoming paths respectively. Remember from Proposition 3 that any state $\sigma \in \Sigma_i$ has only one outgoing path. We get that

$$g_i = \sum_{\sigma \in \Sigma_i} \frac{P(\sigma)}{M} h(P(\sigma)) = M^{-1} \sum_{\sigma \in \Sigma_t} P(\sigma) \log P(\sigma), \tag{1}$$

or equivalently that

$$M = g_i^{-1} \sum_{\sigma \in \Sigma_t} P(\sigma) \log P(\sigma). \tag{2}$$

Setting $i = t$, we get that

$$M \leq \sum_{\sigma \in \Sigma_t} P(\sigma) \log P(\sigma). \tag{3}$$

Theorem 12. *A $(2,1)$-separating code has size $M \leq 2^t$.*

Proof. Let Σ_i be the set of states at time i with multiple incoming paths as before, and let Σ'_i be the set of states at time $i \leq t$ with a unique incoming path, and a path leading to a state in Σ_t. Obviously, we have $\#\Sigma_i + \#\Sigma'_i \leq 2^i$. Also note that a state in Σ_i (for $i < t$) must have a (unique) outgoing path leading to a state in Σ_t. Observe that $\Sigma'_t = \emptyset$.

We will prove that for $i = 0, \ldots, t$, we have

$$M \leq 2^i \left(\sum_{\sigma \in \Sigma_{t-i}} P(\sigma)\log P(\sigma) + \#\Sigma'_{t-i} \right). \qquad (4)$$

This holds for $i = 0$ by (3), so it is sufficient to show that

$$\sum_{\sigma \in \Sigma_i} P(\sigma)\log P(\sigma) + \#\Sigma'_i \leq 2 \left(\sum_{\sigma \in \Sigma_{i-1}} P(\sigma)\log P(\sigma) + \#\Sigma'_{i-1} \right). \qquad (5)$$

for $0 < i \leq t$. Since $\Sigma_0 = \emptyset$ and Σ'_0 is the singleton set containing the initial state, (4) implies $M \leq 2^t$ by inserting $i = t$, which will prove the theorem.

Each state σ in Σ_i must have paths from one or two states $\Sigma_{i-1} \cup \Sigma'_{i-1}$. If there is only one such state σ', then we have $P(\sigma) = P(\sigma')$ and $\sigma' \in \Sigma_{i-1}$.

If there are two such states σ_1 and σ_2, we get that $P(\sigma_1) + P(\sigma_2) = P(\sigma)$. If $P(\sigma_j) = 1$, we have $\sigma_j \in \Sigma'_{i-1}$, otherwise $\sigma_j \in \Sigma_{i-1}$. Observe that

$$P(\sigma) \log P(\sigma) \leq 2(P(\sigma_1) \log P(\sigma_1) + P(\sigma_2) \log P(\sigma_2)),$$
$$\text{if } \sigma_1, \sigma_2 \in \Sigma_{i-1}, \qquad (6)$$

$$P(\sigma) \log P(\sigma) \leq 2P(\sigma_1) \log P(\sigma_1) + 1,$$
$$\text{if } \sigma_1 \in \Sigma_{i-1}, \sigma_2 \in \Sigma'_{i-1}, \qquad (7)$$

$$P(\sigma) \log P(\sigma) = 2,$$
$$\text{if } \sigma_1, \sigma_2 \in \Sigma'_{i-1}. \qquad (8)$$

Each of these three equations describes one type of state $\sigma \in \Sigma_i$. Recall that $\sigma_j \in \Sigma_{i-1}$ can have but one outgoing edge. For any state $\sigma \in \Sigma'_i$ there is one state $\sigma' \in \Sigma'_{i-1}$, and each such state σ' has a path to one or two states in $\Sigma_i \cup \Sigma'_i$.

We note that each $\sigma \in \Sigma_i$ in (5) contributes to the right hand side with the maximum amount from the bounds (6) to (8). The term $\#\Sigma'_{i-1}$ is multiplied by two to reflect the fact that each $\sigma' \in \Sigma'_{i-1}$ can have an edge to two different states in $\Sigma_i \cup \Sigma'_i$. This proves the bound.

Proposition 13. *If Conjecture 10 is true, then any asymptotical $(2,1)$-SS has rate $R \leq 1/3$. Similarly, if Conjecture 11 is true, then any asymptotical $(2,1)$-SS has rate $R \leq 0.28$.*

The proof is is similar to the ones used to prove upper bounds on linear $(2,1)$-SS in past, see e.g. [15,6].

Proof. From the Plotkin-type bound on τ, we get $\tau \le \frac{1}{2}(1 - R)$, and from Theorem 12 we thus get $R \le \frac{1}{2}(1 - R)$ which proves the result. The proof of the second sentence is similar, replacing the Plotkin bound by the MRRW bound.

Remark 14. *Theorem 12 combined with the tistance Singleton bound, $R \le 1 - \tau$, implies that $R \le 0.5$ for any $(2, 1)$-SS by the proof above, providing a new proof for the old bound. Any stronger bound on R in terms of τ for non-linear codes, will improve the rate bound for $(2, 1)$-separating codes.*

Remark 15. *By using (2), we get for any i that*

$$M \le h_i^{-1} 2^i,$$

by a proof similar to that of Theorem 12.

6   Balance

From Table 1, we know that an asymptotic upper bound of the rate of a $(1, 2)$-separating code is $R \le 1/2$. Starting with an asymptotic family with rate close to $1/2$, we construct a family with the same rate and only codewords with weight close to $n/2$. Let C be a $(1, 2)$-SS of rate $R = 1/2 - \alpha$, where $\alpha > 0$ is a sufficiently small constant.

Consider a partition (P_1, P_2) of the coordinates with $|P_1| = \lceil (1/2 + 1.1\alpha)n \rceil =: n_1$. Let $U_i \subseteq C$ be the set of codewords matching no other codeword on P_i. It is easy to check that $C \subset U_1 \cup U_2$. (Otherwise, some codeword would be matched by at most two others on P_1 and P_2, thus not separated). Since $|U_2| \le 2^{|P_2|} = o(|C|)$, we get $|U_1| = (1 - o(1))|C|$.

Projecting C on P_1 gives a code $C_1(n_1, 2^{(1/2-\alpha)n}(1 - o(1)))$ of rate $R_1 \approx (1/2 - \alpha)/(1/2 + 1.1\alpha) \approx 1 - 4.2\alpha$. Thus, the relative dominating weight ω_1 in C_1 must be close to $1/2$.

Now, we expurgate by keeping only codewords of C which get relative weight ω_1 when projected on P_1. Thus we get a code C' with rate asymptotically equal to that of C.

We repeat the procedure with a new partition (P_1', P_2'), almost disjoint from the previous one (i.e., we take $|P_1 \cap P_1'| = \lceil 2.2\alpha n \rceil$). The code C'' obtained after the second expurgation retains both $(1, 2)$-separation and rate $\approx 1/2$. Its codewords, being balanced on P_1 and P_1', are 'almost' balanced, as the following theorem states.

Theorem 16. *For all $c'' \in C''$, we have $|w(c'')/n - 1/2| = o(1)$.*

Remark 17. *This result generalises easily to $(1, t)$-separation. Any such code with rate close to the optimal rate of $1/t$ is almost balanced.*

We have translated the old combinatorial question of separating codes into the language of trellises. This has enabled us to shed new light on the matter, by putting to use concepts like entropy and higher weights.

References

1. L. A. Bassalygo. Supports of a code. In G. Cohen, M. Giusti, and T. Mora, editors, *Applied Algebra, Algebraic Algorithms and Error-Correcting Codes*, volume 948 of *Springer Lecture Notes in Computer Science*. Springer-Verlag, 1995.
2. Simon R. Blackburn. Frameproof codes. *SIAM J. Discrete Math.*, 16(3):499–510, 2003.
3. Dan Boneh and James Shaw. Collusion-secure fingerprinting for digital data. *IEEE Trans. Inform. Theory*, 44(5):1897–1905, 1998. Presented in part at CRYPTO'95.
4. Gérard Cohen, Simon Litsyn, and Gilles Zémor. Upper bounds on generalized distances. *IEEE Trans. Inform. Theory*, 40(6):2090–2092, 1994.
5. Gérard Cohen and Gilles Zémor. Intersecting codes and independent families. *IEEE Trans. Inform. Theory*, 40:1872–1881, 1994.
6. Gérard D. Cohen, Sylvia B. Encheva, Simon Litsyn, and Hans Georg Schaathun. Intersecting codes and separating codes. *Discrete Applied Mathematics*, 128(1):75–83, 2003.
7. Gérard D. Cohen, Sylvia B. Encheva, and Hans Georg Schaathun. More on (2, 2)-separating codes. *IEEE Trans. Inform. Theory*, 48(9):2606–2609, September 2002.
8. Gérard D. Cohen and Hans Georg Schaathun. Asymptotic overview on separating codes. Technical Report 248, Dept. of Informatics, University of Bergen, May 2003. Available at `http://www.ii.uib.no/publikasjoner/texrap/index.shtml`.
9. Tor Helleseth, Torleiv Kløve, and Johannes Mykkeltveit. The weight distribution of irreducible cyclic codes with block lengths $n_1((q^l - 1)/n)$. *Discrete Math.*, 18:179–211, 1977.
10. János Körner. On the extremal combinatorics of the Hamming space. *J. Combin. Theory Ser. A*, 71(1):112–126, 1995.
11. A. Krasnopeev and Yu. L. Sagalovich. The Kerdock codes and separating systems. In *Eight International Workshop on Algebraic and Combinatorial Coding Theory*, 2002.
12. Robert J. McEliece, Eugene R. Rodemich, Howard Rumsey, Jr., and Lloyd R. Welch. New upper bounds on the rate of a code via the Delsarte-MacWilliams inequalities. *IEEE Trans. Inform. Theory*, IT-23(2):157–166, 1977.
13. Ilan Reuven and Yair Be'ery. Entropy/length profiles, bounds on the minimal covering of bipartite graphs, and the trellis complexity of nonlinear codes. *IEEE Trans. Inform. Theory*, 44(2):580–598, March 1998.
14. Ilan Reuven and Yair Be'ery. Generalized Hamming weights of nonlinear codes and the relation to the Z_4-linear representation. *IEEE Trans. Inform. Theory*, 45(2):713–720, March 1999.
15. Yu. L. Sagalovich. Separating systems. *Problems of Information Transmission*, 30(2):105–123, 1994.
16. Jessica N. Staddon, Douglas R. Stinson, and Ruizhong Wei. Combinatorial properties of frameproof and traceability codes. *IEEE Trans. Inform. Theory*, 47(3):1042–1049, 2001.
17. Victor K. Wei. Generalized Hamming weights for linear codes. *IEEE Trans. Inform. Theory*, 37(5):1412–1418, 1991.
18. Chaoping Xing. Asymptotic bounds on frameproof codes. *IEEE Trans. Inform. Theory*, 40(11):2991–2995, November 2002.

Tessellation Based Multiple Description Coding

Canhui Cai and Jing Chen

Institute of Information Science and Technology, Huaqiao University,
Quanzhou, Fujian, 362011, China
{chcai, jingzi}@hqu.edu.cn

Abstract. A novel multiple description coding framework, Tessellation
Based Multiple Description Coding (TMDC), is proposed in this paper.
In this work, we first decompose each wavelet coefficient into two parts by
bit-wise-down sampling, generating odd and even wavelet transformed
images. Then, we restructure these two images in wavelet blocks and
group the wavelet blocks to form the row and column packets. These
packets are separately dispatched through diverse channels. Since the
row and the column packets have limited intersection, even with the
loss of row and column packets simultaneously, the proposed system
exhibits good error resilient ability. Experimental results have verified
the improved performance of the proposed algorithm.

1 Introduction

The transmission of image and video information over today's heterogeneous,
and unreliable networks presents new challenges for research in image coding.
Since long burst errors in wireless networks due to the fading effect and network
congestion or occasional link failures in the Internet, robust coding mechanisms
are necessary to deliver acceptable reconstruction quality in such environments.
Conventional approaches for dealing with packet loss and transmission errors
for static data, such as retransmission may not be applicable in such condition
due to real time nature of the video. As a possible solution to minimize the
effect of packet loss and transmission errors, Multiple Description Coding (MDC)
approach has recently received considerable attentions.

The basic idea of the multiple description coding is to encode a single source
into several mutually refining, self-decodable, and equal important bit streams
called descriptions, then deliver them through different channels to the receiver.
The reconstruction quality at the decoder depends on the number of descriptions
received. If all the descriptions are received correctly, the decoder can have a high
fidelity reconstruction quality. If only one description is available, the decoder is
still able to retrieve some information of the lost parts of signals and reconstruct
the source with an acceptable quality. Due to its flexibility, multiple description
coding has emerged as an attractive framework for robust information transmission over unreliable channels. Many MDC schemes have been proposed under
different application scenarios [1].

The first MDC framework, Multiple Description Scalar Quantizer (MDSQ)
is due to Vaishampayan [2]. In MDSQ, each sample is quantized to two indexes.

N.P. Smart (Ed.): Cryptography and Coding 2005, LNCS 3796, pp. 68–77, 2005.

If both indexes are available in the receiver, the sample is reconstructed with a small quantization error. If only one index is received, the reconstruction error is larger. Servetto introduced a wavelet based MDC scheme by combining the multiple description scalar quantizers and a subband codec [3]. Jiang and Ortega proposed a zero-tree based MDC framework, called the Polyphase Transform and Selective Quantization (PTSQ), where zero-trees are separated into multiple parts (called phases in the paper), and a description is built from the data of one phase and the protection information of another [4]. Miguel et al. developed another zero-tree based multiple protection scheme, MD-SPIHT, with more protection information in the description, receiving better error resilient results in 16 descriptions [5]. In our earlier paper, Structure Unanimity Base Multiple Description Subband Coding (SUMDSC) [6], we split each significant DWT coefficient into two coefficients by separating the odd number and the even number location bits of the coefficient, forming two descriptions with similar tree structures. For notational convenience, we call the coefficient made of odd bits the odd coefficient and that made of the even bits the even coefficient. Similar to the PTSQ, a description in SUMDSC includes data from one type of coefficients and protection from the other. Since two types of coefficients share the same hierarchical tree structures, no tree structures in the protection part need to be coded, resulting in an improvement in the coding efficiency.

In this paper, a novel MDC framework, called Tessellation Based Multiple Description Coding (TMDC), is proposed. Here the wavelet coefficients of an image are first split into two parts: odd coefficients and even coefficients. The coefficients in each part are then separately grouped to form the odd image and the even image. These two images are next decomposed into hierarchical quad-tree blocks. The blocks from the odd and even images are grouped in the horizontal and vertical directions, respectively, generating the row packets and the column packets. Because each row packet and each column packet have limited intersection, the proposed MDC framework has very good error resilient ability.

The rest of this paper is organized as follows: Section 2 describes the proposed MDC framework, including a brief introduction to the SUMDSC, the hierarchical tree block, generation of the packets, and the error resilient image reconstruction. Section 3 shows some simulation results to illustrate the improved performances of the proposed algorithm. Some concluding remarks are presented in Section 4.

2   Tessellation Based Multiple Description Coding Framework

2.1   Structure Unanimity Base Multiple Description Subband Coding (*SUMDSC*)

The block diagram of the *SUMDSC* framework is shown in the Fig. 1. Here the wavelet coefficients and their delayed versions are bitwise-down-sampled into even position bits and odd position bits, creating even and odd coefficients. Let x be a significant coefficient with $n + 1$ bits,

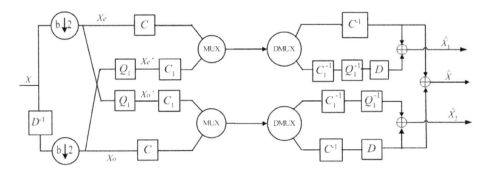

Fig. 1. Block Diagram of Structure Unanimity Base Multiple Description Subband Coding

$$x = b_n b_{n-1} \cdots b_1 b_0 = \sum_{k=even} b_k 2^k + \sum_{k=odd} b_k 2^k = x_e + x_o \tag{1}$$

Where x_e and x_o denote the odd and the even coefficients, respectively. Without loss of generality, set n to be an odd number, $n = 2m + 1$. Then,

$$x_o = b_{2m+1} \cdots b_1 = \sum_{k=0}^{m} b_{2k+1} 2^k \tag{2}$$

$$x_e = b_{2m} \cdots b_0 = \sum_{k=0}^{m} b_{2k} 2^k \tag{3}$$

Where $b_{2m+1} = 1$. However, if $b_{2m} = 0$, the reconstructed image from the even coefficients will produce serious side distortions. For instance, if $x = 100000011$, then $x_e = 10001$ and $x_o = 0001$. If only x_o is available, the reconstructed value will be $\hat{x}_0 = 1$ To avoid such catastrophic situation, the most significant bit (MSB) of x, b_{2m+1}, should be included in both x_e and x_o, and the above equations results in

$$x_o = 1 b_{2m-1} \cdots b_1 \tag{4}$$

$$x_e = 1 b_{2m} \cdots b_0 \tag{5}$$

The odd and the even coefficients are then separately grouped into even and odd images X_e and X_o respectively. Each image is coded independently by a codec C, forming the primary part of the description. In case of channel erasure, the counterpart component is coarsely re-quantized and coded as a redundant part of the description. Because both X_e and X_o have the same zero mapping, the redundant part need only to code the absolute value of the re-quantized coefficients and the coding efficiency is improved. To simplify the expression, we will call these two descriptions later odd description and even description, respectively.

2.2   Tessellation Based Multiple Description Encoding

The $SUMDSC$ works quite well in the simple erasure channel model. However, as most communication channels used nowadays are mainly packet exchange networks, not all packets in a description but only part of them are lost due to the network congestion. Moreover, the packet loss may occur in both descriptions even we transport them in different channels. To provide the best quality of the reconstructed image, all available packets in the receiver should be used in the image reconstruction. To this end, the descriptions must be well packetized before dispatched.

A natural idea is to segment the transformed image into N sub-images and apply $SUMDSC$ to each sub-image to generate sub-odd-description and sub-even-description, and transport all sub-odd-descriptions from one channel and all sub-even-descriptions from the other. However, if two sub-descriptions of a sub-image are totally lost, the reconstructed image will be severely degraded. To minimize the maximum degradation, we propose a novel packetized multiple description coding algorithm called the Tessellation Based Multiple Description Coding ($TMDC$).

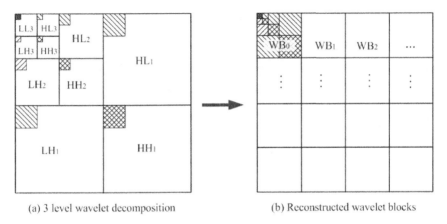

(a) 3 level wavelet decomposition   (b) Reconstructed wavelet blocks

Fig. 2. Rearrangement of a wavelet transformed image into wavelet blocks

Similar to the $SUMDSC$, the wavelet coefficients of a transformed image are bitwise-down-sampled, forming two wavelet transformed images, even image X_e and odd image X_o. Wavelet coefficients in both images are then rearrange into wavelet blocks as shown in Fig. 2.

Fig. 2 demonstrates how wavelet coefficients are shaped into wavelet blocks in a three level dyadic wavelet decomposition image with 10 subband, LL_3, HL_i, LH_i, and HH_i, $i = (1, 2, 3)$. Fig. 2(a) illustrates a zerotree [7] rooted in the LL_3 band. Fig. 2 (b) shows how coefficients in the zerotree in Fig. 2(a) are rearranged to form a wavelet block. From this picture, one can see that a wavelet block is in fact a integrated zero-tree rooted in the lowest frequency band.

Wavelet blocks in X_e are then grouped row by row to form row packets (Fig. 3(a)), and the wavelet blocks in X_o are grouped column by column to form column packets (Fig. 3(b)). The number of block per packet depends on the bit rate and the size of MTU (maximum transfer unit).

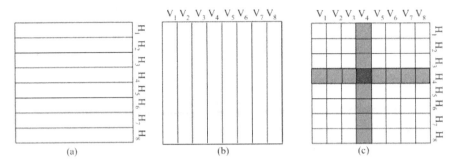

Fig. 3. (a)grouped wavelet blocks in X_e row by row to form row packets; (b)grouped wavelet blocks in X_o column by column to form column packets; (c)reconstruct lost coefficients in LL band from its neighborhood packets

Thus, the encoding process is as follows:

1. Form row and column packets:

   1.1 Group the wavelet blocks from X_e horizontally and code them to form the main parts of row packets;

   1.2 Group the wavelet blocks from X_o vertically and code them to form the main parts of column packets;

   1.3 Re-quantize coefficients in the wavelet blocks of X_e, group and code them vertically to produce the redundant parts of the column packets;

   1.4 Re-quantize coefficients in the wavelet blocks of X_o, group and code them horizontally to produce the redundant parts of the row packets;

2. The row packets and the column packets are respectively formed row and column descriptions, coded by an X-tree codec [9] and dispatch through two different channels to the receiver. For expression convenience, row description and column description will be shortened to description in the rest of the paper.

2.3   Tessellation Based Multiple Description Decoding

The decoder makes use of all available packets to reconstruct the source. Since each row packet and each column packet have only limited intersections, a loss of row packets along with few column packets will not introduce serious distortions, and vice versa. Moreover, since the lowest frequency band of an image is quite smooth, the difference between the lost coefficient and its neighbor is small. Thus, it is possible to estimate the lost lowest band coefficients from those of the received signal with small error. In this context, we can reconstruct the lowest frequency band (LL band for short) of the lost sub-image from its neighborhood packets.

Let H_i and V_j denote the i-th row packet and the j-th column packet. For instance, H_4 stands for the fourth row packet, and V_4, the fourth column packet. If both H_4 and V_4 are unavailable, their intersection will be totally lost (Fig. 3(c)). However, the lowest frequency band of the lost section can be estimated from its neighboring pixels V_3, V_5, H_3, and H_5 by linear interpolation.

In this way, the decoding algorithm in the *TMDC* is as follows:

1) group all available packets to reconstruct as many as possible coefficients,

2) estimate the missing pixels in the lowest band by their neighbors and set all other missing pixels to zero, and

3) reconstruct the image from the wavelet coefficients.

3 Simulation Results and Discussions

We test the proposed method on the common used 512×512 gray image "Lena". In all our experiments, the five level pyramid wavelet decomposition with 7/9 bi-orthogonal wavelet filters [8] are used. The transformed image is partitioned into 16 description packets, and an X-tree codec along with an arithmetic coding have been used to code these descriptions. The computer simulation results on the image at total bit rate 0.5 bpp and 0.1 bpp by the proposed algorithm are shown in Fig. 4 and 5, where TMDC1 denotes the experimental results by the proposed algorithm if packet loss only happens on either transmission channels due to network congestion, and TMDC2 indicates the simulation results by the proposed algorithm if packets loss occurs on both channels.

PSNRs illustrated in these pictures are means of experimental results for all possible conditions. For instance, in the case of one description lost, there are 16 possibilities, so the PSNR shown in the picture is the average of PSNRs from

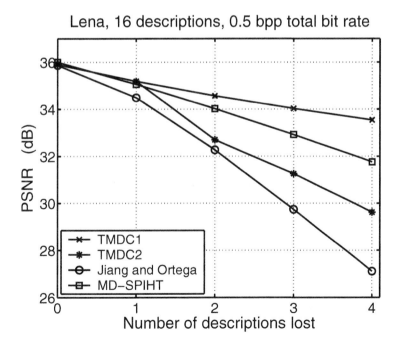

Fig. 4. Experimental results on "Lena" (0.5bpp)

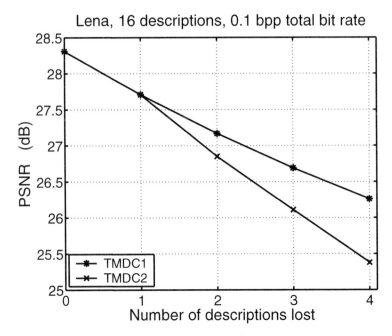

Fig. 5. Experimental results on "Lena" (0.1bpp)

Fig. 6. Reconstruction "Lena" at 0.5bpp by TMDC1 : (a) with all descriptions; (b) with 4 column descriptions (3-6) lost; (c) with 4 row descriptions (3-6) lost

related 16 simulation results. As references, we also present the experimental results from literature [4] and [5] by PTSQ and MD-SPIHT in 16 descriptions with 0.5 bpp total bit rate in Fig. 4. From this figure, it can be seen that if only one channel failure during the image transmission, the outcome of the proposed algorithm, TMDC1 has the superior results. Even if packet loss happens in both channels, the outcome TMDC2 is still better than that of PTSQ.

Fig. 6 are reconstructed images with all descriptions, lost 4 row descriptions, and lost 4 column descriptions at total bit rate 0.5 bpp by TMDC1. From this

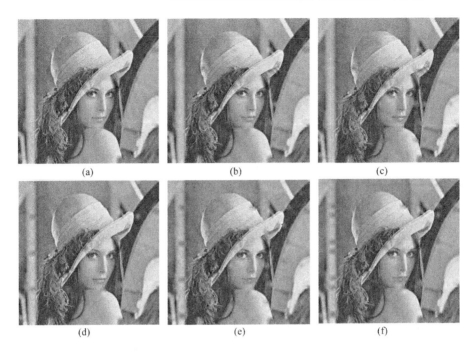

Fig. 7. Reconstruction "Lena" at 0.5bpp by TMDC2 : (a) loss of 4th row description and 4th column description; (b) loss of 4th column description and 4-5th row descriptions; (c) loss of 4th row description and 4-5th column descriptions; (d) loss of 4th row description and 3-5th column descriptions; (e) loss of 4-5th column description and 4-5th column descriptions; (f)loss of 4th column description and 3-5th row descriptions

Fig. 8. Reconstruction "Lena" at 0.1bpp by TMDC1 : (a) with all descriptions; (b) with 2 row descriptions (4-5) lost; (c) with 4 row descriptions (3-6) lost

picture, we can see little difference among these three images. The reconstructed images illustrate their good quality at such bit rate even 4 descriptions have been lost. Fig. 8 are rebuilt images with all descriptions, lost 2 row descriptions, and lost 4 row descriptions at total bit rate 0.1 bpp by TMDC1. Since the rate

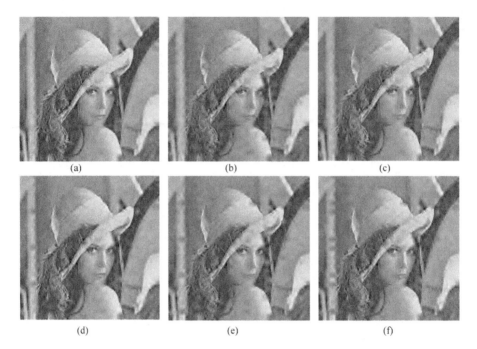

Fig. 9. Reconstruction "Lena" at 0.51pp by TMDC2 : (a) loss of 4th row description and 4th column description; (b) loss of 4th column description and 4-5th row descriptions; (c) loss of 4th row description and 4-5th column descriptions; (d) loss of 4th row description and 3-5th column descriptions; (e) loss of 4-5th column description and 4-5th column descriptions; (f)loss of 4th column description and 3-5th row descriptions

budget is so limited, any additional information will be helpful to upgrade the quality of the reconstructed image. The quality promotion as the increment of the number of available descriptions is clearly showed up in this picture.

Fig. 7 and 9 are reconstructed images with 2 descriptions (4th row and 4th column) lost, 3 descriptions (4th row, 4th and 5th columns / 4th column, 4th and 5th rows) lost, and 4 descriptions (4th row, 3-5th columns / 4th column, 3-5th rows / 4-5th columns and 4-5th rows) lost at total bit rate 0.5 bpp and 0.1 bpp by TMDC2. From Fig. 7(a) and 9(a), we can see that if the neighborhood packets of the lost packet are all available, the reconstruction of lost LL band coefficients by interpolation is very effective. Fig. 9 also show that even at low bit rate and 25 percent packet lost, the quality of reconstructed images is still acceptable.

4   Concluding Remarks

This paper has developed a new MDC framework, called Tessellation Based Packetized Multiple Description Coding (TMDC) for image transmission in the packet erasure networks. In this work, we presented the measure to split a wavelet coefficient into one odd-bit-coefficient and one even-bit-coefficient, forming two descriptions of the source, and discussed how to form row packets and column packets from the two descriptions in detail. Combined with a X-tree based codec, the proposed MDC framework was developed into a new MDC system. Experimental results have proven the error resilient ability of our proposed MDC system. If only row packets or column packets are lost, the proposed MDC framework has very good quality of reconstruction. Since the row packet and the column packet have limited intersection, even packet loss happens on both horizontal and vertical directions, the outcomes of the proposed system still steadily outperform PTSQ and are comparable with MD-SPIHT.

Acknowledgements

This work is partially supported by the National Natural Science Foundation of China under grant 60472106 and the Fujian Province Natural Science Foundation under Grant A0410018.

References

1. Goyal, V.K.: Multiple Description Coding: Compression Meets the Network. IEEE Signal Processing Magazine 18 (2001) 74–93.
2. Vaishampayan, V.A.: Design of Multiple Description Scalar Quantizers. IEEE Trans. on Information Theory 39 (1993) 821–834.
3. Servetto, S.D.: Multiple Description Wavelet Based Image Coding. IEEE Trans. on Image Processing 9 (2000) 813-26.
4. Jiang, W., Ortega, A.: Multiple Description Coding via Polyphase Transform and Selective Quantization. In: Proc. of Visual Communications and Image Processing (1999) 998–1008.
5. Miguel, A.C., Mohr, A.E., Riskin, E.A.: SPIHT for Generalized Multiple Description Coding. In: Proc. of ICIP (1999) 842–846.
6. Cai, C., Chen, J.: Structure Unanimity Based Multiple Description Subband Coding. In: Proc. of ICASSP2004, vol. 3 (2004) 261–264.
7. Shapiro, J.M.: Embedded Image Coding Using Zerotrees of Wavelet Coefficients. IEEE Trans. on Signal Processing 41 (1993) 3445–3463.
8. Antonini, M., Barlaud, M., Mathieu. P, Daubechies, I.: Image Coding Using Wavelet Transform. IEEE Trans. on Image Processing 4 (1992) 205–221.
9. Cai, C., Mitra, S.K., Ding, R.: Smart Wavelet Image Coding: X-Tree Approach. Signal Processing 82 (2002) 239–249

Exploiting Coding Theory for Collision Attacks on SHA-1*

Norbert Pramstaller, Christian Rechberger, and Vincent Rijmen

Institute for Applied Information Processing and Communications (IAIK),
Graz University of Technology, Austria
{Norbert.Pramstaller, Christian.Rechberger,
Vincent.Rijmen}@iaik.tugraz.at

Abstract. In this article we show that coding theory can be exploited efficiently for the cryptanalysis of hash functions. We will mainly focus on SHA-1. We present different linear codes that are used to find low-weight differences that lead to a collision. We extend existing approaches and include recent results in the cryptanalysis of hash functions. With our approach we are able to find differences with very low weight. Based on the weight of these differences we conjecture the complexity for a collision attack on the full SHA-1.

Keywords: Linear code, low-weight vector, hash function, cryptanalysis, collision, SHA-1.

1 Introduction

Hash functions are important cryptographic primitives. A hash function produces a hash value or message digest of fixed length for a given input message of arbitrary length. One of the required properties for a hash function is collision resistance. That means it should be practically infeasible to find two messages m and $m^* \neq m$ that produce the same hash value.

A lot of progress has been made during the last 10 years in the cryptanalysis of dedicated hash functions such as MD4, MD5, SHA-0, SHA-1 [1,5,6,12]. In 2004 and 2005, Wang *et al.* announced that they have broken the hash functions MD4, MD5, RIPEMD, HAVAL-128, SHA-0, and SHA-1 [14,16]. SHA-1, a widely used hash function in practice, has attracted the most attention and also in this article we will mainly focus on SHA-1.

Some of the attacks on SHA-1 exploit coding theory to find characteristics (propagation of an input difference through the compression function) that lead to a collision [10,12]. The basic idea is that the set of collision-producing differences can be described by a linear code. By applying probabilistic algorithms the attacker tries to find low-weight differences. The Hamming weight of the resulting low-weight differences directly maps to the complexity of the collision

---

* The work in this paper has been supported by the Austrian Science Fund (FWF), project P18138.

N.P. Smart (Ed.): Cryptography and Coding 2005, LNCS 3796, pp. 78–95, 2005.

attack on SHA-1. Based on [10,12] we present several different linear codes that we use to search for low-weight differences. Our new approach is an extension of the existing methods and includes some recent developments in the cryptanalysis of SHA-1. Furthermore, we present an algorithm that reduces the complexity of finding low-weight vectors for SHA-1 significantly compared to existing probabilistic algorithms. We are able to find very low-weight differences within minutes on an ordinary computer.

This article is structured as follows. In Section 2, we present the basic attack strategy and review recent results on the analysis of SHA-1. How we can construct linear codes to find low-weight collision-producing differences is shown in Section 3. Section 4 discusses probabilistic algorithms that can be used to search for low-weight differences. We also present an algorithm that leads to a remarkable decrease of the search complexity. The impact of the found low-weight differences on the complexity for a collision attack on SHA-1 is discussed in Section 5. In Section 6 we compare our low-weight difference with the vectors found by Wang *et al.* for the (academical) break of SHA-1. Finally, we draw conclusions in Section 7.

2   Finding Collisions for SHA-1

In this section we shortly describe the hash function SHA-1. We present the basic attack strategy and review recent results in the analysis of SHA-1. For the remainder of this article we use the notation given in Table 1. Note that addition modulo 2 is denoted by '+' throughout the article.

Table 1. Used notation

notation	description
$A + B$	addition of A and B modulo 2 (XOR)
$A \vee B$	logical OR of two bit-strings A and B
M_t	input message word t (32-bits), index t starts with 0
W_t	expanded input message word t (32-bits), index t starts with 0
$A \ll n$	bit-rotation of A by n positions to the left
$A \gg n$	bit-rotation of A by n positions to the right
step	the SHA-1 compression function consists of 80 steps
round	the SHA-1 compression function consists of 4 rounds = 4×20 steps
A_j	bit value at position j
$A_{t,j}$	bit value at position j in step t
A'_j	bit difference at position j
$A'_{t,j}$	bit difference at position j in step t

2.1   Short Description of SHA-1

The SHA family of hash functions is described in [11]. Briefly, the hash functions consist of two phases: a message expansion and a state update transformation.

These phases are explained in more detail in the following. SHA-1 is currently the most commonly used hash function. The predecessor of SHA-1 has the same state update but a simpler message expansion. Throughout the article we will always refer to SHA-1.

Message Expansion. In SHA-1, the message expansion is defined as follows. The input message is split into 512-bit message blocks (after padding). A single message block is denoted by a row vector m. The message is also represented by 16 32-bit words, denoted by M_t, with $0 \le t \le 15$.

In the message expansion, this input is expanded linearly into 80 32-bit words W_t, also denoted as the 2560-bit expanded message row-vector w. The words W_t are defined as follows:

$$W_t = M_t, \qquad\qquad\qquad\qquad 0 \le t \le 15 \qquad (1)$$
$$W_t = (W_{t-3} + W_{t-8} + W_{t-14} + W_{t-16}) \lll 1, \quad 16 \le t \le 79 . \qquad (2)$$

Since the message expansion is linear, it can be described by a 512×2560 matrix $\mathbf{M}$ such that $w = m\mathbf{M}$. The message expansion starts with a copy of the message, cf. (1). Hence, there is a $512 \times 32(80 - 16)$ matrix $\mathbf{F}$ such that $\mathbf{M}$ can be written as:

$$\mathbf{M}_{512 \times 2560} = [\mathbf{I}_{512}\ \mathbf{F}_{512 \times 2048}] . \qquad (3)$$

State Update Transformation. The state update transformation starts from a (fixed) initial value for 5 32-bit registers (referred to as iv) and updates them in 80 steps $(0,\ldots,79)$ by using the word W_t in step t. Figure 1 illustrates one step of the state update transformation. The function f depends on the step number: steps 0 to 19 (round 1) use the *IF-function* and steps 40 to 59 (round 3) use the *MAJ-function*:

$$f_{IF}(B, C, D) = BC + \overline{B}D \qquad (4)$$
$$f_{MAJ}(B, C, D) = BC + BD + CD . \qquad (5)$$

The remaining rounds 2 and 4, use a 3-input XOR referred to as f_{XOR}. A step constant K_t is added in every step. There are four different constants; one for each round. After the last application of the state update transformation, the initial register values are added to the final values (feed forward), and the result is either the input to the next iteration or the final hash value.

We linearize the state update by approximating f_{IF} and f_{MAJ} by a 3-input XOR. The linear state update can then be described by a 2560×160 matrix $\mathbf{S}$, a 160×160 matrix $\mathbf{T}$, and a vector k that produce the output vector o from the input message vector m:

$$o = m\mathbf{MS} + iv\mathbf{T} + k . \qquad (6)$$

The (linear) transformation of the initial register value iv is described by the matrix $\mathbf{T}$. The constant k includes the step constants.

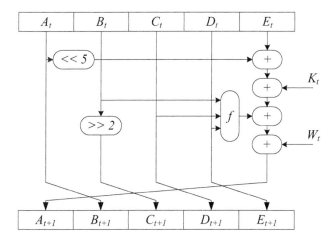

Fig. 1. One step of the linearized state update transformation of SHA-1

2.2 The Basic Attack Strategy

Recent collision attacks on SHA-1 use the following strategy. Firstly, a characteristic, *i.e.* the propagation of an input difference through the compression function of the hash function, is constructed. Secondly, messages are constructed, which follow the characteristic. This strategy is based on the attack on SHA-0 by Chabaud and Joux [5].

They observed that every collision for the linearized compression function of SHA (SHA-0, SHA-1) can be written as a linear combination of local collisions. These local collisions consist of a *perturbation* and several *corrections*. Rijmen and Oswald [12] described the first extension of this attack to SHA-1. Their method extends the Chabaud-Joux method and works with any characteristic that produces output difference zero.

Since a characteristic propagates in a deterministic way through a linear function, the characteristic is determined completely by the choice of the input difference. Hence, there are 2^{512} different characteristics. A fraction of 2^{-160} of these, results in a zero output difference (a collision). A difference corresponding to a characteristic is called a collision-producing difference.

Two messages m_1 and $m_2 = m_1 + \delta$ collide if

$$(m_1 + \delta)\mathbf{MS} - m_1\mathbf{MS} = 0 \iff \delta\mathbf{MS} = 0, \tag{7}$$

where δ is the input difference. Therefore, we are interested in solutions of the following equation:

$$v\mathbf{S} = 0, \tag{8}$$

whereas $v = \delta\mathbf{M}$ represents a collision-producing difference. Among the set of 2^{352} solutions we are searching for a small subset where

– v has a low Hamming weight
– the probability for the characteristic to hold is maximal.

There is a strong correlation between these two requirements, which will be explained in Section 3. Using a suitable low-weight difference, the attack proceeds as follows:

– conditions for following the characteristic are derived
– some conditions are pre-fulfilled by setting certain bits in the message
– during the final search, the most naive approach to fulfill all remaining conditions is to preform random trials. The time-complexity of this final search is determined by the number of conditions which are not pre-fulfilled.

The problem of finding low-weight difference vectors is the main topic of this article. We present efficient algorithms to cover this search space in Section 4. Using the found low-weight difference, we describe a general way to derive conditions that need to hold in order to follow this difference in Section 5.

2.3   Overview of Existing Attacks on SHA-1

We now review recent advances in the analysis of SHA-1. The conclusions drawn in this section will be used in subsequent sections.

Using More Than One Message Block. In multi-block collisions, we can also use characteristics that do not result in a zero output. For instance, for a two-block collision, all we require is that the output differences in both blocks are equal, because then, the final feed-forward will result in cancelation of the differences (with a certain probability). For an i-block collision, we get $512i - 160$ free bits ($512i - 320$ if we require that the perturbation pattern is a valid expanded message).

Easy Conditions. For the second step of the attack, constructing a pair of messages that follows this characteristic, a number of conditions on message words and intermediate chaining variables need to be fulfilled. As already observed in [5], conditions on the first steps can be pre-fulfilled. Using the idea of *neutral bits*, this approach was extended to cover the first 20 steps of the compression function of SHA-0 [1]. Wang *et al.* and Klima do something similar for MD4 and MD5 to pre-fulfill conditions, which is there called message modification [8,15,17]. For this reason, whenever we refer to a weight of a characteristic (collision-producing difference), we omit the weight of the first 20 words, unless stated otherwise.

Exploiting Non-Linearity. The state update is a non-linear transformation, and this can be exploited during the construction of the characteristic. While for a linear transformation the characteristic is determined completely by the input difference, in a non-linear transformation, one input difference can correspond to several characteristics.

Using a characteristic different from the one constructed from the linear approximation results in an overall increase of the number of equations. However, as explained before, the conditions in the first 15 steps are easy to fulfill. A good strategy is to look for a characteristic that has low weight and follows the linear approximation after the first 15 steps. This appears to be the strategy followed in [16]. A similar observation is made in [2,7]. We will use this strategy in Section 3.3 and Section 3.4.

3   From a Set of Collision-Producing Differences to a Linear Code

With the message expansion described by the matrix $\mathbf{M}_{512 \times 2560} = [\mathbf{I}_{512} \times \mathbf{F}_{512 \times 2048}]$ and the linearized state update described by the matrix $\mathbf{S}_{2560 \times 160}$, the output (hash value) of one SHA-1 iteration is $o = m\mathbf{MS} + iv\mathbf{T} + k$ (cf. Section 2.1). Two messages m_1 and $m_1^* = m_1 + m_1'$ collide if:

$$o_1^* - o_1 = (m_1 + m_1')\mathbf{MS} + k - (m_1\mathbf{MS} + k) = m_1'\mathbf{MS} = 0 \ . \tag{9}$$

Hence, the set of collision-producing differences is a linear code with check matrix $\mathbf{H}_{160 \times 512} = (\mathbf{MS})^t$. The dimension of the code is $512 - 160 = 352$ and the length of the code is $n = 512$.

Observation 1. *The set of collision-producing differences is a linear code. Therefore, finding good low-weight characteristics corresponds to finding low-weight vectors in a linear code.*

Based on this observation we can now exploit well known and well studied methods of coding theory to search for low-weight differences. We are mainly interested in the existing probabilistic algorithms to search for low-weight vectors, since a low-weight difference corresponds to a low-weight codeword in a linear code. In the remainder of this section we present several linear codes representing the set of collision-producing differences for the linearized model of SHA-1 as described in Section 2. Note that if we talk about SHA-1 in this section, we always refer to the linearized model. For the different linear codes we also give the weights of the found differences. How the low-weight vectors are found is discussed in Section 4.

As described in Section 2, we are interested in finding low-weight differences that are valid expanded messages and collision producing. Later on, we apply the strategy discussed in Section 2.3, *i.e.* we do not require the difference to be collision-producing. With this approach we are able to find differences with lower weight. The found weights are summarized in Table 4.

3.1   Message Expansion and State Update—Code C_1

For our attack it is necessary to look at the expanded message words and therefore we define the following check matrix for the linear code C_1 with dimension $dim(C_1) = 352$ and length $n = 2560$:

$$\mathbf{H1}_{2208 \times 2560} = \begin{bmatrix} \mathbf{S}^t{}_{160 \times 2560} \\ \mathbf{F}^t_{2048 \times 512} \ \mathbf{I}_{2048} \end{bmatrix} \ . \tag{10}$$

This check matrix is derived as following. Firstly, we want to have a valid expanded message. Since $m\mathbf{M} = w = m_{1 \times 512}[\mathbf{I}_{512}\mathbf{F}_{512 \times 2048}]$ and $\mathbf{M}$ is a systematic generator matrix, we immediately get the check matrix $[\mathbf{F}_{2048 \times 512}^t \mathbf{I}_{2048}]$. If a codeword w fulfills $w[\mathbf{F}_{2048 \times 512}^t \mathbf{I}_{2048}]^t = 0$, w is a valid expanded message. Secondly, we require the codeword to be collision-producing. This condition is determined by the state update matrix $\mathbf{S}$. If $w\mathbf{S} = 0$ then w is collision-producing. Therefore, we have the check matrix $\mathbf{S}^t$. Combining these two check matrices leads to the check matrix $\mathbf{H1}$ in (10).

The resulting codewords of this check matrix are valid expanded messages and collision-producing differences. When applying a probabilistic algorithm to search for codewords in the code C_1 (see Section 4) we find a lowest weight of 436 for 80 steps. The same weight has also been found by Rijmen and Oswald in [12]. As already described in Section 2.2, we do not count the weight of the first 20 steps since we can pre-compute these messages such that the conditions are satisfied in the first 20 steps. The weights for different number of steps are listed in Table 2.

Table 2. Lowest weight found for code C_1

steps $0, \ldots, 79$	steps $15, \ldots, 79^*$	steps $20, \ldots, 79$
436	333	293

*weight also given in [12]

A thorough comparison of these results with the weights given by Matusiewicz and Pieprzyk in [10] is not possible. This is due to the fact that in [10] *perturbation* and *correction* patterns have to be valid expanded messages. Furthermore, Matusiewicz and Pieprzyk give only the weights for the *perturbation* patterns.

3.2　Message Expansion Only and Multi-block Messages—Code C_2

Instead of working with a single message block that leads to a collision, we can also work with multi-block messages that lead to a collision after i iterations (cf. Section 2.3). For instance take $i = 2$. After the first iteration we have an output difference $o_1' \neq 0$ and after the second iteration we have a collision, *i.e.* $o_2' = 0$. The hash computation of two message blocks is then given by

$$o_1 = m_1\mathbf{MS} + iv\mathbf{T} + k$$
$$o_2 = m_2\mathbf{MS} + o_1\mathbf{T} + k$$
$$= m_2\mathbf{MS} + m_1\mathbf{MST} + \underbrace{iv\mathbf{T}^2 + k\mathbf{T} + k}_{constant} .$$

Based on the same reasoning as for the check matrix $\mathbf{H1}$ in Section 3.1, we can construct a check matrix for two block messages as follows:

$$\mathbf{H2}_{4256 \times 5120} = \begin{bmatrix} \mathbf{ST}_{160 \times 2560}^t & \mathbf{S}_{160 \times 2560}^t \\ \mathbf{F}_{2048 \times 512}^t \mathbf{I}_{2048} & \mathbf{0}_{2048 \times 2560} \\ \mathbf{0}_{2048 \times 2560} & \mathbf{F}_{2048 \times 512}^t \mathbf{I}_{2048} \end{bmatrix} . \quad (11)$$

The same can be done for i message blocks that collide after i iterations. The output in iteration i is given by

$$o_i = \sum_{j=0}^{i-1} m_{i-j} \mathbf{MST}^j + \underbrace{iv\mathbf{T}^i + k \sum_{l=0}^{i-1} \mathbf{T}^l}_{constant} . \tag{12}$$

Searching for low-weight vectors for a two-block collision in C_2 with **H2** and a three-block collision with the check matrix **HM2** given in Appendix A, leads to the weights listed in Table 3.

Table 3. Weight for two and three message blocks

weight of collision-producing differences for steps 20-79				
two-block collision		three-block collision		
exp. message 1	exp. message 2	exp. message 1	exp. message 2	exp. message 3
152	198	107	130	144

As it can be seen in Table 3, using multi-block collisions results in a lower weight for each message block. The complexity for a collision attack is determined by the message block with the highest weight. Compared to the weight for a single-block collision in Table 2 (weight = 293), we achieve a remarkable improvement. However, as shown in Table 4, the weight of the chaining variables is very high. Why this weight is important and how we can reduce the weight of the chaining variables is presented in the following section.

3.3 Message Expansion and State Update—Code C_3

For deriving the conditions such that the difference vector propagates for the real SHA-1 in the same way as for the linearized model, we also have to count the differences in the chaining variables (see Section 5). That means that for the previously derived collision-producing differences we still have to compute the weight in the chaining variables. It is clear that this leads to an increase of the total weight (see Table 4). Therefore, our new approach is to define a code that also counts in the chaining variables and to look for low-weight vectors in this larger code. This leads to lower weights for the total.

Furthermore, we now apply the strategy discussed in Section 2.3. In terms of our linear code, this means that we only require the codewords to be valid expanded messages and no longer to be collision-producing, *i.e.* they correspond to characteristics that produce zero ouput difference in the fully linearized compression function. This can be explained as follows. Our code considers only 60 out of 80 steps anyway. After 60 steps, we will have a non-zero difference. For a collision-producing difference, the 'ordinary' characteristic over the next 20 steps would bring this difference to zero. But in fact, for any difference after step 60 we will later be able to construct a special characteristic that maps the resulting difference to a zero difference in step 79. Hence, we can drop the requirement that the difference should be collision-producing. If we place the 20 special steps at the beginning, then the number of conditions corresponding to the special steps can be ignored.

Searching for the codeword producing the lowest number of conditions in the last 60 steps, we will work backwards. Starting from a collision after step 79 (chaining variables $A_{80}, \ldots, E_{80}$), we will apply the inverse linearized state update transformation to compute the chaining variables for step 78,77,...,20. We obtain a generator matrix of the following form:

$$\mathbf{G3}_{512 \times 11520} = [\mathbf{M}_{j \times n} \mathbf{A}_{j \times n} \mathbf{B}_{j \times n} \mathbf{C}_{j \times n} \mathbf{D}_{j \times n} \mathbf{E}_{j \times n}], \qquad (13)$$

where $j = 512$ and $n = 1920$. The matrices $\mathbf{A}_{j \times n}, \ldots, \mathbf{E}_{j \times n}$ can easily be constructed by computing the state update transformation backwards starting from step 79 with $A'_{80} = B'_{80} = \cdots = E'_{80} = 0$ and ending at step 20. The matrix $\mathbf{M}_{j \times n}$ is defined in Section 2.1.

The matrix defined in (13) is a generator matrix for code C_3 with $dim(C_3) = 512$ and length $n = 11520$. The lowest weight we find for code C_3 is 297. Note, that this low-weight vector now also contains the weight of the chaining variables $A'_t, \ldots, E'_t$. The weight for the expanded message is only 127. Compared with the results of the previous sections (code C_1) we achieve a remarkable improvement by counting in the weight of the chaining variables and by only requiring that the codewords are valid expanded messages.

3.4   Message Expansion, State Update, and Multi-block Messages—Code C_4

As shown in Section 3.2, we are able to find differences with lower weight if we use multi-block messages. We will do the same for the code C_4. A multi-block collision with $i = 2$ is shown in Figure 2.

As it can be seen in Figure 2, if we have the same output difference for each iteration we have a collision after the second iteration due to the feed forward.

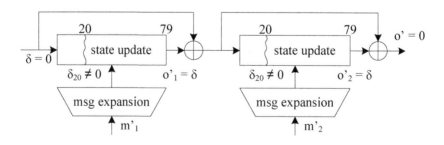

Fig. 2. Multi-block collision for SHA-1

We can construct a generator matrix as in Section 3.3 but we have to extend it such that we do not require a collision after the first iteration, *i.e.* we want an output difference of $o'_1 = o'_2 = \delta$. Therefore, we add 160 rows to the generator matrix in (13) that allow an output difference $o'_1 = o'_2 = \delta$. For the code C_4 we get a generator matrix

$$\mathbf{G4}_{672 \times 11520} = \begin{bmatrix} \mathbf{M}_{j \times n} & \mathbf{A}_{j \times n} & \mathbf{B}_{j \times n} & \mathbf{C}_{j \times n} & \mathbf{D}_{j \times n} & \mathbf{E}_{j \times n} \\ \mathbf{0}_{l \times n} & \mathbf{A}'_{l \times n} & \mathbf{B}'_{l \times n} & \mathbf{C}'_{l \times n} & \mathbf{D}'_{l \times n} & \mathbf{E}'_{l \times n} \end{bmatrix}, \qquad (14)$$

where $j = 512$, $l = 160$, and $n = 1920$. The matrix in (14) is a generator matrix for code C_4 with $dim(C_4) = 672$ and $n = 11520$. Searching for low-weight vectors in C_4 results in a smallest weight of 237. As we will show in Section 4, for this code it is infeasible to find codewords with weight 237 by using currently known algorithms (the same holds for code C_3). We found this difference vector by using an efficient way to reduce the search space as will be discussed in Section 4.2. Again, this weight includes also the weight of the chaining variables. For the message expansion only we have a weight of 108 (for one block we had 127). The difference vector is shown in Table 7, Appendix B.

3.5 Summary of Found Weights

To give an overview of the improvements achieved by constructing different codes we list the weights of the found codewords in Table 4.

Table 4. Summary of found weights

	Code C_1	Code C_2		Code C_3	Code C_4	
	single-block	two-block		single-block	two-block	
		msg 1	msg 2		msg 1	msg 2
weight expanded message	293	152	198	127	108	108
weight state update	563	4730	4817	170	129	129
total weight	856	4882	5015	297	237	237

4 Finding Low-Weight Vectors for Linear Codes Representing the Linearized SHA-1

In this section we describe different probabilistic algorithms that can be used to find low-weight vectors in linear codes. We describe the basic idea of these algorithms and present an algorithm that improves the low-weight vector search for SHA-1 significantly.

4.1 Probabilistic Algorithms to Find Low-Weight Vectors

We will briefly discuss some probabilistic algorithms presented by Leon [9] and modified by Chabaud [4], Stern [13], and by Canteaut and Chabaud [3]. The ·

basic approach of these algorithms is to take a (randomly permuted) subset of a code C and to search for low-weight vectors in this punctured code $C^{\bullet}$. A found low-weight codeword in the punctured code is a good candidate for a low-weight codeword in the initial code C.

A modified variant of Leon's algorithm [9] was presented by Chabaud [4]. It is applied to the generator matrix $\mathbf{G}_{k \times n}$ of a code C and defines the parameters p and s. The length of the punctured code $C^{\bullet}$ with generator matrix $\mathbf{Z} = \mathbf{Z}_{k \times (s-k)}$ is defined by s, where $s > dim(C^{\bullet}) = k$. For computing the codewords in $C^{\bullet}$ all linear combinations of at most p rows of $\mathbf{Z}$ are computed. The parameter p is usually 2 or 3. Values for the parameter s are $k+13, \ldots, k+20$ (see for instance [4]).

Stern's algorithm [13] is applied to the check matrix $\mathbf{H}_{(n-k) \times n}$. The parameters of the algorithm are l and p. The generator matrix $\mathbf{Z} = \mathbf{Z}_{l \times k}$ for the punctured code $C^{\bullet}$ is determined by k and l. The columns of $\mathbf{Z}$ are further split into two sets $\mathbf{Z}_1$ and $\mathbf{Z}_2$. Then the linear combinations of at most p columns are computed for both $\mathbf{Z}_1$ and $\mathbf{Z}_2$ and their weight is stored. Then searching for a collision of both weights allows to search for codewords of weight $2p$. Usually, the parameter p is 2 or 3 and l is at most 20 (see for instance [13]).

To compare these two algorithms we used the work-factor estimations to find an existing codeword with weight wt given by Chabaud [4]. For the comparison we used code C_4 (cf. Section 3.4) with $dim(C_4) = 672$, length $n = 11520$, and the weight $wt = 237$. The optimal parameters for Stern's algorithm are $p = 3$ and $l = 20$ for C_4. To find a codeword with $wt = 237$ in C_4 requires approximately 2^{50} elementary operations. Leon's algorithm, with parameters $p = 3$ and $s = dim(C_4) + 12$, requires approximately 2^{43} elementary operations.

Canteaut and Chabaud [3] have presented a modification of these algorithms. Instead of performing a Gaussian elimination after the random permutation in each iteration, Canteaut and Chabaud use a more efficient updating algorithm. More precisely, only two randomly selected columns are interchanged in each iteration, that is, only one step of a Gaussian elimination has to be performed. Even if this reduces the probability of finding a 'good' subset of the code, this approach leads to considerable improvements as they have shown for several codes in [3].

4.2   Improving Low-Weight Search for SHA-1

During our research on the different codes we observed that the found low-weight vectors all have in common that the ones and zeroes occur in bands. More precisely, the ones in the expanded message words usually appear in the same position (see also Tables 6 and 7). This observation has also been reported by Rijmen and Oswald in [12]. This special property of the low-weight differences for SHA-1 can be used to improve the low-weight vector search as follows. By applying Algorithm 1 to the generator matrix we force certain bits in the codewords to zero. With this approach we are able to reduce the search space significantly. As already mentioned, the basic idea of the probabilistic algorithms described in the beginning of this section, is to use a randomly selected set of columns of the generator matrix $\mathbf{G}$ to construct the punctured code. This corresponds to a reduction of the search space. If we apply Algorithm 1 to $\mathbf{G}$, we actually do the same but we do not have any randomness in constructing the punctured code. Algorithm 1 shows the pseudo-code.

Algorithm 1 Forcing certain bits of the generator matrix to zero

---

Input: generator matrix $\mathbf{G}$ for code C, integer r defining the minimum $rank$ of $\mathbf{Z}$
Output: generator matrix $\mathbf{Z}$ for punctured code $C^{\bullet}$ with $rank(\mathbf{Z}) = r$
1: $\mathbf{Z} = \mathbf{G}$
2: **while** $rank(\mathbf{Z}) > r$ **do**
3:     search in row x $(0 \le x < rank(\mathbf{Z}))$ for a one in column y $(0 \le y < length(\mathbf{Z}))$
4:     add row x to all other rows that have a one in the same column
5:     remove row x
6: **end while**
7: return $\mathbf{Z}$

Prior to applying the probabilistic search algorithms we apply Algorithm 1 to reduce the search space of the code. Since we force columns of the codewords to zero, we do not only reduce the dimension of the code but also the length. For the low-weight search we remove the zero-columns of $\mathbf{G}$. Computing the estimations for the complexities of this 'restricted code' shows that the expected number of operations decreases remarkably. For instance, applying Algorithm 1 to the generator matrix for code C_4 with $r = 50$ leads to the following values for the punctured code $C_4^{\bullet}$: $dim(C_4^{\bullet}) = 50$ and length $n = 2327$ (zero-columns removed). Stern's algorithm with optimal parameter $p = 2$ and $l = 4$ requires approx. 2^{37} elementary operations. For Leon's algorithm we get a work factor of approx. 2^{25} with $p = 3$ and $s = dim(C_4^{\bullet}) + 8$. With all the above-described algorithms we find the 237-weight difference within minutes on an ordinary computer by using Algorithm 1.

5   Low-Weight Vectors and Their Impact on the Complexity of the Attack

In this section we show how we can derive conditions for the low-weight differences found in Section 3. Based on the low-weight difference of code C_4, we will show some example conditions. The complexity for a collision attack depends on the number of conditions that have to be fulfilled. Since the number of conditions directly depends on the weight of the difference vector we see the correlation between weight and complexity: the lower the weight of the difference the lower the complexity for the attack.

The low-weight difference found for code C_4 leads to a collision after step 79 of the second compression function for the linearized SHA-1. Now, we want to define conditions such that the propagation of this difference is the same for the real SHA-1. In other words the conditions ensure that for this difference the real SHA-1 behaves like the linearized model.

As already mentioned in Section 2, the non-linear operations are f_{IF}, f_{MAJ}, and the addition modulo 2^{32}. Since we pre-compute message pairs such that all conditions in the first 20 steps are fulfilled, we only have to deal with f_{MAJ} and with the modular addition. For the addition we have to ensure that no carry occurs in the difference. For f_{MAJ}, we have to define conditions such that the differential behavior is the same as for f_{XOR}. Table 5 shows these conditions. For the sake of completeness also the conditions for f_{IF} are listed. Depending on the input difference we get conditions for the bit values of the inputs. For instance, if the input difference is $B_j'C_j'D_j' = 001$ then B_j and C_j have to be opposite, i.e. $B_j + C_j = 1$. The differential behavior of f_{MAJ} and f_{XOR} is the same if this condition is satisfied.

Table 5. Conditions that need to be fulfilled in order to have a differential behavior identical to that of an XOR

input difference $B_j'C_j'D_j'$	$f_{XOR}(B_j', C_j', D_j')$	$f_{IF}(B_j', C_j', D_j')$	$f_{MAJ}(B_j', C_j', D_j')$
000	0	always	always
001	1	$B_j = 0$	$B_j + C_j = 1$
010	1	$B_j = 1$	$B_j + D_j = 1$
011	0	never	$C_j + D_j = 1$
100	1	$C_j + D_j = 1$	$C_j + D_j = 1$
101	0	$B_j + C_j + D_j = 0$	$B_j + D_j = 1$
110	0	$B_j + C_j + D_j = 0$	$B_j + C_j = 1$
111	1	$C_j + D_j = 0$	always

Now, we show an example how to derive conditions for f_{XOR} and f_{MAJ}. Firstly, we take from Table 7 the difference corresponding to step $t = 28$ and bit position $j = 30$. We obtain the following:

$$A_{t,j}' = 0, B_{t,j}' = 0, C_{t,j}' = 1, D_{t,j}' = 0, E_{t,j}' = 0, A_{t+1,j}' = 0, W_{t,j}' = 1 .$$

For the following description we denote the output of f_{XOR} and f_{MAJ} by $F_{t,j}$.

Since $20 \leq t < 40$, the function f is f_{XOR}. Due to the input difference $C_{t,j}' = 1$ we always have $F_{t,j}' = 1$. Also $W_{t,j}' = 1$, and we have to ensure that there is no difference in the carry. This can be achieved by requiring that $F_{t,j}$ and $W_{t,j}$ have opposite values, i.e. $F_{t,j} + W_{t,j} = 1$. With $F_{t,j} = B_{t,j} + C_{t,j} + D_{t,j}$ we get $B_{t,j} + C_{t,j} + D_{t,j} + W_{t,j} = 1$. Since $B_t = A_{t-1}$, $C_t = A_{t-2} \lll 2$, $D_t = A_{t-3} \lll 2$, and $E_t = A_{t-4} \lll 2$, the condition for this example is:

$$A_{t-1,j} + A_{t-2,j+2} + A_{t-3,j+2} + W_{t,j} = 1 .$$

Secondly, we consider the difference for $t = 46$ and $j = 31$. This is the same difference as before but now f is f_{MAJ}, and therefore we have to ensure that

f_{MAJ} behaves like f_{XOR}. For the input difference $B'_{t,j}C'_{t,j}D'_{t,j} = 010$, we first get the following condition (cf. Table 5): $B_{t,j} + D_{t,j} = 1$. If this condition is satisfied we have the same situation as for f_{XOR}, namely $F'_{t,j} = C'_{t,j}$. Different to the previous example we do not get any further condition because the difference occurs in bit-position 31. The difference for this example is:

$$A_{t-1,j} + A_{t-3,j+2} = 1 .$$

If we derive the equations (conditions) for the complete low-weight vector in Table 7 we get a set of 113 equations. The equations are either in A only or in A and W. We can rework some of the equations to get (linear) equations involving bits of the expanded message words W only. This equations can easily be solved since they can directly be mapped to conditions on the message words. After reworking the 113 equations, we get 31 in W only and 82 equations in A, and in A and W. The overall complexity of the attack is determined by the (nonlinear) equations involving bits of the chaining variables and/or expanded message words. This is due to the fact that after pre-fulfilling the conditions for the first 20 steps the remaining conditions are fulfilled using random trials. Hence, solving this 82 (nonlinear) equations takes at most 2^{82} steps.

6   Comparison with Results of Wang *et al.*

In this section we compare the results of Wang *et al.* in [16] with the found low-weight difference given in Table 7. The difference in Table 7 is the lowest weight we found. The next higher weight we found (weight $= 239$) with the probabilistic search algorithms can also be constructed directly from the vector in Table 7. This is done by computing another iteration (see (2) and Figure 1) at the end and omitting the values of the first row such that we have again 60 steps. Since it is only a shifted version of the vector in Table 7 we can still use this vector for the comparison. The difference in Table 7, chaining variable A'_{t+1}, is the same disturbance vector as the one used by Wang *et al.* for near-collisions given in [16, Table 5] (italicized indices 20,...,79). To compare the two tables consider that Wang *et al.* index the steps from 1,...,80 (we from 0,...,79) but because Wang *et al.* use the shifted version the indices are the same except that the last pattern (index 80 for Wang *et al.*) is missing in Table 7. Also the Hamming weight for round 2-4 given in [16, Table 6] for 80 steps is the same. In [16, Table 7] one can find the difference vectors and the according number of conditions. The number of conditions and the conjectured attack complexity we stated in the previous section is remarkable higher than the values from [16]. However, no details on the exact way to derive conditions are given in [16].

7   Conclusions

In this article we have shown how coding theory can be exploited efficiently for collision attacks on the hash function SHA-1. We gave an overview of existing attack strategies and presented a new approach that uses different linear codes for

finding low-weight differences that lead to a collision. We also presented an algorithm that allows to find the low-weight differences very efficiently. Furthermore,
we gave an outline on how we can derive conditions for the found low-weight difference. We have shown that the number of conditions and hence the complexity
for a collision attack on SHA-1, directly depends on the Hamming weight of the
low-weight differences found.

Currently we are still working on improving the condition generation phase
to reduce the overall complexity of the collision attack on SHA-1. We will also
extend our approach such that we can perform similar analyses of alternative
hash functions such as the members of the SHA-2 family and RIPEMD-160.

Acknowledgements

We would like to thank Mario Lamberger for fruitful discussions and comments
that improved the quality of this article.

References

1. Eli Biham and Rafi Chen. Near-Collisions of SHA-0. In *Proccedings of CRYPTO
   2004*, volume 3152 of *LNCS*, pages 290–305. Springer, 2004.
2. Eli Biham, Rafi Chen, Antoine Joux, Patrick Carribault, Christophe Lemuet, and
   William Jalby. Collisions of SHA-0 and Reduced SHA-1. In *Proceedings of EU-
   ROCRYPT 2005*, volume 3494 of *LNCS*, pages 36–57. Springer, 2005.
3. Anne Canteaut and Florent Chabaud. A New Algorithm for Finding Minimum-
   Weight Words in a Linear Code: Application to McEliece's Cryptosystem and
   to Narrow-Sense BCH Codes of Length 511. *IEEE Transactions on Information
   Theory*, 44(1):367–378, 1998.
4. Florent Chabaud. On the Security of Some Cryptosystems Based on Error-
   correcting Codes. In *Proceedings of EUROCRYPT '94*, volume 950 of *LNCS*,
   pages 131–139. Springer, 1995.
5. Florent Chabaud and Antoine Joux. Differential Collisions in SHA-0. In *Proceedings of CRYPTO '98*, volume 1462 of *LNCS*, pages 56–71. Springer, 1998.
6. Hans Dobbertin. Cryptanalysis of MD4. In *Proceedings of Fast Software Encryption*, volume 1039 of *LNCS*, pages 53–69. Springer, 1996.
7. Antoine Joux, Patrick Carribault, William Jalby, and Christophe Lemuet. Full
   iterative differential collisions in SHA-0, 2004. Preprint.
8. Vlastimil Klima. Finding MD5 Collisions on a Notebook PC Using Multi-message
   Modifications, 2005. Preprint, available at http://eprint.iacr.org/2005/102.
9. Jeffrey S. Leon. A probabilistic algorithm for computing minimum weights of large
   error-correcting codes. *IEEE Transactions on Information Theory*, 34(5):1354–
   1359, 1988.
10. Krystian Matusiewicz and Josef Pieprzyk. Finding good differential patterns
    for attacks on SHA-1. In *Proccedings of WCC 2005*. Available online at
    http://www.ics.mq.edu.au/~kmatus/FindingGD.pdf.
11. National Institute of Standards and Technology (NIST). FIPS-180-2: Se-
    cure Hash Standard, August 2002. Available online at http://www.itl.nist.
    gov/fipspubs/.

12. Vincent Rijmen and Elisabeth Oswald. Update on SHA-1. In *Proceedings of CT-RSA 2005*, volume 3376 of *LNCS*, pages 58–71. Springer, 2005.
13. Jacques Stern. A method for finding codewords of small weight. In *Proccedings of Coding Theory and Applications 1988*, volume 388 of *LNCS*, pages 106–113. Springer, 1989.
14. Xiaoyun Wang, Dengguo Feng, Xuejia Lai, and Xiuyuan Yu. Collisions for Hash Functions MD4, MD5, HAVAL-128 and RIPEMD, August 2004. Preprint, available at `http://eprint.iacr.org/2004/199`, presented at the Crypto 2004 rump session.
15. Xiaoyun Wang, Xuejia Lai, Dengguo Feng, Hui Chen, and Xiuyuan Yu. Cryptanalysis for Hash Functions MD4 and RIPEMD. In *Proceedings of EUROCRYPT 2005*, volume 3494 of *LNCS*, pages 1–18. Springer, 2005.
16. Xiaoyun Wang, Yiqun Lisa Yin, and Hongbo Yu. Finding Collisions in the Full SHA-1. In *Proceedings of CRYPTO 2005*, volume 3621 of *LNCS*, pages 17–36. Springer,2005.
17. Xiaoyun Wang and Hongbo Yu. How to Break MD5 and Other Hash Functions. In *Proceedings of EUROCRYPT 2005*, volume 3494 of *LNCS*, pages 19–35. Springer, 2005.

A   Check Matrix for 3-Block collision

The hash output for three message blocks, is given by

$$o_3 = m_3\mathbf{MS} + m_2\mathbf{MST} + m_1\mathbf{MST}^2 + \underbrace{iv\mathbf{T}^3 + k\mathbf{T}^2 + k\mathbf{T} + k}_{constant} \ .$$

The set of collision-producing differences is a linear code with check matrix:

$$\mathbf{HM2}_{6304\times7680} = \begin{bmatrix} (\mathbf{ST}^2)^t_{160\times2560} & \mathbf{ST}^t_{160\times2560} & \mathbf{S}^t_{160\times2560} \\ \mathbf{F}^t_{2048\times512}\mathbf{I}_{2048} & \mathbf{0}_{2048\times2560} & \mathbf{0}_{2048\times2560} \\ \mathbf{0}_{2048\times2560} & \mathbf{F}^t_{2048\times512}\mathbf{I}_{2048} & \mathbf{0}_{2048\times2560} \\ \mathbf{0}_{2048\times2560} & \mathbf{0}_{2048\times2560} & \mathbf{F}^t_{2048\times512}\mathbf{I}_{2048} \end{bmatrix} \ . \quad (15)$$

B   Found Low-Weight Differences

Table 6. Lowest weight found for code C_2 — weight = 436. Note that the ones and zeroes appear in bands

step	W_t	step	W_t
t=0	06e00000	t=40	1a780000
t=1	d9000000	t=41	f5000000
t=2	a2e00000	t=42	b7700000
t=3	82e00000	t=43	06800000
t=4	cd580000	t=44	78b00000
t=5	57500000	t=45	00000000
t=6	9b660000	t=46	6a900000
t=7	c0ce0000	t=47	60f00000
t=8	c0b20000	t=48	6c200000
t=9	d1f00000	t=49	e7100000
t=10	7d980000	t=50	8bc00000
t=11	c3bc0000	t=51	85d00000
t=12	3a500000	t=52	08000000
t=13	54c00000	t=53	80100000
t=14	bd840000	t=54	35000000
t=15	47bc0000	t=55	25900000
t=16	60e40000	t=56	82700000
t=17	6f280000	t=57	23200000
t=18	ab380000	t=58	c3200000
t=19	edd00000	t=59	02400000
t=20	068c0000	t=60	b2000000
t=21	d0cc0000	t=61	47800000
t=22	17000000	t=62	63e00000
t=23	501c0000	t=63	20e00000
t=24	1a040000	t=64	44200000
t=25	d4c80000	t=65	84000000
t=26	99d80000	t=66	c0000000
t=27	c1500000	t=67	87400000
t=28	ab200000	t=68	16000000
t=29	b4d00000	t=69	44000000
t=30	16600000	t=70	a7a00000
t=31	47500000	t=71	50a00000
t=32	ca100000	t=72	82e00000
t=33	80a00000	t=73	c5800000
t=34	e6780000	t=74	23000000
t=35	6cb80000	t=75	80c00000
t=36	74180000	t=76	04c00000
t=37	44f00000	t=77	00c00000
t=38	efb80000	t=78	01400000
t=39	8f380000	t=79	01000000

Table 7. Lowest weight found for code C_4 — weight = 237

step	W'_t	A'_{t+1}	B'_{t+1}	C'_{t+1}	D'_{t+1}	E'_{t+1}
t=20	80000040	00000000	00000002	00000000	a0000000	80000000
t=21	20000001	00000003	00000000	80000000	00000000	a0000000
t=22	20000060	00000000	00000003	00000000	80000000	00000000
t=23	80000001	00000002	00000000	c0000000	00000000	80000000
t=24	40000042	00000002	00000002	00000000	c0000000	00000000
t=25	c0000043	00000001	00000002	80000000	00000000	c0000000
t=26	40000022	00000000	00000001	80000000	80000000	00000000
t=27	00000003	00000002	00000000	40000000	80000000	80000000
t=28	40000042	00000002	00000002	00000000	40000000	80000000
t=29	c0000043	00000001	00000002	80000000	00000000	40000000
t=30	c0000022	00000000	00000001	80000000	80000000	00000000
t=31	00000001	00000000	00000000	40000000	80000000	80000000
t=32	40000002	00000002	00000000	00000000	40000000	80000000
t=33	c0000043	00000003	00000002	00000000	00000000	40000000
t=34	40000062	00000000	00000003	80000000	00000000	00000000
t=35	80000001	00000002	00000000	c0000000	80000000	00000000
t=36	40000042	00000002	00000002	00000000	c0000000	80000000
t=37	40000042	00000000	00000002	80000000	00000000	c0000000
t=38	40000002	00000000	00000000	80000000	80000000	00000000
t=39	00000002	00000002	00000000	00000000	80000000	80000000
t=40	00000040	00000000	00000002	00000000	00000000	80000000
t=41	80000002	00000000	00000000	80000000	00000000	00000000
t=42	80000000	00000000	00000000	00000000	80000000	00000000
t=43	80000002	00000002	00000000	00000000	00000000	80000000
t=44	80000040	00000000	00000002	00000000	00000000	00000000
t=45	00000000	00000002	00000000	80000000	00000000	00000000
t=46	80000040	00000000	00000002	00000000	80000000	00000000
t=47	80000000	00000002	00000000	80000000	00000000	80000000
t=48	00000040	00000000	00000002	00000000	80000000	00000000
t=49	80000000	00000002	00000000	80000000	00000000	80000000
t=50	00000040	00000000	00000002	00000000	80000000	00000000
t=51	80000002	00000000	00000000	80000000	00000000	80000000
t=52	00000000	00000000	00000000	80000000	00000000	00000000
t=53	80000000	00000000	00000000	00000000	00000000	80000000
t=54	80000000	00000000	00000000	00000000	00000000	00000000
t=55	00000000	00000000	00000000	00000000	00000000	00000000
t=56	00000000	00000000	00000000	00000000	00000000	00000000
t=57	00000000	00000000	00000000	00000000	00000000	00000000
t=58	00000000	00000000	00000000	00000000	00000000	00000000
t=59	00000000	00000000	00000000	00000000	00000000	00000000
t=60	00000000	00000000	00000000	00000000	00000000	00000000
t=61	00000000	00000000	00000000	00000000	00000000	00000000
t=62	00000000	00000000	00000000	00000000	00000000	00000000
t=63	00000000	00000000	00000000	00000000	00000000	00000000
t=64	00000000	00000000	00000000	00000000	00000000	00000000
t=65	00000004	00000004	00000000	00000000	00000000	00000000
t=66	00000080	00000000	00000004	00000000	00000000	00000000
t=67	00000004	00000000	00000000	00000001	00000000	00000000
t=68	00000009	00000008	00000000	00000000	00000001	00000000
t=69	00000101	00000000	00000008	00000000	00000000	00000001
t=70	00000009	00000000	00000000	00000002	00000000	00000000
t=71	00000012	00000010	00000000	00000000	00000002	00000000
t=72	00000202	00000000	00000010	00000000	00000000	00000002
t=73	0000001a	00000008	00000000	00000004	00000000	00000000
t=74	00000124	00000020	00000008	00000000	00000004	00000000
t=75	0000040c	00000000	00000020	00000002	00000000	00000004
t=76	00000026	00000000	00000000	00000008	00000002	00000000
t=77	0000004a	00000040	00000000	00000000	00000008	00000002
t=78	0000080a	00000000	00000040	00000000	00000000	00000008
t=79	00000060	00000028	00000000	00000010	00000000	00000000
weight	108	26	25	25	26	27

Hash Based Digital Signature Schemes

C. Dods, N.P. Smart, and M. Stam

Dept. Computer Science, University of Bristol,
Merchant Venturers Building, Woodland Road,
Bristol, BS8 1UB, United Kingdom
chris@rydertech.co.uk, {nigel, stam}@cs.bris.ac.uk

Abstract. We discuss various issues associated with signature schemes based solely upon hash functions. Such schemes are currently attractive in some limited applications, but their importance may increase if ever a practical quantum computer was built. We discuss issues related to both their implementation and their security. As far as we are aware this is the first complete treatment of practical implementations of hash based signature schemes in the literature.

1 Introduction

Digital signature schemes based on non-number theoretic assumptions are interesting for a number of reasons. Firstly they extend the number of underlying problems on which we base our security, secondly all number theoretic assumptions currently used in cryptography will be insecure if a quantum computer is ever built. Of particular interest are those schemes based solely on the security of cryptographic hash functions; such schemes are not only of historical interest but are also relatively easy to implement. See [6] for a survey of other possible signature schemes in a post-quantum computing world.

However, systems based on hash functions have a number of drawbacks. The main problem is that they usually arise in the context of one-time signatures. One-time signatures are public key signature schemes which have the property that they can only be used to sign one single message. On the other hand such schemes are usually highly efficient and easy to implement in very constrained devices, since they only require the implementation of hash functions and not any advanced arithmetic operations.

Usually, however, one requires multi-time signatures. A one-time signature can be turned into a multi-time signature in one of two ways. The first way allows one to sign as many messages as one wishes over time, however this comes at the cost of our signatures growing in size with every signature produced. This not only gives efficiency problems, but also reveals extra information since a signature size reveals information about the prior number of signatures produced.

Another way of turning a one-time signature into a multi-time signature is to fix beforehand the total number of signatures which will ever be produced and then to use a Merkle tree to authenticate the different one-time signatures.

N.P. Smart (Ed.): Cryptography and Coding 2005, LNCS 3796, pp. 96–115, 2005.

Despite the drawback of needing to fix the number of signatures beforehand it is this latter approach which is used when hash based one-time signatures are implemented.

There are a number of hash based one-time signature schemes available. Of particular interest is the Winternitz scheme which is easy to understand, relatively efficient and has been used in practical situations. Also of interest is the construction of Bleichenbacher and Maurer which is more complicated but which is theoretically the most efficient construction known.

If one wished to protect digital signatures for a long time, in particular against the possible advent of quantum computers, one is currently restricted to the use of hash based one-time signature schemes with Merkle trees. In this paper we investigate what is the most efficient construction for such schemes, by comparing the Winternitz based approach with the Bleichenbacher–Maurer approach. These are combined with a method for constructing authentication paths in Merkle trees due to Szydlo.

We present our experiments with a combination of Szydlo's methods and the Winternitz and Bleichenbacher–Maurer constructions. In addition we present various security results which generalise on the results in the literature. In particular our security proofs are built upon the assumption that any practical scheme will be implemented using only one type of hash function, hence we focus on what properties this function should have rather than requiring different properties and hence possibly different hash functions for different parts of the construction.

2   Efficient One-Time Signature Schemes

In this section we describe two schemes. The first, due to Winternitz, is a generalisation of the scheme of Merkle [16], which itself is based on the scheme of Lamport [11] (although this latter scheme is highly inefficient). In the second scheme we present the scheme of Bleichenbacher and Maurer which attempts to achieve the best possible efficiency in terms of signature size and number of hash function evaluations per bit of the message. We present a full explanation of the Bleichenbacher and Maurer construction since the original papers [4,5] concentrated on the combinatorial constructions and did not cover the issue of how this is actually used in practice to construct a one-time signature scheme.

We let f denote a hash function with domain $\{0,1\}^*$ and codomain $\{0,1\}^n$. Repeated application of f is denoted by superscripts, i.e., $f^2(x) = f(f(x))$. We assume the actual message to be signed, say M, has first been hashed via f to obtain an n-bit hash value m which will be signed by the one-time signature scheme.

In all schemes we assume that from a single n-bit key x one can derive a set of pseudo-random bit strings x_i all of length n. This can be done, for example, by using f in a key derivation function such as ANSI-X9.63-KDF [1].

Following [4] we define the efficiency of a graph based one-time signature scheme, which signs n bits, to be equal to $\Gamma = \frac{n}{v+1}$ where v is number of vertices in the associated graph. This is essentially equal to the number of message bits which can be signed per hash function evaluation during the key generation phase. It also assumes a binary tree is used to hash the public key into a single vertex. Whilst this is a crude measure, in practice it does give some guide as to the practical efficiency of such schemes, but we shall see that it does not tell the whole truth in a practical situation.

2.1   Winternitz Scheme

The Winternitz scheme is parametrised by a value w, typically chosen to be a small power of two. For convenience, the explanation below is tailored for the case where $w = 2^t$. The generalization to arbitrary w is straightforward.

One defines $N = \lceil n/t \rceil + \lceil (\lceil \log_2 n - \log_2 t \rceil + t)/t \rceil$.

- **Key Generation:** The private key is given by x_i for $i = 1, \ldots, N$. The public key is then computed by first computing $y_i = f^{2^t-1}(x_i)$ and then computing $Y = f(y_1 \| y_2 \| \cdots \| y_N)$.
- **Signing:** The hash of the message m is split into $\lceil n/t \rceil$ segments $b_1, \ldots, b_{\lceil n/t \rceil}$ of t-bits in length (using padding with zero if necessary). Treating the m values of b_i as integers, we form the check symbol

$$C = \sum_{i=1}^{\lceil n/t \rceil} 2^t - b_i.$$

  Note, that $C \leq \lceil n/t \rceil 2^t$ and so we can write in base 2^t as $b_{\lceil n/t \rceil+1}, \ldots, b_N$. The signature is given by $s_i = f^{b_i}(x_i)$, for $i = 1, \ldots, N$.
- **Verification:** From m generate the b_i as in the signing process and compute $v_i = f^{2^t-b_i-1}(s_i)$. The signature is said to verify if and only if $Y = f(v_1 \| v_2 \| \cdots \| v_N)$
- **Efficiency:** The number of vertices in the associated graph is easily seen to be equal to

$$v = 2^t \cdot N = 2^t \left(\lceil n/t \rceil + \lceil (\lceil \log_2 n - \log_2 t \rceil + t)/t \rceil \right) \approx 2^t \left(n + \log_2 n \right)/t.$$

  Hence, $\Gamma \approx t/2^t$ as $n \longrightarrow \infty$. So we expect the scheme to be most efficient when $t = 2$.

2.2   Bleichenbacher–Maurer Scheme

In this section we describe the best known graph construction of a one-time signature scheme [5]. We present the construction in full generality and explain how one obtains from the combinatorial construction a one-time signature scheme, by using an analogue of the check symbol from above. This contrasts with the original description [5], where a non-constructive argument is given based on the pigeon hole principle.

The scheme is parametrised by an integer w (originally chosen to be three) and an integer B, the value of B is dependent on n and w and will be given below. We define the scheme via a set of B blocks, each block is an array of width w and height $w+1$. There is also an additional 0th block (not included in the B above) which consists of a single row of w entries. We use the subscripts $z_{b,r,c}$ to refer to the entry in the bth block and in the rth row and cth column, where rows and columns are numbered from zero. The entries are assumed to hold values, and they are inferred from the following computational rule:

$$z_{b,r,c} = \begin{cases} f(z_{b,r-1,c}\|z_{b-1,w,(c+r)} \pmod w) & \text{If } r > 0 \text{ and } b > 1, \\ f(z_{b,r-1,c}\|z_{b-1,0,(c+r)} \pmod w) & \text{If } r > 0 \text{ and } b = 1, \\ x_{bw+c} & \text{If } r = 0. \end{cases} \quad (1)$$

To define a signature we first need to define a signature pattern. This is an ordered list of w numbers $\mathfrak{p} = (r_0, r_1, \ldots, r_{w-1})$, i.e. one height or row per column. We select the set of patterns $\mathfrak{p}$ such that

$$\bigcup_{i \in \{0,\ldots,w-1\}} \{i + j \pmod w : r_i \leq j < w\} = \{0, \ldots, w-1\}.$$

We let p denote the number of such signature patterns, which depends on the choice of w. For example when $w = 2$ one has $p = 6$, when $w = 3$ one has $p = 51$ and when $w = 4$ one has $p = 554$. To each pattern we assign a weight given by

$$wt(\mathfrak{p}) = \sum_{i=0}^{w-1} (w + 1 - r_i).$$

Note that $w^w < p < (w+1)^w$ and $wt(\mathfrak{p}) \leq w(w+1) \leq p$. For asymptotic purposes we will use $\log_2 p = \Theta(w \log_2 w)$.

- **Key Generation:** The secret key is assumed to consist of $N = (B+1)w$ values $x_0, \ldots, x_{N-1}$ which are placed in the bottom row of each block as above. The public key is produced by

$$Y = f(z_{B,w,0}\|z_{B,w,1}\| \cdots \|z_{B,w,w-1}),$$

  i.e., by hashing together the values in the top row of the final block.
- **Signing:** The (hash of the) message m is first written in base p, as $m = \sum_{i=0}^{l-1} b_i p^i$ where $l = \lceil n/\log_2 p \rceil$. We then compute the check symbol via

$$C = \sum_{i=0}^{l-1} wt(\mathfrak{p}_{b_i}) = b_l + b_{l+1}p \ldots + b_{l+l'-1}p^{l'-1}.$$

If we set $l' = \lceil 1 + \log_p l \rceil$ then the above base p expansion of C is valid. To sign the n-bit message we will require B blocks where $B = l + l' = \lceil n/\log_2 p \rceil + \lceil \log_p \lceil n/\log_2 p \rceil \rceil$. The signature is computed by, for each $0 \leq i \leq B-1$, taking the value b_i and taking the signature pattern $\mathfrak{p}_{b_i} = (r_{i,0}, \ldots, r_{i,w-1})$. Then releasing the values of the entries $\{z_{i+1,r_{i,j},j}\}$, for $0 \leq i \leq B-1$ and $0 \leq j \leq w-1$, plus the values of $z_{0,0,j}$, for $j = 0, \ldots, w-1$.

- **Verification:** Verification proceeds by again writing m in base p. The check symbol is computed and also written in base p. The released entries are then taken, and assumed to correspond, to the patterns given by the message and the check symbol. Using the computation rule (1) the values of $z_{B,w,0}, z_{B,w,1}, \ldots, z_{B,w,w-1}$ are computed and the public key Y is checked to be equal to

$$Y' = f(z_{B,w,0} \| z_{B,w,1} \| \cdots \| z_{B,w,w-1}).$$

- **Efficiency:** We have

$$\Gamma = \frac{B \log_2 p}{1 + w(w+1)B} \longrightarrow \frac{\log_2 p}{w(w+1)} \approx \frac{\log_2 w}{w} \quad \text{as } n \longrightarrow \infty.$$

We expect therefore the scheme to be most efficient when $w = 3$.

3   One-Time to Many-Time

As mentioned in the introduction there are two mechanisms to convert a one-time signature into a many-time signature. The first technique, see [15–Section 11.6.3], allows an arbitrary number of signatures to be signed, however as the number of signatures grows so do the signature size and the verification time. A second approach is to use Merkle authentication trees. Here one bounds the total number of signatures which can be issued at key generation time. The benefit is that signature size and verification time are then constant.

A Merkle tree in the context of a one-time signature scheme is defined as follows. It is a complete binary tree equipped with a function h, with codomain the set of bit strings of length k, and an assignment ϕ which maps the set of nodes to the set of k-length bit strings. For two interior child nodes, n_l and n_r, of any interior node, n_p, the assignment satisfies

$$\phi(n_p) = h(\phi(n_l) \| \phi(n_r)).$$

For each leaf l_i the value of $\phi(l_i)$ is defined to be the public-key of a one-time signature scheme. From these values the values of the interior nodes can be computed. The value of $\phi(n)$ for the root node n is defined to be the public key of the many-time signature scheme. The leaf nodes are numbered from left to right and if the tree has height H, then we can sign at most 2^H messages.

For a tree of height H and the ith node l_i, we define $A_{h,i}$ to be the value of the sibling of the height h node on the path from the leaf l_i to the root. The values $A_{h,i}$ are called the authentication path for the leaf l_i.

To sign a message one signs a message with the next unused one-time signature, and then one authenticates the public key of this one-time signature using the Merkle tree. This is done by revealing the authentication path from the used leaf up-to the public key at the root. To verify the signature one verifies both the one-time signature and the authentication path. The current best algorithm for creating authentication paths in Merkle trees is described by Szydlo [23]. This algorithm creates the paths in a sequential manner, i.e., the path for leaf l_{i+1} is created after the path for leaf l_i. Since to verify the authentication path one also needs to know the value of i, a one-time signature implemented using Szydlo's algorithm will reveal the time ordering of different issued signatures. In some applications this may be an issue.

4   Provable Security of One-Time Signatures

Prior practical constructions of one-time signatures do not come with security proofs. Hevia and Micciancio [10] prove the security of a different class of graph based signatures, to those considered above. Unfortunately, most known schemes as they are implemented in real life, do not seem to be covered by the proof directly. In particular, Hevia and Micciancio do not include the use of length-preserving one-way functions, even though these are most commonly used in proposals for practical one-time signature schemes. In theory, one could simulate a length-preserving function by composing hash function and pseudorandom generators, but this would incur a considerable performance loss. A second problem is that Hevia and Micciancio use a slightly different definition of graph based signatures. In particular, Bleichenbacher and Maurer allow the output of a hash function to be the input to several other hashes, whereas Hevia and Micciancio enforce the use of a pseudorandom generator to perform fanout. Moreover, Hevia and Micciancio pose a strong restriction on the type of signatures that is allowed. Consequently, there is a gap between what Hevia and Micciancio prove secure and what Bleichenbacher and Maurer describe as a graph based one-time signature scheme.

In this section we generalise Hevia and Micciancio's result to a broader class of graphs based on weaker, more natural assumptions, all in the standard model. However, even this broad class does not include the construction of Bleichenbacher and Maurer. To prove a security result for this latter construction we need to apply to the Random Oracle Model.

4.1   Definitions

Building signature schemes requires composing the functions in non-trivial ways and the properties of the original function need not hold for the composition. For a length-preserving one-way function, the most obvious composition is repeated application, for instance $f^2(x) = f(f(x))$. Composition for a collection of functions is done pointwise.

The properties of the functions we require in our proofs are one-wayness, collision resistance and undetectability. We intuitively describe these properties here, leaving the technical definitions to Appendix A.

- **One wayness:** A function is one-way (or preimage resistant) if it is easy to compute but hard to invert.
- **Collision resistance:** A function is collision resistant if it is infeasible to find any pair (x, x') in the domain that map to the same value.

- **Second pre-image resistance:** A function is second pre-image resistant if, given some x in the domain, it is hard to find some x' unequal to x that maps to the same value.
- **Undetectable:** A function is called undetectable if an adversary cannot distinguish the output from f with a uniform distribution over its range. In practice, this means that, given some n-bit element, an adversary is not able to tell whether f was applied to it or not.

4.2   Generalities on Graph Based One-Time Signatures

Graph based signatures, such as that of Winternitz or the Bleichenbacher–Maurer construction, are based on directed acyclic graphs. The use of these graphs for one-time signature schemes was first described by Bleichenbacher and Maurer [3]. Later a different model was given by Hevia and Micciancio [10]. There are some subtle but important differences between these two models. We give a unifying framework, mainly based on Bleichenbacher and Maurer's terminology, although we will follow Hevia and Micciancio's suggestion to swap the meaning of nodes and edges.

Let (V, E) be a directed acyclic graph. The sources and sinks of the graph correspond to the secret keys and public keys respectively. The graph denotes how the public key is computed from the secret key in the following way. The internal nodes correspond to functions, the edges to input and output values associated to those functions. There are two basic types of internal nodes: those with indegree one and those with indegree two. A node with indegree two represents a call to a hash function h. Some convention has to be used to distinguish between the first and second input for h. A node with indegree one represents a call to a length-preserving one-way function f_i. Here $i \in \{1, \ldots, w\}$ needs to be specified for the node at hand. If a node has several outgoing arcs, these all represent copies of the same output value. We assume that the graph is optimised in the sense that there are no two nodes representing the same function that have the same input (one could make an exception for hash functions, where the two possible input orderings presumably give different outcomes, but we do not).

As in any directed acyclic graph, the graph defines a partial ordering on the nodes and allows us to talk of predecessors, successors, parents, children, sinks and sources. We also fix some total ordering that is consistent with the induced partial ordering. For instance, the private key is smaller than the public key.

A signature corresponds to the release of the values associated with a certain set of edges. We will also call this set of edges a signature or sometimes a signature-set or signature-pattern. Given a set of edges S, the set of directly computable nodes is defined as the set of nodes all of whose incoming arcs are in S. The set of computable nodes is defined recursively as the set of nodes computable from the union of S with the outgoing edges of the directly computable nodes.

We will impose three conditions on a set of edges representing a signature:

1. It is verifiable: the entire public key is in the set of computable nodes.

2. It is consistent: either all arcs originating from a node are in a signature or none is. (This requirement is needed because all arcs originating from a single node represent the same output value.)

3. It is minimal: leaving out any edge would destroy verifiability or consistency.

Intuitively, the set of computable nodes from a signature denote the functions that the signer, or a forger, has to invert given the public key. Two signatures are called compatible if neither of the sets of computable nodes is a subset of the other. As a result, given one signature forging a second, compatible one still requires finding some non-trivial preimage. A signature scheme is a collection $\mathcal{S}$ of mutually compatible sets $S \subseteq V$ together with an efficiently computable collision-resistant mapping from the message space into $\mathcal{S}$.

The total number of nodes $|V|$ in the graph corresponds to the total number of operations during the key-generation, taking into account that for the leaves a call to the pseudorandom function is needed (and handing out a penalty for too many public keys).

Since all nodes have only one distinct output Bleichenbacher and Maurer identify (the value of) the output with the function-node itself. Hence, they describe a signature in terms of nodes and not edges (as a result, consistency is immediate). Moreover, Bleichenbacher and Maurer's main focus is on using only one type of one-way function. We will refer to a scheme with multiple one-way functions as a *generalised BM scheme*.

Hevia and Micciancio consider a restricted model by allowing only certain types of graphs and certain types of signature sets given the graph. Their graphs consist of two types of internal nodes: nodes with indegree two and outdegree one (representing a hash function), and nodes with indegree one and outdegree two (representing a pseudorandom generator). An important difference with our model is that fanout is only allowed by using pseudorandom generators, with the result that the arcs leading out of this node represent *different* values instead of copies of the same output value (as is the case in our model). An easy consequence of this model is that consistency is immediate.

One can transform a Hevia-Miccancio graph into a generalised BM-graph. To do this, simply replace every pseudorandom node in the Hevia-Micciancio graph with an f_1- and an f_2-node in the transformed graph, where both nodes have the same input and f_1 takes care of the first output of the generator and f_2 of the second. This transformation increases the number of nodes, which influences the efficiency measure. We believe it is reasonable to assume that a pseudorandom generator will take twice the time of a length preserving function or a hash function (on compatible input/output lengths).

Another restriction of Hevia and Micciancio's approach is that they only consider cuts for their signature sets. Given a graph, a cut consists of the arcs between the two parts of a non-trivial partition $(S, \tilde{S})$ of the set of nodes. It is a requirement that all arcs point from S to $\tilde{S}$ (or equivalently, that S is predecessor closed or that $\tilde{S}$ is successor closed). It is easy to see that there is a one-to-one correspondence between the cuts in a Hevia-Micciancio graph and the consistent cuts in the corresponding generalised BM-graph.

As mentioned before, most proposals for one-time signature schemes do not actually fit into the Hevia-Micciancio framework. A small number of one-time signature schemes do not fit into the generalised Bleichenbacher–Maurer framework [17,19,20] either. Although the relation between the public key and the secret key can still be expressed in terms of a graph and a signature still corresponds to a set of edges, it is no longer a requirement that all possible signatures are incompatible. Instead of using a mapping from the message space to the signature space $\mathcal{S}$ that is collision resistant, a stronger mapping is used, such that it is hard to find "incompatible collisions".

4.3   Security Proofs

We generalise Hevia and Micciancio's work in three ways. We allow more diverse graphs, we allow more diverse signature patterns and our requirements on the functions with which the scheme will be instantiated are more lenient.

Hevia and Micciancio gave a proof of security for the signature schemes that fit into their model (all fanout is provided by pseudorandom generators and signatures are based on cuts) under the assumption that the hash function h is regular and collision-resistant and that an injective pseudorandom generator is used. Regularity of h is a structural requirement, as is injectivity of g. We show that these structural requirements can be replaced by more lenient and to-the-point security properties.

We also generalise Hevia and Micciancio's work by allowing more one-way functions f_i and several private and public key nodes. The latter has a slight advantage even if all public key nodes are subsequently hashed together (since then the final hash only needs to be collision-resistant). Consequently, our results directly apply to schemes using primarily a single length-preserving function (such as Winternitz' scheme), whereas in Hevia and Micciancio's work one seemingly needed to replace this with hashing the output of a pseudorandom generator (to fit in what was proven). We do however still assume that if a node has several arcs leading out, they all lead to different f_i nodes (and not to h-nodes). This opens the door for compressing the graph into something more akin to a Hevia-Micciancio graph by replacing all the f_i-nodes serviced from the same node by a single new node of indegree one, but a larger outdegree. Let $W \subseteq \{1, \ldots, w\}$ be the set of indices for which f_i was replaced. We will model the new node as representing the function $f_W : \{0,1\}^n \to \{0,1\}^{|W|n}$. Each outgoing arc corresponds to one of the f_i that has been replaced. We will refer to the graph that arises from this reinterpretation (for all nodes) as the compressed graph $(\tilde{V}, \tilde{E})$.

Finally, we relax the requirement that signatures are based on cuts. Suppose two different signatures s and $\tilde{s}$ are given (i.e., two sets of arcs). Consider the set of nodes computable from s but not from $\tilde{s}$ (this set is non-empty by definition of compatibility). Call a node in this set allowable if the intersection of the predecessor edges with $\tilde{s}$ is empty. Call s and $\tilde{s}$ strongly compatible if the set of allowable nodes is non-empty (note that, unlike compatibility, strong compatibility is not a symmetric relation). We require from our signature scheme that all pairs of signatures are strongly compatible. It is clear that if all signatures are cuts, the set of allowable nodes is exactly the same as the set of nodes computable from s but not from $\tilde{s}$, thus the schemes by Hevia and Micciancio form a (proper) subset of our schemes.

We show that any compressed graph based one-time signature scheme with strongly compatible signature sets is secure if h and all the relevant f_W are collision-resistant, one-way and undetectable.

Theorem 1. *Let* $\Pr[\text{FORGE}]$ *be the success probability of any forger of the signature scheme. Let* $\Pr[\text{DETECT}]$ *be the success probability for detecting using algorithms running in polynomial time with oracle access to the signature forger and let* $\Pr[\text{INVERT or COLLISION}]$ *be the maximum of the probabilities of finding a collision or preimage for some* f_W *or* h, *using algorithms as above. Let* $|\tilde{V}|$ *be the number of nodes in the compressed graph and let* $\alpha = |\tilde{V}|^2$. *Then the following holds*

$$\Pr[\text{FORGE}] \leq |\tilde{V}|(\Pr[\text{INVERT or COLLISION}] + \alpha \Pr[\text{DETECT}]) .$$

Proof. The proof is given in Appendix B.

Our result implies Hevia and Micciancio's result [10–Theorem 1], apart from the tightness regarding detection. The proofs are similar, but we make the hybrids more explicit. This gives a shorter proof that is hopefully closer to intuition and easier to understand. The price we have to pay is that our reduction is less tight.

On the other hand the above theorem does not apply to the types of graphs used in the Bleichenbacher–Maurer construction, whilst it does apply to the Winternitz construction. The problem with the Bleichenbacher–Maurer schemes is that there might be two edges in a single signature that are comparable under the partial ordering induced by the graph. Hence to prove a result for such graphs we are required to resort to the random oracle model.

Theorem 2. *Let* $\Pr[\text{FORGE}]$ *be the success probability of any forger of the signature scheme. Let* $\Pr[\text{INVERT}]$ *be the maximum success probabilities for inversion of some* f_i *using algorithms running in polynomial time with oracle access to the signature forger, let* $\Pr[\text{COLLISION}]$ *be the maximum probability of finding a collision for* f_i *and let* $\Pr[\text{DETECT}]$ *be the maximum probability for detecting a relevant* f_W, *using algorithms as above. Furthermore, let* $|V|$ *be the number of nodes in the graph, let* $\alpha = |V|^2$, *and let* δ *be a neglible quantity that upper bounds the probability of inverting a random oracle or finding a collision for it in polynomial time. Then the following holds*

$$\Pr[\text{FORGE}] \leq |V|(\Pr[\text{COLLISION}] + \Pr[\text{INVERT}] + \alpha \Pr[\text{DETECT}] + \delta) .$$

Proof. The proof is given in Appendix C.

5   Practical Considerations

A theoretical analysis shows that the Bleichenbacher–Maurer and the Winternitz scheme have similar asymptotics. That is, for both schemes key generation, signature generation and signature verification take time $O(nw/\log w)$ with signature size $O(n/\log w)$. Hence any theoretical difference is hidden in the constants. Table 1 gives a theoretical comparison of the two one-time signature schemes. Here we adhere to Bleichenbacher and Maurer's assumption that both hashing two n-bit values into a single n-bit value and hashing a single n-bit value cost unit time, whereas hashing a dn-bit value for $d > 1$ takes $d - 1$ units.

Table 1. Comparison of Winternitz and Bleichenbacher–Maurer, in the number of signature bits, respectively number of hash-operations per message bit

	Winternitz				Bleichenbacher–Maurer		
	$w = 2$	$w = 3$	$w = 4$	$w = 5$	$w = 2$	$w = 3$	$w = 4$
Signature Size	1.00	0.69	0.50	0.43	0.77	0.53	0.44
Key Generation	3.00	2.52	2.50	2.58	2.32	2.11	2.19
Signature Generation	1.50	1.26	1.25	1.29	2.32	2.11	2.19
Signature Verification	1.50	1.26	1.25	1.29	1.16	1.05	1.09

Note that the numbers for key generation are the reciprocals of the efficiency measure as introduced by Bleichenbacher and Maurer. Based on these figures they concluded that their scheme with $w = 3$ fixed, is the most efficient. However, the Bleichenbacher–Maurer scheme requires more complicated code and bookkeeping, seems to rely on the assumption about the cost of hash functions, and its benefits are mainly based on key generation and to a lesser extent signature verification.

To test the relative merits of the different schemes we implemented both, and the associated Merkle trees. The various timings, in milli-seconds, can be found in Tables 2 and 3. Surprisingly we found that the Winternitz scheme was more efficient, regardless of the output size of the hash function used.

Our experiments show that the Winternitz scheme is significantly faster than the Bleichenbacher–Maurer scheme. Moreover, Winternitz with $w = 3$ has slightly cheaper key generation than $w = 4$; Winternitz signatures and verification do no take the same time, contrary to the theoretical expectation; for the Bleichenbacher–Maurer scheme, signing is slightly more expensive than keygeneration.

There are a couple of causes for the differences between the theory put forward by Bleichenbacher and Maurer and practice as we implemented it. In the theory only calls to the hash functions are counted, however in reality mapping the message to the signature set is not for free. Thus, in the Bleichenbacher-Maurer scheme, where key generation and signing take almost the same number of hash calls, signature creation also requires the additional creation of a radix

p representation of the message and mapping from this to signature sets, so one would expect signing to be slower than key generation as shown in the table. Another example is the Winternitz scheme for $w = 4$, where we implemented both a radix 4 representation using modular reductions (timings for these are within parentheses) and a direct bitwise method. The latter gave slightly faster times as expected.

Our experiments also show that the assumption on the cost of hashing different size messages is inaccurate. For instance, in the Winternitz scheme the signing and verification equally divide the number of hash calls made within the rakes. However, when signing the private keys need to be generated whereas in the verification the public keys need to be hashed together. In the theoretical model we assume that the costs for these two operations are equal, but in practice hashing many short values is clearly more expensive than hashing one long value. In the extreme and unrealistic case—where hashing one long value takes unit time, so hashing the public key is essentially ignored—the theoretical advantage of Bleichenbacher–Maurer dissipates and Winternitz is most efficient, in particular with $w = 3$.

Finally, one sees that for the Bleichenbacher-Maurer scheme the time for key generation increases from $w = 2$ to $w = 4$ (skipping $w = 3$), whereas that of signature verification decreases. This actually does follow from a theoretically more thorough analysis, where one takes into account the forbidden signature patterns. The patterns that would have been cheapest to verify are banned, hence increasing the cost of verification to just over half the cost of key generation. However, for small w this unbalancing effect weighs in stronger than for larger w.

To conclude, the theoretical advantage promised by Bleichenbacher and Maurer may not be a real advantage for real-life hash functions and practical values of n. For larger values of n, it is possible that Bleichenbacher and Maurer's scheme starts performing better, but the likely cause of the discrepancy between theory and practice carries through in the asymptotic case. It is recommended to use the Winternitz' scheme with $w = 4$, since it is very fast, easy to implement and provides short signatures.

6   Key Sizes

In this section we examine what the security reductions of our proofs and generic attacks on one-way functions imply for the recommended security parameters. In Table 4 we summarize the best known black-box attacks against different functions given different goals. Here $f : \{0,1\}^n \rightarrow \{0,1\}^n$ is length-preserving, $g : \{0,1\}^n \rightarrow \{0,1\}^{dn}$ can be a function with any expansion factor d and $h : \{0,1\}^{2n} \rightarrow \{0,1\}^n$ is a hash function.

In the security reduction, an inverter, collision finder or sample-distinguisher is described that basically runs in the same time as a potential signature forger, but with decreased success probability. By running the simulator multiple times, the probability of (say) inverting can be boosted. Assuming indepedence, this boost

Table 2. Timings (in ms) of Operations for One-Time Signature Schemes

	f	Params	Key Gen	Sign	Verify
Winternitz	SHA-1	$w = 2$	1.2	0.9	0.5
Winternitz	SHA-1	$w = 3$	1.1	0.7	0.5
Winternitz	SHA-1	$w = 4$	1.1 (1.1)	0.7 (0.7)	0.5 (0.5)
Winternitz	SHA-1	$w = 5$	1.2	0.7	0.6
BM	SHA-1	$w = 2$	1.4	1.5	0.7
BM	SHA-1	$w = 3$	1.3	1.4	0.6
BM	SHA-1	$w = 4$	1.4	1.4	0.7
Winternitz	SHA-256	$w = 2$	2.1	1.4	1.0
Winternitz	SHA-256	$w = 3$	1.8	1.2	0.9
Winternitz	SHA-256	$w = 4$	1.9 (1.9)	1.1 (1.1)	0.9 (1.0)
Winternitz	SHA-256	$w = 5$	2.0	1.2	1.0
BM	SHA-256	$w = 2$	3.3	3.5	1.9
BM	SHA-256	$w = 3$	3.2	3.3	1.7
BM	SHA-256	$w = 4$	3.5	3.6	1.8
Winternitz	SHA-512	$w = 2$	14.7	9.3	6.1
Winternitz	SHA-512	$w = 3$	13.1	7.9	5.9
Winternitz	SHA-512	$w = 4$	13.1 (13.4)	7.6 (7.7)	5.9 (6.3)
Winternitz	SHA-512	$w = 5$	14.1	8.1	7.0
BM	SHA-512	$w = 2$	22.8	23.1	12.4
BM	SHA-512	$w = 3$	21.8	21.9	11.0
BM	SHA-512	$w = 4$	23.9	24.0	11.7

Table 3. Timings (in ms) of Operations for Many-Time Signature Schemes. Note, maximum number of signatures is 2^R.

	f	Params	R	Key Gen	Sign	Verify
Winternitz	SHA-1	$w = 4$	8	300	8	1
Winternitz	SHA-1	$w = 4$	14	18700	8	1
BM	SHA-1	$w = 3$	8	350	9	1
BM	SHA-1	$w = 3$	14	22100	12	1
Winternitz	SHA-256	$w = 4$	8	500	14	2
Winternitz	SHA-256	$w = 4$	14	30800	12	2
BM	SHA-256	$w = 3$	8	800	24	2
BM	SHA-256	$w = 3$	14	56400	24	2

is linear, so to achieve an overwhelming probability of inverting we would have to invest at most $|V|/\Pr[\text{FORGE}]$ time. Our assumption is that inverting using the signature forger takes at least as much time as a generic attack, so the time taken by a generic attack should in any case be less than $|V|/\Pr[\text{FORGE}]$, which implies an upper bound in the probability of a signature forger expressed in $|V|$ and the runtime of a generic inverter. Since this runtime is a function of the security parameter, i.e., the length of the bitstrings that f works on, it is now possible to determine a value for the security parameter that is guaranteed to provide security (based on

Table 4. Complexity of generic attacks on one-way functions

	Classical			Quantum		
	f	g	h	f	g	h
one-wayness	2^n	2^n	2^n	$2^{n/2}$	$2^{n/2}$	$2^{n/2}$
collision-resistance	2^n	2^n	$2^{n/2}$	2^n	2^n	$2^{n/3}$
undetectability	2^n	2^n	2^{2n}	$2^{n/2}$	$2^{n/2}$	2^n

the assumption that generic attacks are the best available). These are necessarily conservative estimates: the reduction might not be tight. In concreto, finding collisions for the hash function is going to be the dominating factor in determining the minimum recommended security parameter.

In the classical world, we have that $\Pr[\text{FORGE}] \leq |V|/2^{n/2}$, where the number of nodes $|V|$ is still to be determined. It is fair to assume that the messages will be hashed, so we need to sign $2k$-bit hash to achieve a security of at least 2^{-k}. Using the Winternitz scheme processing two bits at a time, this results in approximately $|V| = 3k$ (effectively), so we recommend a keysize $n \geq 2k + 2\lg k + 2\lg 3$.

In the quantum world it is easier to find collisions [2]. Indeed, we need $\Pr[\text{FORGE}] \leq |V|/2^{n/3}$, where the number of nodes $|V|$ is approximately $4.5k$ using a Winternitz scheme processing two bits at a time (we need to sign a $3k$-bit hash to achieve 2^{-k}-security). The recommended keysize then becomes $n \geq 3k + 3\lg k + 3\lg 4.5$.

To conclude, the security parameter should be 1.5 times as long in the quantum setting as in the classical setting. Note that the tightness in the reduction (for instance of simulator failure) or the fact that we need to sign longer hashes in the quantum setting hardly influences the recommendations for the security parameter.

References

1. ANSI X9.63. *Public Key Cryptography for the Financial Services Industry: Key Agreement and Key Transport Using Elliptic Curve Cryptography.* October 1999, Working Draft.
2. G. Brassard, P. Høyer, and A. Tapp. Quantum cryptanalysis of hash and claw-free functions. In C. L. Lucchesi and A. V. Moura, editors, *LATIN'98*, volume 1380 of *Lecture Notes in Computer Science*, pages 163–169. Springer-Verlag, 1998.
3. D. Bleichenbacher and U. M. Maurer. Directed acyclic graphs, one-way functions and digital signatures. In Y. Desmedt, editor, *Advances in Cryptography— Crypto'94*, volume 839 of *Lecture Notes in Computer Science*, pages 75–82. Springer-Verlag, 1994.
4. D. Bleichenbacher and U. Maurer. Optimal tree-based one-time digital signature schemes. *Proc. Symp. Theoretical Aspects of Comp. Sci. – STACS '96*, Springer-Verlag LNCS 1046, 363–374, 1996.
5. D. Bleichenbacher and U. Maurer. On the efficiency of one-time digital signature schemes. *Advances in Cryptology – ASIACRYPT '96*, Springer-Verlag LNCS 1163, 145–158, 1996.

6. J. Buchmann, C. Coronado, M. Döring, D. Engelbert, C. Ludwig, R. Overbeck, A. Schmidt, U. Vollmer and R.-P. Weinmann. Post-quantum signatures. Preprint, 2004.
7. S. Even, O. Goldreich, and S. Micali. On-line/off-line digital signatures. *Journal of Cryptology*, 9(1):35–67, 1996.
8. J. Håstad and M. Näslund. Practical construction and analysis of pseudo-randomness primitives. In C. Boyd, editor, *Advances in Cryptography—Asiacrypt'01*, volume 2248 of *Lecture Notes in Computer Science*, pages 442–459. Springer-Verlag, 2001.
9. R. Hauser, M. Steiner, and M. Waidner. Micro-payments based on iKP. Technical report, IBM Research, 1996.
10. A. Hevia and D. Micciancio. The provable security of graph-based one-time signatures and extensions to algebraic signature schemes. In Y. Zheng, editor, *Advances in Cryptography—Asiacrypt'02*, volume 2501 of *Lecture Notes in Computer Science*, pages 379–396. Springer-Verlag, 2002.
11. L. Lamport. Constructing digital signatures from a one-way function. *SRI International, CSL-98*, 1979.
12. L. A. Levin. One-way functions and pseudorandom generators. *Combinatorica*, 7(4):357–363, 1987.
13. R. J. Lipton and R. Ostrovsky. Micro-payments via efficient coin-flipping. In R. Hirschfeld, editor, *FC'98*, volume 1465 of *Lecture Notes in Computer Science*, pages 1–15. Springer-Verlag, 1998.
14. M. Lomas, editor. *Security Protocols*, volume 1189 of *Lecture Notes in Computer Science*. Springer-Verlag, 1996.
15. A.J. Menezes and P.C. van Oorschot and S.A. Vanstone. *CRC-Handbook of Applied Cryptography*, CRC Press, 1996.
16. R. Merkle. A certified digital signature. *Advances in Cryptology – CRYPTO '89*, Springer-Verlag LNCS 435, 218–238, 1990.
17. M. Mitzenmacher and A. Perrig. Bounds and improvements for BiBa singature schemes. Technical report, Harvard University, Cambridge, Massachusetts, 2002.
18. T. P. Pedersen. Electronic payments of small amount. In Lomas [14], pages 59–68.
19. A. Perrig. The BiBa one-time signature and broadcast authentication protocol. In M. Reiter and P. Samarati, editors, *CCS'01*, pages 28–37. ACM Press, 2001.
20. L. Reyzin and N. Reyzin. Better than BiBa: Short one-time signatures with fast signing and verifying. In L. M. Batten and J. Seberry, editors, *ACISP'02*, volume 2384 of *Lecture Notes in Computer Science*, pages 144–153. Springer-Verlag, 2002.
21. R. L. Rivest and A. Shamir. PayWord and MicroMint: Two simple micropayment schemes. In Lomas [14], pages 69–78.
22. D. R. Simon. Finding collisions on a one-way street: Can secure hash functions be based on general assumptions? In K. Nyberg, editor, *Advances in Cryptography—Eurocrypt'98*, volume 1403 of *Lecture Notes in Computer Science*, pages 334–345. Springer-Verlag, 1998.
23. M. Szydlo. Merkle tree traversal in log space and time. *Advances in Cryptology – EUROCRYPT 2004*, Springer-Verlag LNCS 3027, 541–554, 2004.

A   Properties of One-Way Functions

In this section we will review some well-known properties of one-way functions. Our starting point will be a collection $\mathcal{F}$ of length-regular functions acting on bitstrings, together with a probabilistic polynomial time (ppt) algorithm I that samples elements from $\mathcal{F}$. On input a string of ones I returns a description of some $f \in \mathcal{F}$ of length polynomially related to the length of I's input. Let $m(f)$ denote the input length of the function indexed by f and let $n(f)$ denote the corresponding output length, so we can write $f : \{0,1\}^m \to \{0,1\}^n$ (omitting the dependence of n and m on f). We will assume that an efficient algorithm is given that, given f and $x \in \{0,1\}^m$ computes $f(x)$. We denote the uniform distribution over all strings of length n by U_n.

Definition 1 (One wayness). *Let $\mathcal{F}$ and I be given as above. Then $(\mathcal{F}, I)$ is one-way if and only if (iff) for all ppt adversaries A and all polynomials p eventually (there is an integer N such that for all $k > N$)*

$$\Pr[f(x') = f(x) \wedge |x'| = m : x' \leftarrow A(f,y), y \leftarrow f(x), x \leftarrow U_m, f \leftarrow I(1^k)] < \frac{1}{p(k)}.$$

Unfortunately, a function f can be one-way but f^2 not. Vice versa, a function f^2 can be one-way without f being one-way. However, if f is one-way and undetectable, then so is f^2 and similarly, if f is one-way and collision resistant, then so is f^2.

Definition 2 (Collision Resistance). *Let $\mathcal{F}$ and I be given as above. Then $(\mathcal{F}, I)$ is collision resistant iff for all ppt adversaries A and all polynomials p eventually (in k)*

$$\Pr[f(x') = f(x) \wedge x \neq x' \wedge |x| = |x'| = m : (x, x') \leftarrow A(f), f \leftarrow I(1^k)] < \frac{1}{p(k)}.$$

It is clear that collision-resistance implies second pre-image resistance, but the converse is unlikely to hold. Any injective function is collision-resistant and second pre-image resistant. It is also well known that collision-resistance implies one-wayness for compressing functions that are regular or for which the compression ratio can be asymptotically bounded away from 1. It can be easily shown that a function f is collision resistant if and only if f^2 is collision resistant.

Definition 3 (Undetectability). *Let $\mathcal{F}$ be a collection of functions with sampling algorithm I. Then we say $(\mathcal{F}, I)$ is undetectable iff for all ppt adversaries A and all polynomials p eventually (in k)*

$$|\Pr[b = b'|b' \leftarrow A(f,y), y \leftarrow y_b, b \leftarrow \{0,1\}, y_0 \leftarrow U_n, y_1 \leftarrow f(x),$$
$$x \leftarrow U_m, f \leftarrow I(1^k)] - \frac{1}{2}| < \frac{1}{p(k)}.$$

In the literature the security requirement undetectability is usually associated with the morphological property of f being an expanding function. In this case f is known as a pseudorandom generator. We deviate from the literature by considering the two separately, especially so since undetectability is a useful property even for non-expanding functions in certain proofs. Note that naming this property pseudorandomness would have conflicted with the existing notion of pseudorandom functions.

If f is undetectable, then so is f^2 by a straightforward hybrid argument. Suppose we would have a distinguisher A_2 for f^2. Since A_2 distinguishes $f^2(U_n)$ and U_n by assumption, it must also distinguish between at least one of the pairs $f^2(U_n), f(U_n)$ and $f(U_n), U_n$. In the latter case A_2 distinguishes f, in the former case $A_2 \circ f$ does the trick. It is possible however to construct a function that is detectable applied only once, but undetectable when applied twice.

For the security of hash-chains in authentication schemes it usually suffices if the function is one-way on its iterates [8,9,18,12,21,13], but for signature schemes something potentially stronger is needed. The relevant problem has been defined by Even, Goldreich and Micali [7] (under the name quasi-inversion). We give a slightly different definition with different terminology.

Definition 4 (One-deeper preimage resistance). *Let $l_1 \geq q$ and $l_0 \geq 0$. A collection of length-preserving functions $(\mathcal{F}, I)$ is called (l_0, l_1)-one-deeper preimage resistant if for all probabilistic polynomial time adversaries A, all polynomials p and all sufficiently large k the following holds:*

$$\Pr[f^{l_0}(y) = f^{l_0+1}(x') : x' \leftarrow A(f, y), y \leftarrow f^{l_1}(w), w \leftarrow U_n, f \leftarrow I(1^k)] < \frac{1}{p(k)}.$$

If $l_0 = 0$ with ranging l_1, the definition reduces to that of one-wayness on its iterates. If the property holds for all (l_0, l_1) polynomial in k we call the function chainable. It is well known and easy to verify that one-way permutations are chainable. We give sufficient conditions for a length-preserving one-way function to be chainable.

Lemma 1. *If a family of length-preserving functions $(\mathcal{F}, I)$ is one-way, undetectable and collision resistant, then it is chainable.*

If a function f is secure for some l_0 and l_1, this does imply f to be one-way, but it does not necessitate f to be either undetectable or second-preimage resistant. These properties are needed in the proof to amplify the onewayness of f. Our result should not be confused with that of Even et al. [7] who remark that the existence of one-way functions suffices for the existence of functions that are one-deeper preimage resistant (for polynomially bounded l_0 and l_1). This is interesting, since the existence of one-way functions is not known to be sufficient for the existence of one-way permutations or collision-resistant functions in general [22].

If our goal is to prove that a function is one-way on its iterates, it suffices that f is one-way and undetectable (which is hardly surprising). If a function is one-way, undetectable and second preimage-resistant, then it is $(1, l_1)$-one-deeper preimage resistant for all (polynomially bounded) l_1.

B   Proof of Theorem 1

The idea of the proof is simple and will take place in the compressed graph $(\tilde{V}, \tilde{E})$. The forger asks for a signature and then has to produce another one.

The signature scheme is set up in such a way that there are nodes computable from the forged signature that were not computable from the queried one. We will say that the forger inverts on these nodes. If we can guess beforehand which node the forger is going to invert, we can put the value we wish to invert in that node. The problem is that this action changes the distribution of the public key (and the queried signature), so we need to show that an adversary capable of spotting the difference between the two distribution can be used to detect the application of some combination of functions. Using hybrids, this is fairly easy to show. We are helped in this task by the fact that if an adversary inverts a node, the predecessor edges are completely unknown (this is where we need that the signatures are partitions), so the simulator does not need to know or have consistent values for these edges, and by the fact that all fanout is the work of pseudorandom generators.

Let y_h be generated according to $h(U_n)$ and y_W according to $f_W(U_n)$, for $W \subseteq \{1, \ldots, w\}$. We will describe three games, show that a success in Game 3 leads to either a collision (for h or some f_W) or a preimage (for some y_W or y_h) and that if the success probability of Game 3 is negligible but an efficient signature forger exists, some function is detectable.

Game 1. The original game. The simulator creates the public key as in the original key generation and feeds it to the forger. Because the simulator knows the entire secret key, he can easily provide any one-time signature the forger requests. The success probability of this game is by definition $\Pr[\text{FORGE}]$, so

$$\Pr[\text{SUCCESS}_1] = \Pr[\text{FORGE}] . \tag{2}$$

Game 2. As Game 1, but with the modification that the simulator picks an internal node (the target) at random and aborts if the forger does not invert on this node or if the forger's signature query includes an arc predecessing the target node. Since the actions of the simulator are independent of the chosen target, the forger behaves as in Game 1. The success probability is at least $\Pr[\text{SUCCESS}_1]/|\tilde{V}|$, which, combined with (2), gives

$$\Pr[\text{SUCCESS}_2] \geq \Pr[\text{FORGE}]/|\tilde{V}| . \tag{3}$$

Game 3. As Game 2, but with the following modification to the key generation. The simulator replaces the output value of the targeted node with whichever he wishes to invert (obviously, this value should be from a suitable distribution for the node at hand). The simulator also replaces all arcs leading out of the predecessor graph of the target node with random values. The simulator recomputes the public key.

The simulator can produce for all arcs a value, except for those within the predecessor graph of the target. If the target node is allowable, the adversary's query will not contain any arcs within the predecessor graph (by definition). Thus for allowable targets, the simulator will be able to answer the signature query correctly.

We claim the following, from which the theorem follows:

$$\varepsilon = \Pr[\text{SUCCESS}_2] - \Pr[\text{SUCCESS}_3] \leq |\tilde{V}| \Pr[\text{DETECT}]$$
$$\Pr[\text{SUCCESS}_3] \leq \Pr[\text{INVERT or COLLISION}]$$

First consider the case in which Game 3 is successful. For the targeted node the forger has thus supplied a value y' and its preimage. If $y' = y$, we have found a preimage to the target and are done. On the other hand, if $y' \neq y$, it is possible to find a collision somewhere in the path to the root.

Now consider the difference ε in success probability between Game 2 and Game 3. Remember that a full ordering on the nodes of the graph is fixed consistent with the partial ordering induced by the graph.

Game 3'. Consider Game 3', parametrised by an internal node t of the graph (we also allow a node at infinity). The simulator picks a random internal node: if it is less than t it plays Game 2, otherwise it plays Game 3. If t is the minimal element, Game 3' is identical to Game 3, whereas the other extreme, t is the node at infinity, corresponds to Game 2. A standard hybrid argument implies the existence of a node t^* such that the difference in success probability between Game 3' with $t = t^*$ and with t the next node (in the full ordering) is at least $\varepsilon/|\tilde{V}|$.

Game 3''. Now define Game 3'', parametrised by an internal node r that is a predecessor of t^*. The simulator picks an internal node in the full graph at random. If it is less than t^* the simulator plays Game 2, if it is larger than t^* it plays Game 3. If it equals t^*, the simulator assigns random values to all edges emanating from a node smaller than r and recomputes the public key.

If r is minimal, then Game 3'' equals Game 3' with $t = t^*$, but if $r = t^*$, then Game 3'' equals Game 3' with t the next node. As a result, there must be some node r^* such that the difference in success probability between Game 3'' with $r = r^*$ and r being the next node is at least $\varepsilon/|\tilde{V}|^2$. The only difference between these two instantiations of Game 3'' can be traced back to node r^*. If $r = r^*$, it is computed based on an input computed uniformly at random. If r is the next node, then the output of r^* will be made uniformly random by the simulator. This finally leads to a distinguisher for the function that node r^* represents.

C   Proof of Theorem 2

The idea of the proof is similar to the previous proof. If we can guess which node the forger is going to invert, we can put the value we wish to invert in that node, provided it is an f_i-node. An adversary that does not invert any f-node, is forced to invert the random oracle, which can only be done with negligible probability given only a polynomial number of oracle queries.

Let y_i be generated according to $f_i(U_m)$ for $i = 1, \ldots, w$. We will describe three games, show that a success in Game 3 leads to either a collision (for some f_i or the random oracle), or an inversion of f_i and that the success probability of Game 3 cannot be negligible if an efficient signature forger exists and f_W are undetectable.

Game 1. The original game, where calls to the hash function are replaced by oracle queries. The simulator creates the public key as in the original key generation and feeds it to the forger. Because the simulator knows the entire secret key, he can easily provide any one-time signature the forger requests. The success probability of this game is by definition $\Pr[\text{FORGE}]$.

Game 2. As Game 1, but with the modification that the simulator picks an internal node at random and aborts if the forger does not invert on this node. Although the simulator makes up his mind for several oracle queries, he does not make the calls anymore (the forger can of course query based on the public key and signature he receives). Since the actions of the simulator are independent of the node he picked, the forger behaves as in Game 1. The success probability is at least $\Pr[\text{SUCCESS}_1]/|V|$.

Game 3. As Game 2, but with the modification that if the simulator has picked an f_i-node it replaces its output value with y_i. Moreover, it looks at f_i's predecessor nodes until it hits an h-node. The simulator replaces the output of the f_i-children of the f_i-predecessors in a direct line with random values and recomputes the public key. The only arcs for which the simulator does not know a value are the arcs directly leading out of the f_i-predecessors in a direct line of the target node. From any of these arcs the target is computable and hence, if the forger inverts on the target, by minimality of signature patterns these arcs cannot have been queried.

It is easy to see that an adversary that distinguishes between Game 2 and Game 3 also distinguishes between a correct evaluation of the tree consisting of the f_i-predecessors in a direct line of the target together with their f_i children (which will include the target itself) and an equal number of independently chosen uniform random values. ¿From the result in the previous section it already follows that if all the f_W involved are undetectable, this probability is negligible. In fact, $|\Pr[\text{SUCCESS}_3] - \Pr[\text{SUCCESS}_2]| \leq \alpha \Pr[\text{DETECT}]$.

If the forger inverts some f-node, the above leads either to a one-deeper inversion, a collision on f or a collision on the random oracle. The probability of finding a collision on the random oracle in polynomial time is negligible.

If the forger does not invert any f-node, he will have to invert an h-node, corresponding to inverting the random oracle. Since it is an inversion, not both edges can be known from the signature-query allowed to the forger. So one edge is still unknown. Presumably, the simulator already knows the value of this edge and copies of it might be used elsewhere in the signature scheme. However, no information can be known to the adversary if it is solely used as input to other h-nodes (by virtue of the random oracle modelling these nodes). Moreover, it cannot be the input to an f-node, since in that case the adversary would also invert an f-node, which it does not (by assumption). Hence, the adversary only has a negligible guessing probability in succeeding.

A General Construction for Simultaneous Signing and Encrypting

John Malone-Lee

University of Bristol, Department of Computer Science,
Woodland Road, Brsitol, BS8 1UB, UK
malone@cs.bris.ac.uk

Abstract. In this paper we present a very efficient, general construction for simultaneous signing and encrypting data. Our construction uses the KEM-DEM methodology of Cramer and Shoup combined with a secure signature scheme.

We describe an instantiation of our construction that provides all the functionality of a signature scheme and of an encryption scheme. This instantiation is more efficient than similar constructions proposed to-date.[1]

1 Introduction

Encryption schemes and signature schemes are the basic tools offered by public key cryptography for providing privacy and authenticity respectively. Originally they were always viewed as important, but distinct, primitives for use in higher level protocols; however, there are many settings where the services of both are needed, perhaps the most obvious being secure e-mailing. In this scenario messages should be encrypted to ensure confidentiality and signed to provide authentication. In this case it is of course possible to use an encryption scheme combined with a digital signature scheme. It has been observed in earlier work by An et al. [1] that there are often subtleties when using such combinations. Moreover, it may be possible to use features particular to the case where one wants both authentication and encryption to gain in efficiency and functionality. Motivated by such considerations there has been much recent research into schemes and methods for simultaneous signing and encrypting.

The first proposed construction combining the functionality of an encryption scheme with that of a signature scheme appeared in a paper by Zheng [22]. The motivation behind this work was to obtain some efficiency benefit when compared to encrypting and signing separately. Zheng called schemes designed to achieve this goal *signcryption* schemes. Many schemes have subsequently been designed with Zheng's original motivation in mind [2,3,13,14,15,18]. Some of these have been formally analysed using complexity-theoretic reductions [2,13,14,15]; however, all this analysis relies on the random oracle model [5].

---

[1] Many thanks to Liqun Chen for discussions and suggestions that lead to this work.

N.P. Smart (Ed.): Cryptography and Coding 2005, LNCS 3796, pp. 116–135, 2005.
© Springer-Verlag Berlin Heidelberg 2005

The first time that a formal security treatment was applied to signcryption schemes was the work of An et al. [1]. Unlike the work of Zheng, the purpose here was not simply to achieve efficiency; the goal was to provide a rigorous framework to analyse any scheme or composition paradigm used to achieve the combined functionality of encryption and signature. Several security notions were introduced: *insider* and *outsider* security; and *two-user* and *multi-user* security. We will discuss these notions further in Section 7. In addition to providing a security framework, several composition paradigms are also proposed and analysed in [1].

Here we continue research in this area. We propose a general construction that can be used whenever one requires a message to be signed and encrypted. Our construction uses the KEM-DEM methodology of Cramer and Shoup [9,10] combined with a secure signature scheme. We prove a result that tightly relates the security of our construction to the security of the atomic components that it is composed from. Our construction is similar to the *sign-then-encrypt* method proposed in [1]; however, we show how, in certain situations, one can produce a more efficient solution by using a KEM-DEM approach.

We also describe an instantiation of our scheme based on a signature scheme of Boneh and Boyen [6] and a KEM of Cramer and Shoup [9]. If one takes Zheng's original definition of signcryption – to achieve an efficiency gain over combined encrypting and signing – we argue that our instantiation can be viewed as the first signcryption scheme that is provably secure without appealing to the random oracle model [5]. This is particularly significant when one considers the recent separation results between this model and the standard model [4,8,16].

The paper proceeds as follows. In Section 2 we describe the primitives that our construction is built from. We present the construction itself in Section 3. The security notion that we will be using is discussed in Section 4, and we present our security result in Section 5. In Section 6 we describe how our construction could be instantiated. The final section of the paper gives a detailed discussion of what we have done relative to existing research in the area.

2   Preliminaries

In this section we discuss the primitives from which our construction is built. Before doing this we define some notational conventions that will be used here and throughout the paper.

2.1   Notation

Let $\mathbf{Z}$ denote the ring of integers and let $\mathbf{Z}_{\geq 0}$ denote the non-negative integers. For a positive integer a let $\mathbf{Z}_a$ denote the ring of integers modulo a and let $\mathbf{Z}_a^*$ denote the corresponding multiplicative group of units.

If S is a set then we write $v \leftarrow S$ to denote the action of sampling from the uniform distribution on S and assigning the result to the variable v. If S contains one element s we use $v \leftarrow s$ as shorthand for $v \leftarrow \{s\}$.

We shall be concerned with probabilistic polynomial-time (PPT) algorithms. If A is such an algorithm we denote the action of running A on input I and assigning the resulting output to the variable v by $v \leftarrow \mathsf{A}(I)$. Note that since A is probabilistic, $\mathsf{A}(I)$ is a probability space and not a value. We denote the set of possible outputs of $\mathsf{A}(I)$ by $[\mathsf{A}(I)]$.

If E is an event defined in some probability space, we denote the probability that E occurs by $\Pr[E]$ (assuming the probability space is understood from the context).

2.2   Signature Schemes

A signature scheme SIG consists of the following three algorithms.

- A PPT *key generation algorithm* SIG.KeyGen that takes as input 1^κ for $\kappa \in \mathbf{Z}_{\geq 0}$. It outputs a public/secret key pair $(\mathsf{PK}_s, \mathsf{SK}_s)$. The structure of PK_s and SK_s depends on the particular scheme.
- A polynomial-time *signing algorithm* SIG.Sig that takes as input 1^κ for $\kappa \in \mathbf{Z}_{\geq 0}$, a secret key SK_s, and a message $m \in \{0, 1\}^*$. It outputs a signature σ. Algorithm SIG.Sig may be probabilistic or deterministic.
- A polynomial-time *verification algorithm* SIG.Ver that takes as input 1^κ for $\kappa \in \mathbf{Z}_{\geq 0}$, a public key PK_s, a message m, and a purported signature σ on m. The verification algorithm outputs $\top$ if σ is indeed a signature on m under PK_s; it outputs $\bot$ otherwise.

Security of Signature Schemes: Strong existential unforgeability The standard notion of security for a signature scheme is *existential unforgeability under adaptive chosen message attack* [11]. In this section we recall the slightly stronger notion of *strong existential unforgeability* for signature schemes as defined in [1]. To describe this notion we consider an adversary A that is a probabilistic, polynomial-time oracle query machine that takes as input a security parameter 1^κ for $\kappa \in \mathbf{Z}_{\geq 0}$. The attack game for this security notion proceeds as follows.

Stage 1: The adversary queries a *key generation oracle* with 1^κ. The key generation oracle computes $(\mathsf{PK}_s, \mathsf{SK}_s) \leftarrow \mathsf{SIG.KeyGen}(1^\kappa)$ and responds with PK_s.

Stage 2: The adversary makes a series of at most n_s queries to a *signing oracle*. For each query m_i, for $i \in \{1, \dots, n_s\}$, the signing oracle computes $\sigma_i \leftarrow \mathsf{SIG.Sig}(1^\kappa, \mathsf{SK}_s, m)$ and responds with σ_i. The adversary may choose its queries m_i adaptively based on the responses to previous queries.

Stage 3: The adversary attempts to output a pair (m, σ) such that
  1. $(m, \sigma) \notin \{(m_1, \sigma_1), \dots, (m_{n_s}, \sigma_{n_s})\}$ and
  2. $x = \top$ where $x \leftarrow \mathsf{SIG.Ver}(1^\kappa, \mathsf{PK}_s, m, \sigma)$.
  We say that the adversary *wins* if it succeeds in doing this.

We define $\mathsf{AdvEF}_{\mathsf{SIG},\mathsf{A}}(\kappa)$ to be $\Pr[\mathsf{A}\ \text{wins}]$ where wins is defined in the above game. The probability is taken over the random choices of A and those of A's oracles.

2.3 Key Encapsulation Mechanisms

A *key encapsulation mechanism* KEM [9] consists of the following algorithms.

- A PPT *key generation algorithm* KEM.KeyGen that takes as input 1^λ for $\lambda \in \mathbf{Z}_{\geq 0}$. It outputs a public/secret key pair $(\mathsf{PK}_r, \mathsf{SK}_r)$. The structure of PK_r and SK_r depends on the particular scheme.
- A PPT *encryption algorithm* KEM.Enc that takes as input 1^λ for $\lambda \in \mathbf{Z}_{\geq 0}$, and a public key PK_r. It outputs a pair (K, ψ) where K is a key and ψ is a ciphertext.
  For any $(\mathsf{PK}_r, \mathsf{SK}_r)$ output by KEM.KeyGen, we assume that

$$\max_{\psi} \Pr[\psi^* = \psi :\ (K^*, \psi^*) \leftarrow \mathsf{KEM.Enc}(1^\lambda, \mathsf{PK}_r)] \leq \frac{1}{2^{\lambda-1}}$$

  where $1/2^{\lambda-1}$ can be replaced with any negligible function of λ and an appropriate adjustment made to our security analysis.
  A key K is a bit string of length $\mathsf{KEM.KeyLen}(\lambda)$, where $\mathsf{KEM.KeyLen}(\lambda)$ is another parameter of KEM.
- A deterministic, polynomial-time *decryption algorithm* KEM.Dec that takes as input 1^λ for $\lambda \in \mathbf{Z}_{\geq 0}$, a secret key SK_r, and a ciphertext ψ. It outputs either a key K or the symbol $\perp$ to indicate that the ciphertext was invalid.

Soundness. We also require a notion of *soundness* for KEM. Let us say that a public/secret key pair $(\mathsf{PK}_r, \mathsf{SK}_r)$ is *bad* if, for some $(K, \psi) \in [\mathsf{KEM.Enc}(1^\lambda, \mathsf{PK}_r)]$, we have $x \neq K$ where x is the output of $\mathsf{KEM.Dec}(1^\lambda, \mathsf{SK}_r, \psi)$. Let $\mathsf{BadKP}_{\mathsf{KEM}}(\lambda)$ denote the probability that the key generation algorithm outputs a bad key pair for a given value of λ. Our requirement on KEM is that $\mathsf{BadKP}_{\mathsf{KEM}}(\lambda)$ grows negligibly in λ.

Real or Random Security Against Passive Attack. Here we introduce a weak notion of security for KEM: *real or random security against passive attack*. Note that, although this weak notion is sufficient for our application, we are categorically not suggesting that it should replace stronger notions proposed for other applications [9].

An adversary A that mounts a passive attack against KEM is a PPT algorithm that takes as input 1^λ, for security parameter $\lambda \in \mathbf{Z}_{\geq 0}$. Below we describe the attack game used to define security against passive attack.

Stage 1: The adversary queries a *key generation oracle* with 1^λ. The key generation oracle computes $(\mathsf{PK}_r, \mathsf{SK}_r) \leftarrow \mathsf{KEM.KeyGen}(1^\lambda)$ and responds with PK_r.

Stage 2: The adversary makes a query to a *challenge encryption oracle*. The challenge encryption oracle does the following.

$$(K^*, \psi^*) \leftarrow \mathsf{KEM.Enc}(1^\lambda, \mathsf{PK}_r);\ K^+ \leftarrow \{0,1\}^{l_k};\ b \leftarrow \{0,1\};$$
$$\text{if } b = 0,\ K^\dagger \leftarrow K^*;\ \text{else } K^\dagger \leftarrow K^+;$$

where $l_k = \mathsf{KEM.KeyLen}(\lambda)$. Finally, it responds with $(K^\dagger, \psi^*)$.

Stage 3: The adversary outputs $b' \in \{0, 1\}$.

If A is playing the attack game above, we define

$$\mathsf{AdvRR}_{\mathsf{KEM},\mathsf{A}}(\lambda) := |\mathsf{Pr}[b' = 1 | b = 1] - \mathsf{Pr}[b' = 1 | b = 0]|. \qquad (1)$$

The probability is taken over the random choices of A and those of A's oracles.

2.4   One-Time Symmetric-Key Encryption

A *one-time symmetric-key encryption scheme* [9] SKE consists of the following two algorithms.

- A deterministic, polynomial-time *encryption algorithm* SKE.Enc that takes as input 1^λ for $\lambda \in \mathbf{Z}_{\geq 0}$, a key K, and a message $m \in \{0, 1\}^*$. It outputs a ciphertext $\chi \in \{0, 1\}^*$.
  The key K is a bit string of length SKE.KeyLen(λ) where SKE.KeyLen(λ) is a parameter of the encryption scheme.
- A deterministic, polynomial-time *decryption algorithm* SKE.Dec that takes as input 1^λ for $\lambda \in \mathbf{Z}_{\geq 0}$, a key K, and a ciphertext χ. It outputs either a message $m \in \{0, 1\}^*$ or the symbol $\perp$ to indicate that the ciphertext was invalid.
  The key K is a bit string of length SKE.KeyLen(λ).

Soundness. We require our symmetric-key encryption scheme SKE to satisfy the following soundness condition: For all $\lambda \in \mathbf{Z}_{\geq 0}$, for all $\kappa \in \{0, 1\}^{\mathsf{SKE.KeyLen}(\lambda)}$, and for all $m \in \{0, 1\}^*$, we have

$$x = m \text{ where } x \leftarrow \mathsf{SKE.Dec}(1^\lambda, K, \mathsf{SKE.Enc}(1^\lambda, K, m)).$$

Indistinguishability of Encryptions Under Passive Attack. Here we introduce a weak notion of security for SKE: *indistinguishability of encryptions under passive attack*. An adversary A that mounts a passive attack against SKE is a PPT algorithm that takes as input 1^λ, for security parameter $\lambda \in \mathbf{Z}_{\geq 0}$. Below we describe the attack game used to define this security notion.

Stage 1: The adversary chooses two messages, m_0 and m_1, of equal length. It gives these to an *encryption oracle*.

Stage 2: The encryption oracle does the following.

$$K \leftarrow \{0, 1\}^{l_s}; \; b \leftarrow \{0, 1\}; \; \chi^* \leftarrow \mathsf{SKE.Enc}(1^\lambda, K, m_b),$$

where $l_s = \mathsf{SKE.KeyLen}(\lambda)$. Finally, it responds with χ^*.

Stage 3: The adversary outputs $b' \in \{0, 1\}$.

If A is playing the attack game above, we define

$$\mathsf{AdvIND}_{\mathsf{SKE},\mathsf{A}}(\lambda) := |\mathsf{Pr}[b' = 1 | b = 1] - \mathsf{Pr}[b' = 1 | b = 0]|.$$

The probability is taken over the random choices of A and those of A's oracles.

3   Simultaneous Signing and Encrypting

A construction SSE for simultaneous signing and encrypting (SSE) consists of the following four algorithms.

- A PPT *sender key generation algorithm* SSE.SKeyGen that takes as input 1^κ for $\kappa \in \mathbf{Z}_{\geq 0}$, and outputs a public/secret key pair $(\mathsf{PK}_s, \mathsf{SK}_s)$. The structure of PK_s and SK_s depends on the particular scheme.
- A PPT *receiver key generation algorithm* SSE.RKeyGen that takes as input 1^λ for $\lambda \in \mathbf{Z}_{\geq 0}$, and outputs a public/secret key pair $(\mathsf{PK}_r, \mathsf{SK}_r)$. The structure of PK_r and SK_r depends on the particular scheme.
- A PPT *sign/encrypt algorithm* SSE.SigEnc that takes as input 1^κ for $\kappa \in \mathbf{Z}_{\geq 0}$, 1^λ for $\lambda \in \mathbf{Z}_{\geq 0}$, a sender secret key SK_s, a receiver public key PK_r and a message $m \in \{0,1\}^*$. It outputs a ciphertext C.
- A deterministic, polynomial-time *decrypt/verify algorithm* SSE.DecVer that takes as input 1^κ for $\kappa \in \mathbf{Z}_{\geq 0}$, 1^λ for $\lambda \in \mathbf{Z}_{\geq 0}$, a sender public key PK_s, a receiver secret key SK_r and a purported ciphertext C. It outputs either a message m or a symbol $\perp$ to indicate that C is invalid.

SKeyGen: On input 1^κ for $\kappa \in \mathbf{Z}_{\geq 0}$:

    1. $(\mathsf{PK}_s, \mathsf{SK}_s) \leftarrow \mathsf{SIG.KeyGen}(1^\kappa)$
    2.   Return the public key PK_s and the secret key SK_s

RKeyGen: On input 1^λ for $\lambda \in \mathbf{Z}_{\geq 0}$:

    1. $(\mathsf{PK}_r, \mathsf{SK}_r) \leftarrow \mathsf{KEM.KeyGen}(1^\lambda)$
    2.   Return the public key PK_r and the secret key SK_r

SigEnc: On input 1^κ for $\kappa \in \mathbf{Z}_{\geq 0}$, 1^λ for $\lambda \in \mathbf{Z}_{\geq 0}$, secret key PK_s, public key PK_r and message $m \in \{0,1\}^*$:

    SE1: $(K, \psi) \leftarrow \mathsf{KEM.Enc}(1^\lambda, \mathsf{PK}_r)$
    SE2: $\sigma \leftarrow \mathsf{SIG.Sig}(1^\kappa, \mathsf{SK}_s, m \| \psi)$
    SE3: $\chi \leftarrow \mathsf{SKE.Enc}(1^\lambda, K, m \| \sigma)$
    SE4: Return the ciphertext (ψ, χ)

DecVer: On input 1^κ for $\kappa \in \mathbf{Z}_{\geq 0}$, 1^λ for $\lambda \in \mathbf{Z}_{\geq 0}$, public key PK_s, secret key SK_r and ciphertext (ψ, χ):

    DV1: $K \leftarrow \mathsf{KEM.Dec}(1^\lambda, \mathsf{SK}_r, \psi)$
    DV2: $m \| \sigma \leftarrow \mathsf{SKE.Dec}(1^\lambda, K, \chi)$
    DV3: $x \leftarrow \mathsf{SIG.Ver}(1^\kappa, \mathsf{PK}_s, m \| \psi, \sigma)$
    DV4: If $x = \top$, return m otherwise return $\perp$

Fig. 1. A general construction for simultaneous signing and encrypting

In Figure 1 we describe how our general construction works. We use primitives SIG, KEM and SKE as defined in Section 2. We assume for simplicity that KEM.KeyLen(λ) = SKE.KeyLen(λ).

4 Security Notions for Simultaneous Signing and Encrypting

4.1 Indistinguishability of Encryptions

In this section we describe the notion of *indistinguishability of encryptions under an adaptive chosen-plaintext and chosen-ciphertext attack* (ICPCA) for a SSE scheme SSE. This is a natural analogue of IND-CCA2 security for standard public key encryption [19]. To describe this notion we consider an adversary A that is a PPT oracle query machine that takes as input a security parameters 1^κ for $\kappa \in \mathbf{Z}_{\geq 0}$, and 1^λ for $\lambda \in \mathbf{Z}_{\geq 0}$. The attack game for this security notion proceeds as follows.

Stage 1: The adversary queries a *sender key generation oracle* with 1^κ. The sender key generation oracle computes $(\mathsf{PK}_s, \mathsf{SK}_s) \leftarrow \mathsf{SSE.SKeyGen}(1^\kappa)$ and responds with PK_s.

Stage 2: The adversary queries a *receiver key generation oracle* with 1^λ. The receiver key generation oracle computes $(\mathsf{PK}_r, \mathsf{SK}_r) \leftarrow \mathsf{SSE.RKeyGen}(1^\lambda)$ and responds with PK_r.

Stage 3: The adversary makes a series of queries to two oracles: a *sign/encrypt oracle* and a *decrypt/verify oracle*. (During Stage 3 and Stage 5 the adversary makes at most n_s queries to the sign/encrypt oracle and at most n_d queries to the decrypt/verify oracle.)
For each query m, the sign/encrypt oracle computes

$$C \leftarrow \mathsf{SSE.SigEnc}(1^\kappa, 1^\lambda, \mathsf{SK}_s, \mathsf{PK}_r, m)$$

and responds with C.
For each query C, the decrypt/verify oracle computes

$$x \leftarrow \mathsf{SSE.DecVer}(1^\kappa, 1^\lambda, \mathsf{PK}_s, \mathsf{SK}_r, C)$$

and responds with x.
The adversary may choose its queries to these oracles adaptively based on the responses to previous queries.

Stage 4: The adversary chooses two messages m_0 and m_1 of equal length. It gives these to a *challenge oracle*. The challenge oracle does the following.

$$b \leftarrow \{0, 1\}; \ C^* \leftarrow \mathsf{SigEnc}(1^\kappa, 1^\lambda, \mathsf{SK}_s, \mathsf{PK}_r, m_b)$$

Finally, it returns C^* to the adversary.

Stage 5: The adversary continues to query the oracles of Stage 3 subject to the condition that it does not query the decrypt/verify oracle with C^*

Stage 6: The adversary outputs $b' \in \{0, 1\}$.

If A is playing the above attack game we define

$$\mathsf{AdvICPCA}_{\mathsf{SSE},\mathsf{A}}(\kappa, \lambda) := |\Pr[b' = 1|b = 1] - \Pr[b' = 1|b = 0]|.$$

The probability is taken over the random choices of A and those of A's oracles.

4.2   Unforgeability

As our construction is using a signature scheme that is assumed to be existentially unforgeable under adaptive chosen message attack, we do not treat unforgeability explicitly here; our contribution is to demonstrate how to use a weak encryption scheme and maintain strong security. Suffice to say that, using the assumed unforgeability of the signature scheme, unforgeability results for our construction can easily be proved using the techniques of [1].

5   Security Result

Here we state our security result pertaining to the security of our construction under the definition given in Section 4.1. The proof may be found in Appendix A.

Theorem 1. *Let SSE be an instance of our construction that uses SIG, KEM and SKE. Let A be an adversary of SSE that uses an adaptive chosen-plaintext and chosen-ciphertext attack to attempt to distinguish encryptions produced using SSE. Suppose that A makes at most n_s sign/encrypt queries and at most n_d decrypt/verify queries. From this adversary it is possible to construct adversaries A_1, A_2 and A_3, whose running times are essentially the same as that of A, such that*

$$\mathsf{AdvICPCA}_{\mathsf{SSE},\mathsf{A}}(\kappa, \lambda) \leq 2\mathsf{AdvEF}_{\mathsf{SIG},\mathsf{A}_1}(\kappa) + 2\mathsf{BadKP}_{\mathsf{KEM}}(\lambda)$$
$$+ 2\mathsf{AdvRR}_{\mathsf{KEM},\mathsf{A}_2}(\lambda) + 2\mathsf{AdvIND}_{\mathsf{SKE},\mathsf{A}_3}(\lambda) + \frac{n_s}{2^{\lambda-2}}.$$

6   An Instantiation

In this section we briefly describe a very efficient instantiation of our construction. This instantiation could be viewed as the first signcryption scheme in the literature that is provably secure without appealing to the random oracle model. We justify this claim in the subsequent discussion section.

   The signature scheme that we recommend for our construction is one proposed by Boneh and Boyen [6]. This scheme uses a bilinear group, such as those frequently used to construct identity-based cryptosystems such as those proposed in [7,12]. It is strongly existentially unforgeable under adaptive chosen message attack and thus satisfies the security criterion for our construction. This can be proved without using the random oracle heuristic under an assumption dubbed the *strong Diffie-Hellman assumption*. The scheme is very efficient: signing requires one group exponentiation and verifying requires two group exponentiations and two pairing computations.

To instantiate the key encapsulation mechanism we suggest the Hashed El-Gamal (HEG) scheme proposed by Cramer and Shoup (see Section 10.1 of [9]). In [9] this is proved to be secure under an adaptive chosen ciphertext attack in the random oracle model provided that the gap Diffie Hellman problem [17] is hard. As you will recall from sections 2.3 and 5, our construction does not require such a strong notion of security from its KEM; moreover, it follows from the results of Tsiounis and Yung [21] that such ElGamal based KEMs are secure against passive attack – as we require – under the decisional Diffie-Hellman assumption. This result holds without using the random oracle model.

The beauty of using such a KEM is that the security of our construction holds, under the definition that we have used so far, without the random oracle model. If we require stronger security notions, such as those that we discuss in the next section, then we can simply apply the result of Cramer and Shoup that holds in the random oracle model and achieve this security without changing the KEM (as long as we are content to make do with a random oracle model proof).

To instantiate the symmetric-key encryption scheme we recommend a block-cipher such as AES used in an appropriate mode. Although it is not possible to prove anything about the security of AES, it has been subject to thorough analysis and is widely believed to be sound.

7   Discussion

In this final section we discuss some related work in this area so that we are able to put our own work in to context and to justify some of the decisions that we have made.

7.1   Other Security Notions

As we mentioned in the introduction, An et al. defined several notions of security for schemes providing both signing and encryption [1]. We describe these below.

Outsider Security. The definition that we have given so far is an instance of outsider security: an adversary is assumed to have access to a sign/encrypt oracle for SK_s/PK_r and to a decrypt/verify oracle for PK_s/SK_r; however, it is an *outsider* of the system in that it does not know SK_s. (If it knew SK_r it could perform decryption itself and so indistinguishability of encryptions would be impossible!)

Insider Security. Note that if we surrender SK_s to an adversary in our construction we obtain a standard public-key encryption scheme. This is so because SK_s can be used by the adversary to produce ciphertexts. This resulting scheme is called the *induced* encryption scheme in [1]. The construction is said to offer insider security if the induced encryption scheme is IND-CCA2 secure [19].

An et al. state that in many situations outsider security may be sufficient; however, they do point out that it guarantees that ciphertexts produced by the owner of SK_s for the owner of PK_r remain confidential even if SK_s is subsequently compromised. Many signcryption schemes, such as those proposed in [3,14,22], do not have this property.

Here we sketch what we need to do to ensure that our construction offers insider security. The first point to note is that we cannot use the strong existential unforgeability of the signature scheme in our security result as we did for Theorem 1. The reason for this is that, under the definition of insider-security, we are going to provide an adversary with SK_s. If an adversary has SK_s it can produce signatures on any messages that it likes without forging anything!

We need a method of simulating the decrypt/verify oracle that does not not exploit the existential unforgeability of the signature scheme used in the construction. To do this we require stronger security notions from KEM and SKE used in our construction.

Consider the real or random security notion for the security of KEM, as introduced in Section 2.3. We require hat KEM is secure in this sense even when an adversary is given access to a decryption oracle that uses SK_r. This is the notion of adaptive chosen ciphertext attack for KEMs that we mentioned in Section 6. We again remark that the HEG KEM of Cramer and Shoup recommended for our instantiation already has a proof of security under this stronger notion, albeit in the random oracle model [9].

The notion of security that we require of SKE is similar to that presented in Section 2.4. The difference is that we allow an adversary access to a decryption oracle that works with the key K. Cramer and Shoup show how an encryption scheme that is secure against passive attack – as in Section 2.4 – can be made secure under this stronger notion by using a message authentication code (see Theorem 4 of [9] for details).

Suppose that we have primitives KEM and SKE that satisfy these stronger notions that we have just described. A simulation for an insider adversary of our construction would work as follows. One first runs SIG.KeyGen and gives the resulting signing key to the adversary. Using the decryption oracles that would granted to adversaries of KEM and SKE it is possible to simulate the decrypt/verify oracle as required. The resulting proof is very similar to Theorem 5 of [9].

Having discussed outsider and insider security, the following two notions remain.

Two-User Setting. The definition that we have used thus far is an instance of security in a two-user setting: the adversary wishes to distinguish encryption created using SK_s and PK_r using access to a sign/encrypt oracle for $\mathsf{SK}_s/\mathsf{PK}_r$ and to a decrypt/verify oracle for $\mathsf{PK}_s/\mathsf{SK}_r$. The point to observe here is that these oracles are fixed for PK_r and PK_s respectively.

Multi-User Setting. Unlike in the two-user setting, in a multi-user setting the adversary is able to choose the public keys to input to its oracles. For example, it can obtain encryptions produced using SK_s and an arbitrary public key of its choosing.

An et al. argue that considering security in the multi-user setting is crucial to avoid problems of identity fraud. To motivate this example, consider a construction where a message is encrypted with Bob's public key and the resulting ciphertext is signed using Alice's secret key. This construction may be secure in the two-user setting if both its components are secure; however, suppose that there is a user Charlie who can intercept data and alter it. Charlie can remove Alice's signature on the ciphertext and replace it with his own. In this scenario Bob will be convinced that the encrypted message that he receives came from Charlie when it really came from Alice. This is done without violating the security of any of the components.

In the multi-user setting we are unable to use the strong existential unforgeability of the signature scheme in our security proof, but for an altogether different reason from the insider security case discussed above. The decrypt/verify oracle in the proof of Theorem 1 uses the fact that, if it is presented with a valid ciphertext (ψ, χ), then, within this ciphertext, there must be a signature on ψ. Since the simulator knows the keys K corresponding to all the ψs that have been signed, it is able to decrypt and verify. This will only hold if ciphertexts are decrypted and verified with respect to a specific sender public key PK_r - not the case in the multi-user setting.

It turns out that the very security requirements that we outlined above for KEM and SKE to give security in the insider-case also suffice to give security in the multi-user case: simulating the decrypt/verify oracle only requires decryption oracles for KEM and SKE; moreover, both these are independent of any keys belonging to the sender. To simulate the sign/encrypt oracle one only requires a signing oracle valid for PK_s; generation of ψ and the final encryption can be done using by the public key supplied by the adversary.

7.2 The Instantiation

Until now, any construction for simultaneous signing and encrypting offering provable security without the random oracle model would have to use an encryption scheme that is fully IND-CCA2 secure and a signature scheme that is existentially unforgeable under adaptive chosen message attack [1]. The results for the encryption scheme and the signature scheme would clearly have to hold in the standard model. All other solutions rely on the random oracle model [2,13,14,15] for their security proofs.

Although our framework still requires a signature scheme that is existentially unforgeable, we can greatly improve on efficiency when it comes to encryption: the HEG KEM that fits our security notion for Theorem 1 requires only two group exponentiations for encrypting and one for decrypting. This compares with five for encrypting and four for decrypting when using the analogous method to achieve full IND-CCA2 security [9].

We can therefore claim to have designed the first signcryption scheme offering provable security without the random oracle model – when one takes Zheng's original motivation as discussed in the introduction. At present the random oracle free version of our result only holds for outsider security in the two-user setting. It is an interesting open question to extend our results to the other models proposed in [1].

7.3   The Security Requirement on SIG

We have used the notion of strong existential unforgeability [1] for our scheme rather than the standard existential unforgeability [11]. Here we provide some intuition for this.

The challenge ciphertext, (ψ^*, χ^*), is such that χ^* is the encryption of $m_b||\sigma^*$, where σ^* is a signature on $m_b||\psi^*$. Now, if the signature scheme is not strongly existentially unforgeable, the possibility that the adversary finds $\sigma' \neq \sigma^*$ such that σ' is a valid signature on $m_b||\psi^*$ is not ruled out; moreover, if the adversary could somehow come up with $\chi' \neq \chi^*$ such that χ' decrypts to give $m_b||\sigma'$ then it could break the scheme by submitting (ψ^*, χ') to the decrypt/verify oracle.

The property of strong existential unforgeability is used in the proof of Lemma 2 in the appendix.

We note that An et al. state in the full version of their paper that strong existential unforgeability can also be used in their encrypt-then-sign construction to weaken the security requirement on the encryption scheme [1]. The method that one would use to prove this is similar to the methods used in [1] rather than the method here and, moreover, in the encrypt-then-sign scenario of [1] one ends up with a signature on a ciphertext rather than on a message. Non-repudiation of the message is not then necessarily straightforward as it is in our construction.

References

1. J. H. An, Y. Dodis, and T. Rabin. On the security of joint signature and encryption. In *Advances in Cryptology - EUROCRYPT 2002*, volume 2332 of *Lecture Notes in Computer Science*, pages 83–107. Springer-Verlag, 2002. Full version available at http://eprint.iacr.org/2002/046.
2. J. Baek, R. Steinfeld, and Y. Zheng. Formal proofs for the security of signcryption. In *Public Key Cryptography - PKC 2002*, volume 2274 of *Lecture Notes in Computer Science*, pages 80–98. Springer-Verlag, 2002.
3. F. Bao and R. H. Deng. A signcryption scheme with signature directly verifiable by public key. In *Public Key Cryptography - PKC '98*, volume 1431 of *Lecture Notes in Computer Science*, pages 55–59. Springer-Verlag, 1998.

4. M. Bellare, A. Boldyreva, and A. Palacio. An uninstantiable random-oracle-model scheme for a hybrid-encryption problem. In *Advances in Cryptology - EURO-CRYPT 2004*, volume 3027 of *Lecture Notes in Computer Science*, pages 171–188. Springer-Verlan, 2004.
5. M. Bellare and P. Rogaway. Random oracles are practical: A paradigm for designing efficient protocols. In 1st *ACM Conference on Computer and Communications Security*, pages 62–73, 1993.
6. D. Boneh and X. Boyen. Short signatures without random oracles. In *Advances in Cryptology - EUROCRYPT 2004*, volume 3027 of *Lecture Notes in Computer Science*, pages 56–73. Springer-Verlan, 2004. Full version available at `http://crypto.stanford.edu/~dabo/abstracts/bbsigs.html`.
7. D. Boneh and M. Franklin. Identity-based encryption from the Weil pairing. In *Advances in Cryptology - CRYPTO 2001*, volume 2139 of *Lecture Notes in Computer Science*, pages 213–229. Springer-Verlag, 2001.
8. R. Canetti, O. Goldreich, and S. Halevi. The random oracle model, revisited. In 30th *ACM Symposium on Theory of Computing*, pages 209–218. ACM Press, 1998.
9. R. Cramer and V. Shoup. Design and analysis of practical public-key encryption schemes secure against adaptive chosen ciphertext attack. *SIAM Journal on Computing*, 33(1):167–226, 2003.
10. A. W. Dent. A designer's guide to KEMs. In *Cryptography and Coding*, volume 2898 of *Lecture Notes in Computer Science*, pages 133–151. Springer-Verlag, 2003.
11. S. Goldwasser, S. Micali, and R. Rivest. A digital signature scheme secure against adaptive chosen-message attacks. *SIAM Journal on Computing*, 17(2):281–308, 1988.
12. F. Hess. Efficient identity based signature schemes based on pairings. In *Selected Areas in Cryptography (2002)*, volume 2595 of *Lecture Notes in Computer Science*, pages 310–324. Springer-Verlag, 2003.
13. B. Libert and J. J. Quisquater. New identity-based signcryption schemes from pairings. In *IEEE Information Theory Workshop 2003*. Full version available at `http://eprint.iacr.org/2003/023/`.
14. J. Malone-Lee. Signcryption with non-interactive non-repudiation. Technical Report CSTR-02-004, Department of Computer Science, University of Bristol, 2004. Available at `http://www.cs.bris.ac.uk/Publications/index.jsp`.
15. J. Malone-Lee and W. Mao. Two birds one stone: Signcryption using RSA. In *Topics in Cryptology - CT-RSA 2003*, volume 2612 of *Lecture Notes in Computer Science*, pages 211–226. Springer-Verlag, 2003.
16. J. B. Nielsen. Separating random oracle proofs from complexity theoretic proofs: The non-committing encryption case. In *Advances in Cryptology - CRYPTO 2002*, volume 2442 of *Lecture Notes in Computer Science*, pages 111–126. Springer-Verlag, 2002.
17. T. Okamoto and D. Pointcheval. The gap-problems: A new class of problems for the security of cryptographic schemes. In *Public Key Cryptography - PKC 2001*, volume 1992 of *Lecture Notes in Computer Science*, pages 104–118. Springer-Verlag, 2001.
18. H. Petersen and M. Michels. Cryptanalysis and improvement of signcryption schemes. *IEE Proceedings - Computers and Digital Techniques*, 145(2):149–151, 1998.
19. C. Rackoff and D. Simon. Non-interactive zero-knowledge proof of knowledge and chosen ciphertext attack. In *Advances in Cryptology - CRYPTO '91*, volume 576 of *Lecture Notes in Computer Science*, pages 433–444. Springer-Verlag, 1992.
20. V. Shoup. OAEP reconsidered. In *Advances in Cryptology - CRYPTO 2001*, volume 2139 of *Lecture Notes in Computer Science*, pages 239–259. Springer-Verlag, 2001.

21. Y. Tsiounis and M. Yung. On the security of ElGamal based encryption. In *Public Key Cryptography - PKC '98*, volume 1431 of *Lecture Notes in Computer Science*, pages 117–134. Springer-Verlag, 1998.
22. Y. Zheng. Digital signcryption or how to achieve cost(signature & encryption) << cost(signature) + cost(encryption). In *Advances in Cryptology - CRYPTO '97*, volume 1294 of *Lecture Notes in Computer Science*, pages 165–179. Springer-Verlag, 1997.

A   Proof of Theorem 1

Our proof strategy is as follows. We define a sequence $\mathbf{G}_0, \dots, \mathbf{G}_7$ of modified attack games. The only difference between games is how the environment responds to A's oracle queries. For any $0 \le i \le 7$, we let S_i be the event that $b' = b$ in game $\mathbf{G}_i$, where b is the bit chosen by A's challenge oracle and b' is the bit output by A. This probability is taken over the random choices of A and those of A's oracles.

We will make frequent of the following useful lemma from [20].

Lemma 1. *Let E, F and G be events defined on a probability space such that* $Pr[E \wedge \neg G] = Pr[F \wedge \neg G]$. *We have*

$$|Pr[E] - Pr[F]| \le Pr[G].$$

Game $\mathbf{G}_0$. We simulate the view of an adversary in a real attack by running the appropriate key generation algorithms and using the resulting keys to respond to A's queries. The view of A is therefore the same as it would be in a real attack and so

$$Pr[S_0] = \frac{1}{2}\left(\mathsf{AdvICPCA}_{\mathsf{SSE,A}}(\kappa, \lambda) + 1\right). \tag{2}$$

Game $\mathbf{G}_1$. In this game we do not modify the responses of the oracles with which A interacts; the only change is to add some bookkeeping that will be used in subsequent games. To this end a list L_s is maintained. This list is initially empty. On each query to the sign/encrypt oracle, the appropriate $(K, m||\psi, \sigma)$ is added to L_s. When the challenge oracle is called, the list L_s is also updated accordingly.

Clearly this change has no impact on the adversary and so

$$Pr[S_1] = Pr[S_0]. \tag{3}$$

Game $\mathbf{G}_2$. We now modify how the challenge ciphertext is generated. At the beginning of the simulation, once A has made its receiver key generation query, we compute $(K, \psi) \leftarrow \mathsf{KEM.Enc}(1^\lambda, \mathsf{PK}_r)$, we set $(K^*, \psi^*) \leftarrow (K, \psi)$, and we use (K^*, ψ^*) in the generation of the challenge ciphertext. Again, this change does not alter the view of A meaning that

$$Pr[S_2] = Pr[S_1]. \tag{4}$$

Game G_3. In this game we slightly modify how the sign/encrypt oracle responds to queries from A; specifically, we add a rule **SE1'** that we apply between **SE1** and **SE2**.

SE1': If $\psi = \psi^*$, abort the simulation.

The chances that the sign/encrypt oracle generates ψ^* when responding to a sign/encrypt query are $q_s/2^{\lambda-1}$; moreover, G_2 and G_3 proceed in an identical manner until this happens. Therefore, by Lemma 1

$$|\Pr[S_3] - \Pr[S_2]| \leq \frac{q_s}{2^{\lambda-1}}. \tag{5}$$

Game G_4. We now modify how the decrypt/verify oracle responds to queries from A. We do this by adding a rule **DV2'** that we apply between **DV2** and **DV3**. Let (ψ, χ) be the ciphertext that has been sent to the oracle. The new rule is as follows.

DV2': If $(\hat{K}, \hat{m}||\psi, \hat{\sigma}) \notin L_s$ for any $\hat{K}, \hat{m}, \hat{\sigma}$, return $\perp$. Else if $\psi = \psi^*$, return $\perp$. Else proceed to apply rule **DV3**.

Now, let R_4 be the event that, in game G_4, some ciphertext is submitted to the decrypt/verify oracle that is rejected, but is such that it would not have been rejected in G_3. Since the events $S_4 \wedge \neg R_4$ and $S_3 \wedge \neg R_4$ are identical, Lemma 1 tells us that

$$|\Pr[S_4] - \Pr[S_3]| \leq \Pr[R_4]. \tag{6}$$

Therefore we have to bound $\Pr[R_4]$ which we do in the lemma below.

Lemma 2. *There exists a PPT algorithm A_1, whose running time is essentially the same as that of A, such that*

$$\Pr[R_4] \leq \mathsf{AdvEF}_{\mathsf{SIG},A_1}(\kappa). \tag{7}$$

We leave the proof of all lemmas to the end of the main proof.

Game G_5. In this game we further modify how the decrypt/verify oracle responds to a query (ψ, χ) from A; we replace **DV1** with the following.

DV1: Search L_s until an entry $(\hat{K}, \hat{m}||\psi, \hat{\sigma})$ is found for some $\hat{K}, \hat{m}, \hat{\sigma}$ or until the end of the list is reached; if such entry is found , use $\hat{K}$ in **DV2**; else, return $\perp$.

Note that if no entry $(\hat{K}, \hat{m}||\psi, \hat{\sigma})$ appears in L_s then any ciphertext whose first component is ψ would be rejected in game G_4 in any case; therefore, this modification introduces no change from the point of view of A unless $\hat{K}' \neq \hat{K}$ where $\hat{K}'$ is the result of running $\mathsf{KEM.Dec}(1^\lambda, \mathsf{SK}_r, \psi)$. Therefore, by Lemma 1

$$|\Pr[S_5] - \Pr[S_4]| \leq \mathsf{BadKP}_{\mathsf{KEM}}(\lambda). \tag{8}$$

Game G_6. In this game we modify how the challenge ciphertext is generated. For the challenge ciphertext only we replace step **SE3** with step **SE3*** below.

SE3* $K^* \leftarrow \{0,1\}^{l_k}; \; \chi^* \leftarrow \mathsf{SKE.Enc}(1^\lambda, K^*, m_b || \sigma^*)$

Since rule **DV2′** was introduced in G_4, we do not need to know K^* for the remainder of the simulation and so we add $(-, m_b || \psi^*, \sigma^*)$ to L_s rather than $(K^*, m_b || \psi^*, \sigma^*)$. (Where $-$ denotes the fact that we leave a blank space.) Also, since the rule **DV1** was re-defined in G_5, we no longer require knowledge of SK_r to simulate A.

Lemma 3. *There exists a PPT algorithm A_2, whose running time is essentially the same as that of A, such that*

$$|Pr[S_6] - Pr[S_5]| = AdvRR_{KEM,A_2}(\lambda). \tag{9}$$

Game G_7. In this, the final game, we once more modify how the challenge ciphertext is generated. We replace the step **SE3*** introduced in game G_6 with the following.

SE3* $K^* \leftarrow \{0,1\}^{l_k}; \; str \leftarrow \{0,1\}^{l_m}$ (where l_m is the bit length of $m_b || \sigma^*$)
$\chi \leftarrow \mathsf{SKE.Enc}(1^\lambda, K^*, str)$

Now, since the challenge ciphertext, and everything else in the simulation, is entirely independent of b we have

$$Pr[S_7] = \frac{1}{2}. \tag{10}$$

To complete the proof we require the following lemma.

Lemma 4. *There exists a PPT algorithm A_3, whose running time is essentially the same as that of A, such that*

$$|Pr[S_7] - Pr[S_6]| \leq AdvIND_{SKE,A_3}(\lambda). \tag{11}$$

The result now follows from (2), (3), (4), (5), (6), (7), (8), (9), (10) and (11).

Proof of Lemma 2. To prove this we demonstrate how to construct an adversary A_1 of SIG to violate the assumed strong existential unforgeability. This adversary satisfies the bound (7).

Adversary A_1 is constructed by running adversary A. We respond to A's queries as follows.

- When A makes its signer key generation oracle query, A_1 calls its sender key generation oracle to obtain PK_s. It returns PK_s to A.
- When A makes its receiver key generation oracle query, A_1 runs KEM.KeyGen to obtain $(\mathsf{PK}_r, \mathsf{SK}_r)$. It keeps SK_r secret and return PK_r to A. At this point A_1 also runs $\mathsf{KEM.Enc}(1^\lambda, \mathsf{PK}_r)$ to obtain (K^*, ψ^*) which it keeps secret.

- To respond to A's sign/encrypt oracle query for a message m, A_1 first obtains (K, ψ) by running KEM.Enc under PK_r. It checks whether $\psi = \psi^*$ and if so it aborts the simulation. Assuming it did not abort, it then makes a signing query to its signing oracle to obtain a signature σ on $m||\psi$. After that, A_1 obtains χ by running SKE.Enc on $m||\sigma$ under key K. It responds to A with (ψ, χ) and it adds $(K, m||\psi, \sigma)$ to a list L_s that is initially empty.
- To respond to A's decrypt/verify query on a ciphertext (ψ, χ), A_1 uses the key SK_r generated earlier in the simulation to obtain $m||\sigma$ by running KEM.Dec, SKE.Dec. For each decrypt/verify query that results in a valid message, before A_1 responds to A it does the following.
  First it checks to see whether or not $(\hat{K}, \hat{m}||\psi, \hat{\sigma})$ appears in L_s for any $\hat{K}, \hat{m}, \hat{\sigma}$; there are two cases to consider.
    1. If it does not appear, A_1 halts the simulation and outputs $(m||\psi, \sigma)$.
    2. Otherwise, if $(\hat{K}, \hat{m}||\psi^*, \hat{\sigma})$ appears in L_s then A_1 halts the simulation and outputs $(\hat{m}||\psi^*, \hat{\sigma})$.
- Adversary Adv_1 responds to Adv's challenge query (m_0, m_1) by first choosing $b \leftarrow \{0, 1\}$. It then queries its signing oracle to obtain a signature σ^* on $m_b||\psi^*$ (recall that (K^*, ψ^*) was generated earlier in the simulation). It obtains χ^* by encrypting $m_b||\sigma^*$ using SKE.Enc with key K^*. Finally it returns (ψ^*, χ^*) to A.

Now, A is run in the above simulation by A_1 in exactly the same way that it is run in both game G_3 and G_4 until the simulation halts as in one of the cases that we have described above.

We will now argue that in both cases A_1 outputs a valid forgery. The first case is obvious. Let us now consider the second case. Suppose that such a query is made before the challenge ciphertext is given to A. In this case no signing query has been made by A_1 involving ψ^* and so $(\hat{m}||\psi^*, \hat{\sigma})$ is clearly a valid forgery. Suppose now that this query occurs after the challenge ciphertext is given to A. Then either $\hat{m} \neq m_b$ or $\hat{\sigma} \neq \sigma^*$ since otherwise it would be that challenge ciphertext itself submitted to the decrypt/verify oracle; moreover, only one signature query is ever made by A_1 involving ψ^*. We conclude that $(\hat{m}||\psi^*, \hat{\sigma})$ is a valid forgery.

Putting all this together we have

$$Pr[R_4] \leq AdvEF_{SIG,A_1}(\kappa)$$

as required.

Proof of Lemma 3. To prove this lemma we show how A can be used to construct an adversary A_2 of KEM that breaks its real or random indistinguishability. Algorithm A_2 works by defining an environment in which to run A. It responds to the queries of A as follows.

- The first thing that A_2 does is to call its receiver key generation oracle to obtain PK_r. Once it has done this it calls its challenge oracle to obtain $(K^\dagger, \psi^*)$.

- When A makes its sender key generation oracle query, A_2 runs SIG.KeyGen to obtain (PK_s, SK_s). It returns PK_s to A and keeps SK_s secret.
- When A makes its receiver key generation oracle query, A_2 returns PK_r to A.
- Adversary A_2 responds to the sign/encrypt and decrypt/verify queries made by A in very much the same way that these queries would be responded to in game G_5; the only difference is that signatures are now produced using SK_s rather than using a signing oracle. Note that, as in G_5, the secret key corresponding to PK_r is not necessary to run A_2.
- When A submits messages (m_0, m_1) to its challenge oracle, A_2 chooses $b \leftarrow \{0, 1\}$; it uses SK_s to produce a signature σ^* on $m_b \| \psi^*$; it uses SKE.Enc to obtain the encryption χ^* of $m_b \| \sigma^*$ under $K^\dagger$; finally it returns the ciphertext (ψ^*, χ^*) to A.
- At the end of the simulation, A outputs a bit b'. If $b' = b$, A_2 outputs 1, otherwise it outputs 0.

Let d be the internal bit of A_2's challenge oracle which A_2 seeks to determine and let d' be the bit output by A_2. By construction we have

$$\Pr[S_5] = \Pr[b' = b | d = 1] = \Pr[d' = 1 | d = 1],$$
$$\Pr[S_6] = \Pr[b' = b | d = 0] = \Pr[d' = 1 | d = 0]$$

and so by definition we have

$$|\Pr[S_6] - \Pr[S_5]| = \mathsf{AdvRR}_{\mathsf{KEM}, A_2}(\lambda).$$

Proof of Lemma 4. The proof of this lemma proceeds in two stages. We first show that we can construct an adversary A_3' that breaks the security of SKE under a new definition that we describe below. We then show how such an adversary can be used to construct the requisite A_3 to break the assumed indistinguishability of encryptions of SKE.

To begin with we introduce a new notion of security for SKE: *real or random indistinguishability under passive attack*. An adversary A_3' that mounts a passive attack against SKE is a PPT algorithm that takes as input 1^λ, for security parameter $\lambda \in \mathbf{Z}_{\geq 0}$. Below we describe the attack game used to define this security notion.

Stage 1: The adversary queries chooses a messages m of length l_m. It gives this message to an *encryption oracle*.

Stage 2: The encryption oracle does the following.

$$K \leftarrow \{0, 1\}^{\mathsf{SKE.KeyLen}(\lambda)}; \quad str \leftarrow \{0, 1\}^{l_m}; \quad b \leftarrow \{0, 1\};$$
$$\text{if } b = 1, \ \chi^* \leftarrow \mathsf{SKE.Enc}(1^\lambda, K, m);$$
$$\text{if } b = 0, \ \chi^* \leftarrow \mathsf{SKE.Enc}(1^\lambda, K, str).$$

Finally, it responds with χ^*.

Stage 3: The adversary outputs $b' \in \{0, 1\}$.

If A'_3 is playing the attack game above, we define

$$\mathsf{AdvRR}_{\mathsf{SKE},\mathsf{A}'_3}(\lambda) := |\Pr[b' = 1|b = 1] - \Pr[b' = 1|b = 0]|.$$

The probability is taken over the random choices of A'_3 and those of A'_3's oracles.
    We now show how to construct such an adversary A'_3 using A. The simulation of A'_3 works as follows.

- The first thing that A'_3 does is to run KEM.KeyGen to obtain $(\mathsf{PK}_r, \mathsf{SK}_r)$. It keeps PK_r and discards SK_r. Once it has done this it runs KEM.Enc to obtain (K^*, ψ^*). It keeps ψ^* and discards K^*.
- When A makes its sender key generation oracle query, A'_3 runs SIG.KeyGen to obtain $(\mathsf{PK}_s, \mathsf{SK}_s)$. It returns PK_s to A and keeps SK_s secret.
- When A makes its receiver key generation oracle query, A'_3 returns PK_r to A.
- Adversary A'_3 responds to the sign/encrypt and decrypt/verify queries made by A in very much the same way that these queries would be responded to in game $\mathbf{G}_6$; the only difference is that signatures are now produced using SK_s rather than using a signing oracle.
- When A submits messages (m_0, m_1) to its challenge oracle, A'_3 chooses $b \leftarrow \{0, 1\}$; it uses SK_s to produce a signature σ^* on $m_b||\psi^*$; it calls its challenge encryption oracle on $m_b||\sigma^*$ to obtain χ^*; finally it returns the ciphertext (ψ^*, χ^*) to A.
- At the end of the simulation, A outputs a bit b'. If $b' = b$, A'_3 outputs 1, otherwise it outputs 0.

Let d be the internal bit of A'_3's challenge oracle which A'_3 seeks to determine and let d' be the bit output by A'_3. By construction we have

$$\Pr[S_6] = \Pr[b' = b|d = 1] = \Pr[d' = 1|d = 1],$$
$$\Pr[S_7] = \Pr[b' = b|d = 0] = \Pr[d' = 1|d = 0]$$

and so by definition we have

$$|\Pr[S_7] - \Pr[S_6]| = \mathsf{AdvRR}_{\mathsf{SKE},\mathsf{A}'_3}(\lambda).$$

To complete the proof we demonstrate how a real or random distinguishing adversary such as A'_3 above can be used to build an adversary A_3 in the sense of indistinguishability of encryptions. To construct A_3 proceed as follows.

- Run A'_3 until it outputs a message m of length l_m.
- Choose $str \leftarrow \{0, 1\}^{l_m}$.
- Set $m_0 \leftarrow str$ and $m_1 \leftarrow m$.
- Send (m_0, m_1) to Adv_3's encryption oracle and receive ciphertext χ^* in response.

- Return χ^* to Adv_3'.
- At the end of the simulation Adv_3' outputs a bit b', return b'.

From the above construction it is clear that any advantage A_3' has translates directly into advantage for A_3. Therefore

$$\mathsf{AdvRR}_{\mathsf{SKE},\mathsf{A}_3'}(\lambda) \leq \mathsf{AdvIND}_{\mathsf{SKE},\mathsf{A}_3}(\lambda)$$

as required.

Non-interactive Designated Verifier Proofs and Undeniable Signatures

Caroline Kudla[*] and Kenneth G. Paterson

Information Security Group,
Royal Holloway, University of London, UK
{c.j.kudla, kenny.paterson}@rhul.ac.uk

Abstract. Non-interactive designated verifier (NIDV) proofs were first introduced by Jakobsson *et al.* and have widely been used as confirmation and denial proofs for undeniable signature schemes. There appears to be no formal security modelling for NIDV undeniable signatures or for NIDV proofs in general. Indeed, recent work by Wang has shown the original NIDV undeniable signature scheme of Jakobsson *et al.* to be flawed. We argue that NIDV proofs may have applications outside of the context of undeniable signatures and are therefore of independent interest. We therefore present two security models, one for general NIDV proof systems, and one specifically for NIDV undeniable signatures.

We go on to repair the NIDV proofs of Jakobsson *et al.*, producing secure NIDV proofs suited to combination with Chaum's original undeniable signature scheme resulting in a secure and efficient concrete NIDV undeniable signature scheme.

1 Introduction

Undeniable signatures were first presented in 1989 by Chaum and van Antwerpen [7], and were designed to have the property that signatures could be freely distributed, but were not self-authenticating. In other words, signatures cannot be verified without the cooperation of the signer. However, any party wrongly accused of having produced the signature can deny having produced the signature. The true signer may prove his authorship of an undeniable signature by running a confirmation protocol with a verifier, and a falsely implicated signer may deny his involvement by running a denial protocol with a verifier. Obviously, only the true signer should be able to successfully complete a confirmation protocol. Moreover the true signer should be unable to successfully complete a denial protocol for any of his signatures. Therefore the true signer cannot deny having produced his signatures.

The confirmation and denial protocols for the undeniable signature scheme of [7] were made zero-knowledge in [5], and this goes some way to ensuring that the signer has control over who can verify an undeniable signature. However, as was pointed out in [10], even though the confirmation protocol is zero-knowledge, the

---

[*] This author is funded by Hewlett-Packard Laboratories.

N.P. Smart (Ed.): Cryptography and Coding 2005, LNCS 3796, pp. 136–154, 2005.

signer may still not always be able to control who is able to verify the validity of a signature if a group of verifiers cooperate. The undeniable signature scheme in [7] is also vulnerable to a blackmailing attack [12]. Jakobsson et al. [13] provide a solution, called designation of verifiers, to ensure that only a specified verifier can confirm an undeniable signature.

Informally, a designated verifier (DV) proof is a proof of correctness of some "statement" that either the prover or some designated verifier could have produced. If the prover created the proof, then the "statement" is correct, however a designated verifier could simulate a valid proof without a correct "statement". A secure DV proof should convince the designated verifier of the correctness of the "statement" since the designated verifier knows that he did not create the proof himself. But no other party will be convinced of the validity of the proof since the designated verifier could have created it. In the context of undeniable signatures, a DV proof can be used to convince (only) the designated verifier of the validity of the undeniable signature.

Although the authors of [13] did not give a formal definition of DV proofs, they provided concrete examples of such proofs. The construction of DV proofs in [13] used trapdoor commitment schemes [2]. It was also shown there that the DV proofs could be made non-interactive. Such proofs are called NIDV proofs. However an attack on the concrete scheme of [13] was recently discovered by Wang [23], whereby a cheating signer can create a "non-standard" undeniable signature which the signer can prove valid via the NIDV confirmation proof and later deny via the denial proof. Wang proposed two ways to repair the scheme of [13], but did not offer any proofs of security. In fact, prior to this work, no formal definitions for NIDV proofs or for their security have ever been proposed, nor has a security model for NIDV undeniable signatures ever been developed.

For normal undeniable signature schemes, unforgeability and invisibility (or anonymity) are usually considered to be the key notions of security. As for the security of confirmation and denial proofs, the literature suggests that most authors have been content to simply prove that they are zero-knowledge and sound. This may suffice for the case where zero-knowledge confirmation and denial proofs are used, but it is unclear whether these notions of security are satisfactory for NIDV proofs. In fact we argue here that they are not.

To summarise, little work has been done on security for NIDV undeniable signatures, and the earliest scheme [13] is now known to be insecure.

1.1   Our Contribution

We present a formal definition for NIDV proof systems which is compatible with the concrete schemes of [13]. We then propose a model of security for NIDV proof systems which we believe are of independent interest.

We then present a formal definition for NIDV undeniable signature schemes as well as a security model which models the security of both the core signature scheme and the NIDV confirmation and denial proof systems with which it is composed. Essentially, two NIDV proof systems are required to construct an NIDV undeniable signature scheme. The NIDV proof systems are for complementary languages, one providing confirmation proofs and the other denial proofs for the core signature scheme.

Our work represents the first time that a formal security model for NIDV undeniable signatures has been developed. The model does not require the signature scheme to be randomized. It is also a multi-party model, reflecting the fact that designated verifier proofs naturally involve more than one party, and that a party may play different roles at different times.

We consider NIDV proofs and NIDV undeniable signatures separately. One reason for doing so is that, in any application, signatures may exist independently of proofs. For example, a prover may generate a signature as a commitment to a message but only later provide an NIDV proof of its correctness, or a prover may generate many proofs for different designated verifiers on the same signature. A second reason is that NIDV proofs may also be useful in contexts other than undeniable signatures. One possible application of NIDV proofs is to deniable proofs of knowledge, in particular, proofs of knowledge of a private key. For example, when registering a public key with certification authority C, A could demonstrate knowledge of the appropriate private key by presenting C with an NIDV proof of knowledge of the private key of A.

We go on to repair the NIDV proofs of [13], producing secure NIDV proof systems suited to combination with the full domain hash variant of Chaum's undeniable signature scheme [5,7]. The NIDV proofs we obtain are actually a little shorter than those in [13]. Our work confirms that one of the fixes proposed by Wang [23] does indeed repair the NIDV proofs of [13]. The result is a concrete and efficient NIDV undeniable signature scheme. Our paper concludes with some open problems and ideas for further extensions of our work.

1.2   Related Notions and Work

In parallel work, Lipmaa *et al.* [15] examine the security properties of what they refer to as designated verifier signature (DVS) schemes. In common with other authors [14,19], they ascribe the term DVS to [13] and describe the concrete NIDV undeniable signature scheme of [13] as a DVS. In fact, this terminology was *never* used in [13]. Examination of [15] reveals that a DVS effectively combines an undeniable signature and an NIDV proof of correctness for that signature into a single entity. DVS, then, are closely related to (two-party) ring signatures. In contrast, the authors of [13] did not explicitly define such an object, preferring to keep undeniable signatures and their proofs separate. Our formal definition of an NIDV undeniable signature scheme also maintains the separation of signatures and proofs, and so is closer in spirit to the informal definitions in [13]. We have given several reasons why this separation is appropriate in the preceding section. It is unfortunate that this confusion over nomenclature has arisen in the literature, and we hope our work can help to clarify the situation.

Lipmaa *et al.* [15] go on to formalize and extend the attack of Wang [23] on [13] to DVS. They then propose two new security properties which are required for secure DVS, namely non-delegatability and disavowability. Although these notions are presented in the context of DVS, they could also be applied in the context of NIDV undeniable signatures. Our security model for NIDV undeniable signatures does in fact capture the notion of non-delegatability, but we do not consider disavowability to be a feature of NIDV proofs, and our concrete examples do not have this property since they are unconditionally non-transferable.

Another related notion is that of strong designated verifier (SDV) proofs [13,19,22], which provide stronger security guarantees than NIDV proofs. SDV proofs provide similar properties to NIDV proofs except that only the designated verifier is able to verify the proofs produced, since the verification algorithm requires the private key of the designated verifier. By contrast, NIDV proofs are universally verifiable, but only convincing to the designated verifier.

Also related to NIDV proofs (or more specifically to NIDV undeniable signatures) are universal designated verifier (UDV) signatures [14,20,21]. Although at first glance these appear to be similar to NIDV undeniable signatures, they are quite different. UDV signature schemes produce signatures which are universally verifiable. However any party in possession of a valid UDV signature on message M from signer S can provide (to any verifier of their choice) a designated verifier proof that they possess a valid UDV signature on M by S. On the other hand, NIDV undeniable signatures are not universally verifiable, and only the signer is able to produce designated verifier proofs of a signature's validity. We do not consider UDV signatures in this paper.

2   Preliminaries

Let G be a finite, multiplicative group of prime order q, and let g be a generator of G. We denote by $DL(g, h)$ the discrete logarithm of h with respect to base g in group G. So $g^{DL(g,h)} = h$ in G. We also informally define the following problems in G:

Discrete Logarithm (DL) Problem: Given $g, g^a \in G$, $a \in_R \mathbb{Z}_q$, compute a.
Computational Diffie-Hellman (CDH) Problem: Given $g, g^a, g^b \in G$, $a, b \in_R \mathbb{Z}_q$, compute g^{ab}.
Decisional Diffie-Hellman (DDH) Problem: Given $g, g^a, g^b, g^c \in G$, $a, b \in_R \mathbb{Z}_q$, determine whether $c = ab \bmod q$.

3   Non-interactive Designated Verifier Proof Systems

We now present a formal definition for NIDV proof systems. This formal definition was previously lacking in [13] but our definitions are compatible with the concrete scheme of [13].

A non-interactive designated verifier (NIDV) proof system is defined with respect to some family of languages $\mathcal{L}$. The goal of an NIDV proof system is to prove the membership of elements e in a language $L \in \mathcal{L}$. An NIDV proof system consists of the following algorithms:

- A probabilistic *Setup* algorithm which takes a security parameter l as input and returns the system parameters *params* and a description of a family of languages $\mathcal{L}$. Amongst the public parameters *params* are descriptions of the following spaces: a public key space $\mathcal{PK}$, a private key space $\mathcal{SK}$, an element space $\mathcal{E}$ and a proof space $\mathcal{P}$.
- A probabilistic *KeyGen* algorithm which takes as input the public parameters *params* and returns a key pair (x, X) where $x \in \mathcal{SK}$ is a private key and $X \in \mathcal{PK}$ is the corresponding public key.
- A proof generation algorithm *PGen* which takes as input $\langle x_P, X_V, e \rangle$ where $x_P \in \mathcal{SK}$, $X_V \in \mathcal{PK}$, and $e \in \mathcal{E}$ with $e \in L(X_P)$, and produces an NIDV proof $\pi \in \mathcal{P}$ for e.
- A verification algorithm *PVerify* which takes as input $\langle X_P, X_V, e, \pi \rangle$ where $X_P, X_V \in \mathcal{PK}, e \in \mathcal{E}$ and $\pi \in \mathcal{P}$, and outputs Accept or Reject.

Note that we parameterize the languages $L \in \mathcal{L}$ by public keys, and for any public key $X \in \mathcal{PK}$, $L(X) \in (E)$. This parametrization will be needed for the proof systems required for use with undeniable signatures but is not necessary in general.

4   Security of NIDV Proof Systems

We say that an NIDV proof system is secure if it satisfies the notions of correctness, non-transferability and soundness. These are defined next.

4.1   Correctness

An NIDV proof system is *correct* if when PGen is run on any input $x_P \in \mathcal{SK}, X_V \in \mathcal{PK}, e \in \mathcal{E}$ and outputs some $\pi \in \mathcal{P}$, then PVerify on input $\langle X_P, X_V, e, \pi \rangle$ outputs Accept.

4.2   Non-transferability

We say that an NIDV proof system is *non-transferable* if there exists a polynomial time algorithm A that on input tuples $\langle X_P, x_V, e \rangle$, where $X_P \in \mathcal{PK}, x_V \in \mathcal{SK}, e \in \mathcal{E}$, but where e is not necessarily in $L(X_P)$, produces proofs $\pi \in \mathcal{P}$ such that $\langle X_P, X_V, e, \pi \rangle$ is accepted by PVerify and the distribution of proof π is polynomially indistinguishable from proof π' produced by PGen when run on inputs $\langle x_P, X_V, e' \rangle$ where $e' \in \mathcal{E}$ and $e' \in L(X_P)$.

4.3   Soundness

Soundness of an NIDV proof system is defined via the following game between a challenger C and an adversary E:

Setup: C runs Setup and KeyGen for a given security parameter l to obtain the public parameters $params$, a description of a family of languages $\mathcal{L}$, as well as a set of public and private key pairs (X_i, x_i). C sets up each participant oracle I with its public and private keys X_I, x_I. E is given $params, \mathcal{L}$ and the public keys $\{X_i\}$ of all participants while C retains the private keys $\{x_i\}$. We define the set of all participants' public keys to be $\mathcal{X}$.

E can make the following types of query to the participant oracles:

EGen Queries: E can make an EGen query to an oracle P with public key X_P (possibly with input some $seed$). The oracle outputs an element $e \in L(X_P)$.

PGen Queries: E can make a PGen query to oracle P with public key X_P on input $\langle X_V, e \rangle$ where $e \in \mathcal{E}$, $X_V \in \mathcal{X}$. If $e \in L(X_P)$, the oracle produces an NIDV proof $\pi \in \mathcal{P}$ from P to V for e. If $e \notin L(X_P)$ then the oracle outputs "invalid".

FakePGen Queries: E can make a FakePGen query to oracle V with public key X_V on input $\langle X_P, e \rangle$ where $e \in \mathcal{E}$, $X_P \in \mathcal{X}$. The oracle runs algorithm A to produce an NIDV proof $\pi \in \mathcal{P}$ for e. The oracle outputs π.

Corrupt Queries: E can request the private key x_I of any oracle I.

Output: Finally E outputs $\langle X_P^*, X_V^*, e^*, \pi^* \rangle$, where $X_P^*, X_V^* \in \mathcal{X}$ and X_V^* is uncorrupted, $e^* \in \mathcal{E}$ and $\pi^* \in \mathcal{P}$. E wins if $\langle X_P^*, X_V^*, e^*, \pi^* \rangle$ is accepted by PVerify, π^* was not the output of some PGen query to P^* on $\langle X_V^*, e^* \rangle$ or some FakePGen query to V^* on $\langle X_P^*, e^* \rangle$, and either:
1. X_P^* is uncorrupted, or
2. $e^* \notin L(X_P^*)$.

Definition 1. *We say that an NIDV proof system is* sound *if the probability of success of any polynomially bounded adversary in the above game is negligible (as a function of the security parameter l).*

4.4   Notes on the Security Definitions for NIDV Proof Systems

Soundness: The soundness definition guarantees that if an uncorrupted verifier receives a valid NIDV proof, then it was created using the private key x_P and $e \in L(X_P)$. So a prover cannot cheat. Soundness also guarantees that no-one other than P can convince an uncorrupted designated verifier that $e \in L(X_P)$. In the context of undeniable signatures, this means that no-one other than the real signer is able to produce an NIDV proof of a valid undeniable signature that will be accepted by an uncorrupted designated verifier. This is essential for the security of undeniable signatures. The model for soundness is multiparty, reflecting the fact that NIDV proofs naturally involve more than one party, and that a party may play different roles at different times.

Non-transferability: The existence of algorithm A that can be run by V to create an NIDV proof for any element e (not necessarily in L) ensures that no-one besides V will be convinced by an NIDV proof for e.

We note that we do not require the elements e and e' to be indistinguishable for non-transferability, rather only the proofs π and π' are required to be indistinguishable. If an outside party can already distinguish elements in L from elements not in L, then they have no need of NIDV proofs for L. However an NIDV proof for an element e should not give any extra information regarding e to parties other than the designated verifier.

FakePGen queries: The existence of algorithm A from Section 4.2 also enables oracles to answer FakePGen queries. We consider it important to model such queries since an adversary may have access to such "faked" NIDV proofs that are produced by dishonest verifiers using algorithm A.

5   NIDV Undeniable Signature Schemes

As mentioned earlier, the main application of NIDV proofs has historically been in undeniable signatures, even though the current security models for undeniable signatures do not seem to support NIDV proofs. We now present a formal definition for NIDV undeniable signature schemes.

Definition 2. *An NIDV undeniable signature scheme consists of a core signature scheme as well as NIDV confirmation and denial proof systems. The core signature scheme consists of the following algorithms:*

- A probabilistic *Setup* algorithm which takes a security parameter l as input and returns the system parameters *params*. Amongst the public parameters are descriptions of the following spaces: a public key space $\mathcal{PK}$, a private key space $\mathcal{SK}$, a message space $\mathcal{M}$ and a signature space $\mathcal{S}$.
- A probabilistic *KeyGen* algorithm which takes as input the public parameters *params* and returns a key pair (x, X) where $x \in \mathcal{SK}$ and $X \in \mathcal{PK}$.
- A (possibly probabilistic) signature generation algorithm *Sign* which on input $\langle x, m \rangle$ where $x \in \mathcal{SK}, m \in \mathcal{M}$, produces an undeniable signature $\sigma \in \mathcal{S}$.

The core signature scheme defines a language $L(X)$ for each public key X, where $L(X) = \{(m, \sigma) : \sigma = Sign(x, m)\}$. In other words, $L(X)$ is the language of all possible valid message and signature pairs for public key X and $\overline{L(X)}$ is the language of all invalid message and signature pairs for public key X. The family of languages $\mathcal{L}$ is defined as $\mathcal{L} = \{L(X) : X \in \mathcal{PK}\}$ and $\overline{\mathcal{L}}$ is defined as $\overline{\mathcal{L}} = \{\overline{L(X)} : X \in \mathcal{PK}\}$.

The families of languages $\mathcal{L}$ and $\overline{\mathcal{L}}$ parameterize the confirmation and denial proofs. The confirmation proof is an NIDV proof system $\mathcal{C}$ for $\mathcal{L}$, and the denial proof is an NIDV proof system $\mathcal{D}$ for $\overline{\mathcal{L}}$. The setup algorithms for $\mathcal{C}$ and $\mathcal{D}$ use the public key space $\mathcal{PK}$, the private key space $\mathcal{SK}$, and set the element space $\mathcal{E}$ to be $\mathcal{M} \times \mathcal{S}$. The proof spaces for $\mathcal{C}$ and $\mathcal{D}$ are denoted $\mathcal{P}_C$ and $\mathcal{P}_D$ respectively. The following algorithms then make up the confirmation and denial proofs.

- A confirmation proof generation algorithm *ConfGen* which, on input $\langle x_P, X_V, m, \sigma \rangle$ where $x_P \in \mathcal{SK}, X_V \in \mathcal{PK}, (m, \sigma) \in L(X_P)$ runs *PGen* of $\mathcal{C}$ on $\langle x_P, X_V, (m, \sigma) \rangle$.

- A confirmation proof verification algorithm *ConfVerify* which, on input $\langle X_P, X_V, m, \sigma, \pi_C \rangle$ where $X_P, X_V \in \mathcal{PK}$, $(m, \sigma) \in \mathcal{E}$ and $\pi_C \in \mathcal{P}_C$ runs *PVerify* of $\mathcal{C}$ on $\langle X_P, X_V, (m, \sigma), \pi_C \rangle$.
- A denial proof generation algorithm *DenyGen* which, on input $\langle x_P, X_V, m, \sigma \rangle$ where $x_P \in \mathcal{SK}$, $X_V \in \mathcal{PK}$, $(m, \sigma) \in \overline{L(X_P)}$ runs *PGen* of $\mathcal{D}$ on $\langle x_P, X_V, (m, \sigma) \rangle$.
- A denial proof verification algorithm *DenyVerify* which, on input $\langle X_P, X_V, m, \sigma, \pi_D \rangle$ where $X_P, X_V \in \mathcal{PK}$, $(m, \sigma) \in \mathcal{E}$ and $\pi_D \in \mathcal{P}_C$ runs *PVerify* of $\mathcal{D}$ on $\langle X_P, X_V, (m, \sigma), \pi_D \rangle$.

6 Security of NIDV Undeniable Signatures

In analyzing the security of NIDV undeniable signatures, we consider the security of the confirmation and denial proofs being used, and their composition with the core signature scheme.

Unless explicitly stated, we will represent the public key of a participant I by X_I, and the private key as x_I. P will in general represent a prover, and V a verifier.

Definition 3. *The confirmation (denial) proof of an NIDV undeniable signature is secure if $\mathcal{C}$ ($\mathcal{D}$) is a secure NIDV proof system for $\mathcal{L}$ ($\overline{\mathcal{L}}$). That is, $\mathcal{C}$ ($\mathcal{D}$) is correct, non-transferable and sound.*

6.1 The Security of the Core Signature Scheme

The security of the core signature scheme is defined via the following notions:

Unforgeability: Unforgeability of an undeniable signature scheme is defined via the following game between a challenger C and an adversary E:

Setup: C runs the Setup and KeyGen algorithms for a given security parameter l to obtain the public parameters *params* as well as a set of public and private key pairs (X_i, x_i). E is given *params* and the set of public keys $\{X_i\}$ of all participants while C retains the private keys $\{x_i\}$. C sets up each participant oracle I with its public and private keys X_I, x_I. We define the set of all participants' public keys to be $\mathcal{X}$.

E can make the following types of query to the participant oracles:

Sign Queries: E can make a Sign query to any participant I with public key X_I on input m where $m \in \mathcal{M}$, and the oracle runs Sign on $\langle x_I, m \rangle$ to produce a signature $\sigma \in \mathcal{S}$. The oracle outputs σ.

Conf/Deny Queries: E can make a Conf/Deny query to any participant P with public key X_P on input $\langle X_V, m, \sigma \rangle$ where $X_V \in \mathcal{X}$, $m \in \mathcal{M}$ and $\sigma \in \mathcal{S}$. If $(m, \sigma) \in L(X_P)$ then the oracle runs ConfGen on $\langle x_P, X_V, m, \sigma \rangle$ to produce an NIDV proof $\pi_C \in \mathcal{P}_C$ which it outputs, otherwise it runs DenyGen on $\langle x_P, X_V, m, \sigma \rangle$ to produce an NIDV proof $\pi_D \in \mathcal{P}_D$ which it outputs.

FakeConf Queries: E can make a FakeConf query to any participant V with public key X_V on input $\langle X_P, m, \sigma \rangle$ where $X_V \in \mathcal{X}$, $m \in \mathcal{M}$ and $\sigma \in \mathcal{S}$. The oracle runs the algorithm A of the NIDV proof system $\mathcal{C}$ to produce an NIDV proof $\pi_C \in \mathcal{P}_C$ which it outputs.

FakeDeny Queries: E can make a FakeDeny query to any participant V with public key X_V on input $\langle X_P, m, \sigma \rangle$ where $X_V \in \mathcal{X}$, $m \in \mathcal{M}$ and $\sigma \in \mathcal{S}$. The oracle runs the algorithm A of the NIDV proof system $\mathcal{D}$ to produce an NIDV proof $\pi_D \in \mathcal{P}_D$ which it outputs.

Corrupt Queries: E can make a corrupt query to any participant I with public key X_I, and the oracle outputs x_I.

Output: Finally E produces $X_P^* \in \mathcal{X}$, $m^* \in \mathcal{M}$ and $\sigma^* \in \mathcal{S}$, where X_P^* is uncorrupted and σ^* was not the output of some previous Sign query to P^* on m^*. E wins the game if $(m^*, \sigma^*) \in L(X_P^*)$.

Definition 4. *We say that an undeniable signature scheme is* unforgeable *if the probability of success of any polynomially bounded adversary in the above game is negligible in l.*

Invisibility: Invisibility of an undeniable signature scheme is defined via the following game between a challenger C and an adversary E:

Setup: This is as in the Unforgeability game above.

Phase 1: The adversary can make Sign, Conf/Deny, FakeConf, FakeDeny and Corrupt queries, and these are all answered as in the Unforgeability game.

Challenge: E produces $m^* \in \mathcal{M}$, $X_P^* \in \mathcal{X}$, where X_P^* is uncorrupted. In addition, if the Sign algorithm is deterministic, then E should not have previously made a Sign query to P^* on m^* in Phase 1. C chooses a random bit b and if $b = 0$, C sets $\sigma^* = r$ where r is randomly chosen from $\mathcal{S}$, otherwise C sets $\sigma^* = Sign(x_P^*, m^*)$. C gives σ^* to E.

Phase 2: Again E can make queries as in Phase 1, except that E cannot make a Conf/Deny query to P^* on $\langle X_V, m^*, \sigma^* \rangle$ for any X_V. If the signature algorithm Sign is deterministic, E is also forbidden from making a Sign query to P^* on m^*.

Output: Finally E outputs a bit b' and wins the game if $b' = b$.

Definition 5. *We say that an undeniable signature scheme is* invisible *if the probability of success of any polynomially bounded adversary in the above game is negligible in l.*

6.2   Notes on the Security Definitions for Undeniable Signatures

Correctness: Although we do not explicitly define correctness for NIDV undeniable signatures, correctness is handled by the correctness of proof systems $\mathcal{C}$ and $\mathcal{D}$.

Unforgeability: Our model of unforgeability differs from security models for normal undeniable signatures in two main ways. Firstly it is multiparty due to the multiparty nature of the NIDV confirmation and denial proofs. Secondly, we allow the adversary to make FakeConf and FakeDeny queries. We consider these to be necessary since an adversary may conceivably have access to such "fake" proofs produced by dishonest designated verifiers.

Invisibility: We include the notion of invisibility in our model of security rather than anonymity. Anonymity, which could be defined in a similar way to [11], captures the notion that an adversary cannot determine which of two possible signers created a given signature. Analogous results to those in [11] could be used to show that invisibility is the stronger notion and implies anonymity. However we feel that the stronger definition of invisibility is appropriate for NIDV undeniable signatures since we model the existence of fake NIDV proofs and their corresponding (possibly fake) signatures, and these should be indistinguishable from true signatures and NIDV proofs.

Determinism: Our definitions of unforgeability and invisibility encompass both non-deterministic and deterministic undeniable signatures. We assume that signers can identify their own valid signatures. For deterministic undeniable signatures this is trivial because signers can just re-sign a message, but in the case of non-deterministic (or randomized) undeniable signatures this may be non-trivial.

We note that in the deterministic case, invisibility actually implies unforgeability, since an adversary who can forge signatures can trivially win the invisibility game. However for randomized signatures, these properties are distinct. We keep the properties distinct when proving the security of a deterministic undeniable signature scheme later in the paper because it makes the proofs easier.

7   A Concrete NIDV Undeniable Signature Scheme

We present the full domain hash variant of the undeniable signature scheme of Chaum [5] with NIDV confirmation and denial proofs.

7.1   The Core Signature

Setup: For some security parameter l, let p and q be large primes, where $q|(p-1)$. Let G be a multiplicative subgroup of $\mathbb{Z}_p^*$ of order q and let g be a generator of G. We also assume that $H_1 : \{0,1\}^* \to G$ is a cryptographic hash function. For example, such a hash function may be constructed by using a standard hash function to map the input to a bitstring representing an integer, reducing that integer modulo p and then exponentiating the result to the power $(p-1)/q$ modulo p. We set $\mathcal{PK} = \mathcal{S} = G$, $\mathcal{M} = \{0,1\}^*$ and $\mathcal{SK} = \mathbb{Z}_q$. The public parameters are $params = (p, q, g, H_1, \mathcal{PK}, \mathcal{SK}, \mathcal{M}, \mathcal{S})$.

KeyGen: To set up a user I's public and private keys, the private key x_I is chosen at random from $\mathbb{Z}_q$, and the public key is $X_I = g^{x_I} \bmod p$.

Sign: On input $\langle x_I, m \rangle$ where $x_I \in \mathbb{Z}_q$, $m \in \{0,1\}^*$, compute $\sigma = H_1(m)^{x_I} \bmod p$. Output σ.

7.2   The Confirmation and Denial Proofs

The Sign algorithm defines a language $L(X_I) = \{(m, \sigma) : \sigma = \mathrm{Sign}(x_I, m)\}$ for each public key X_I. For the above signature scheme, we can write $L(X_I) = \{(m, \sigma) : DL(\sigma, H_1(m)) = DL(x_I, g)\}$. In other words, $L(X_I)$ is the language of all possible message and signature pairs (m, σ) where the discrete logarithm of σ to the base $H_1(m)$ equals the discrete logarithm of X_I to the base g modulo p. The family of languages $\mathcal{L}$ is defined as $\mathcal{L} = \{L(X_I) : X_I \in G\}$.

We can now define confirmation and denial proofs with respect to the languages $L(X_I)$.

Confirmation proof: The confirmation proof requires a secure NIDV proof system $\mathcal{C}$ for $\mathcal{L}$. Informally, $\mathcal{C}$ must prove the equality of two discrete logarithms (EDL).

Denial proof: The denial proof requires a secure NIDV proof system $\mathcal{D}$ for $\overline{\mathcal{L}}$. Informally, $\mathcal{D}$ must prove the inequality of two discrete logarithms (IDL).

7.3   A Concrete NIDV EDL Proof System

The NIDV proof we present is a slight modification of the scheme of Jakobsson et al. [13] since the original proof was shown to be insecure by Wang [23].

Since our NIDV EDL proof will be used with the above undeniable signature scheme, the Setup algorithm will be identical to that in Section 7.1 except that in addition we require another cryptographic hash function $H_2 : \{0,1\}^* \rightarrow \mathbb{Z}_q$, and we define the spaces $\mathcal{E} = \mathcal{M} \times \mathcal{S}$ and $\mathcal{P} = \mathbb{Z}_q^4$. KeyGen will be exactly as in Section 7.1. The family of languages will be defined by the Sign algorithm of the concrete scheme as described above in Section 7.2. We still need to define the PGen and PVerify algorithms.

NIDV EDL PGen: On input $\langle x_P, X_V, m, \sigma \rangle$ where $x_P \in \mathcal{SK}$, $X_V \in \mathcal{PK}$, $m \in \mathcal{M}$ and $\sigma \in \mathcal{S}$, the algorithm picks random $w, r, t \in \mathbb{Z}_q$ and computes:

$$c = g^w X_V^r \bmod p$$
$$G = g^t \bmod p$$
$$M = H_1(m)^t \bmod p$$
$$h = H_2(c, G, M, m, \sigma, X_P)$$
$$d = t - x_P(h + w) \bmod q$$

The algorithm outputs $\pi = \langle w, r, h, d \rangle$.

NIDV EDL PVerify: On input $\langle X_P, X_V, m, \sigma, \pi \rangle$ where $X_P, X_V \in \mathcal{PK}$, message $m \in \mathcal{M}$, signature $\sigma \in \mathcal{S}$, and proof $\pi = \langle w, r, h, d \rangle \in \mathcal{P}$, the algorithm computes:

$$c = g^w X_V^r \bmod p$$
$$G = g^d X_P^{(h+w)} \bmod p$$
$$M = H_1(m)^d \sigma^{(h+w)} \bmod p$$

and verifies that $h = H_2(c, G, M, m, \sigma, X_P)$. If the last equation holds, then the algorithm outputs Accept, otherwise it outputs Reject.

Comparison to the scheme of Jakobsson et al. The main difference is that we include the values σ and X_P in the input of H_2. Our proof π also has one less element than in the NIDV EDL proof of [13].

7.4   A Concrete NIDV IDL Proof System

The denial proof is an NIDV version of the proof of inequality of discrete logarithms in [4].

Our Setup and KeyGen algorithms are as above in Section 7.3 for the NIDV EDL proof scheme except that now $\mathcal{P} = G \times \mathbb{Z}_q^4$.

NIDV IDL PGen: On input $\langle x_P, X_V, m, \sigma \rangle$ where $x_P \in \mathcal{SK}$, $X_V \in \mathcal{PK}$, $m \in \mathcal{M}$, and $\sigma \in \mathcal{S}$, the algorithm picks random $r \in \mathbb{Z}_q$ and computes $C = (\frac{H_1(m)^{x_P}}{\sigma})^r \bmod p$.
The algorithm then constructs a designated verifier proof to demonstrate knowledge of some α and β such that $C = H_1(m)^\alpha \sigma^{-\beta} \bmod p$ and $1 = g^\alpha X_P^{-\beta} \bmod p$. The algorithm sets $\alpha = x_P r \bmod q$ and $\beta = r$ for some random $r \in \mathbb{Z}_q$, picks random $r_1, r_2 \in \mathbb{Z}_q$ and computes:

$$c = g^w X_V^r \bmod p$$
$$G = g^{r_1}(X_P)^{-r_2} \bmod p$$
$$M = H_1(m)^{r_1}(\sigma)^{-r_2} \bmod p$$
$$h = H_2(C, c, G, M, m, \sigma, X_P)$$
$$d_1 = r_1 - \alpha(h + w) \bmod q$$
$$d_2 = r_2 - \beta(h + w) \bmod q$$

The algorithm outputs $\pi = \langle C, w, r, h, d_1, d_2 \rangle$ as the NIDV proof to verifier V that $\mathrm{DL}(\sigma, H_1(m)) \neq \mathrm{DL}(X_P, g)$.
NIDV IDL PVerify: On input $\langle X_P, X_V, m, \sigma, \pi \rangle$ where $X_P, X_V \in \mathcal{PK}$, $m \in \mathcal{M}$, $\sigma \in \mathcal{S}$, and $\pi = \langle C, w, r, h, d_1, d_2 \rangle \in \mathcal{P}$, the algorithm first checks that $C \neq 1$ and then computes:

$$c = g^w X_V^r \bmod p$$
$$G = g^{d_1}(X_P)^{-d_2} \bmod p$$
$$M = C^{h+w} H_1(m)^{d_1}(\sigma)^{-d_2} \bmod p$$

and verifies that $h = H_2(C, c, G, M, m, \sigma, X_P)$. If the last equation holds, then the algorithm outputs Accept, otherwise it outputs Reject.

8   Security of the Concrete Scheme

8.1   Security of the NIDV EDL and IDL Proof Systems

Theorem 1. *The NIDV EDL proof system of Section 7.3 is* correct.

Theorem 2. *The NIDV EDL proof system of Section 7.3 is* non-transferable.

Theorem 3. *The NIDV EDL proof system of Section 7.3 is* sound *in the random oracle model assuming the hardness of the discrete logarithm problem in* G.

The proof of correctness is trivial and is therefore omitted. The proofs of the other two theorems appear in the Appendix. The proofs of correctness, non-transferability and soundness for the NIDV IDL proof system of Section 7.4 are similar to those for the NIDV EDL proof system, so we omit the details.

8.2   Application to the Core Signature Scheme

Since our NIDV EDL and IDL proof systems are secure, they can be composed with our concrete scheme to form secure NIDV confirmation and denial proofs for the NIDV undeniable signature scheme. All that remains for the whole NIDV undeniable signature scheme to be secure is to show that the core signature scheme satisfies the unforgeability and invisibility properties.

Theorem 4. *The core signature scheme of Section 7.1 is* unforgeable *in the random oracle model assuming the hardness of the Computational Diffie-Hellman problem in* G.

Theorem 5. *The core signature scheme of Section 7.1 has* invisibility *in the random oracle model assuming the hardness of the Decision Diffie-Hellman problem in* G.

The proof of Theorem 4 is similar to the proof in [17] (corrected in [16]), although we use a slightly different security model. The details are left to the reader. The proof of Theorem 5 is fairly simple and is therefore also left to the reader. Both proofs will appear in the full version of the paper.

Alternative constructions for NIDV EDL and IDL proofs may be possible. For example, it may be the case that the techniques of [8] could yield more general constructions of such NIDV proofs, although we believe that such general constructions are unlikely to be more efficient than the concrete examples presented here.

9   Conclusions and Open Problems

We have presented models of security for NIDV proof systems and NIDV undeniable signatures and argued that NIDV proofs can have applications outside of the context of undeniable signatures such as in deniable proofs of knowledge or possession. We then repaired the original NIDV undeniable signature scheme of [13], producing a concrete scheme that is efficient and proven secure.

In future work, it would be interesting to investigate how to extend our model to include strong designated verifier proofs [19,22,13] and DVS schemes [15]. It would also be interesting to provide models of security for NIDV versions of confirmer signatures [1,6,9,3] and other signature schemes closely related to undeniable signatures.

Acknowledgements

We would like to thank Steven Galbraith for valuable comments on this work.

References

1. J. Boyar, D. Chaum, I. Damgård, and T. P. Pedersen. Convertible undeniable signatures. In A. Menezes and S.A. Vanstone, editors, *Advances in Cryptology – CRYPTO '90*, volume 537 of *LNCS*, pages 189–205. Springer-Verlag, 1991.
2. G. Brassard, D. Chaum, and C. Crépeau. Minimum disclosure proofs of knowledge. *Journal of Computer and System Sciences*, 37(2):156–189, 1988.
3. J. Camenisch and M. Michels. Confirmer signature schemes secure against adaptive adversaries. In B. Preneel, editor, *Advances in Cryptology – EUROCRYPT 2000*, volume 1807 of *LNCS*, pages 243–258. Springer-Verlag, 2000.
4. J. Camenisch and V. Shoup. Practical verifiable encryption and decryption of discrete logarithms. In D. Boneh, editor, *Advances in Cryptology – CRYPTO 2003*, volume 2729 of *LNCS*, pages 126–144. Springer-Verlag, 2003.
5. D. Chaum. Zero-knowledge undeniable signatures. In I.B. Damgård, editor, *Advances in Cryptology – EUROCRYPT '90*, volume 473 of *LNCS*, pages 458–464. Springer-Verlag, 1990.
6. D. Chaum. Designated confirmer signatures. In A. De Santis, editor, *Advances in Cryptology – EUROCRYPT '94*, volume 950 of *LNCS*, pages 86–91. Springer-Verlag, 1994.
7. D. Chaum and H. van Antwerpen. Undeniable signatures. In G. Brassard, editor, *Advances in Cryptology – CRYPTO '89*, volume 435 of *LNCS*, pages 212–216. Springer-Verlag, 1990.
8. R. Cramer, I. Damgard, and B. Schoenmakers. Proofs of partial knowledge and simplified design of witness hiding protocols. In Y. Desmedt, editor, *Advances in Cryptology - CRYPTO '94*, volume 893 of *Lecture Notes in Computer Science*, pages 174–187. Springer-Verlag, 1995.
9. I. Damgård and T. Pedersen. New convertible undeniable signature schemes. In U.M. Maurer, editor, *Advances in Cryptology – EUROCRYPT '96*, volume 1070 of *LNCS*, pages 372–386. Springer-Verlag, 1996.

10. Y. Desmedt and M. Yung. Weakness of undeniable signature schemes. In D.W. Davies, editor, *Advances in Cryptology – EUROCRYPT '91*, volume 547 of *LNCS*, pages 205–220. Springer-Verlag, 1991.
11. S. D. Galbraith and W. Mao. Invisibility and anonymity of undeniable and confirmer signatures. In M. Joye, editor, *CT-RSA 2003*, volume 2612 of *LNCS*, page 8097. Springer-Verlag, 2003.
12. M. Jakobsson. Blackmailing using undeniable signatures. In A. De Santis, editor, *Advances in Cryptology – EUROCRYPT '94*, volume 950 of *LNCS*, pages 425–427. Springer-Verlag, 1994.
13. M. Jakobsson, K. Sako, and R. Impagliazzo. Designated verifier proofs and their applications. In U.M. Maurer, editor, *Advances in Cryptology – EUROCRYPT '96*, volume 1070 of *LNCS*, pages 143–154. Springer-Verlag, 1996.
14. F. Laguillaumie and D. Vergnaud. Designated verifier signatures: Anonymity and efficient construction from *any* bilinear map. In C. Blundo and S. Cimato, editors, *SCN 2004*, volume 3352 of *LNCS*, pages 105–119. Springer-Verlag, 2005.
15. H. Lipmaa, G. Wang, and F. Bao. Designated verifier signature schemes: Attacks, new security notions and a new construction. In L. Caires et al., editor, *Automata, Languages and Programming, ICALP 2005*, volume 3580 of *LNCS*, pages 459–471. Springer-Verlag, 2005.
16. W. Ogata, K. Kurosawa, and S. Heng. The security of the FDH variant of Chaum's undeniable signature scheme. Cryptology ePrint Archive, Report 2004/290, 2004. Available from http://eprint.iacr.org/2004/290.
17. W. Ogata, K. Kurosawa, and S. Heng. The security of the FDH variant of Chaum's undeniable signature scheme. In S. Vaudenay, editor, *Public Key Cryptography - PKC 2005*, volume 3386 of *LNCS*, pages 328–345. Springer-Verlag, 2005.
18. D. Pointcheval and J. Stern. Security proofs for signature schemes. In U.M. Maurer, editor, *Advances in Cryptology – EUROCRYPT '96*, volume 1070 of *LNCS*, pages 387–398. Springer-Verlag, 1996.
19. S. Saeednia, S. Kremer, and O. Markowitch. An efficient strong designated verifier signature scheme. In J.I. Lim and D.H. Lee, editors, *Information Security and Cryptology - ICISC 2003*, volume 2971 of *LNCS*, pages 40–54. Springer-Verlag, 2003.
20. R. Steinfeld, L. Bull, H. Wang, and J. Pieprzyk. Universal designated-verifier signatures. In C.S. Laih, editor, *Advances in Cryptology - ASIACRYPT 2003*, volume 2894 of *Lecture Notes in Computer Science*, pages 523–542. Springer-Verlag, 2003.
21. R. Steinfeld, H. Wang, and J. Pieprzyk. Efficient extension of standard Schnorr/RSA signatures into universal designated-verifier signatures. In F. Bao et al., editor, *PKC 2004*, volume 2947 of *Lecture Notes in Computer Science*, pages 86–100. Springer-Verlag, 2004.
22. W. Susilo, F. Zhang, and Y. Mu. Identity-based strong designated verifier signature schemes. In H. Wang et al., editor, *ACISP 2004*, volume 3108 of *LNCS*, pages 313–324. Springer-Verlag, 2004.
23. G. Wang. An attack on not-interactive designated verifier proofs for undeniable signatures. Cryptology ePrint Archive, Report 2003/243, 2003. Available from http://eprint.iacr.org/.

Appendix

Proof of Theorem 2. We define algorithm A as follows. On input $\langle X_P, x_V, m, \sigma \rangle$, where $X_P \in \mathcal{PK}$, $x_V \in \mathcal{SK}$, $(m, \sigma) \in \mathcal{E}$,

A chooses random $d, \alpha, \beta \in \mathbb{Z}_q$ and calculates:

$$c = g^\alpha \bmod p$$
$$G = g^d X_P^{-\beta} \bmod p$$
$$M = H_1(m)^d \sigma^{-\beta} \bmod p$$
$$h = H_2(c, G, M, m, \sigma, X_P)$$
$$w = \beta - h \bmod q$$
$$r = (\alpha - w)x_V^{-1} \bmod q$$

A outputs $\pi = \langle w, r, h, d \rangle$. It is easy to check that $\langle X_P, X_V, n, \sigma, \pi \rangle$ will be accepted by PVerify and that π is indistinguishable from any $\pi' = \langle w', r', h', d' \rangle$ produced by running PGen on input $\langle x_P, X_V, m', \sigma' \rangle$. □

Proof of Theorem 3. Suppose that H_1 and H_2 are random oracles and there exists an algorithm E that makes at most μ_i queries to the random oracles $H_i, i = \{1, 2\}$, at most μ_s Sign queries, and wins the soundness game of Section 4.3 in time at most τ with probability at least $\eta = 10(\mu_s + 1)(\mu_s + \mu_2)/q$, where q is exponential in security parameter l.

We show how to construct an algorithm B that uses E to solve the discrete logarithm problem in G. B will simulate the random oracles and the challenger C in a game with E. B's goal is to solve the discrete logarithm problem on input $\langle g, X, p, q \rangle$, that is to find $x \in \mathbb{Z}_q$ such that $g^x = X \bmod p$, where g is of prime order q modulo prime p and generates group G.

Simulation:

B uses the parameters $\langle g, p, q \rangle$ to run Setup, and gives all the public parameters to E. B generates a set of participants U, where $|U| = \rho(l)$ and ρ is a polynomial function of the security parameter l. For some participant J, B sets $X_J = X$, and for each $I \neq J$, x_I is chosen randomly from $\mathbb{Z}_q$, and B sets $X_I = g^{x_I} \bmod p$. E is given all the public keys X_i. We define the set of all participants' public keys to be $\mathcal{X}$. B now simulates the challenger by simulating all the oracles which E can query as follows:

H_1-Queries: B simulates the random oracle by keeping a list of tuples $\langle M_i, r_i \rangle$ which is called L_{H_1}. When the oracle is queried with an input $M \in \{0, 1\}^*$, B responds as follows:
1. If the query M is already in L_{H_1} in the tuple $\langle M, r_i \rangle$, then B outputs $g^{r_i} \bmod p$.
2. Otherwise B selects a random $r \in \mathbb{Z}_q$, outputs $g^r \bmod p$ and adds $\langle M, r \rangle$ to L_{H_1}.

H_2-Queries: B simulates the H_2 oracle in the same way as H_1 by keeping a list of tuples L_{H_2}, but the tuples are of the form $\langle M, s \rangle$, where s is chosen randomly from $\mathbb{Z}_q$, and the output to a query on M is s.

EGen Queries: E can make EGen queries to any oracle I with public key X_I on input $\langle m \rangle$ where $m \in \mathcal{M}$. If $X_I \neq X_J$ then B runs Sign(x_I, m) to produce a signature $\sigma \in \mathcal{S}$ such that $(m, \sigma) \in L(X_I)$. If $X_I = X_J$ then B

queries m on the H_1 oracle and receives some g^{r_i} as response. B then sets $\sigma = X_I^{r_i} \bmod p$. B outputs $\langle m, \sigma \rangle$.

PGen Queries: E can make PGen queries to any oracle P with public key X_P on input $\langle X_V, m, \sigma \rangle$. If $X_P \neq X_J$ then B runs PGen on $\langle x_P, X_V, m, \sigma \rangle$ and outputs the response. If $X_P = X_J$ then B queries m on H_1 and receives some g^{r_i}. If $X_P^{r_i} \bmod p \neq \sigma$ then B outputs "invalid". Otherwise, B picks random $w, r, t, h \in \mathbb{Z}_q$ and computes:

$$c = g^w X_V^r \bmod p$$
$$G = g^d X_P^{(h+w)} \bmod p$$
$$M = H_1(m)^d \sigma^{(h+w)} \bmod p$$

If the H_2 oracle has previously been queried on input c, G, M, m, σ, X_P, then B starts again by picking new w, r, t, h. Otherwise B sets $M = "c, G, M, m, \sigma, X_P"$, adds the tuple $\langle M, h \rangle$ to L_{H_2} and outputs $\pi = \langle w, r, h, d \rangle$.

FakePGen Queries: E can make FakePGen queries to any oracle V with public key X_V on input $\langle X_P, m, \sigma \rangle$. If $X_V \neq X_J$ then B runs Algorithm A defined in the proof of Theorem 2 on $\langle X_P, x_V, m, \sigma \rangle$ and outputs the response. Otherwise B picks random $w, r, t, h \in \mathbb{Z}_q$ and computes:

$$c = g^w X_V^r \bmod p$$
$$G = g^d X_P^{(h+w)} \bmod p$$
$$M = H_1(m)^d \sigma^{(h+w)} \bmod p$$

If the H_2 oracle has previously been queried on input c, G, M, m, σ, X_P, then B starts again by picking new w, r, t, h. Otherwise B sets $M = "c, G, M, m, \sigma, X_P"$, adds the tuple $\langle M, h \rangle$ to L_{H_2} and outputs $\pi = \langle w, r, h, d \rangle$.

Corrupt Queries: E can make a Corrupt query to any oracle I with public key X_I. If $X_I = X_J$, then B aborts and terminates E. Otherwise B returns the appropriate private key x_I.

Output: On termination, E outputs $\langle X_P^*, X_V^*, m^*, \sigma^*, \pi^* \rangle$, where $X_P^*, X_V^* \in \mathcal{X}$, X_V^* is uncorrupted, $m^* \in \mathcal{M}$, $\sigma^* \in \mathcal{S}$ and $\pi^* \in \mathcal{P}$. E wins if $\langle X_P^*, X_V^*, m^*, \sigma^*, \pi^* \rangle$ is accepted by PVerify, $\pi^* = \langle w^*, r^*, h^*, d^* \rangle$ was not the output of some PGen query to P^* on $\langle X_V^*, m^*, \sigma^* \rangle$ or some FakePGen query to V^* on $\langle X_P^*, m^*, \sigma^* \rangle$, and either:
1. X_P^* is uncorrupted, or
2. $(m^*, \sigma^*) \notin L(X_P^*)$.

Case 1. Suppose that X_P^* is uncorrupted. If h^* was not output by any previous PGen or FakePGen query, then by the forking lemma of [18], with a certain probability B can repeat its simulation so that E outputs another tuple $\langle X_P^*, X_V^*, m^*, \sigma^*, \pi \rangle$ where proof $\pi = \langle w, r, h, d \rangle$ and $h \neq h^*$. We then get the equations

$$g^{w^*} X_V^{*\,r^*} = g^w X_V^r \bmod p \tag{1}$$

$$g^{d^*} X_P^{*\,(h^*+w^*)} = g^d X_P^{*\,(h+w)} \bmod p \tag{2}$$

$$H_1(m^*)^{d^*} \sigma^{*(h^*+w^*)} = H_1(m^*)^d \sigma^{*(h+w)} \bmod p. \tag{3}$$

Now if $X_V^* \neq X_V$ and $X_V \neq X_J$, or $r^* \neq r$ then B can solve (1) for the discrete logarithm of X_V^*. The probability that $X_V^* = X_J$ is $\frac{1}{\rho}$. If $X_V^* \neq X_V$ and $X_V = X_J$ then again B can solve (1) for the discrete logarithm of $X_V = X_J$.

If $X_V^* = X_V$ and $r^* = r$, then $w^* = w$, and since $h^* \neq h$ we have that $h^* + w^* \neq h + w$ so B can solve (2) for the discrete logarithm of X_P^*. The probability that $X_P^* = X_J$ is $\frac{1}{\rho}$.

If h^* was output by some previous PGen query to P^* on $\langle X_V, m^*, \sigma^* \rangle$ which produced proof $\pi = \langle w, r, h^*, d \rangle$, then since h and h^* were outputs from H_2, with overwhelming probability the inputs to H_2 were identical and we obtain the equations

$$g^{w^*} X_V^{*\,r^*} = g^w X_V^r \bmod p \tag{4}$$

$$g^{d^*} X_P^{*\,(h^*+w^*)} = g^d X_P^{*\,(h^*+w)} \bmod p \tag{5}$$

$$H_1(m^*)^{d^*} \sigma^{*(h^*+w^*)} = H_1(m^*)^d \sigma^{*(h^*+w)} \bmod p. \tag{6}$$

As for equation (1), if $X_V^* \neq X_V$ or $r^* \neq r$, then B can solve (4) for the discrete logarithm of X_J with probability $\frac{1}{\rho}$.

If $X_V^* = X_V$ and $r^* = r$, then $w^* = w$, so $h^* + w^* = h^* + w$ and therefore $d^* = d$. But this means that $\pi^* = \langle w^*, r^*, h^*, d^* \rangle$ was the output of some PGen query to P^* on $\langle X_V^*, m^*, \sigma^* \rangle$, contradicting our assumption.

If h^* was output by some previous FakePGen query to V^* on $\langle X_P, m^*, \sigma^* \rangle$ which produced proof $\pi = \langle w, r, h^*, d \rangle$, then again with overwhelming probability we obtain the equations

$$g^{w^*} X_V^{*\,r^*} = g^w X_V^{*\,r} \bmod p \tag{7}$$

$$g^{d^*} X_P^{*\,(h^*+w^*)} = g^d X_P^{(h^*+w)} \bmod p \tag{8}$$

$$H_1(m^*)^{d^*} \sigma^{*(h^*+w^*)} = H_1(m^*)^d \sigma^{*(h^*+w)} \bmod p. \tag{9}$$

Now if $r^* \neq r$ then B can solve (7) for the discrete logarithm of X_V^*. The probability that $X_V^* = X_J$ is $\frac{1}{\rho}$. If $X_P^* \neq X_P$ or $w^* \neq w$ then B can solve (8) for the discrete logarithm of X_P^* (or X_P). The probability that $X_P^* = X_J$ (or $X_P = X_J$) is $\frac{1}{\rho}$.

If $r^* = r$, $X_P^* = X_P$ and $w^* = w$, then $d^* = d$. But this means that $\pi^* = \langle w^*, r^*, h^*, d^* \rangle$ was the output of some FakePGen query to V^* on $\langle X_P^*, m^*, \sigma^* \rangle$, contradicting our assumption.

Case 2. Suppose now that $(m^*, \sigma^*) \notin L(X_P^*)$. If h^* was not output by any previous PGen or FakePGen query, then by the forking lemma of [18], with a certain probability B can repeat its simulation so that E outputs another tuple

$\langle X_P^*, X_V, m^*, \sigma^*, \pi \rangle$ where proof $\pi = \langle w, r, h, d \rangle$ and $h \neq h^*$. We then get the equations

$$g^{w^*} X_V^{* \, r^*} = g^w X_V^r \bmod p \tag{10}$$

$$g^{d^*} X_P^{* \, (h^*+w^*)} = g^d X_P^{* \, (h+w)} \bmod p \tag{11}$$

$$H_1(m^*)^{d^*} \sigma^{*(h^*+w^*)} = H_1(m^*)^d \sigma^{*(h+w)} \bmod p. \tag{12}$$

As for equation (1), if $X_V^* \neq X_V$ or $r^* \neq r$ then B can solve (10) for the discrete logarithm of X_J with probability $\frac{1}{\rho}$.

If $X_V^* = X_V$ and $r^* = r$, then $w^* = w$. But since $h^* \neq h$ we can rewrite equations (11) and (12) as $X_P^* = g^{\frac{d-d^*}{h^*-h}}$ and $\sigma^* = H_1(m^*)^{\frac{d-d^*}{h^*-h}}$. But we assumed that $(m^*, \sigma^*) \notin L(X_P^*)$, contradicting our assumption.

Now h^* cannot have been output by some PGen query to P^* on $\langle X_V, m^*, \sigma^* \rangle$ since $(m^*, \sigma^*) \notin L(X_P^*)$ and the PGen query would have output "invalid" in this case.

If h^* was output by some previous FakePGen query to V^* on $\langle X_P, m^*, \sigma^* \rangle$ which produced proof $\pi = \langle w, r, h^*, d \rangle$, then again with overwhelming probability we obtain the equations

$$g^{w^*} X_V^{* \, r^*} = g^w X_V^{* \, r} \bmod p \tag{13}$$

$$g^{d^*} X_P^{* \, (h^*+w^*)} = g^d X_P^{(h^*+w)} \bmod p \tag{14}$$

$$H_1(m^*)^{d^*} \sigma^{*(h^*+w^*)} = H_1(m^*)^d \sigma^{*(h^*+w)} \bmod p. \tag{15}$$

Now if $r^* \neq r$ then B can solve (13) for the discrete logarithm of X_V. $X_V = X_J$ with probability $\frac{1}{\rho}$. As for (8), if $X_P^* \neq X_P$ or $w^* \neq w$ then B can solve (14) for the discrete logarithm of X_J with probability $\frac{1}{\rho}$.

If $r^* = r$, $X_P^* = X_P$ and $w^* = w$, then $d^* = d$. But this means that $\pi^* = \langle w^*, r^*, h^*, d^* \rangle$ was the output of some FakePGen query to V^* on $\langle X_P^*, m^*, \sigma^* \rangle$, contradicting our assumption.

Since the forking lemma produces a second appropriate signature with expected time at most $\tau' = 120686\mu_s\tau$, we find that B can solve the discrete logarithm problem in time at most τ' and with probability at least η/ρ, which is non-negligible, contradicting the hardness of the discrete logarithm problem. $\square$

Partial Key Recovery Attacks on XCBC, TMAC and OMAC

Chris J. Mitchell

Information Security Group, Royal Holloway, University of London
c.mitchell@rhul.ac.uk
http://www.isg.rhul.ac.uk/~cjm

Abstract. The security provided by the XCBC, TMAC and OMAC schemes is analysed and compared with other MAC schemes. In particular, 'partial' key recovery attacks against all three of these schemes are described, yielding upper bounds on the effective security level. The results imply that there is relatively little to be gained practically through the introduction of these schemes by comparison with other well-established MAC functions.

1 Introduction

In this paper the security of three related methods for computing Message Authentication Codes (MACs) is analysed and compared with the level of security provided by other, more well-established, MACing techniques. The security analysis is given in terms of the most efficient (known) forgery and key recovery attacks that can be launched against the schemes.

The three MAC schemes considered here are known as XCBC [1], OMAC [2] and TMAC [3] (see also [4]). These three schemes are all examples of CBC-MACs, i.e. they are all based on the use of a block cipher in Cipher Block Chaining Mode — see, for example, [5]. Various CBC-MAC schemes have been in wide use for many years for protecting the integrity and guaranteeing the origin of data.

Note that all three of these new schemes have been specifically designed for use with messages of variable length, with the goal of minimising the number of block cipher operations required to compute a MAC. We compare the efficiency and security of these MAC schemes with two other schemes also designed for messages of arbitrary length, namely EMAC [6] and the ANSI retail MAC [7], also known as MAC algorithms 2 and 3 (respectively) from ISO/IEC 9797-1 [8].

2 Key Recovery and Forgery Attacks

There are two main classes of attack on a MAC scheme, namely *key recovery* attacks, in which an attacker is able to discover the secret key used to compute the MACs, and *forgery attacks* in which an attacker is able to determine the correct MAC for a message (without a legitimate key holder having generated

N.P. Smart (Ed.): Cryptography and Coding 2005, LNCS 3796, pp. 155–167, 2005.

it). Key recovery attacks are clearly more powerful than forgery attacks since once the key is known arbitrary forgeries are possible. We also consider *partial key recovery attacks* in which an attacker is able to obtain part of the secret key.

Using a simplified version of the approach of [8], we use a three-tuple $[a, b, c]$ to quantify the resources needed for an attack, where a denotes the number of off-line block cipher encipherments (or decipherments), b denotes the number of known data string/MAC pairs, and c denotes the number of chosen data string/MAC pairs. In each case, the resources given are those necessary to ensure that an attack has a probability of successful completion greater than 0.5.

3   XCBC and Some Simple Attacks

The XCBC scheme was originally proposed by Black and Rogaway in 2000 [1], with the objective of providing a provably secure CBC-MAC scheme which minimises the number of block cipher encryptions and decryptions.

3.1   Definition

The XCBC scheme operates as follows. First (as throughout) suppose that the underlying block cipher transforms an n-bit block of plaintext into an n-bit block of ciphertext (i.e. it is an n-bit block cipher), and that it uses a key of k bits. If X is an n-bit block then we write $e_K(X)$ (or $d_K(X)$) for the block cipher encryption (or decryption) of the n-bit block X using key K.

The XCBC MAC scheme uses a triple of keys (K_1, K_2, K_3) where K_1 is a block cipher key, i.e. it contains k bits, and K_2, K_3 are both n-bit strings. The XCBC MAC computation is as follows.

The message D on which the MAC is to be computed is padded and split into a sequence of q n-bit blocks: $D_1, D_2, \ldots, D_q$. Note that there are two possibilities for the padding process. If the bit-length of the message is already an integer multiple of n then no padding is performed. However, if the bit-length of the message is not a multiple of n then the padded message consists of the message concatenated with a single one bit followed by the minimal number of zeros necessary to make the bit-length of the padded message a multiple of n. (Note that this padding strategy is not a 1-1 mapping of messages to padded messages; however, problems are avoided by the use of two different MAC computation strategies, as described immediately below).

The computation of the MAC depends on whether or not padding has been necessary. In the first case, i.e. where no padding is necessary, the MAC computation is as follows:

$$H_1 = e_{K_1}(D_1),$$
$$H_i = e_{K_1}(D_i \oplus H_{i-1}), \quad (2 \leq i \leq q - 1), \text{ and}$$
$$\text{MAC} = e_{K_1}(D_q \oplus H_{q-1} \oplus K_2).$$

In the second case, i.e. where padding is applied, the MAC computation is as follows:

$$H_1 = e_{K_1}(D_1),$$
$$H_i = e_{K_1}(D_i \oplus H_{i-1}), \quad (2 \le i \le q-1), \text{ and}$$
$$\text{MAC} = e_{K_1}(D_q \oplus H_{q-1} \oplus K_3).$$

That is, the keys K_2 and K_3 are ex-ored with the final plaintext block depending on whether or not padding is necessary.

Note that the MAC used will be truncated to the left-most m bits of the MAC value given in the above equation, where $m \le n$. In this paper we only consider the case where $m = n$, i.e. where no truncation is performed.

Before proceeding we observe that other authors have also considered the security of XCBC and related schemes. In particular, Furuya and Sakurai [9] have considered various attacks against 2-key variants of XCBC. However, some of the previous work (including that in [9]) has focussed on weaknesses arising from particular choices for the underlying block cipher. This contrasts with the approach followed in this paper which considers attacks independent of the block cipher. Note also that the attacks described below do not contradict the proofs of security for XCBC, TMAC and OMAC — they simply establish the tightness of the results; nevertheless, the existence of 'partial key recovery attacks', as described herein, is something that is both undesirable and not evident from theoretical analysis of the schemes.

3.2 Forgery Attacks on XCBC

Suppose a fixed key triple (K_1, K_2, K_3) is in use for computing XCBC-MACs. Let $D_1, D_2, \ldots, D_q$ be any sequence of n-bit blocks, where $q \ge 0$ is arbitrary. Suppose (by some means) an attacker learns the MACs for $2^{n/2}$ different messages which, after padding and splitting into a sequence of n-bit blocks, all have $q+1$ blocks, and whose first q blocks are $D_1, D_2, \ldots, D_q$; i.e. all the padded messages have the form $D_1, D_2, \ldots, D_q, X$, for some n-bit block X. Because padding has been applied the MACs will all be computed using the key K_3. (Note that, since padding is applied, all of these messages must have unpadded bit-length ℓ satisfying $qn < \ell < (q+1)n$).

Suppose that the attacker also has the MACs for a further $2^{n/2}$ different messages to which padding is not applied and which, after division into a sequence of n-bit blocks, have the form $D_1, D_2, \ldots, D_q, Y$, for some n-bit block Y. Because padding has not been applied, the MACs will all be computed using the key K_2. Note that, since no padding is applied, all these messages will have length precisely $(n+1)q$ bits.

The total number of message/MAC pairs required is clearly $2^{n/2+1}$. In the discussion below we 'cheat' slightly and refer to these as 'known MACs' rather than 'chosen MACs'. The justification for this is that if $q = 0$ then we do not impose any conditions on the messages for which MACs are required (except for their lengths). Also, there may be applications where the first part of a message is fixed, and only the last block is variable — again in such a case the required message/MAC pairs can be obtained without choosing the messages.

By the usual birthday paradox probability arguments (see, for example, Sect. 2.1.5 of [5] or [10]), with a probability of approximately $1 - e^{-1} \simeq 0.63$ one of the MACs from the first set of messages will equal one of the MACs from the second set. Suppose the pair of messages concerned are respectively $D_1, D_2, \ldots, D_q, X^*$ and $D_1, D_2, \ldots, D_q, Y^*$ for some n-bit blocks X^* and Y^*.

Before proceeding, suppose that Q is the 'simple' CBC-MAC for the q-block message $D_1, D_2, \ldots, D_q$, i.e. if

$$H_1 = e_{K_1}(D_1), \text{ and}$$
$$H_i = e_{K_1}(D_i \oplus H_{i-1}), \quad (2 \leq i \leq q)$$

then $Q = H_q$.

Then, by definition, we immediately have that

$$e_{K_1}(Q \oplus X^* \oplus K_3) = e_{K_1}(Q \oplus Y^* \oplus K_2).$$

Hence, since encryption with a fixed key is a permutation of the set of all n-bit blocks, we have $Q \oplus X^* \oplus K_3 = Q \oplus Y^* \oplus K_2$, i.e. $X^* \oplus Y^* = K_2 \oplus K_3$. That is, the attacker has learnt the value of $K_2 \oplus K_3$. Knowledge of this value immediately enables forgeries to be computed.

Specifically, suppose $(D_1, D_2, \ldots, D_q)$ is the padded version of a message (of unpadded length ℓ satisfying $(q-1)n < \ell < qn$) for which the MAC M is known. Then the **unpadded** message $(D_1, D_2, \ldots, D_q \oplus K_2 \oplus K_3)$ also has MAC M. The overall complexity of this forgery attack is $[0, 2^{n/2+1}, 0]$.

Note that the above is, in some sense, also a partial key recovery attack, since the attacker has reduced the number of unknown key bits from $k + 2n$ to $k + n$. However, to simplify the presentation below we do not consider this further here.

3.3   Key Recovery Attacks on XCBC

We describe two main types of key recovery attack. The first attack (essentially based on the Preneel-van Oorschot attack [11,12]) requires a significant number of known MACs and 'only' 2^k block cipher operations. The second attack (a 'meet-in-the-middle' attack) requires minimal numbers of known MACs, but potentially larger numbers of block cipher operations (and more storage).

The first attack is as follows. Suppose an attacker knows the MACs for $2^{n/2}$ different messages of length less than n bits. Thus, after padding and division into n-bit blocks, all these messages will consist of one block. Suppose the attacker also knows the MACs for a further $2^{n/2}$ different messages of bit-length ℓ satisfying $n < \ell < 2n$, i.e. messages which, after padding, contain two blocks.

Exactly as above, there is a good chance (probability $\simeq 0.63$) that a message from the first set will have the same MAC as a message from the second set. Suppose that the one-block and two-block messages concerned are X and $(Y,$

Z) respectively. Since both messages involve padding, key K_3 is used in both cases. Then we know that

$$e_{K_1}(K_3 \oplus X) = e_{K_1}(K_3 \oplus Z \oplus e_{K_1}(Y)).$$

Hence, since e_{K_1} is a permutation on the set of all n-bit blocks, we have $K_3 \oplus X = K_3 \oplus Z \oplus e_{K_1}(Y)$, i.e. $X \oplus Z = e_{K_1}(Y)$. It is now possible (at least in principle) to perform an exhaustive search through all possible values for K_1, and as long as $k < n$ it is likely that only the correct value will satisfy this equation. If $k \geq n$ then a number of 'false' matches will be found — however, these can be eliminated in the next stage with minimal effort.

Given a candidate for K_1, any of the known MACs for one-block messages can be decrypted using this value of K_1 to reveal the value of K_3. This candidate key pair can then be tested on a further known MAC, and all false keys can quickly be eliminated (the complexity of this step does not affect the overall attack complexity since it is only conducted when a candidate for K_1 is found, which will only happen occasionally). The total expected complexity of this first key recovery attack is thus $[2^k, 2^{n/2+1}, 0]$, since the correct key will certainly be found by the time an exhaustive search of the key space is complete (given that a MAC collision has been found).

For the second attack we present just one variant (many other meet-in-the-middle variants exist). Suppose that the attacker has access to $\lceil (k+n)/n \rceil$ known single-block message/MAC pairs all of which involve no padding (and hence K_2 is used), together with a single known message/MAC pair for which padding is applied, i.e. a total of $\lceil (k+2n)/n \rceil$ known MACs. One interesting point regarding the attack we now describe is that negligible storage is required, unlike similar attacks on EMAC (although this would no longer be true if the messages contained more than one block).

The attacker chooses a single-block message for which padding is not used — suppose the message is D and the MAC is M. The attacker first goes through all 2^k possible values for the key K_1 and computes $d_{K_1^*}(M) \oplus D = K_2^*$ for each candidate value K_1^*. The pair (K_1^*, K_2^*) is then tested as a candidate for (K_1, K_2) using a second known message/MAC pair for which padding was not used. This will require a single block cipher operation, and almost all incorrect candidate key pairs will be eliminated. By means of further tests against known message/MAC pairs (from the set of $\lceil (k+n)/n \rceil$), with high probability all but the correct key pair can be eliminated. The single remaining known MAC can be used to derive K_3. The total complexity of this attack is thus $[2^{k+1}, \lceil (k+2n)/n \rceil, 0]$. (Note that this attack requires only four times as many block cipher operations as the previous attack, and requires only a handful of known MACs compared to a very large number for the previous attack).

4   TMAC and Its Security

The TMAC scheme, a simple variant of XCBC, was proposed by Kurosawa and Iwata [2] with the goal of reducing the number of required keys from three to two.

4.1   Definition of TMAC

The TMAC scheme operates in exactly the same way as XCBC except that it only uses a key pair (K, K') instead of a key triple, where K is a k-bit block cipher key and K' contains n bits. A key triple (K_1, K_2, K_3), as used by XCBC, is then derived from (K, K') by setting $K_1 = K$, $K_2 = u.K'$ and $K_3 = K'$, where u is a constant (defined in [2]) and multiplication by u takes place in a specific representation of the finite field of 2^n elements (also specified in [2]).

4.2   Forgery Attacks on TMAC

Clearly the forgery attack on XCBC described in Sect. 3.2 will also apply to TMAC. There does not appear to be any obvious way in which to take advantage of the added structure in TMAC to make such an attack more efficient.

4.3   Key Recovery Attacks on TMAC

Again, both the key recovery attacks on XCBC described in Sect. 3.3 will also apply to TMAC.

There also exists a *partial* key recovery attack which will yield the key K' rather more simply than the entire key can be obtained — most importantly this attack does not require a search through the entire key space.

Suppose that the attacker performs the forgery attack described in Sect. 3.2. Then, the attacker will learn the value of $K_2 \oplus K_3 = S$, say. However, in the case of TMAC, we also know that $K_2 = u.K_3$, where multiplication by the public constant u is defined over the finite field of 2^n elements. Thus $K_3 = S.(u+1)^{-1}$, and hence the attacker can learn the values of both K_2 and K_3. The total complexity of this partial key recovery attack is thus the same as that of the forgery attack, i.e. $[0, 2^{n/2+1}, 0]$. (Note that this yields another full key recovery attack with complexity $[2^{k-1}, 2^{n/2+1}, 0]$).

Before proceeding we consider the implications of knowledge of K_2 and K_3. At first glance it is not obvious that this is any worse than knowing the value of $K_2 \oplus K_3$, which already enables simple forgeries. However, it is more serious since it enables a far wider range of forgeries to be performed. For example, suppose the (unpadded) message $D_1, D_2, \ldots, D_q$ has MAC M, i.e. $M = e_{K_1}(K_2 \oplus D_q \oplus H_{q-1})$, where H_{q-1} is defined as above. Then, if message $E_1, E_2, \ldots, E_r$ has MAC N, it is not hard to see that the message

$$D_1, D_2, \ldots, D_{q-1}, D_q \oplus K_2, E_1 \oplus M, E_2, E_3, \ldots, E_r$$

also has MAC N.

Finally note that this partial key recovery attack against TMAC has previously been described by Sung, Hong and Lee [13].

4.4   Improving TMAC

The main reason that TMAC is significantly weaker than XCBC is the fact that a simple algebraic relationship exists between K_2 and K_3. This not only enables

K_2 to be trivially deduced from K_3 (and vice versa), it also enables a second linear equation in K_2 and K_3 to be used to deduce both K_2 and K_3.

However, there is no reason for such a simple relationship to exist between K_2 and K_3. One way of avoiding this would be to cryptographically derive both K_2 and K_3 from the single key K'. One way in which this could be done would be to define two different fixed n-bit strings, S_2 and S_3 say, and to put $K_2 = e_{K'}(S_2)$ and $K_3 = e_{K'}(S_3)$. With this definition, knowledge of one of K_2 (or K_3) will not enable K_3 (or K_2) to be deduced, as long as the block cipher e resists known ciphertext attacks. Also, knowledge of $K_2 \oplus K_3$ will also not enable K_2 and K_3 to be deduced (again assuming that e resists known ciphertext attacks). This change would, however, mean that K' contains k rather than n bits.

Of course, this change invalidates the security proof for TMAC. Moreover, an analogous change proposed for OMAC (see Sect. 5.4) has been criticised by Iwata and Kurosawa [14,15].

5   The Security of OMAC

The OMAC scheme, a further simple variant of XCBC, was proposed by Iwata and Kurosawa [3] with the goal of further reducing the number of required keys from three to one. This scheme has recently been adopted by NIST under the title of CMAC [16].

5.1   Definition of OMAC

The OMAC scheme operates in exactly the same way as XCBC except that it only uses a single key K instead of a key triple, where K is a k-bit block cipher key. A key triple (K_1, K_2, K_3), as used by XCBC, is then derived from K by setting $L = e_K(0^n)$, $K_1 = K$, $K_2 = u.L$ and $K_3 = u^2.L$, where 0^n is the n-bit block of all zeros, and u is a constant (defined in [3]) and multiplication by u and u^2 takes place in a specific representation of the finite field of 2^n elements (also specified in [3]).

Note that there are, in fact, two different variants of OMAC, known as OMAC1 and OMAC2. The version defined above is OMAC1, and is the one analysed here. However the analysis is almost identical for OMAC2, which is identical to OMAC1 except that $K_3 = u^{-1}.L$.

5.2   Forgery Attacks on OMAC

Just as for TMAC, the forgery attack on XCBC described in Sect. 3.2 will also apply to OMAC. There does not appear to be any obvious way in which to take advantage of the added structure in OMAC to make such an attack more efficient.

5.3   Key Recovery Attacks on OMAC

Both the key recovery attacks on XCBC described in Sect. 3.3 will also apply to OMAC, as will a simple variant of the partial key recovery attack described in Sect. 4.3. In this case however, a second partial key recovery attack exists, which we now describe. This attack is designed to enable L to be determined, knowledge of which immediately enables both K_2 and K_3 to be determined. The attack is similar to that described in Sect. 3.3.

Suppose an attacker knows the MACs for $2^{n/2}$ different messages of length less than n bits. Thus, after padding and division into n-bit blocks, all these messages will consist of one block. Suppose the attacker also knows the MACs for a further $2^{n/2}$ different messages of bit-length ℓ satisfying $n < \ell < 2n$, i.e. messages which, after padding, contain two blocks, and for which the first n bits are all zero.

Exactly as above, there is a good chance (probability $\simeq 0.63$) that a message from the first set will have the same MAC as a message from the second set. Suppose that the one-block and two-block messages concerned are X and $(0^n, Z)$ respectively (recall that the second message must begin with n zeros). Since both messages involve padding, key K_3 is used in both cases. Then we know that

$$e_{K_1}(K_3 \oplus X) = e_{K_1}(K_3 \oplus Z \oplus e_{K_1}(0^n)).$$

Hence, since e_{K_1} is a permutation on the set of all n-bit blocks, we have $K_3 \oplus X = K_3 \oplus Z \oplus e_{K_1}(0^n)$, i.e. $X \oplus Z = e_{K_1}(0^n)$. But $K_1 = K$, and thus we know that $L = X \oplus Z$. Thus L, and hence K_2 and K_3, are immediately available to the attacker. This latter attack has complexity $[0, 2^{n/2}, 2^{n/2}]$.

5.4   Improving OMAC

Analogously to the proposed improvements to TMAC (given in Sect. 4.4), one possibility would be to derive K_2 and K_3 from K using the following process: put $K_2 = e_K(S_2)$ and $K_3 = e_K(S_3)$, where S_2 and S_3 are fixed and distinct n-bit strings. This will avoid the attack described in Sect. 4.3. In addition, in order to avoid the OMAC-specific attack described in Sect. 5.3, it is suggested that the key K_1 used in MAC computations should not be the same as the key used to derive K_2 and K_3, to prevent MAC computations accidentally revealing K_2 and/or K_3. This is simple to achieve by setting $K_1 = K \oplus S_1$ for a fixed k-bit string S_1.

However, we note that the above simple approach to modifying OMAC has been criticised by Iwata and Kurosawa [14,15], who show that a proof of security cannot be obtained for the modified scheme using the 'standard' assumptions about the underlying block cipher. This suggests that it would be interesting to develop variants of TMAC and OMAC for which proofs of security can be readily developed, and which nevertheless do not permit 'partial key recovery attacks' of the type described.

6   Benchmark Results and Comparisons

We next consider the security provided by two well-known and standardised CBC-MAC schemes, namely EMAC and the ANSI retail MAC. Note that, unlike XCBC, TMAC and OMAC, these schemes operate independently of whether or not a message is padded and how padding is performed.

6.1   EMAC

EMAC is standardised as MAC algorithm 2 in ISO/IEC 9797-1 [8], and has been proven secure by Petrank and Rackoff [17]. EMAC uses a pair of keys (K_1, K_2) where K_1 and K_2 are both block cipher keys, i.e. they contain k bits. A message D is first padded and split into a sequence of q n-bit blocks: $D_1, D_2, \ldots, D_q$.

The EMAC computation, which essentially involves double encrypting the final block, is as follows:

$$H_1 = e_{K_1}(D_1),$$
$$H_i = e_{K_1}(D_i \oplus H_{i-1}), \quad (2 \leq i \leq q - 1), \text{ and}$$
$$\text{MAC} = e_{K_2}(e_{K_1}(D_q \oplus H_{q-1})).$$

As summarised in [8], the most effective (known) forgery attack against EMAC has complexity $[0, 2^{n/2}, 1]$ and the best key recovery attacks have complexity either $[2^{k+1}, 2^{n/2}, 0]$ or $[s.2^k, \lceil 2k/n \rceil, 0]$ (for some small value of s), where the second attack requires $O(2^k)$ storage. Note that the second attack is a meet-in-the-middle attack.

6.2   ANSI Retail MAC

The ANSI retail MAC (abbreviated as ARMAC below) is standardised as MAC algorithm 3 in ISO/IEC 9797-1 [8]. Note that this scheme is widely used with the block cipher DES (which has $n = 64$ and $k = 56$) in environments where obtaining $2^{n/2} = 2^{32}$ message/MAC pairs is deemed infeasible. However, it seems that a security proof for this scheme does not exist. Nevertheless, since it closely resembles EMAC, heuristically one might expect a similar level of provable security.

This MAC scheme again uses a pair of keys (K_1, K_2) where K_1 and K_2 are both block cipher keys, i.e. they contain k bits. A message D is first padded and split into a sequence of q n-bit blocks: $D_1, D_2, \ldots, D_q$. The MAC computation is as follows:

$$H_1 = e_{K_1}(D_1),$$
$$H_i = e_{K_1}(D_i \oplus H_{i-1}), \quad (2 \leq i \leq q - 1), \text{ and}$$
$$\text{MAC} = e_{K_1}(d_{K_2}(e_{K_1}(D_q \oplus H_{q-1}))).$$

As summarised in [8], the best known forgery attack against the ANSI retail MAC has complexity $[0, 2^{n/2}, 1]$ and the best-known key recovery attack has complexity $[2^{k+1}, 2^{n/2}, 0]$ — note that one attraction of ARMAC is that it does not appear to be subject to meet-in-the-middle attacks.

6.3   Comparisons

We compare the three 'new' MAC algorithms, i.e. XCBC, OMAC and TMAC, with the two longer-established schemes with respect to two different criteria: efficiency and security.

Efficiency can be further sub-divided into two different categories: key length, and the number of block cipher operations required to compute the MAC for a message. The key lengths for the five MAC schemes considered here are given in Table 1.

Table 1. Key lengths

XCBC	TMAC	OMAC	EMAC	ARMAC
$k + 2n$	$k + n$	k	$2k$	$2k$

The number of block cipher operations (encryptions or decryptions) required to compute the MAC for a message is specified in Table 2. Note that it is assumed that EMAC and ARMAC are used with the "always add a '1' and then as many zeros as necessary" padding method, which is standardised as padding method 2 in ISO/IEC 9797-1 [8].

Table 2. Computational complexity (block cipher operations)

No. of data bits (ℓ)	XCBC	TMAC	OMAC	EMAC	ARMAC
$(t - 1)n < \ell < tn$	t	t	t	$t + 1$	$t + 2$
$\ell = tn$	t	t	t	$t + 2$	$t + 3$

From Table 2 it should be clear that XCBC, TMAC and OMAC all offer workload advantages over EMAC and ARMAC. This workload advantage is slightly increased by the fact that XCBC, TMAC and OMAC only require one block cipher key 'set up' per MAC computation, whereas EMAC and OMAC require two — the difference this makes depends on the block cipher in use. However, these advantages are probably insignificant for messages that are more than a few blocks long, and even for short messages they are unlikely to be a major issue; banking networks have been using ARMAC with a relatively slow block cipher such as DES for many years for very large numbers of messages, using relatively primitive hardware.

We sub-divide the security comparison into three sub-categories, covering forgery attacks, key recovery attacks and partial key recovery attacks. The complexities of forgery attacks against the five MAC schemes considered here are specified in Table 3.

The complexities of key recovery attacks are specified in Table 4. Note that this table does not take account of the fact that the complexities of the second

Table 3. Forgery attack complexities

XCBC	TMAC	OMAC	EMAC	ARMAC
$[0,2^{n/2+1},0]$	$[0,2^{n/2+1},0]$	$[0,2^{n/2+1},0]$	$[0,2^{n/2},1]$	$[0,2^{n/2},1]$

Table 4. Key recovery attack complexities

XCBC	TMAC	OMAC	EMAC	ARMAC
$[2^k,2^{n/2+1},0]$	$[2^k,2^{n/2+1},0]$	$[2^k,2^{n/2+1},0]$	$[2^{k+1},2^{n/2},0]$	$[2^{k+1},2^{n/2},0]$
$[2^{k+1},\lceil(k+2n)/n\rceil,0]$	$[2^{k+1},\lceil(k+n)/n\rceil,0]$	$[2^{k+1},\lceil k/n\rceil,0]$	$[s.2^k,\lceil 2k/n\rceil,0]$	

attacks for XCBC, TMAC and OMAC require no significant storage, whereas the second attack against EMAC requires around $O(2^k)$ storage.

Finally, the complexities of partial key recovery attacks (where they exist) are specified in Table 5.

Table 5. Partial key recovery attack complexities

XCBC	TMAC	OMAC	EMAC	ARMAC
—	$[0,2^{n/2+1},0]$	$[0,2^{n/2+1},0]$	—	—
		$[0,2^n,2^n]$		

7 Conclusions

It should be clear from the analysis above that, in terms of security, XCBC, TMAC and OMAC offer no significant advantage by comparison with EMAC and ARMAC. Moreover, in some cases, they would appear to be weaker, although the most significant weaknesses of TMAC and OMAC might be avoided by changing the key derivation procedure[1]. Unfortunately, such changes invalidate the proofs of security for these schemes, and a simple proposed change of this type has been heavily criticised. It would be interesting to investigate other possible changes to the key derivation process.

Nevertheless, XCBC, TMAC and OMAC do offer a small practical advantage in terms of a modest reduction in the number of block cipher operations, although this is unlikely to be significant in most applications. In summary, there does not appear to be a compelling case for standardising these new CBC-MAC schemes.

---

[1] Note that the reference to weaknesses should not be interpreted as implying that these schemes are 'weak' — the existing security proofs establish their robustness as long as $2^{n/2}$ MACs are not available to an attacker and k is sufficiently large (e.g. $k \geq 128$).

References

1. Black, J., Rogaway, P.: CBC-MACs for arbitrary length messages: The three-key constructions. In Bellare, M., ed.: Advances in Cryptology — Crypto 2000. Volume 1880 of Lecture Notes in Computer Science., Springer-Verlag, Berlin (2000) 197–215

2. Kurosawa, K., Iwata, T.: TMAC: Two-key CBC MAC. In Joye, M., ed.: Topics in Cryptology — CT-RSA 2003. Volume 2612 of Lecture Notes in Computer Science., Springer-Verlag, Berlin (2003) 33–49

3. Iwata, T., Kurosawa, K.: OMAC: One-key CBC MAC. In Johansson, T., ed.: Fast Software Encryption, 10th International Workshop, FSE 2003, Lund, Sweden, February 24-26, 2003, Revised Papers. Volume 2889 of Lecture Notes in Computer Science., Springer-Verlag, Berlin (2003) 129–153

4. Iwata, T., Kurosawa, K.: Stronger security bounds for OMAC, TMAC and XCBC. In Johansson, T., Maitra, S., eds.: Progress in Cryptology — INDOCRYPT 2003, 4th International Conference on Cryptology in India, New Delhi, India, December 8-10, 2003, Proceedings. Volume 2904 of Lecture Notes in Computer Science., Springer-Verlag, Berlin (2003) 402–415

5. Menezes, A.J., van Oorschot, P.C., Vanstone, S.A.: Handbook of Applied Cryptography. CRC Press, Boca Raton (1997)

6. Berendschot, A., den Boer, B., Boly, J.P., Bosselaers, A., Brandt, J., Chaum, D., Damgard, I., Dichtl, M., Fumy, W., van der Ham, M., Jansen, C.J.A., Landrock, P., Preneel, B., Roelofsen, G., de Rooij, P., Vandewalle, J.: Integrity primitives for secure information systems. Volume 1007 of Lecture Notes in Computer Science. Springer-Verlag, Berlin (1995)

7. American Bankers Association Washington, DC: ANSI X9.19, Financial institution retail message authentication. (1986)

8. International Organization for Standardization Genève, Switzerland: ISO/IEC 9797-1, Information technology — Security techniques — Message Authentication Codes (MACs) — Part 1: Mechanisms using a block cipher. (1999)

9. Furuya, S., Sakurai, K.: Risks with raw-key masking — The security evaluation of 2-key XCBC. In Deng, R.H., Qing, S., Bao, F., Zhou, J., eds.: Information and Communications Security, 4th International Conference, ICICS 2002. Volume 2513 of Lecture Notes in Computer Science., Springer-Verlag, Berlin (2002) 327–341

10. Girault, M., Cohen, R., Campana, M.: A generalized birthday attack. In Guenther, C., ed.: Advances in Cryptology — EUROCRYPT '88, Workshop on the Theory and Application of Cryptographic Techniques, Davos, Switzerland, May 25-27, 1988, Proceedings. Volume 330 of Lecture Notes in Computer Science., Springer-Verlag, Berlin (1988) 129–156

11. Preneel, B., van Oorschot, P.C.: A key recovery attack on the ANSI X9.19 retail MAC. Electronics Letters **32** (1996) 1568–1569

12. Preneel, B., van Oorschot, P.C.: On the security of iterated Message Authentication Codes. IEEE Transactions on Information Theory **45** (1999) 188–199

13. Sung, J., Hong, D., Lee, S.: Key recovery attacks on the RMAC, TMAC, and IACBC. In Safavi-Naini, R., Seberry, J., eds.: ACISP 2003. Volume 2727 of Lecture Notes in Computer Science., Springer-Verlag, Berlin (2003) 265–273

14. Iwata, T., Kurosawa, K.: On the security of a new variant of OMAC. In Lim, J.I., Lee, D.H., eds.: Information Security and Cryptology — ICISC 2003, 6th International Conference, Seoul, Korea, November 27-28, 2003, Revised Papers. Volume 2971 of Lecture Notes in Computer Science., Springer-Verlag, Berlin (2003) 67–78

15. Iwata, T., Kurosawa, K.: On the security of a MAC by Mitchell. IEICE Trans. Fundamentals **E88-A** (2005) 25–32
16. National Institute of Standards and Technology (NIST): NIST Special Publication 800-38B, Recommendation for Block Cipher Modes of Operation: The CMAC Mode for Authentication. (2005)
17. Petrank, E., Rackoff, C.: CBC MAC for real-time data sources. Journal of Cryptology **13** (2000) 315–338

Domain Expansion of MACs: Alternative Uses of the FIL-MAC[*]

Ueli Maurer and Johan Sjödin

Department of Computer Science,
Swiss Federal Institute of Technology (ETH), Zurich,
CH-8092 Zurich, Switzerland
{maurer, sjoedin}@inf.ethz.ch

Abstract. In this paper, a study of a paradigm for domain expansion of MACs is generalized. In particular, a tradeoff between the efficiency of a MAC and the tightness of its security reduction is investigated in detail. Our new on-line single-key AIL-MAC construction, the PDI-construction, transforms any FIL-MAC into an AIL-MAC and is superior to all previous AIL-MAC constructions given in the literature (taking the tradeoff into account). It appears obvious that this construction is essentially optimal.

Keywords: Message authentication code (MAC), arbitrary-input-length (AIL), variable-input-length (VIL), fixed-input-length (FIL).

1   Introduction

1.1   Motivation: Data Integrity

A message authentication code (MAC) is a function family

$$H := \{h_k : \mathcal{M} \to \mathcal{T}\}_{k \in \mathcal{K}},$$

where $\mathcal{M}$ is the message space, $\mathcal{T}$ the tag space, and $\mathcal{K}$ the key space. It is the most commonly used method for assuring the integrity of data communicated between two parties sharing a secret key k. A party authenticates a message m by computing a tag $\tau = h_k(m)$ which is sent along with m to the other party. A party receiving (m', τ') accepts the message m' if (m', τ') is valid, i.e., satisfies $\tau' = h_k(m')$. Of course, it should be infeasible for a party not in possession of k to be able to generate a valid message-tag pair (which is new), since this would contradict data integrity. The function h_k is referred to as an *instantiation* of the MAC H.

---

[*] This work was partially supported by the Zurich Information Security Center. It represents the views of the authors.

N.P. Smart (Ed.): Cryptography and Coding 2005, LNCS 3796, pp. 168–185, 2005.

1.2   Domain Expansion of MACs

Cryptographic primitives can be classified according to their domain. We refer to a primitive with domain:

- $\{0,1\}^L$, i.e., the set of all bitstrings of length L, as a *fixed-input-length* (FIL) primitive.
- $\{0,1\}^*$, i.e., the set of all bitstrings of finite length, as a *arbitrary-input-length* (AIL) primitive.
- $\{0,1\}^{\leq N}$, i.e., the set of all bitstrings of length at most N, as a *variable-input-length* primitive.

VIL- and AIL-primitives are often constructed by iterating applications of some FIL-primitive.

In the context of constructing VIL- or AIL-MACs, a natural and weak assumption on the FIL-primitive is that of being a MAC. This was first studied by An and Bellare in [1], who proposed and proved the security of the NI-construction, the first VIL-MAC based on a FIL-MAC. Domain expansion of MACs was further studied in [8], where a general paradigm for constructing VIL- and AIL-MACs by iterating applications of a FIL-MAC was proposed. Several improvements on the NI-construction and two single-key AIL-MAC constructions, Chain-Shift (CS) and Chain-Rotate (CR), were presented. While the CS-construction transforms FIL-MACs with input-length/output-length ratio at least 2, the CR-construction transforms any FIL-MAC (irrespectively of its input-length/output-length ratio) at the cost of a less tight security reduction (by a factor of roughly 5). In this paper the paradigm is generalized and analyzed further.

Domain expansion is well studied for many cryptographic primitives such as collision resistant hash function [6,10], pseudo-random functions (PRFs) [2,3,11,7], universal one-way hash functions [4,12], and random oracles [5]. Since a PRF is (trivially) also a MAC, many VIL- and AIL-PRFs based on a FIL-PRF are widely used as VIL- and AIL-MACs, respectively. However, these MACs are only guaranteed to be secure under the (relative strong) assumption that the FIL-primitive is a PRF. For instance the CBC-MAC [3] is not secure under the assumption that the FIL-primitive is a secure MAC [1]. In cryptography a central goal is to prove the security of cryptographic schemes under as weak assumptions as possible. Demanding that the FIL-primitive is a MAC (rather than a PRF) is a more cautious cryptographic assumption.

1.3   The Construction Paradigm

Let us briefly recall the construction paradigm of [8] as a reference for our contributions. Throughout this paper, the function family

$$G := \{g_k : \{0,1\}^L \to \{0,1\}^\ell\}_{k \in \{0,1\}^\kappa}$$

(with $L > \ell$) denotes a FIL-MAC with *compression parameter*

$$b := L - \ell.$$

The paradigm considers a type of construction $C^{\cdot}$, which uses G to construct an AIL-MAC[1]

$$C^G := \{C^{g_k} : \{0,1\}^* \to \{0,1\}^\ell\}_{k \in \{0,1\}^\kappa}.[2]$$

More precisely, the computation of the tag $\tau = C^{g_k}(m)$ for an n-bit message m can be described as follows. In a pre-processing step m is encoded into a bit string m', for instance by padding m and appending information about its length. The processing step is best described with a buffer initialized with m', where each call to g_k fetches (and deletes) some L bits and writes back the ℓ-bit result to the buffer (for instance by concatenating it at the end of the buffer). This reduces the number of bits in the buffer by b with each call to g_k. The output of the last (possible) call to g_k is returned as the tag (instead of being written back to the buffer). The length of m' is appropriately chosen to be $t(n) \cdot b + \ell$ for some $t(n) \in \mathbb{N}$, to leave the buffer empty after the computation. We stress that $t(n)$ is exactly the number of calls to g_k before the tag is returned. The function $t(\cdot)$ is referred to as the *application* function of $C^{\cdot}$. A particular construction can thus be described by the encoding function mapping m to m' and by the scheme by which the L-bit blocks are fetched. The computation process is illustrated in Fig. 1.

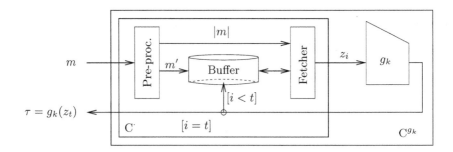

Fig. 1. The construction paradigm

The efficiency of a construction is measured in the number of applications $t(n)$ of the FIL-MAC, or, equivalently, in terms of the *waste*

$$w(n) := t(n) \cdot b + \ell - n,$$

i.e., the amount by which pre-processing expands the message.

---

[1] We consider single-key AIL-MAC constructions in this paper, i.e., AIL-MAC constructions which use one instantiation of the FIL-MAC.

[2] $C^{\cdot}$ denotes the construction, where the dot is the placeholder for G. Actually, $C^{\cdot}$ transforms any FIL-MAC with any compression parameter b and output length ℓ (if not stated otherwise). We describe $C^{\cdot}$ for arbitrary but fixed values of b and ℓ, and let these parameters be implicitly given (if not stated otherwise).

1.4 Our Contribution

In this paper, we generalize the construction paradigm of [8]. The main idea is to comprise constructions which first transform the FIL-MAC G into a FIL-MAC

$$[G]_f := \{[g_k]_f\}_{k\in\{0,1\}^\kappa}$$

defined by

$$[g_k]_f(x) := f(g_k(x)),$$

where $f : \{0,1\}^\ell \to \{0,1\}^{\ell-\delta}$ (with $\delta > 0$) is a key-less compression function, and then the FIL-MAC $[G]_f$ (rather than G) into an AIL-MAC $C^{[G]_f}$ (by some AIL-MAC construction $C^\cdot$). The new type of construction $C^{[\cdot]_f}$ is more efficient than $C^\cdot$, since $[G]_f$ compresses more than G. However, this improvement is at the cost of a worse security reduction. For example if the function f cuts away the δ most significant bits of its input, the security reduction is worsened by a factor of roughly 2^δ.

This tradeoff is investigated in detail. At first sight, a less tight security reduction by some constant factor (as for the CR-construction [8]) seems irrelevant. However, by allowing such a factor, the throughput of other constructions can be improved substantially and result in overall better constructions (see Sect. 4.3).

In this paper, we also propose a new on-line[3] AIL-MAC construction, the PDI-construction, which is superior to all AIL-MAC constructions given in the literature, taking the security/efficiency tradeoff into account.

2 Preliminaries

2.1 Notation

If M is a set, $\#M$ denotes its cardinality. For a sequence S of elements, $|S|$ denotes its length and S_i the sequence of its first $i \leq |S|$ elements. For any $n \in \mathbb{N}_0$, let $[n] := \{1,\ldots,n\}$ (with $[0] := \emptyset$).

For $x, y \in \{0,1\}^*$, let $|x|$ denote the length of x (in bits), $x\|y$ the concatenation of x and y, $\langle n\rangle_b$ a b-bit encoding of a positive integer $n \leq 2^b$, $x[i]$ the ith bit of x, and

$$x[i,j] := x[i]\big\|x[i+1]\big\|\cdots\big\|x[j]$$

for $1 \leq i < j \leq |x|$. Furthermore, let $\mathrm{RR}(\cdot)$ denote the operator on bit strings that rotates the input by one position to the right, i.e.,

$$\mathrm{RR}(x) := x[L]\|x[1, L-1].$$

An encoding $\sigma : \{0,1\}^* \to \{0,1\}^*$ is called *prefix-free* if there are no three strings $x, x', y \in \{0,1\}^*$ such that $x \neq x'$ and $\sigma(x)\|y = \sigma(x')$. A *non-trivial collision* for a function f is a pair $x \neq x'$ of inputs for which $f(x) = f(x')$.

If $\mathcal{E}$ denotes an event, $\bar{\mathcal{E}}$ denotes the complementary event.

---

[3] I.e., the ability to process a message as the message bits arrive, without knowing the message length in advance.

2.2   Security Definition for MACs

A forger F for a MAC $H := \{h_k : \mathcal{M} \to \mathcal{T}\}_{k \in \mathcal{K}}$ has oracle access to $h_k(\cdot)$ (for which k is chosen uniformly at random from $\mathcal{K}$ and kept secret) and can thus learn the tag values for some adaptively chosen messages $m_1, \ldots, m_q$. It then returns a *forgery* (m, τ), i.e., a message m together with a tag τ. The forger F is considered successful if $h_k(m) = \tau$. The only constraint on m is that it must be *new*, i.e., different from all previous messages $m_1, \ldots, m_q$. A forger F is referred to as a (t, q, μ, ε)-forger, if t, q, and μ are upper bounds on the running time, the number of messages (or oracle queries), and the total length (in bits) of the oracle queries including the forgery message m, respectively, and ε is a lower bound on the success probability. Informally, a MAC is considered secure against *existential forgery* under an *adaptive chosen-message attack*, if there is no (t, q, μ, ε)-forger, even for very high values of t, q, and μ, and a very small value of ε.

Definition 1. *A MAC is (t, q, μ, ε)-secure if there exists no (t, q, μ, ε)-forger.*

A forger for a FIL-MAC will be denoted simply as a (t, q, ε)-forger, since the parameter μ is determined by q and the input-length L, i.e., $\mu = (q + 1) \cdot L$.

To prove the security of a MAC, based on a FIL-MAC, one shows that the existence of a (t, q, μ, ε)-forger F for the MAC implies the existence of a (t', q', ε')-forger F' for the FIL-MAC, where t', q', and ε' are functions of t, q, μ, and ε. In all our security proofs F is called only once by F'. Therefore, the running time of F' is essentially that of F, i.e., $t' \approx t$, with some small overhead that is obvious from the construction of F'. We will therefore not bother to explicitly compute the running time of forgers, as this complicates the analysis unnecessarily without providing more insight. Therefore we drop the time parameter t in the sequel.

2.3   Security Reductions

We make use of the proof technique of [8], which we recall for completeness. Let F be a (q, μ, ε)-forger for a MAC $\mathbf{C}^G$ and let

$$F \circ \mathbf{C}^{g_k}$$

denote the process in which F's queries to (its oracle) $\mathbf{C}^{g_k}$ are computed and returned to F, and where F's forgery (m, τ) is verified by computing $\mathbf{C}^{g_k}(m)$. Consider the random variables occurring at the interface to g_k (in the process $F \circ \mathbf{C}^{g_k}$), and let z_i denote the ith input to g_k and $y_i := g_k(z_i)$ the corresponding output. The sequences

$$\mathbf{Z} := (z_1, z_2, \ldots) \quad \text{and} \quad \mathbf{Y} := (y_1, y_2, \ldots)$$

are thus naturally defined. Note that as soon as the key k and the random coins of F are fixed, all values in $\mathbf{Z}$ and $\mathbf{Y}$ are determined, and also whether F is successful or not. Let $\mathcal{E}$ denote the event that F is successful. Without loss of generality we assume that F's forgery message m is distinct from F's oracle queries. Thus $\mathcal{E}$ occurs if and only if $C^{g_k}(m) = \tau$.

A forger F' for the FIL-MAC G simulates $F \circ C^{g_k}$ with the help of F and its oracle access to g_k. At some query z_i to g_k it stops the simulation and returns a forgery (z', τ') for g_k (without making any other oracle queries to g_k). Such a forger is characterized by the time when it stops (i.e., i) and the way it produces its forgery. This is referred to as a *strategy* s of F' and F'_s denotes the corresponding forger.

The most simple strategy is the *naïve* strategy s_{na}. $F'_{s_{na}}$ stops the simulation of $F \circ C^{g_k}$ at the very last query $\mathbf{z}$ to g_k (i.e., $\mathbf{z}$ is the last entry in $\mathbf{Z}$). Then it returns $(\mathbf{z}, \tau)$ as a forgery, where τ is the forgery tag of F's forgery (m, τ) for C^{g_k}. $F'_{s_{na}}$ is successful if the following two conditions hold. First, $\mathcal{E}$ occurs, i.e., $C^{g_k}(m) = \tau$ (and thus $g_k(\mathbf{z}) = \tau$ by definition of $C^{\cdot}$), and second $\mathbf{z}$ is new, i.e., $\mathbf{z}$ is only the last entry in $\mathbf{Z}$. Let $\mathcal{E}_{new}$ denote the event that $\mathbf{z}$ is new. Thus $F'_{s_{na}}$ is successful whenever $\mathcal{E} \wedge \mathcal{E}_{new}$ occurs.

Assume there is a set $\mathcal{S}$ of strategies for a construction with the property that, whenever $\bar{\mathcal{E}}_{new}$ occurs, there exists at least one strategy $s \in \mathcal{S}$ for which F'_s is successful. Such a set is referred to as *complete* for the construction. Obviously, the set $\mathcal{S} \cup \{s_{na}\}$ has the property that whenever $\mathcal{E}$ occurs, there is at least one strategy $s \in \mathcal{S} \cup \{s_{na}\}$ for which F'_s is successful. Thus an overall strategy of F' is to pick its strategy uniformly at random from $\mathcal{S} \cup \{s_{na}\}$. Its success probability is at least the probability that $\mathcal{E}$ occurs, divided by $\#\mathcal{S} + 1$. As F''s number of oracle queries is $|\mathbf{Z}|$, which is a random variable, it is convenient to introduce the following function.

Definition 2. [8] *The* expansion *function* $e : \mathbb{N} \times \mathbb{N} \to \mathbb{N}$ *of a construction* $C^{\cdot}$ *is defined as*

$$e(\tilde{q}, \tilde{\mu}) := \max \left\{ \sum_{i=1}^{\tilde{q}} t(n_i) : n_1, \ldots, n_{\tilde{q}} \in \mathbb{N}_0, \, n_1 + \cdots + n_{\tilde{q}} \leq \tilde{\mu} \right\},$$

where $t(\cdot)$ *is the application function of* $C^{\cdot}$.

It follows that $|\mathbf{Z}| \leq e(q + 1, \mu)$, since there are at most $q + 1$ queries of total length at most μ to C^{g_k} in $F \circ C^{g_k}$. In general, $\#\mathcal{S}$ is a function of $e(q + 1, \mu)$.

Proposition 1. [8] *The existence of a complete set* $\mathcal{S}$ *for a construction* $C^{\cdot}$ *and a* (q, μ, ε)-*forger* F *for* C^G *implies the existence of a* (q', ε')-*forger* F' *for* G, *where* $q' = e(q + 1, \mu)$ *and* $\varepsilon' = \frac{\varepsilon}{\#\mathcal{S} + 1}$.

An important class of strategies for F' are the deterministic strategies. A deterministic strategy s is characterized by a pair (i, f), where $i \in [e(q + 1, \mu)]$ is an index and f a function mapping $(\mathbf{Z}_i, \mathbf{Y}_{i-1})$ to some value $\hat{y}_i \in \{0, 1\}^{\ell}$ (which can be seen as a prediction of y_i). To be more precise, the corresponding forger F'_s stops (the simulation of $F \circ C^{g_k}$) at query z_i and returns $(z_i, \hat{y}_i)$ as a forgery.[4] The forger is successful if $\hat{y}_i = y_i$ and if z_i is new, i.e., not contained in

---

[4] If $i > |\mathbf{Z}|$ the forger aborts.

the sequence $\mathbf{Z}_{i-1}$. In the sequel, we will make use of the following two sets of deterministic forgers (from [8]):

- Let $s_{i,y}$ (for $y \in \{0,1\}^{\ell}$) denote the strategy of stopping at query z_i and returning (z_i, y) as a forgery. Note that whenever the event occurs that an output of g_k is equal to y, i.e., y is an entry in $\mathbf{Y}$, then there exists a strategy $s \in \mathcal{S}_y := \{s_{i,y} | i \in [e(q+1, \mu)]\}$ for which F'_s is successful. We have

$$\#\mathcal{S}_y = e(q+1, \mu). \tag{1}$$

- Let $s_{\mathrm{coll},i,j}$ (for $i > j$) denote the strategy of stopping at query z_i and returning (z_i, y_j) as a forgery. Note that whenever a non-trivial collision for g_k occurs, i.e., $\alpha, \beta \in [|\mathbf{Z}|]$ satisfying $z_\alpha \neq z_\beta$ and $y_\alpha = y_\beta$, then there is a strategy $s \in \mathcal{S}_{\mathrm{coll}} := \{s_{\mathrm{coll},i,j} | i,j \in [e(q+1, \mu)], i > j)\}$ for which F'_s is successful. The cardinality of $\mathcal{S}_{\mathrm{coll}}$ is

$$\#\mathcal{S}_{\mathrm{coll}} = e(q+1, \mu)^2/2 - e(q+1, \mu)/2. \tag{2}$$

3   Concrete AIL-MAC Constructions

In this section we present new on-line AIL-MAC constructions. First, we introduce the Double-Iterated (DI) construction which has constant waste (i.e., $w(n) \in \theta(1)$) and therefore is efficient for long messages. Then, we present the Prefix-Free Iterated (PI) construction which has linear waste (i.e., $w(n) \in \theta(n)$) but is more efficient than the DI-construction for short messages.

Finally, we propose the Prefix-Free Double Iterated (PDI) construction, which depends on some design parameter $r \in \mathbb{N}_0$ and is a hybrid constructions between the DI- and the PI-construction. For $r = 0$ the construction is equivalent to the DI-construction and for $r \to \infty$ to the PI-construction. For values of r between this range the advantages of both the DI- and the PI-construction are exploited. The idea is to simply apply the PI-construction for short messages and the DI-construction for long messages. What short and long means depends on the value of r.

3.1   The Iteration (I) Method

Before the AIL-MAC constructions are presented, we analyze the iteration $I^h_{\mathrm{IV}}(\cdot)$ of a function $h : \{0,1\}^{b+\ell} \to \{0,1\}^{\ell}$, where IV denotes a fixed ℓ-bit initialization value. It is defined as follows and illustrated in Fig. 2 (see Sect. 9.3.1 of [9]).

The value $\tau = I^h_{\mathrm{IV}}(m)$ for a string $m \in (\{0,1\}^b)^*$, i.e., $m_1 \| \cdots \| m_t = m$ for some $t \in \mathbb{N}_0$ and $|m_i| = b$ for $i \in [t]$, is computed as

$$y_0 = \mathrm{IV}; \quad y_i = h(y_{i-1} \| m_i), \ 1 \leq i \leq t; \quad \tau = y_t.$$

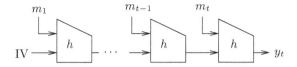

Fig. 2. The iteration (I) method

Lemma 1. *A non-trivial collision in* $I_{IV}^h(\cdot)$ *implies a non-trivial collision in* h *or that an output of* h *is equal to* IV.

Proof. Let $m \neq m'$ and $I_{IV}^h(m) = I_{IV}^h(m')$ denote a non-trivial collision in $I_{IV}^h(\cdot)$. Furthermore, let $(z_1, \ldots, z_t)$ and $(z'_1, \ldots, z'_{t'})$ denote the sequence of inputs to h in the computation of $I_{IV}^h(m)$ and $I_{IV}^h(m')$, respectively. Note that $h(z_t) = I_{IV}^h(m) = I_{IV}^h(m') = h(z'_{t'})$.

Let i denote the smallest index (if any) for which $z_{t-i} \neq z'_{t'-i}$ and $h(z_{t-i}) = h(z'_{t'-i})$. The existence of i directly implies a non-trivial collision in $h(\cdot)$. The non-existence of such an index i implies that one of the sequences $(z_1, \ldots, z_t)$ and $(z'_1, \ldots, z'_{t'})$ is a suffix of the other with $t \neq t'$ since $m \neq m'$. Assume without loss of generality that $t < t'$. In this case we have $IV\|v = z_1 = z'_{t'-t+1} = h(z_{t'-t})\|v$ for some $v \in \{0,1\}^b$, which means that an output of h is equal to IV. □

Lemma 2. $I_{IV}^h(m) = I_{IV'}^h(m')$ *with* $m, m' \in (\{0,1\}^b)^*$ *and* $IV \neq IV'$ *imply a non-trivial collision in* h, *or that an output of* h *is equal to* IV *or* IV'.

Proof. Let $(z_1, \ldots, z_t)$ and $(z'_1, \ldots, z'_{t'})$ denote the sequence of inputs to h in the computation of $I_{IV}^h(m)$ and $I_{IV'}^h(m')$, respectively. Note that $h(z_t) = I_{IV}^h(m) = I_{IV'}^h(m') = h(z'_{t'})$.

Let i denote the smallest index (if any) for which $z_{t-i} \neq z'_{t'-i}$ and $h(z_{t-i}) = h(z'_{t'-i})$. The existence of i directly implies a non-trivial collision in $h(\cdot)$. The non-existence of such an index i implies that one of the sequences $(z_1, \ldots, z_t)$ and $(z'_1, \ldots, z'_{t'})$ is a suffix of the other with $t \neq t'$ since $IV \neq IV'$. If $t < t'$ we have $IV\|v = z_1 = z'_{t'-t+1} = h(z_{t'-t})\|v$ for some $v \in \{0,1\}^b$, which means that an output of h is equal to IV. Analogously, one shows that if $t > t'$ an output of h is equal to IV'. □

Remark 1. The Merkle-Damgård (MD) iteration method [6,10] for collision-resistant hashing is a result of similar nature. The hash value $MD_{IV}^h(m)$, where $m \in \{0,1\}^{\leq 2^b}$ (and $IV \in \{0,1\}^\ell$), is defined by first breaking m into sequence of b-bit blocks $m_1, \ldots, m_t$ (where m_t is padded with zeroes if necessary) and then returning the value $I_{IV}^h(m_1\|\cdots\|m_t\|\langle|m|\rangle_b)$. A non-trivial collision in $MD_{IV}^h(\cdot)$ implies a non-trivial collision in $h(\cdot)$.

3.2 The DI-Construction

The DI-construction is a generalization of the CS-construction [8], which transforms any FIL-MAC (irrespectively of its input-length/output-length ratio) to an AIL-MAC.[5] To be more precise, DI' uses any FIL-MAC G to construct an AIL-MAC $\mathrm{DI}^G := \{\mathrm{DI}^{g_k} : \{0,1\}^* \to \{0,1\}^\ell\}_{k \in \{0,1\}^\kappa}$ as follows.

Break the message $m \in \{0,1\}^*$ (of length n) into a sequence of b-bit blocks $m_1, \ldots, m_{t-1}$ (if $t > 1$) and a $(\lceil \ell/b \rceil b - \ell)$-bit block m_t, where a 1 followed by 0's is used as padding, i.e., $m_1 \| \cdots \| m_t = m \| 10^\nu$ for some $\nu \in \{0, \ldots, b-1\}$. Let

$$\mathrm{DI}^{g_k}(m) := \begin{cases} \mathrm{I}^{g_k}_{1^\ell}\left(\mathrm{I}^{g_k}_{0^\ell}(m_1 \| \cdots \| m_{t-1}) \| m_t\right) & \text{if } t > 1 \\ \mathrm{I}^{g_k}_{1^\ell}(0^\ell \| m_1) & \text{otherwise} \end{cases}.$$

The application function is $t(n) = \lceil \frac{n+1+\ell}{b} \rceil$ (resulting in the waste $w(n) \in \Theta(1)$).

Theorem 1. A (q, μ, ε)-forger F for DI^G implies a (q', ε')-forger F' for G, where $q' = \frac{\mu}{b} + \frac{b+\ell}{b} \cdot (q+1)$ and $\varepsilon' = \frac{\varepsilon}{\frac{1}{2}q'^2 + \frac{3}{2}q' + 1}$.

Proof. We show that $\mathcal{S} := \mathcal{S}_{\mathrm{coll}} \cup \mathcal{S}_{0^\ell} \cup \mathcal{S}_{1^\ell}$ is complete for DI' by proving that, whenever the last input $\mathbf{z}$ to g_k is not new, there is a non-trivial collision in g_k or an output of g_k that is equal to 0^ℓ or 1^ℓ.

Assume that $\mathbf{z}$ is not new. Furthermore, assume that there is no non-trivial collision in g_k and no output of g_k that is equal to 0^ℓ or 1^ℓ. We show that this leads to a contradiction. By Lemma 1, there can not be a non-trivial collision in $\mathrm{I}^{g_k}_{0^\ell}(\cdot)$. Furthermore, no output of $\mathrm{I}^{g_k}_{0^\ell}(\cdot)$ is equal to 0^ℓ, since this would directly imply a non-trivial collision in g_k. As a consequence, the last input $\tilde{m}$ to $\mathrm{I}^{g_k}_{1^\ell}(\cdot)$ is distinct from the other inputs to $\mathrm{I}^{g_k}_{1^\ell}(\cdot)$.[6] Since $\mathbf{z}$ is not new, $\mathbf{z}$ must have been an earlier query to g_k, resulting from some query $m' = m'_1 \| \cdots \| m'_{t'}$ to $\mathrm{I}^{g_k}_{1^\ell}(\cdot)$ with IV $\in \{0^\ell, 1^\ell\}$. Let $z'_1, \ldots, z'_{t'}$ denote the sequence of queries to g_k in the computation of $\mathrm{I}^{g_k}_{\mathrm{IV}}(m')$ and let s be the index for which $z'_s = \mathbf{z}$. Thus, we have $\mathrm{I}^{g_k}_{\mathrm{IV}}(m'_1 \| \cdots \| m'_s) = \mathrm{I}^{g_k}_{1^\ell}(\tilde{m})$. We distinguish two cases:

- If IV $= 0^\ell$, we arrive at a contradiction by Lemma 2.
- If IV $= 1^\ell$, it follows from the construction that $|m'| = |\tilde{m}|$. Thus, we have $m'_1 \| \cdots \| m'_s \neq \tilde{m}$, since $\tilde{m}$ is distinct (from the other queries to $\mathrm{I}^{g_k}_{1^\ell}(\cdot)$). As a consequence, we arrive at a contradiction by Lemma 1.

By definition of $e(q+1, \mu)$, there exist $n_1, \ldots, n_{q+1} \in \mathbb{N}_0$ such that:

$$e(q+1, \mu) = \sum_{i=1}^{q+1} t(n_i) = \sum_{i=1}^{q+1} \left\lceil \frac{n_i + 1 + \ell}{b} \right\rceil \leq \frac{\mu + (b+\ell)(q+1)}{b} =: q'.$$

Thus $\#\mathcal{S} + 1 \leq q'^2/2 + 3q'/2 + 1$ by (1) and (2). Proposition 1 concludes the proof. □

---

[5] The constructions coincide for $b \geq \ell$.
[6] Recall that, with out loss of generality, we assume that the forgery message m of F is distinct from its oracle queries.

Remark 2. The method, used (in [8]) for improving the efficiency of the CS-construction for short messages, can directly be applied for the DI-construction as well. This results in a more efficient construction, which is unfortunately not (completely) on-line.

The DI-construction can also be parallelized in the same way as the CS-construction (see [8]).

3.3 The PI-Construction

The PI-construction uses a prefix-free encoding $\sigma : \{0,1\}^* \to (\{0,1\}^b)^*$, to be defined later, for transforming G into the AIL-MAC $\mathrm{PI}^G := \{\mathrm{PI}^{g_k} : \{0,1\}^* \to \{0,1\}^\ell\}_{k \in \{0,1\}^\kappa}$. It is defined as follows.

For a message $m \in \{0,1\}^*$, let

$$\mathrm{PI}^{g_k}(m) := \mathrm{I}_{0^\ell}^{g_k}(\sigma(m)).$$

Theorem 2. *A (q, μ, ε)-forger for PI^G (with a prefix-free encoding σ) implies a (q', ε')-forger for G, where $q' = e(q+1, \mu)$ and $\varepsilon' = \frac{\varepsilon}{\frac{1}{2}q'^2 + \frac{1}{2}q'+1}$. The expansion function e depends on the concrete choice of σ.*

Proof. We apply Proposition 1 and show that $\mathcal{S} := \mathcal{S}_{\mathrm{coll}} \cup \mathcal{S}_{0^\ell}$ is complete for PI by showing that if $\mathbf{z}$ is not new, then there is a non-trivial collision in g_k or a 0^ℓ-output of g_k. This follows directly from Lemma 1 and the fact that an old $\mathbf{z}$ implies a non-trivial collision in $\mathrm{I}_{0^\ell}^{g_k}(\cdot)$ (due to the prefix-free encoding). □

The on-line property and the efficiency of the construction (hence also the expansion function e) depend on which prefix-free encoding σ is used. It seems obvious that there is no prefix-free encoding for which the construction is on-line and has waste $w(n) \in O(\log(n))$.[7] However, allowing linear waste, i.e., $w(n) \in \theta(n)$, there are prefix-free encodings for which the construction has the on-line property. Throughout this paper, we define σ as follows.

Let
$$\sigma(m) := 0\|m_1\|0\|m_2\|\cdots\|0\|m_{t-1}\|1\|m_t,$$
where $m \in \{0,1\}^*$ and $m_1, \ldots, m_t$ are $(b-1)$-bit blocks such that $m_1\|\cdots\|m_t = m\|10^\nu$ with $\nu \in \{0,\ldots,b-2\}$.

The application function of the PI-construction, with prefix-free encoding σ (as just defined), is $t(n) = \lceil (n+1)/(b-1) \rceil$. This results in waste $w(n) \in \theta(n)$. However, note that the PI-construction is more efficient than the DI-construction if (and only if) the message length is shorter than $\ell(b-1)$.

The following Corollary follows.

---

[7] The prefix-free encoding, described next, has logarithmic waste but is not on-line. Let $\sigma : \{0,1\}^* \to (\{0,1\}^b)^*$ be defined by $r = |\langle|m|\rangle| - 1$ and $\rho(m) := 0^r 1\|\langle|m|\rangle\|m\|0^\nu$, where $\nu \in \{0,\ldots,b-1\}$ is chosen such that the length is a multiple of b.

Corollary 1. *A (q, μ, ε)-forger for PI^G (with σ defined as above) implies a (q', ε')-forger for G, where $q' = \frac{\mu}{b-1} + (q+1)$ and $\varepsilon' = \frac{\varepsilon}{\frac{1}{2}q'^2 + \frac{1}{2}q' + 1}$.*

Proof. The proof follows directly from Theorem 2 and the fact that there exist $n_1, \ldots, n_{q+1} \in \mathbb{N}_0$ such that

$$e(q+1, \mu) = \sum_{i=1}^{q+1} t(n_i) = \sum_{i=1}^{q+1} \left\lceil \frac{n_i + 1}{b-1} \right\rceil \le \sum_{i=1}^{q+1} \frac{n_i + b - 1}{b-1}$$

$$\le \frac{\mu}{b-1} + (q+1) =: q'.$$

As a consequence $\#\mathcal{S} + 1 \le q'^2/2 + q'/2 + 1$ by (1) and (2).   □

3.4   The PDI-Construction

The PDI-construction is an AIL-MAC construction, which is a hybrid construction between the PI- and DI-construction. It exploits the advantages of both constructions as follows.

Let $r \in \mathbb{N}_0$ be a design parameter. The construction PDI_r transforms any FIL-MAC G into the AIL-MAC

$$\mathrm{PDI}_r^G := \{\mathrm{PDI}_r^{g_k} : \{0,1\}^* \to \{0,1\}^\ell\}_{k \in \{0,1\}^\kappa},$$

where $\mathrm{PDI}_r^{g_k}(\cdot)$ is defined as follows.

For a message $m \in \{0,1\}^*$ (of length n), let

$$\mathrm{PDI}_r^{g_k}(m) := \begin{cases} \mathrm{PI}^{g_k}(m) & \text{if } n < r(b-1) \\ \mathrm{DI}^{g_k}(0\|m_1\|0\|m_2\|\cdots\|0\|m_r\|m_{r+1}) & \text{otherwise} \end{cases},$$

where $m_1, \ldots, m_r$ is a sequence of $(b-1)$-bit blocks and m_{r+1} a bitstring such that $m_1\|\cdots\|m_r\|m_{r+1} = m$.

The application function is $t(n) = \begin{cases} \left\lceil \frac{n+1}{b-1} \right\rceil & \text{if } n < r(b-1) \\ \left\lceil \frac{n+1+\ell+r}{b} \right\rceil & \text{otherwise} \end{cases}.$

Although not directly clear from the definition above, this construction is on-line (no matter whether $|m| < r(b-1)$ or not, the processing of m starts out in the same way).

We stress that the PDI-construction is equivalent to the DI-construction for $r = 0$ and to the PI-construction for $r \to \infty$. As is obvious from the definition of PDI_r, the construction is as efficient as PI for messages of shorter length than $r(b-1)$ and slightly less efficient than DI for longer messages.

Theorem 3. *A (q, μ, ε)-forger for PDI_r^G implies a (q', ε')-forger for G, where $q' = \frac{\mu}{b-1} + (q+1) + \frac{\ell+r}{b} \cdot \Lambda - \frac{1}{b \cdot (b-1)} \cdot \Pi$ and $\varepsilon' = \frac{\varepsilon}{\frac{1}{2} \cdot q'^2 + (\frac{1}{2} + \gamma) \cdot q' + 1}$, where*

$$(\Lambda, \Pi) := \begin{cases} (q+1, \mu) & \text{if } r = 0 \\ \left(\left\lfloor \frac{\mu}{r(b-1)} \right\rfloor, 0 \right) & \text{if } \frac{\mu}{q+1} \leq r(b-1) - 1, \\ \left(\min \left(q+1, \left\lfloor \frac{\mu}{r(b-1)} \right\rfloor \right), \mu - q(r(b-1) - 1) \right) & \text{otherwise} \end{cases}$$

and γ takes the value 1 if $\mu \geq r \cdot (b-1)$ and 0 otherwise.

Proof (Sketch). Let γ be an indicator variable that takes the value 1 if $\mu \geq r \cdot (b-1)$ and 0 otherwise. We omit the proof that $\mathcal{S}_{\mathrm{coll}} \cup \mathcal{S}_{0^\ell}$ is complete for the construction if $\gamma = 0$ and that $\mathcal{S}_{\mathrm{coll}} \cup \mathcal{S}_{0^\ell} \cup \mathcal{S}_{1^\ell}$ is complete for the PDI-construction otherwise, since it is similar to the proof of the DI- and PI-construction.[8] Applying Proposition 1 and the following fact concludes the proof.

By definition of $e(q + 1, \mu)$, there exist $n_1, \ldots, n_{q+1} \in \mathbb{N}_0$ such that $e(q + 1, \mu) = \sum_{i=1}^{q+1} t(n_i)$. Let ζ_i be an indicator variable that takes value 1 if $n_i \geq r(b-1)$ and 0 otherwise. We have that

$$\sum_{i=1}^{q+1} t(n_i) \leq \sum_{i=1}^{q+1} \zeta_i \cdot \left\lceil \frac{n_i + 1 + \ell + r}{b} \right\rceil + (1 - \zeta_i) \cdot \left\lceil \frac{n_i + 1}{b - 1} \right\rceil$$

$$\leq \sum_{i=1}^{q+1} \zeta_i \cdot \frac{n_i + b + \ell + r}{b} + (1 - \zeta_i) \cdot \frac{n_i + b - 1}{b - 1}$$

$$\leq \frac{\mu}{b - 1} + (q + 1) + \frac{\ell + r}{b} \cdot \sum_{i=1}^{q+1} \zeta_i - \frac{1}{b \cdot (b - 1)} \sum_{i=1}^{q+1} \zeta_i \cdot n_i.$$

Furthermore, it is straightforward to verify that the following two inequalities hold

$$\sum_{i=1}^{q+1} \zeta_i \leq \begin{cases} q + 1 & \text{if } r = 0 \\ \min \left(q + 1, \left\lfloor \frac{\mu}{r(b-1)} \right\rfloor \right) & \text{otherwise} \end{cases} =: \Lambda$$

$$\sum_{i=1}^{q+1} \zeta_i \cdot n_i \geq \begin{cases} \mu & \text{if } r = 0 \\ 0 & \text{if } \frac{\mu}{q+1} \leq r(b-1) - 1 \\ \mu - q(r(b-1) - 1) & \text{otherwise} \end{cases} =: \Pi.$$

As a consequence $\#\mathcal{S} + 1 \leq q'^2/2 + (1/2 + \gamma) \cdot q' + 1$ by (1) and (2). $\square$

---

[8] Note that if $\mu < r(b-1)$ all queries issued by the forger for $\mathrm{PDI}_r^{g_k}(\cdot)$ (including the forgery message) are shorter than $r(b-1)$ and hence $\mathrm{DI}^{g_k}(\cdot)$ is never invoked.

4   The Generalized Construction Paradigm

In this section, we generalize the construction paradigm to comprise a greater class of constructions. Furthermore, we investigate a tradeoff between the efficiency (of a construction) and the tightness (of its security reduction) in detail.

4.1   An Efficiency/Security Tradeoff

A general design goal of AIL-MAC constructions is to minimize the number of applications $t(n)$ of the FIL-MAC (where n denotes the message length). A natural approach to decrease the number of applications (which is not implied by the type of construction C˙) is to increase the compression parameter of the FIL-MAC before it is transformed by some construction C˙. However, as we will see, this is at the cost of a less tight security reduction.

To be more precise, let $f : \{0,1\}^\ell \rightarrow \{0,1\}^{\ell-\delta}$ be a compression function with compression parameter $\delta > 0$ and let $f^{-1}(y)$ denote the set of all preimages[9] of $y \in \{0,1\}^{\ell-\delta}$. Let $[\cdot]_f$ denote the construction, which transforms G into a FIL-MAC

$$[G]_f := \{[g_k]_f : \{0,1\}^L \rightarrow \{0,1\}^{\ell-\delta}\}_{k \in \{0,1\}^\kappa},$$

defined by

$$[g_k]_f(x) := f(g_k(x)).$$

Lemma 3. *A (q, ε)-forger F for $[G]_f$ implies a $(q, \varepsilon/s)$-forger F' for G, where $s = \max\{\#f^{-1}(y) : y \in \{0,1\}^{\ell-\delta}\}$.*

Proof. The forger F' runs F, answering all its oracle queries with the help of its own oracle. When F returns a forgery (m, τ), F' chooses an element $\hat{\tau}$ uniformly at random from $f^{-1}(\tau)$ and outputs $(m, \hat{\tau})$ as its own forgery. If F' is successful it follows that $\tau = [g_k]_f(m) = f(g_k(m))$. Thus, there is an element $\tau' \in f^{-1}(\tau)$ for which $\tau' = g_k(m)$. The probability that $\hat{\tau} = \tau'$ is

$$1/\#f^{-1}(\tau) \geq 1/s, \quad \text{where} \quad s = \max\{\#f^{-1}(y) : y \in \{0,1\}^{\ell-\delta}\}.$$

Let $\mathcal{E}'$ denote the event that F' is successful and $\mathcal{E}$ the event that F is successful. It follows that

$$\Pr[\mathcal{E}'] \geq \Pr[\mathcal{E}' \mid \mathcal{E}] \cdot \Pr[\mathcal{E}] \geq \underbrace{\Pr[\hat{\tau} = \tau']}_{\geq 1/s} \cdot \underbrace{\Pr[\mathcal{E}]}_{=\varepsilon}.$$

$\square$

---

[9] We assume that, for all $y \in \{0,1\}^{\ell-\delta}$, one can efficiently sample an element uniformly at random from $f^{-1}(y)$.

To get as tight a security reduction in Lemma 3 as possible the largest preimage set of the key-less compression function must be as small as possible. A function achieving this is

$$\Delta_\delta : \{0,1\}^\ell \to \{0,1\}^{\ell-\delta} \ , \quad \text{defined by} \quad x \mapsto x[1, \ell - \delta],$$

which simply cuts off the δ least significant bits of the input. As a consequence Δ_δ can always be chosen as the compression function without loss of generality. To simplify the notation, we write $[\cdot]_\delta$ to denote the construction $[\cdot]_{\Delta_\delta}$.

Corollary 2. *A (q, ε)-forger for $[G]_\delta$ implies a $(q, \varepsilon/2^\delta)$ forger for G.*

Proof. Since each image of $\Delta_\delta(\cdot)$ has equally many preimages, namely 2^δ, the largest preimage set is as small as possible. Apply Lemma 3. □

4.2   The Generalized Paradigm

The AIL-MAC $C^{[G]_\delta}$ is defined by simply letting the construction $C^{\cdot}$ transform the FIL-MAC $[G]_\delta$, which has compression parameter $b' = b + \delta$ and output-length $\ell' = \ell - \delta$. This is illustrated in Fig. 3.

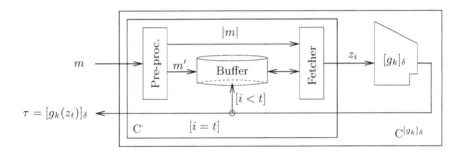

Fig. 3. The generalized construction paradigm

Since $[G]_\delta$ compresses more than G, the number of applications of the FIL-MAC G is in general smaller for $C^{[\cdot]_\delta}$ than for $C^{\cdot}$. However, this is at the cost of having a less tight security reduction for $C^{[G]_\delta}$ by a factor of roughly 2^δ. The tradeoff between the efficiency and the tightness should be taken into account when comparing AIL-MAC constructions with each other (see next section).

Corollary 3. *Let b denote the compression parameter and ℓ the output-length of a FIL-MAC G.[10] If $t_{b,\ell}(n)$ is the application function of $C^{\cdot}$, then $t_{b+\delta,\ell-\delta}(n)$ is the application function of $C^{[\cdot]_\delta}$. If a (q, μ, ε)-forger for C^G implies a (q', ε')-forger for G, where*

$$q' = q'_{b,\ell}(q, \mu, \varepsilon) \quad \text{and} \quad \varepsilon' = \varepsilon'_{b,\ell}(q, \mu, \varepsilon),$$

---

[10] Here we make the parameters b and ℓ explicit.

then a (q, μ, ε)-forger for $C^{[G]_\delta}$ implies a $(q'', \varepsilon''/2^\delta)$-forger for G, where

$$q'' = q'_{b+\delta, \ell-\delta}(q, \mu, \varepsilon) \quad and \quad \varepsilon'' = \varepsilon'_{b+\delta, \ell-\delta}(q, \mu, \varepsilon).$$

Proof. The FIL-MAC $[G]_\delta$ has compression parameter $b' = b + \delta$ and output-length $\ell' = \ell - \delta$. Apply Corollary 2.   □

To prove the security of a construction $C^{[\cdot]_\delta}$, one simply applies Corollary 3 (with the proof technique of Sect. 2.3).

4.3   An Illustrative Example

To illustrate the generalization and the security/efficiency tradeoff, let us first briefly recall the Chain-Rotate construction of [8] (as a reference). The CR-construction transforms any FIL-MAC G into the AIL-MAC $\mathrm{CR}^G := \{\mathrm{CR}^{g_k} : \{0,1\}^* \to \{0,1\}^\ell\}_{k \in \{0,1\}^\kappa}$, as follows.

Parse the message $m \in \{0,1\}^*$ into a sequence $\{m_i\}_{i=1}^t$ of b-bit blocks such that $m_1 \| \cdots \| m_t = m \| 10^\nu$ for a $\nu \in \{0, \ldots, b-1\}$, and let

$$\mathrm{CR}^{g_k}(m) := g_k\left(\mathrm{RR}\left(y \| m_t\right)\right), \quad where \quad y := \begin{cases} \mathrm{I}_{0^\ell}^{g_k}(m_1 \| \cdots \| m_{t-1}) & if \ t > 1 \\ 0^\ell & otherwise \end{cases}.$$

The application function is $t(n) = \left\lceil \frac{n+1}{b} \right\rceil$ (resulting in the waste $w(n) \in \Theta(1)$).

Theorem 4. [8] *A (q, μ, ε)-forger for CR^G implies a (q', ε')-forger for G, where $q' = \frac{\mu}{b} + (q+1)$ and $\varepsilon' = \frac{\varepsilon}{\frac{5}{2}q'^2 + \frac{3}{2}q'+1}$.*

The efficiency of CR is better than for DI and PI (just compare the application functions). However, note that the tightness of the security reduction is roughly a factor 5 worse. At first sight one might be tempted to neglect the factor 5 and consider the CR-construction as the better construction. However, as we show next, the construction $\mathrm{PI}^{[\cdot]_{\delta+1}}$ is as efficient and more secure than $\mathrm{CR}^{[\cdot]_\delta}$ for all δ. This illustrates the importance of taking the security/efficiency tradeoff into account when comparing AIL-MAC constructions.

The following two corollaries follow directly from Corollary 3, using Theorem 4 and Corollary 1, respectively.

Corollary 4. *A (q', ε')-secure FIL-MAC G implies a (q, μ, ε)-secure AIL-MAC $\mathrm{CR}^{[G]_\delta}$, for*

$$\varepsilon \geq 2^\delta \cdot \left(\frac{5}{2} \cdot q'^2 + \frac{3}{2} \cdot q' + 1\right) \cdot \varepsilon' \quad and \quad \frac{\mu}{b+\delta} + (q+1) \leq q'.$$

The application function is $t(n) = \left\lceil \frac{n+1}{b+\delta} \right\rceil$.

Corollary 5. *A (q', ε')-secure FIL-MAC G implies a (q, μ, ε)-secure AIL-MAC $\mathrm{PI}^{[G]_\delta}$, for*

$$\varepsilon \geq 2^\delta \cdot \left(\frac{1}{2} \cdot q'^2 + \frac{1}{2} \cdot q' + 1 \right) \cdot \varepsilon' \quad and \quad \frac{\mu}{b + \delta - 1} + (q+1) \leq q'.$$

The application function is $t(n) = \left\lceil \frac{n+1}{b+\delta-1} \right\rceil$.

It is straightforward to verify that the application function is equivalent for $\mathrm{PI}^{[\cdot]_{\delta+1}}$ and $\mathrm{CR}^{[\cdot]_\delta}$ (and hence the efficiency is the same). Furthermore, the bound for q and μ are also equivalent for the constructions. Since the lower bound for ε is smaller for $\mathrm{PI}^{[\cdot]_{\delta+1}}$, by a factor of roughly 2.5, it follows that $\mathrm{PI}^{[\cdot]_{\delta+1}}$ has a tighter security reduction.

5   Comparisons of AIL-MACs

It is clear from the above that we do not need to consider $\mathrm{CS}^{[\cdot]_\delta}$ and $\mathrm{CR}^{[\cdot]_\delta}$ (for any δ) in our comparison of AIL-MAC constructions, since $\mathrm{DI}^{[\cdot]_\delta}$ is a generalization of the former and $\mathrm{PI}^{[\cdot]_{\delta+1}}$ is more efficient and has a tighter security reduction than the latter. Furthermore, recall that $\mathrm{PDI}_r^{[G]_\delta}$ is equivalent to $\mathrm{DI}^{[G]_\delta}$ for $r = 0$ and to $\mathrm{PI}^{[G]_\delta}$ for $r \to \infty$.

As a consequence, for all AIL-MAC constructions in the literature there is a choice for r and δ for which $\mathrm{PDI}_r^{[\cdot]_\delta}$ is as efficient and secure. The concrete choice for δ and the design parameter r is application dependent. Combining Corollary 3 and Theorem 3, we get:

Corollary 6. *A (q', ε')-secure FIL-MAC G implies a (q, μ, ε)-secure AIL-MAC $\mathrm{PDI}_r^{[G]_\delta}$, for*

$$\varepsilon \geq 2^\delta \cdot \left(\frac{q'^2}{2} + \left(\frac{1}{2} + \gamma \right) \cdot q' + 1 \right) \cdot \varepsilon', \quad and$$

$$\frac{\mu}{b + \delta - 1} + (q+1) + \frac{\ell - \delta + r}{b + \delta} \cdot \Lambda - \frac{1}{(b+\delta)(b+\delta-1)} \cdot \Pi \leq q',$$

where

$$(\Lambda, \Pi) := \begin{cases} (q + 1, \mu) & \text{if } r = 0 \\ \left(\left\lfloor \frac{\mu}{r \cdot (b+\delta-1)} \right\rfloor, 0 \right) & \text{if } \frac{\mu}{q+1} \leq r \cdot (b + \delta - 1) - 1 \\ \left(\min \left(q+1, \left\lfloor \frac{\mu}{r \cdot (b+\delta-1)} \right\rfloor \right), \mu - q \cdot (r \cdot (b + \delta - 1) - 1) \right) & \text{otherwise} \end{cases}$$

and γ equals 1 if $\mu \geq r \cdot (b + \delta - 1)$ and 0 otherwise. The application function is

$$t(n) = \begin{cases} \left\lceil \frac{n+1}{b+\delta-1} \right\rceil & \text{if } n < r(b + \delta - 1) \\ \left\lceil \frac{n+1+\ell-\delta+r}{b+\delta} \right\rceil & \text{otherwise} \end{cases}.$$

Note that Corollary 6 is equivalent to Corollary 5 for $r \to \infty$ and to the following corollary for $r = 0$.

Corollary 7. *A (q', ε')-secure FIL-MAC G implies a (q, μ, ε)-secure AIL-MAC $\mathrm{DI}^{[G]_\delta} (\equiv \mathrm{PDI}_0^{[G]_\delta})$, for*

$$\varepsilon \geq 2^\delta \cdot \left(\frac{1}{2} \cdot q'^2 + \frac{3}{2} \cdot q' + 1 \right) \cdot \varepsilon' \quad and \quad \frac{\mu}{b + \delta} + \frac{b + \ell}{b + \delta} \cdot (q + 1) \leq q'.$$

The application function is $t(n) = \left\lceil \frac{n + 1 + \ell - \delta}{b + \delta} \right\rceil$.

6   Conclusion

In this paper, a study of a paradigm for constructing AIL-MACs by iterating applications of a FIL-MAC was continued. The paradigm was generalized in a natural way and an efficiency/security tradeoff was investigated in detail.

Our new on-line single-key AIL-MAC construction, the PDI-construction, transforms any FIL-MAC into an AIL-MAC with constant waste. It is superior to all constructions given in the literature (taking the tradeoff into account) and it appears obvious that it is essentially optimal.

An open question is whether there exists a prefix-free encoding σ' such that the PI-construction (with encoding σ') is on-line and has logarithmic waste. Our conjecture is that there is no such encoding.

References

1. J. H. An and M. Bellare. Constructing VIL-MACs from FIL-MACs: message authentication under weakened assumptions. In *Advances of Cryptology — CRYPTO 1999*, volume 1666 of *Lecture Notes in Computer Science*, pages 252–269. Springer-Verlag, 1999.
2. M. Bellare, J. Guérin, and P. Rogaway. XOR MACs: new methods for message authentication using finite pseudorandom functions. In *Advances of Cryptology — CRYPTO 1995*, volume 963 of *Lecture Notes in Computer Science*, pages 15–28. Springer-Verlag, 1995.
3. M. Bellare, J. Kilian, and P. Rogaway. The security of the cipher block chaining message authentication code. In *Journal of Computer and System Sciences*, 61(3):362–399, 2000.
4. M. Bellare and P. Rogaway. Collision-resistant hashing: towards making UOWHFs practical. In *Advances in Cryptology — CRYPTO 1997*, volume 1294 of *Lecture Notes in Computer Science*, pages 470–484. Springer-Verlag 1997.
5. J.-S. Coron, Y. Dodis, C. Malinaud, and P. Puniya. Merkle-Damgård revisited: how to construct a hash function. In *Advances of Cryptology — CRYPTO 2005*, volume 3621 of *Lecture Notes in Computer Science*, pages 430–448. Springer-Verlag 2005.
6. I. Damgård. A design principle for hash functions. In *Advances in Cryptology — CRYPTO 1989*, volume 435 of *Lecture Notes in Computer Science*, pages 416–427. Springer-Verlag, 1990.

7. U. Maurer. Indistinguishability of random systems. In *Advances of Cryptology —*
   EUROCRYPT 2002, volume 2332 of *Lecture Notes in Computer Science*, pages
   110–132. Springer-Verlag, 2002.
8. U. Maurer and J. Sjödin. Single-key AIL-MACs from any FIL-MAC. In *Proceedings*
   of ICALP 2005, volume 3580 of *Lecture Notes in Computer Science*, pages 472–484.
   Springer-Verlag, 2005.
9. A. Menezes, P. van Oorschot, and S. Vanstone. *Handbook of applied cryptography.*
   CRC Press, 1997. Available on line at http://www.cacr.math.uwaterloo.ca/hac/.
10. R. Merkle. A certified digital signature. In *Advances in Cryptology — CRYPTO*
    1989, volume 435 of *Lecture Notes in Computer Science*, pages 218–232. Springer-
    Verlag, 1990.
11. E. Petrank and C. Rackoff. CBC MAC for real-time data sources. In *Journal of*
    Cryptology, 13(3):315–338, 2000.
12. V. Shoup. A composition theorem for universal one-way hash functions. In *Ad-*
    vances of Cryptology — EUROCRYPT 2000, volume 1807 of *Lecture Notes in*
    Computer Science, pages 445–452. Springer-Verlag 2000.

Normality of Vectorial Functions

An Braeken, Christopher Wolf, and Bart Preneel

Katholieke Universiteit Leuven,
Dept. Elect. Eng.-ESAT/SCD-COSIC,
Kasteelpark Arenberg 10, 3001 Heverlee, Belgium
{an.braeken, christopher.wolf, bart.preneel}@esat.kuleuven.be

Abstract. The most important building blocks of symmetric cryptographic primitives such as the DES or the AES, are vectorial Boolean functions, also called S-boxes. In this paper, we extend the definition of normality for Boolean functions into several new affine invariant properties for vectorial Boolean functions. We compute the probability of occurrence of these properties and present practical algorithms for each of these new properties. We find a new structural property for the AES S-box, which also holds for a large class of permutation functions when the dimension n is even. Moreover, we prove a relation with the propagation characteristics of a vectorial function and extend the scope of non-APN functions for n even.

Keywords: Cryptography, S-box, AES, normality, algorithm, propagation characteristics, APN.

1   Introduction

The notion of normality was first introduced by Dobbertin in order to construct balanced Boolean functions with high nonlinearity. Normality of Boolean functions is an interesting property since it is affine invariant and it allows us to distinguish different classes of bent functions [4]. In [6], the first non-exhaustive algorithm for computing the normality of Boolean functions was presented. This algorithm was later improved in [2,13]. To the knowledge of the authors, this paper is the first which considers normality of vectorial Boolean functions, also called S-boxes. From a cryptographic point of view, this is far more natural as an attack is usually not launched against a single output bit of an S-box, but against the whole S-box.

In cryptography, vectorial functions are mostly used as building blocks in stream and block ciphers. So the security of the corresponding cipher, *e.g.*, the AES or the DES, strongly depends on the properties of the underlying vectorial Boolean function. Therefore, we believe that it is necessary to generalise the notion of normality for these objects in order to better understand the structure and also the cryptographic security of the vectorial functions involved. We discover a new remarkable structural property for the AES S-boxes and all bijective power functions for n even. This property is also theoretically proven and leads

N.P. Smart (Ed.): Cryptography and Coding 2005, LNCS 3796, pp. 186–200, 2005.

to a new class of functions that are not APN for even n. We leave it as an open question at the moment which degree of normality leads to an efficient attack on stream ciphers or block ciphers.

In Sect. 2, we give some preliminaries and notations. Sect. 3 deals with the new definitions of properties for vectorial functions which are related to the normality of a Boolean function. In Sect. 4, we present counting arguments on the occurrence of these properties for vectorial functions and present the new structural property on all bijective power functions. Then a description of an algorithm for determining these new properties is developed. In Sect. 5, we outline the relation with the propagation characteristics of the function, which measure the resistance of a vectorial function against differential cryptanalysis. This paper concludes with Sect. 6. In particular, this section contains several open problems which we leave to the cryptographic community as interesting research questions.

2   Preliminaries and Notations

Let $f(\overline{x})$ be a Boolean function on $\mathbb{F}_2^n$. Any Boolean function can be uniquely expressed in the algebraic normal form (ANF):

$$f(\overline{x}) = \sum_{(a_1,\ldots,a_n)\in\mathbb{F}_2^n} h(a_1,\ldots,a_n)x_1^{a_1}\cdots x_n^{a_n},$$

with h a function on $\mathbb{F}_2^n$, defined by $h(\overline{a}) = \sum_{\overline{x}\leq\overline{a}} f(\overline{x})$ for any $\overline{a} \in \mathbb{F}_2^n$, where $\overline{x} \leq \overline{a}$ means that $x_i \leq a_i$ for all $i \in \{1,\ldots,n\}$. The *algebraic degree* of f, denoted by $\deg(f)$, is defined as the number of variables in the largest term $x_1^{a_1}\cdots x_n^{a_n}$ in the ANF of f, for which $h(a_1,\ldots,a_n) \neq 0$. An affine Boolean function has degree at most 1. For any $\overline{a} \in \mathbb{F}_2^n, \epsilon \in \mathbb{F}_2$, we can define the affine function $\phi_{\overline{a}} : \mathbb{F}_2^n \to \mathbb{F}_2 : \overline{x} \to \overline{a}\cdot\overline{x} \oplus \epsilon$, where $\overline{a}\cdot\overline{x}$ is the dot product of $\mathbb{F}_2^n$. If $\epsilon = 0$, the function $\phi_{\overline{a}}$ is called a linear function. The following theorem gives a useful property to determine if a function is affine.

Theorem 1. *[15] A Boolean function $f : \mathbb{F}_2^n \to \mathbb{F}_2$ is affine if and only if for all even k with $k \geq 4$ holds that $\forall\overline{x}_1,\ldots,\overline{x}_k$ in $\mathbb{F}_2^n : f(\overline{x}_1) \oplus \cdots \oplus f(\overline{x}_k) = 0$, whenever $\overline{x}_1 \oplus \cdots \oplus \overline{x}_k = 0$.*

A vectorial function F (also called (n,m) S-box) from $\mathbb{F}_2^n$ into $\mathbb{F}_2^m$ can be represented by the vector $(f_1, f_2, \ldots, f_m)$, where f_i are Boolean functions from $\mathbb{F}_2^n$ into $\mathbb{F}_2$. A function $F : \mathbb{F}_2^n \to \mathbb{F}_2^m$ is affine if it can be written as $F(\overline{x}) = A\overline{x} \oplus B$, where A is an $(m \times n)$-matrix and $\overline{b}$ is an m-dimensional vector. Thm. 1 can be generalised for vectorial functions [16].

Theorem 2. *[16] A vectorial function $F : \mathbb{F}_2^n \to \mathbb{F}_2^m$ is affine if and only if for all even k with $k \geq 4$ holds that $\forall\overline{x}_1,\ldots,\overline{x}_k$ in $\mathbb{F}_2^n : F(\overline{x}_1) \oplus \cdots \oplus F(\overline{x}_k) = 0$, whenever $\overline{x}_1 \oplus \cdots \oplus \overline{x}_k = 0$.*

To resist differential attacks, vectorial functions F used in block ciphers need to have good propagation characteristics [14]. The propagation characteristics of F are measured by the differential $\Delta(F)$, defined by

$$\Delta(F) = \max_{\overline{a} \neq 0, \overline{b}} \#\{\overline{x} \in \mathbb{F}_2^n \mid F(\overline{x} \oplus \overline{a}) \oplus F(\overline{x}) = \overline{b}\}. \tag{1}$$

It is obvious that $\Delta(F) \geq \max\{2, 2^{n-m}\}$. If $\Delta(F) = 2^n$, then there exists a *linear structure* $\overline{\alpha} \in \mathbb{F}_2^n$. This is an element for which the function $F(\overline{x}) \oplus F(\overline{x} \oplus \overline{\alpha})$ is constant. If $\Delta(F) = 2$, the function F is called almost perfect nonlinear (APN). It is conjectured that APN permutations only exist in odd dimensions. This statement is proven for some particular cases, most notably power functions [5] and functions of degree 2 [14,9].

3   Normality

3.1   Definitions

A subspace $U \subseteq \mathbb{F}_2^n$ of dimension k can be represented by its basis $U = < \overline{a}_1, \ldots, \overline{a}_k >$, where $\overline{a}_1, \ldots, \overline{a}_k$ are k linearly independent vectors of $\mathbb{F}_2^n$. Moreover, define $\overline{U} := \{\overline{a} \in \mathbb{F}_2^n : \overline{a} \notin U\} \cup \{0\}$ as the complement space which contains all vectors of $\mathbb{F}_2^n$ which are not in U in addition to the zero vector. Now, a coset of the subspace U is represented by $\overline{a} \oplus U$, where $\overline{a} \in \overline{U}$. Apart from the case $\overline{a} = 0$, we have $\overline{a} \notin U$, which in particular implies that $\overline{a}, \overline{a}_1, \ldots, \overline{a}_k$ are linearly independent if $\overline{a}$ is not the zero vector of $\mathbb{F}_2^n$. Instead of speaking about a coset of a subspace, we will speak in the following about a *flat*. We first shortly repeat the definitions of normality and weakly normality for Boolean functions.

Definition 1. *A Boolean function f on $\mathbb{F}_2^n$ is called normal if there exists a flat $V \subset \mathbb{F}_2^n$ of dimension $\lceil n/2 \rceil$ such that f is constant on V.*
   A Boolean function f on $\mathbb{F}_2^n$ is said to be weakly normal if there exists a flat $V \subset \mathbb{F}_2^n$ of dimension $\lceil n/2 \rceil$ such that f is affine on V.

The property of normality is connected with the problem of determining the highest dimension of the flats on which f is constant. As a consequence, a natural generalisation of the previous definitions can be given by:

Definition 2. *A Boolean function f on $\mathbb{F}_2^n$, is said to be k-normal (resp. k-weakly normal) if there exists a flat $V \subset \mathbb{F}_2^n$ of dimension k such that f is constant (resp. affine) on V for $1 \leq k \leq n$.*

Here, we study a similar notion for vectorial functions.

Definition 3. *A vectorial function F from $\mathbb{F}_2^n$ into $\mathbb{F}_2^m$ is called k-normal if there exists a flat $V \subset \mathbb{F}_2^n$ of dimension k such that F is constant on V.*

Therefore, F is normal if and only if all linear combinations of its Boolean components are constant on the same k-dimensional flat, $i.e.$, there exists a flat $V \subset \mathbb{F}_2^n$ of dimension k and a constant $\overline{c} \in \mathbb{F}_2^m$ such that

$$\forall \overline{x} \in V, F(\overline{x}) = \overline{c}$$
$$\Leftrightarrow \forall \overline{x} \in V, \forall \overline{a} \in \mathbb{F}_2^m, \overline{a} \cdot F(\overline{x}) = \overline{a} \cdot \overline{c} \tag{2}$$

It is clear that a sufficient condition for this property is that $\overline{a}_i \cdot F$ is constant on V for $i \in \{1, \dots m\}$, where $(\overline{a}_1, \dots, \overline{a}_m)$ is any basis of $\mathbb{F}_2^m$. Thus, in particular for the standard basis $(\overline{e}_1, \dots, \overline{e}_m)$, with $\overline{e}_i$ the vector of all zeros except on the i-th position, we only have to check the output bit functions f_i of F.

Remark 1. A permutation on $\mathbb{F}_2^n$ (or an injection from $\mathbb{F}_2^n$ into $\mathbb{F}_2^m$) can never be k-normal for any $k > 0$.

To relax the property, we also introduce a weaker version of the original definition.

Definition 4. *A function $F : \mathbb{F}_2^n \to \mathbb{F}_2^m$ is called (k, k')-normal (resp. (k, k')-weakly normal) with $1 \le k \le n$ and $1 \le k' \le m$ if there exists a flat $V \subset \mathbb{F}_2^n$ of dimension k and a flat $W \subset \mathbb{F}_2^m$ of dimension k' such that for all $\overline{a} \in W$, the Boolean functions $\overline{a} \cdot F$ are constant (resp. affine) on V. For $k' = m$, the definition coincides with the definition of k-normality.*

As in the previous case, we can simplify the definition by:

Definition 5. *A function $F : \mathbb{F}_2^n \to \mathbb{F}_2^m$, $n \ge m$, is called (k, k')-normal (resp. (k, k')-weakly normal) if there exists a flat $V \subset \mathbb{F}_2^n$ of dimension k and k' linearly independent vectors $\overline{a}_1, \dots, \overline{a}_{k'}$ of $\mathbb{F}_2^m$ such that for all i, $1 \le i \le k'$, the Boolean functions $\overline{a}_i \cdot F$ are constant (resp. affine) on V.*

From the definitions above, we can conclude that a constant vectorial function from $\mathbb{F}_2^n$ into $\mathbb{F}_2^m$ is n-normal. An affine vectorial function from $\mathbb{F}_2^n$ into $\mathbb{F}_2^m$ is (n, n)-weakly-normal.

The definition of (k, k')-weakly normality can be rephrased into the following definition.

Definition 6. *A function $F : \mathbb{F}_2^n \to \mathbb{F}_2^m$ is said to be (k, k')-weakly normal if there exists a flat $V \subset \mathbb{F}_2^n$ of dimension k such that for all $\overline{x} \in V$, it holds that $F(\overline{x}) = \overline{x}A \oplus \overline{a}$, where A is an $(m \times n)$-matrix with rank less or equal than k' and $\overline{a} \in \mathbb{F}_2^m$.*

It is clear that a (k, k')-weakly normal function only exists if $k' \le 2^k$. Moreover, if $k' = 0$, then the definition coincides with k-normality. A more general definition, where the affine or constant requirement is omitted, is then given by:

Definition 7. *A function $F : \mathbb{F}_2^n \to \mathbb{F}_2^m$ is said to be a (k, k') flat-carrier if there exists a flat $V \subset \mathbb{F}_2^n$ of dimension k and a flat $W \subset \mathbb{F}_2^m$ of dimension k' such that all $\overline{x} \in V$ are mapped on elements of W.*

3.2   Counting Arguments

We now compute the density of the subsets which contain all vectorial Boolean functions from $\mathbb{F}_2^n$ into $\mathbb{F}_2^m$ that are not (k, k')-normal, not (k, k')-weakly normal, and no (k, k') flat-carrier.

Therefore, we make use of the following lemma, concerning the number of subspaces and flats of a certain dimension in a vector space.

Lemma 1. *[11] The number of subspaces $\gamma(n, k)$ and flats $\mu(n, k)$ of dimension k in a vector space of dimension n is given by*

$$\gamma(n, k) = \prod_{i=0}^{k-1} \frac{2^{n-i} - 1}{2^{k-i} - 1};$$

$$\mu(n, k) = 2^{n-k} \prod_{i=0}^{k-1} \frac{2^{n-i} - 1}{2^{k-i} - 1} = 2^{n-k} \gamma(n, k).$$

Lemma 2. *The number of subspaces can be upper bounded as follows:*

$$\gamma(n, k) \leq 2^{nk - k^2 + k} . \tag{3}$$

Proof. We first write $\gamma(n, k)$ as

$$\gamma(n, k) = 2^{nk - k^2 + k} \prod_{i=0}^{k-1} \frac{2^{k-1}}{2^{k-i} - 1} \frac{2^{n-i} - 1}{2^n} = 2^{nk - k^2 + k} p(n, k).$$

We now prove that the function $p(n, k) = \prod_{i=0}^{k-1} p_1(i) p_2(i)$ is smaller than 1 by proving that every factor $p_1(i) p_2(i)$ for $i \in \{0, \ldots, k-1\}$ in the product is smaller than 1. This follows from the fact that for all $0 \leq i \leq k - 1$ holds that

$$p_1(i) = \frac{2^{k-1}}{2^{k-i} - 1} < \frac{2^{k-1}}{2^{k-i-1}} = 2^i,$$

$$p_2(i) = \frac{2^{n-i} - 1}{2^n} < \frac{2^{n-i}}{2^n} = 2^{-i}.$$

$\square$

Theorem 3. *The density of the set of all vectorial functions from $\mathbb{F}_2^n$ into $\mathbb{F}_2^m$ which are*

1. *not (k, k')-normal is greater or equal to*

$$1 - 2^{n(k+1) + m(k' + 2 - 2^k) - k^2 - k'^2} . \tag{4}$$

2. *not (k, k')-weakly normal is greater or equal to*

$$1 - 2^{n(k+1) + m(k' + 2 - 2^k) + kk' - k^2 - k'^2} . \tag{5}$$

3. *no (k, k') flat-carrier is greater or equal to*

$$1 - 2^{n(k+1)+m(k'+2-2^k)+k'(2^k-1)-k^2-k'^2} . \tag{6}$$

These densities tend to zero if n, m tend to infinity for fixed k, k'.

Proof. We start by computing the first density in detail. Therefore, we determine the number $\lambda(n, m, k, k')$ of vectorial Boolean functions from $\mathbb{F}_2^n$ into $\mathbb{F}_2^m$ for which the component functions restricted to a given flat W of dimension k' are constant on a given flat V of dimension k:

$$\lambda(n, m, k, k') = 2^{m(2^n-2^k)}2^m.$$

There exist $\mu(n, k)$ flats of dimension k in $\mathbb{F}_2^n$ and $\mu(m, k')$ flats W of dimension k' in $\mathbb{F}_2^m$. So, the total number $\tau(n, m, k, k')$ of (k, k')-normal vectorial functions is less or equal than

$$\tau(n, m, k, k') \leq \lambda(n, m, k, k')\mu(n, k)\mu(m, k')$$

$$= 2^{m(2^n-2^k+1)}2^{m-k'} \prod_{i=0}^{k'-1} \frac{2^m - 2^i}{2^{k'} - 2^i} 2^{n-k} \prod_{i=0}^{k-1} \frac{2^n - 2^i}{2^k - 2^i}.$$

As a consequence, the density of the set of all vectorial functions which are not k-normal is equal to $1 - \tau(n, m, k, k')2^{-m2^n}$ and thus greater or equal than

$$1 - 2^{m(2^n-2^k+1)}2^{m-k'} \prod_{i=0}^{k'-1} \frac{2^m - 2^i}{2^{k'} - 2^i} 2^{n-k} \prod_{i=0}^{k-1} \frac{2^n - 2^i}{2^k - 2^i} 2^{-m2^n}. \tag{7}$$

By substituting the upperbound from Equation (3) in the formula above, we obtain the first density.

The second and third density can be obtained in a similar way, where only the number $\lambda(n, m, k, k')$ differs. For the second density, this number is equal to $\lambda(n, m, k, k') = 2^{m(2^n-2^k)}\mu(m, k')2^{m-k'2^{k'}(k+1)}$ and for the third density we obtain
$\lambda(n, m, k, k') = 2^{m(2^n-2^k)}\mu(m, k')2^{m-k'2^{k'2^k}}.$ □

Remark 2. We conclude that for same dimensions n, m and fixed k, k' the existence of (k, k')-normality is stronger (less likely) than the existence of a (k, k')-weakly normal function, which are both stronger than the existence of a (k, k') flat-carrier.

We now derive a structural property on the bijective power functions for n even.

Theorem 4. *For n even, every bijective power function on $\mathbb{F}_{2^n}$, i.e., x^r with $\gcd(r, 2^n - 1) = 1$, will be an (k, k) flat-carrier for any divisor k of n. Moreover, the $\frac{2^n-1}{2^k-1}$ input and output subspaces only have the zero-vector in common and thus cover exactly the whole input resp. output space.*

Proof. Recall that $\mathbb{F}_{2^n}^*$ is the cyclic group of order $2^n - 1$. If k is divisor of n, then $2^k - 1$ is divisor of $2^n - 1$. From elementary group theory, every divisor $d = 2^k - 1$ of $2^n - 1$ defines the subgroup G_d of order $2^k - 1$. Moreover, all $\frac{2^n - 1}{2^k - 1}$ cosets of G_d are disjoint and thus partition the whole group $\mathbb{F}_{2^n}^*$.

The group G_d can be represented as $G_d = <g^{\frac{2^n - 1}{d}}>$, where g is a generator of $\mathbb{F}_{2^n}^*$. The group G_d is isomorphic with $\mathbb{F}_{2^k}^*$ and $G_d \cup \{0\}$ is an additive group. Consequently $G_d \cup \{0\}$ can be seen as a subspace of dimension k. Clearly, if $G_d \cup \{0\}$ is a subspace, then also all its disjoint cosets union $\{0\}$ are subspaces, which only have the zero vector in common. The subgroup G_d is mapped on itself and its cosets are mapped on each other under a bijective power function F. Note that they define again additive (subspaces) since F^{-1} is also a bijection.   □

Corollary 1. *For n even, every bijective power function on $\mathbb{F}_{2^n}$, i.e., x^r with $gcd(r, 2^n - 1) = 1$, will be an $(\frac{n}{2}, \frac{n}{2})$ flat-carrier. Moreover, the $2^{\frac{n}{2}} + 1$ input and output subspaces only have the zero-vector in common and thus cover exactly the whole input resp. output space.*

Proof. The proof follows from Theorem 4, together with $2^n - 1 = (2^{n/2} - 1)(2^{n/2} + 1)$.   □

Remark 3. In [3], the normality of highly nonlinear bijective Boolean power functions was studied. It was shown by computer experiments that for high dimensions, n even, the Boolean function was still normal. The previous theorem may give an explanation of this fact.

Corollary 2. *For n even, every bijective power function on $\mathbb{F}_{2^n}$ will be $(2, 2)$-weakly normal. Moreover, the $\frac{2^n - 1}{3}$ input and output subspaces only have the zero-vector in common and thus cover exactly the whole input resp. output space.*

Proof. Since $2^n - 1 = (2^{n/2} - 1)(2^{n/2} + 1)$ and 3 is not a divisor of $2^{n/2}$, we conclude that 3 will divide $2^n - 1$. Therefore, by the proof of the previous theorem, any bijective power function on $\mathbb{F}_{2^n}$ will be a $(2, 2)$ flat-carrier with the disjoint property on the input and output subspaces. The theorem is proven by the fact that a bijective $(2, 2)$ flat-carrier is equivalent with a $(2, 2)$-weakly normal function since the points can only be arranged linearly for flats of dimension 2.   □

Theorem 4 also hold for any permutation on $\mathbb{F}_{2^n}$ which maps the subgroup G_3 to $\mathbb{F}_{2^2}^*$, i.e., for all linear combinations over $\mathbb{F}_2$ of linear functions, bijective power functions, and Dickson permutation polynomials. Similar properties do not hold for power functions in n odd. Therefore, it seems that the power functions for n even have more structure than power functions if n is odd.

In appendix, the input and output flats of the G_3 and G_{15} cosets of the $(8, 8)$ function $x \mapsto x^{-1}$, the S-box of the AES, is presented. Therefore, the AES S-box contains 17 disjoint subspaces, i.e., it is a $(4, 4)$ flat-carrier, and 85 disjoint subspaces for which it is a $(2, 2)$ flat-carrier. We also computed the number of subspaces for which it is a $(3, 3)$ flat-carrier. This number turns out to be equal to 0. These numbers were compared with 10 random $(8, 8)$ S-boxes which are on average a $(3, 3)$ flat-carrier on 5 flats and a $(4, 4)$ flat-carrier on 0 flats. Hence, the AES S-box shows a far stronger structure than we would expect from a random S-box and our theoretical studies from Thm. 3 are confirmed by empirical data.

3.3 Algorithms for Determining the Normality

After introducing the necessary mathematical foundations and definitions, we move on to the presentation of three algorithms which can be used to test for a given cryptographic vector function if the above definitions are fulfilled for given (k, k'). Due to space limitations in this paper, we can only sketch these algorithms and their analysis. In particular we want to remind the reader that [2] needs a whole paper to give the corresponding presentation and analysis for **one** algorithm.

In any case, all algorithms presented here have a rather high computational complexity and are therefore limited to dimensions $n = 8 \ldots 10$. However, in the current constructions of block ciphers, the S-boxes used have a rather small dimension, e.g., $n, m = 8$ for the AES, and even $n = 6, m = 4$ for the DES. Therefore, the algorithms presented in this paper may not be efficient in an algorithmic sense but are certainly practical from a cryptanalytic point of view. Still, we have to state as an open problem if faster algorithms exist for checking these properties.

(k, k') **Flat-Carrier.** In order to determine if a given vectorial function F from $\mathbb{F}_2^n$ into $\mathbb{F}_2^m$ is a (k, k') flat-carrier we suggest the following algorithm: first, perform exhaustive search over all flats of dimension k in the vector space $\mathbb{F}_2^n$. For every flat $a + V$ of dimension k, we construct an $(2^k - 1) \times m$ matrix M which corresponds to a potential flat in the output of the function F. In pseudo-code, we compute

offset $\leftarrow F(a); i \leftarrow 1$
for $b \in (a + V) \backslash \{a\}$ **do**
    RowVector$(M, i) \leftarrow F(b) \oplus$ offset; i ++;

After this, we determine the rank of the matrix M. The minimal value of all such ranks determines the value k' for a given function F and a given parameter k', i.e., if F is a (k, k') flat-carrier.

The analysis of the complexity of the above algorithm can easily be determined using Lem. 1 and is $O(m^2 2^k (2^{n-k} \prod_{n=0}^{k-1} \frac{2^{n-i}-1}{2^{k-i}-1}))$. Note that the constant hidden in the O-notation is in the range of $5 \ldots 20$ and hence negligible for our current purpose. The asymptotic complexities are computed in Table 1 for the AES and DES S-boxes.

In order to speed up the algorithm, we make use of the ideas from previous work computing the normality of a Boolean function, in particular [6,2,13]. In a nutshell, we combine small flats on the input side to derive bigger flats there. This

Table 1. Development of the Asymptotic Function for the AES and the DES in $\log_2$

k	1	2	3	4	5	6	7
$m, n = 8$	22.0	27.4	30.6	31.6	30.6	27.4	22.0
$n = 6,\ m = 4$	16.0	19.3	20.4	19.3			

way, we replace an exhaustive search over all possible input flats of dimension k by the much faster search over all possible input flats of dimension $(k-1)$, cf Lem. 1 for the effect of this change. This idea works as every function F which is a (k, k') flat-carrier, is also a $(k-1, k')$ flat-carrier. Although the contrary is not true, *i.e.*, being a $(k-1, k')$ flat-carrier for several input flats and unknown output flats, does *not* imply that a function is also a (k, k') flat-carrier, it is still a strong indication. Hence, we use the parameter k' as a kind of "filter" to concentrate only on flats which are suitable candidates for the combination step. This way, there are only a few flats to check , so it is computationally much easier to perform the necessary checks. We refer the reader to [6,2,13] for more details on the combination idea, but sketch the idea more detailed here: For a (k, k') flat-carrier, there must exist an input flat V of dimension k and one flat W of dimension k' which contains all points of the output of F when restricted to input from the flat V. Now, all points in all flats V' contained in the input flat V of dimension l with $l \leq k$, will also be contained in the same output flat W. But this also means that if F is a (k, k') flat-carrier with respect to the input flats $\overline{a} \oplus V$, $\overline{b} \oplus V$, both of dimension $k-1$, and the same output flat W, being of dimension k', one can combine the input flats to the flat $\overline{a} \oplus < V, \overline{a} \oplus \overline{b} >$ in order to obtain a (k, k') flat-carrier. Moreover, the idea of random search from [2] instead of exhaustive search over the input flats can be used to achieve an even better running time. However, this is outside the scope of this paper.

(k, k')-Weakly Normality. A (k, k')-weakly normal function for $k' \leq k$ is a special subcase of a (k, k') flat-carrier. Consequently, the same algorithm as for determining the (k, k') flat-carrier can be applied but with the additional check that the affine relations between the vectors in the input flat are maintained in the output (see Thm. 2). At first glance, this looks more difficult to check at a computational level. However, using the idea of Gray codes, we actually can do with a single "gray walk" (cf [2]) and hence obtain a computational complexity of $O(m2^k (2^{n-k} \prod_{n=0}^{k-1} \frac{2^{n-i}-1}{2^{k-i}-1}))$. Moreover, the condition for (k, k')-weakly normality acts as a much stronger filter, so the running time of the corresponding algorithm is even lower in practice. Unfortunately, the combining idea of the previous section may not be used anymore as the affine property does not propagate to flats of smaller dimension.

(k, k')-Normality. The complexity for computing the (k, k')-normality explodes very fast for higher n and m. We propose to compute the k-normality for all the 2^m components of the functions using the efficient algorithms of [2,13].

Then we check if there are components which are k-normal on the same flats taking into account the corresponding constant from the normality in order to obtain the output flat. Note that in most cases, if k is rather high, e.g., $k \approx \lceil \frac{n}{2} \rceil$, the number of flats on which a Boolean function is k-normal is rather small (cf [3] for the computation of the normality for some power functions).

Experiments. At the time of writing, we have a (highly unoptimised) prototype in place for the first and second of our three algorithms. For the third algorithm we used the ideas from [2]. We compared the number of flats for which the DES S-boxes are 2-normal, $(3, 3)$-weakly normal, and $(4, 3)$ flat-carrying with the average results of 10 random $(6, 4)$ S-boxes. The corresponding running times for checking 2-normality is equal to 8.52 s, $(3, 3)$-weakly normality is 83.58 s, and for checking $(4, 3)$ flat-carrying is 229.71 s for random $(6, 4)$ S-boxes; all experiments were carried out on an AMD Athlon XP 2000+. The number of flats for these properties are given in Table 2.

Table 2. Number of 2-normal flats and $(4, 3)$ flat-carrier

	DES1	DES2	DES3	DES4	DES5	DES6	DES7	DES8	Random
2-normal	2	1	0	4	2	2	4	0	0.9
(3,3)-weakly normal	0	0	0	3	0	0	1	0	0.1
(4,3) flat-carrier	2	0	0	12	1	0	0	2	0.6

Note that our results confirm that the fourth S-box can be considered as the weakest one [8]. Moreover, we see that the DES S-boxes have more structure than random S-boxes.

4   Relation with Propagation Characteristic

Let us first show that the new properties are affine invariant, which means that they are invariant for affine equivalent S-boxes. Two S-boxes F_1, F_2 from $\mathbb{F}_2^n$ into $\mathbb{F}_2^m$ are said to be affine equivalent if $F_1 = A_1 \circ F_2 \circ A_2$, where A_1, A_2 are affine transformations on $\mathbb{F}_2^n$ and $\mathbb{F}_2^m$ respectively. Since subspaces and flats are transformed in subspaces and flats by affine transformations, we can conclude that (weakly) normality and (k, k') flat-carrier are affine invariant properties. Affine equivalence classes reduce the total size of (n, m) S-boxes enormously, cf [1] for a practical application of this idea. It is hence interesting to find new properties which can be used for distinguishing different affine equivalence classes. Moreover, these properties lead to a better understanding of the internal structure of vectorial functions. We now show a relation between the (k, k)-weakly normality and the propagation characteristics of a function.

Theorem 5. *Let F be a function from $\mathbb{F}_2^n$ into $\mathbb{F}_2^m$ with $n \geq m$. Then, F is $(2, 2)$-weakly normal if and only if F is not APN ($\Delta(F) \neq 2$).*

Proof. Let us show the direct implication. By definition, F is $(2,2)$-weakly normal if and only if there exists a 2-dimensional flat $\bar{a} \oplus <\bar{a}_i, \bar{a}_j>$, ($\bar{a}_i$ and $\bar{a}_j$ are linearly independent vectors and $\bar{a}$ in the complement vector space of $<\bar{a}_i, \bar{a}_j>$) such that F is affine on this flat. By Theorem 2, this exactly means that

$$F(\bar{a}) \oplus F(\bar{a} \oplus \bar{a}_i) \oplus F(\bar{a} \oplus \bar{a}_j) = F(\bar{a} \oplus \bar{a}_i \oplus \bar{a}_j)$$

So, the equation

$$F(\bar{x} \oplus \bar{\alpha}) \oplus F(\bar{x}) = \bar{\beta} \tag{8}$$

with $\bar{\alpha} = \bar{a}_i, \bar{\beta} = F(\bar{a}) \oplus F(\bar{a} \oplus \bar{a}_i)$ has at least 4 solutions: $\bar{a}, \bar{a} \oplus \bar{a}_i, \bar{a} \oplus \bar{a}_j, \bar{a} \oplus \bar{a}_i \oplus \bar{a}_j$, and thus $\Delta(F) \neq 2$.

We now prove the converse. If there exists $\bar{\alpha} \in \mathbb{F}_2^n \setminus \{0\}$ and $\bar{\beta} \in \mathbb{F}_2^m$, such that the equation $F(\bar{x} \oplus \bar{\alpha}) \oplus F(\bar{x}) = \bar{\beta}$ has more than 2 solutions, e.g., $\bar{a}, \bar{a} \oplus \bar{\alpha}, \bar{b}, \bar{b} \oplus \bar{\alpha}$ where $\bar{a} \neq \bar{b}$. We can write $\bar{b}$ as $\bar{b} = \bar{a} \oplus \bar{a}_i$ for a unique $\bar{a}_i \in \mathbb{F}_2^n$, where $\bar{a}_i \neq \bar{\alpha}$ because of our assumptions. Thus the solutions belong to the flat $\bar{a} \oplus <\bar{a}_i, \bar{\alpha}>$ of dimension 2, and F restricted to this flat is affine by Thm. 2.   □

Combining Theorem 4 with the previous Theorem, we derive a result on the existence of APN functions in even dimension. Other proofs of this fact based on crosscorrelation functions exist in the literature [5].

Corollary 3. *Any bijective power functions on n even is not APN. The number of $(\bar{x}, \bar{a}, \bar{b}) \in \mathbb{F}_2^n \times (\mathbb{F}_2^n \setminus \{0\}) \times (\mathbb{F}_2^n \setminus \{0\})$ with $\bar{a} \neq \bar{b}$ for which $F(\bar{x}) \oplus F(\bar{x} \oplus \bar{a}) \oplus F(\bar{x} \oplus \bar{b}) \oplus F(\bar{x} \oplus \bar{a} \oplus \bar{b}) = 0$ is equal to $24 \times \frac{2^n - 1}{3}$.*

We can generalise Thm. 5 into one direction.

Theorem 6. *Let F be a function from $\mathbb{F}_2^n$ into $\mathbb{F}_2^m$. If F is (k, k)-weakly normal, then $\Delta(F) \geq 2^k$.*

Proof. By definition, F is (k, k)-weakly normal if and only if there exists a k-dimensional flat $V = \bar{a} \oplus <\bar{a}_1, \ldots, \bar{a}_k>$, ($\bar{a}_1, \ldots, \bar{a}_k$ are k linearly independent vectors and $\bar{a}$ is in the complement vector space of $<\bar{a}_1, \ldots, \bar{a}_k>$) such that F is affine on V. By Thm. 2, this exactly means that

$$F(\bar{a}) \oplus F(\bar{a} \oplus lc_1(\bar{a}_1 \ldots, \bar{a}_k)) \oplus F(\bar{a} \oplus lc_2(\bar{a}_1 \ldots, \bar{a}_k)) \oplus$$
$$F(\bar{a} \oplus lc_1(\bar{a}_1 \ldots, \bar{a}_k) \oplus lc_2(\bar{a}_1 \ldots, \bar{a}_k)) = 0,$$

where $lc_i(\bar{a}_1 \ldots, \bar{a}_k)$ for $i \in \{1, 2\}$ represents a linear combination of $\bar{a}_1, \ldots, \bar{a}_k$, *i.e.,* an element of the subspace $<\bar{a}_1, \ldots, \bar{a}_k>$.
Thus, the number of solutions of the equation

$$F(\bar{x} \oplus \bar{\alpha}) \oplus F(\bar{x}) = \bar{\beta}, \tag{9}$$

with $\bar{\alpha} = lc_2(\bar{a}_1 \ldots, \bar{a}_k), \bar{\beta} = F(\bar{a}) \oplus F(\bar{a} \oplus lc_2(\bar{a}_1 \ldots, \bar{a}_k))$ is equal to the number of elements of V, namely 2^k.   □

Remark 4. The opposite of the theorem is not true for $k > 2$. Consider for instance the vectorial function $F : \mathbb{F}_2^n \to \mathbb{F}_2^n$ whose function values are defined by

$$F : \{0, 1, \ldots, 15\} \to \{1, 0, 2, 3, 4, 5, 6, 7, 8, 9, 10, 11, 12, 13, 14, 15\}$$

It can be easily checked that $\Delta(F) = 16$ (the function has a nonzero linear structure, namely the vector 1) and that F is not (4,4)-weakly normal or affine.

Theorem 7. *If a function F from $\mathbb{F}_2^n$ into $\mathbb{F}_2^m$ with $n \geq m$ possesses k linearly independent linear structures, then F will be a (k, k)-weakly normal function.*

Proof. It is well-know that the set of linear structures of a function F forms an affine subspace [7,10]. Let us represent the subspace by $V = < a_1, \ldots, a_k >$, where $a_1, \ldots, a_k$ are the k linearly independent linear structures of F. To show that F is affine when restricted to V, we use the condition of Thm. 2, *i.e.*, we need to prove that $F(a_1) \oplus F(a_2) \oplus F(a_3) \oplus F(a_1 \oplus a_2 \oplus a_3) = 0$ for all $a_1, a_2, a_3 \in V$. As $a_1 \oplus a_2$ is a linear structure, this equation is satisfied by the definition of linear structure. $\qquad\square$

5   Conclusion

We have presented several new affine invariant properties for vectorial Boolean functions which can be seen as generalisations of the notion of normality for Boolean functions. We also have computed the probability of occurrence for each of these properties for randomly chosen functions and presented several practical algorithms for checking them in cryptographically relevant cases. In this context, a new structural property for the bijective power functions is discovered and its existence also proven theoretically. This new property implies that any bijective power function on an even number of variables can be seen as a mapping from $2^{n/2} + 1$ disjoint subspaces into $2^{n/2} + 1$ disjoint subspaces. We applied both this theorem and our algorithms to the S-box of the AES and computed the corresponding subspaces (cf Appendix). Moreover, we established relations with the propagation characteristics of a function.

In addition, we want to point out that our study also has applications in public key cryptography, namely for the C* scheme of Matsumoto and Imai [12]. For finite fields of even dimension n, all our proofs are applicable, and hence we have established the existence of affine subspaces covering completely both the input and the output space of the public key polynomials.

It is an open question and future research will decide if these properties can be exploited in attacks on block ciphers, stream ciphers, or public key schemes. We will now outline some ideas or observations which should be further investigated for possible applications in cryptanalysis.

If the S-box is (k, k)-weakly normal with respect to l flats of dimension k which completely cover the input space of the S-box, then the S-box can be replaced by a set of l affine transformations. Similar, S-boxes which are (k, k')-weakly normal with respect to input flats that cover the input space and output flats which cover the output flats, can be replaced by sets of affine transformations. This can lead to a type of affine approximation attack and hence, we expect the corresponding S-boxes to be rather weak.

The property of (k, k)-normality, (k, k')-normality, and its generalisation of (k, k') flat-carrier can be used in order to obtain a compact hardware design. For instance, it would be interesting to investigate if the AES S-box can be more efficiently implemented using the observations made in this paper.

Acknowledgement

We would like to thank Anne Canteaut for introducing us to this subject. We thank Frederik Vercauteren and the anonymous referees for their useful comments and suggestions. This work was supported in part by the Concerted Research Action (GOA) Ambiorics 2005/11 of the Flemish Government and by the European Commission through the IST Programme under Contract IST2002507932 ECRYPT. An Braeken is an F.W.O. Research Assistant, sponsored by the Fund for Scientific Research - Flanders (Belgium).

References

1. A. Biryukov, C. De Cannière, A. Braeken, B. Preneel, A Toolbox for Cryptanalysis: Linear and Affine Equivalence Algorithms, Eurocrypt, LNCS 2656, Springer-Verlag, pp. 33–50, 2003.
2. A. Braeken, C. Wolf, B. Preneel, A Randomised Algorithm for Checking the Normality of Cryptographic Boolean Functions, 3rd International Conference on Theoretical Computer Science, pp. 51–66, Kluwer, August 2004.
3. A. Braeken, C. Wolf, B. Preneel, Classification of Highly Nonlinear Boolean Power Functions with a Randomised Algorithm for Checking Normality, Cryptology ePrint Archive, Report 2004/214, http://eprint.iacr.org/2004/214/.
4. A. Canteaut, M. Daum, G. Leander, H. Dobbertin. Normal and non normal bent functions, International Workshop on Coding and Cryptography, pp. 91–100, March 2003.
5. P. Charpin, A. Tietäväinen, V. Zonoviev, On Binary Cyclic Codes with $d = 3$, TUCS Technical Report No. 26, June 1996.
6. M. Daum, H. Dobbertin, G. Leander, An Algorithm for Checking Normality of Boolean Functions, International Workshop on Coding and Cryptography, pp. 78–90, March 2003.
7. J.H. Evertse, Linear Structures in Block Ciphers, Eurocrypt, LNCS 304, Springer-Verlag, pp. 249–266, 1987.
8. M.E. Hellman, R. Merkle, R. Schroeppel, L. Washington, W. Die, S. Pohlig, P. Schweitzer, Results of an initial attempt to cryptanalyze the NBS Data Encryption Standard. Technical report, Stanford University, U.S.A., September 1976.
9. X.-D. Hou, Affinity of Permutations, WCC 2003, pp. 273–280, 2003.
10. X. Lai, Additive and Linear Structures of Cryptographic Functions, Eurocrypt, LNCS 1008, Springer-Verlag, pp. 75–85, 1994.
11. F.J. MacWilliams, N.J.A. Sloane, The Theory of Error-Correcting Codes, Elsevier, ISBN 0-444-85193-3, 1991.

12. T. Matsumoto and H. Imai, Public Quadratic Polynomial-Tuples for Efficient Signature Verification and Message-Encryption, EuroCrypt, LNCS 434, Springer-Verlag, pp. 419–545, 1988.
13. K. Nowak, Checking Normality of Boolean functions, 12 pages, Tatra Mountins *to appear*, 2004.
14. K. Nyberg, S-boxes and Round Functions with Controllable Linearity and Differential Cryptanalysis, FSE 1994, LNCS 1008, Springer-Verlag, pp. 11–129, 1995.
15. X.M. Zhang, Y. Zheng, The nonhomomorphicity of Boolean functions, Selected Areas in Cryptography, LNCS 1556, Springer-Verlag, pp. 280–295.
16. Y. Zheng, X.-M. Zhang, The kth-Order Nonhomomorphicity of S-Boxes, Journal of Universal Computer Science, Vol. 6, nr. 8, pp. 830–848, 2000.

A   AES as $(4, 4)$ Flat-Carrier

We now present the 17 disjoint input subspaces of dimension 4 together with the corresponding output subspaces of the AES with field polynomial $x^8 + x^4 + x^3 + x + 1$. Note that the flats are denoted by $< [\bar{a}_1, \bar{a}_2, \bar{a}_3, \bar{a}_4], \bar{b} >$, where $\bar{b}$ represents the coset and $\bar{a}_1, \bar{a}_2, \bar{a}_3, \bar{a}_4$ the four basis vectors of the subspace. Here the vectors are denoted by their radius-2 notation, i.e., $x = x_1 + 2x_2 + \cdots + 2^{n-1}x_n \in \mathbb{Z}$ corresponds with the vector $\bar{x} = (x_1, \ldots, x_n)$.

Input	Output
$< [1, 12, 80, 176], 0 >$	$< [1, 12, 80, 176], 0 >$
$< [2, 24, 97, 160], 0 >$	$< [6, 40, 88, 139], 0 >$
$< [3, 20, 44, 200], 0 >$	$< [4, 48, 91, 153], 0 >$
$< [4, 48, 91, 153], 0 >$	$< [3, 20, 44, 200], 0 >$
$< [5, 11, 50, 67], 0 >$	$< [16, 37, 66, 130], 0 >$
$< [6, 40, 88, 139], 0 >$	$< [2, 24, 97, 160], 0 >$
$< [7, 25, 35, 136], 0 >$	$< [31, 32, 74, 132], 0 >$
$< [8, 33, 65, 151], 0 >$	$< [10, 22, 100, 134], 0 >$
$< [9, 34, 71, 131], 0 >$	$< [21, 38, 79, 128], 0 >$
$< [10, 22, 100, 134], 0 >$	$< [8, 33, 65, 151], 0 >$
$< [15, 18, 68, 129], 0 >$	$< [19, 45, 64, 135], 0 >$
$< [16, 37, 66, 130], 0 >$	$< [5, 11, 50, 67], 0 >$
$< [17, 39, 69, 137], 0 >$	$< [27, 42, 76, 133], 0 >$
$< [19, 45, 64, 135], 0 >$	$< [15, 18, 68, 129], 0 >$
$< [21, 38, 79, 128], 0 >$	$< [9, 34, 71, 131], 0 >$
$< [27, 42, 76, 133], 0 >$	$< [17, 39, 69, 137], 0 >$
$< [31, 32, 74, 132], 0 >$	$< [7, 25, 35, 136], 0 >$

B   AES as $(2, 2)$ Flat-Carrier

The 85 disjoint input subspaces of dimension 2 toghether with the corresponding output subspaces of the AES with field polynomial $x^8 + x^4 + x^3 + x + 1$.

Input	Output	Input	Output
< [1, 188], 0 >	< [1, 188], 0 >	< [2, 97], 0 >	< [94, 141], 0 >
< [3, 220], 0 >	< [107, 157], 0 >	< [4, 194], 0 >	< [47, 203], 0 >
< [5, 122], 0 >	< [82, 130], 0 >	< [6, 163], 0 >	< [123, 184], 0 >
< [7, 25], 0 >	< [63, 209], 0 >	< [8, 151], 0 >	< [114, 154], 0 >
< [9, 34], 0 >	< [21, 79], 0 >	< [10, 244], 0 >	< [41, 65], 0 >
< [11, 67], 0 >	< [103, 167], 0 >	< [12, 81], 0 >	< [92, 176], 0 >
< [13, 224], 0 >	< [80, 177], 0 >	< [14, 50], 0 >	< [119, 146], 0 >
< [15, 129], 0 >	< [126, 185], 0 >	< [16, 37], 0 >	< [57, 77], 0 >
< [17, 137], 0 >	< [42, 158], 0 >	< [18, 68], 0 >	< [45, 135], 0 >
< [19, 234], 0 >	< [75, 156], 0 >	< [20, 231], 0 >	< [52, 153], 0 >
< [21, 79], 0 >	< [9, 34], 0 >	< [22, 134], 0 >	< [96, 190], 0 >
< [23, 44], 0 >	< [48, 95], 0 >	< [24, 162], 0 >	< [46, 88], 0 >
< [26, 193], 0 >	< [40, 213], 0 >	< [27, 102], 0 >	< [54, 204], 0 >
< [28, 100], 0 >	< [73, 182], 0 >	< [29, 197], 0 >	< [64, 148], 0 >
< [31, 164], 0 >	< [61, 143], 0 >	< [32, 74], 0 >	< [58, 145], 0 >
< [33, 214], 0 >	< [110, 140], 0 >	< [35, 150], 0 >	< [117, 132], 0 >
< [36, 136], 0 >	< [85, 155], 0 >	< [38, 207], 0 >	< [78, 168], 0 >
< [39, 84], 0 >	< [76, 133], 0 >	< [40, 213], 0 >	< [26, 193], 0 >
< [41, 65], 0 >	< [10, 244], 0 >	< [42, 158], 0 >	< [17, 137], 0 >
< [45, 135], 0 >	< [18, 68], 0 >	< [46, 88], 0 >	< [24, 162], 0 >
< [47, 203], 0 >	< [4, 194], 0 >	< [48, 95], 0 >	< [23, 44], 0 >
< [49, 210], 0 >	< [69, 174], 0 >	< [51, 128], 0 >	< [108, 131], 0 >
< [52, 153], 0 >	< [20, 231], 0 >	< [54, 204], 0 >	< [27, 102], 0 >
< [55, 70], 0 >	< [66, 183], 0 >	< [56, 200], 0 >	< [91, 169], 0 >
< [57, 77], 0 >	< [16, 37], 0 >	< [58, 145], 0 >	< [32, 74], 0 >
< [61, 143], 0 >	< [31, 164], 0 >	< [62, 83], 0 >	< [89, 147], 0 >
< [63, 209], 0 >	< [7, 25], 0 >	< [64, 148], 0 >	< [29, 197], 0 >
< [66, 183], 0 >	< [55, 70], 0 >	< [69, 174], 0 >	< [49, 210], 0 >
< [71, 138], 0 >	< [105, 149], 0 >	< [73, 182], 0 >	< [28, 100], 0 >
< [75, 156], 0 >	< [19, 234], 0 >	< [76, 133], 0 >	< [39, 84], 0 >
< [78, 168], 0 >	< [38, 207], 0 >	< [80, 177], 0 >	< [13, 224], 0 >
< [82, 130], 0 >	< [5, 122], 0 >	< [85, 155], 0 >	< [36, 136], 0 >
< [87, 175], 0 >	< [98, 191], 0 >	< [89, 147], 0 >	< [62, 83], 0 >
< [91, 169], 0 >	< [56, 200], 0 >	< [92, 176], 0 >	< [12, 81], 0 >
< [94, 141], 0 >	< [2, 97], 0 >	< [96, 190], 0 >	< [22, 134], 0 >
< [98, 191], 0 >	< [87, 175], 0 >	< [101, 161], 0 >	< [124, 166], 0 >
< [103, 167], 0 >	< [11, 67], 0 >	< [105, 149], 0 >	< [71, 138], 0 >
< [107, 157], 0 >	< [3, 220], 0 >	< [108, 131], 0 >	< [51, 128], 0 >
< [110, 140], 0 >	< [33, 214], 0 >	< [112, 139], 0 >	< [121, 160], 0 >
< [114, 154], 0 >	< [8, 151], 0 >	< [117, 132], 0 >	< [35, 150], 0 >
< [119, 146], 0 >	< [14, 50], 0 >	< [121, 160], 0 >	< [112, 139], 0 >
< [123, 184], 0 >	< [6, 163], 0 >	< [124, 166], 0 >	< [101, 161], 0 >
< [126, 185], 0 >	< [15, 129], 0 >		

Related-Key Differential Attacks on Cobra-H64 and Cobra-H128

Changhoon Lee[1], Jongsung Kim[2], Jaechul Sung[3],
Seokhie Hong[1], Sangjin Lee[1], and Dukjae Moon[4]

[1] Center for Information Security Technologies(CIST),
Korea University, Anam Dong, Sungbuk Gu, Seoul, Korea
{crypto77, hsh, sangjin}@cist.korea.ac.kr
[2] Katholieke Universiteit Leuven, ESAT/SCD-COSIC, Belgium
Kim.Jongsung@esat.kuleuven.be
[3] Department of Mathematics, University of Seoul,
90 Cheonnong Dong, Dongdaemun Gu, Seoul, Korea
jcsung@uos.ac.kr
[4] National Security Research Institute, 161 Gajeong-dong,
Yuseong-gu, Daejeon, 305-350, Korea
djmoon@etri.re.kr

Abstract. Cobra-H64 and Cobra-H128, which use data-dependent permutations as a main cryptographic primitive, are 64-bit and 128-bit iterated block ciphers with 128-bit and 256-bit keys, respectively. Since these ciphers use very simple key scheduling and controlled permutation (CP) for fast hardware encryption, they are suitable for wireless communications networks which require high-speed networks. Actually, these ciphers have better hardware performances than other ciphers used in security layers of wireless protocols (Wap, OMA, UMTS, IEEE 802.11 and so on). In this paper, however, we show that Cobra-H64 and Cobra-H128 are vulnerable to related-key differential attacks. We first describe how to construct full-round related-key differential characteristics of Cobra-H64 and Cobra-H128 with high probabilities and then we exploit them to attack full-round Cobra-H64 with a complexity of $2^{15.5}$ and Cobra-H128 with a complexity of 2^{44}.

Keywords: Block Ciphers, Cobra-H64, Cobra-H128, Related-Key Attacks, Data-Dependent Permutations.

1 Introduction

Many network applications of encryption require low power devices and fast computation components which imply that the number and complexity of the encryption operations should be kept as simply as possible. Recently, data-dependent permutations (DDP) have been introduced as one of cryptographic primitives suitable to attain such goal and the various DDP-based ciphers have been proposed for hardware implementation with low cost, such as CIKS-1 [16],

N.P. Smart (Ed.): Cryptography and Coding 2005, LNCS 3796, pp. 201–219, 2005.
© Springer-Verlag Berlin Heidelberg 2005

SPECTR-H64 [2]. Since all of them use very simple key scheduling in order to have no time consuming key preprocessing, they are suitable for the applications of many networks requiring high speed encryption in the case of frequent change of keys. However, the simply designed key scheduling algorithms make to help the cryptanalysts apply related-key attacks to such kinds of block ciphers [12,13].

Cobra-H64 and Cobra-H128 [17], use the switchable operations to prevent weak keys, are 64-bit and 128-bit block ciphers with simple linear key scheduling algorithms, respectively. These ciphers have better hardware implementations and performances (FPGA and ASIC) than other ciphers used in security layers of most of wireless protocols, WAP, OMA, UMTS, IEEE 802.11 and so on.

Table 1. Summary of our attacks on Cobra-H64 and Cobra-H128

Block Cipher	Number of Rounds	Complexity Data / Time
Cobra-H64	10 (full)	$2^{15.5}$RK-CP / $2^{15.5}$
Cobra-H128	12 (full)	2^{44}RK-CP / 2^{44}

RK-CP: Related-Key Chosen Plaintexts, Time: Encryption units

In this paper, we first present the structural properties of new CP-boxes used in the round function of Cobra-H64 and Cobra-H128, which allow us to make full-round related-key differential characteristics with high probabilities. Finally, we present two related-key differential attacks on full-round Cobra-H64 and Cobra-H128, which require about $2^{15.5}$ and 2^{44} data and time complexity, respectively. Table 1 summarizes our results.

This paper is organized as follows; In Section 2, we introduce some notations and properties of the used controlled permutations. Section 3 briefly describes two block ciphers, Cobra-H64, Cobra-H128, and their structural properties, and Sections 4 and 5 present related-key differential attacks of Cobra-H64 and Cobra-H128. Finally, we conclude in Section 6.

2   Preliminaries

In this section, we introduce notations used in this paper and some properties of controlled permutations which are the components of Cobra-H64 and Cobra-H128. The following notations are used throughout the paper. Bits will be numbered from left to right, starting with bit 1. If $P = (p_1, p_2, \cdots, p_n)$ then p_1 is the most significant bit and p_n is the least significant bit.

- $e_{i,j}$: A binary string in which the i-th and j-th bits are one and the others are zeroes, e.g., $e_{1,2} = (1,1,0,\cdots,0)$.
- $\oplus$: Bitwise-XOR operation
- $\lll$ ($\ggg$) : Left (Right) cyclic rotation
- $\cap$: Logical AND

2.1   Controlled-Permutations

In general, controlled permutation (CP) box used in DDP-based ciphers is defined as follows.

Definition 1. *Let $F(X, V)$ be a function $F : \{0, 1\}^n \times \{0, 1\}^m \to \{0, 1\}^n$. F is called a CP-box, if $F(X, V)$ is a bijection for any fixed V.*

We denote the above CP-box $F(X, V)$ by $P_{n/m}$, performing permutations on n-bit binary vectors X depending on some controlling m-bit vector V. The $P_{n/m}$-box is constructed by using elementary switching elements $P_{2/1}$ as elementary building blocks performing controlled transposition of two input bits x_1 and x_2. Here, $P_{2/1}$-box is controlled with one bit v and outputs two bits y_1 and y_2, where $y_1 = x_{1+v}$ and $y_2 = x_{2-v}$, i.e., if $v = 1$, it swaps two input bits otherwise (if $v = 0$), does not.

In other words, $P_{n/m}$-box can be represented as a superposition of the operations performed on bit sets :

$$P_{n/m} = L^{V_1} \circ \pi_1 \circ L^{V_2} \circ \pi_2 \circ \cdots \circ \pi_{s-1} \circ L^{V_s}$$

where L is an active layer composed of $n/2$ $P_{2/1}$ parallel elementary boxes, $V_1, V_2, \cdots V_s$ are controlling vectors of the active layers from 1 to $s = 2m/n$, and $\pi_1, \pi_2, \cdots, \pi_{s-1}$ are fixed permutations (See Fig. 1). Fig. 2 shows structure of the $P_{32/96}$ $(P_{32/96}^{-1})$ and $P_{64/192}$ $(P_{64/192}^{-1})$ used in Cobra-H64 and Cobra-H128. Due to the symmetric structure, the mutual inverses, $P_{n/m}$ and $P_{n/m}^{-1}$, differ only with the distribution of controlling bits over the boxes $P_{2/1}$, e.g., $P_{32/96}^{V}$ and $P_{32/96}^{V'}$ are mutually inverse when $V = (V_1, V_2, \cdots, V_6)$ and $V' = (V_6, V_5, \cdots, V_1)$.

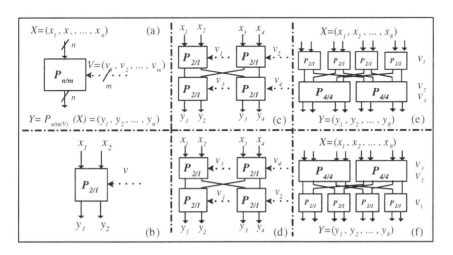

Fig. 1. *CP*-boxes : (a) $P_{n/m}$, (b) $P_{2/1}$, (c) $P_{4/4}$, (d) $P_{4/4}^{-1}$, (e) $P_{8/12}$, (f) $P_{8/12}^{-1}$

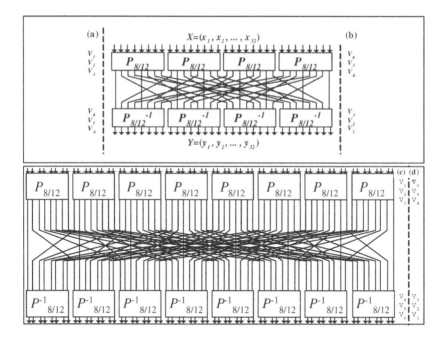

Fig. 2. CP-boxes : (a) $P_{32/96}$, (b) $P^{-1}_{32/96}$, (c) $P_{64/192}$, (d) $P^{-1}_{64/192}$

Now, we present general properties of CP-boxes which can induce properties of operations used in the round function of Cobra-H64 and Cobra-H128.

Property 1. [12,13] Let an input and controlling vector differences of $P_{2/1}$-box be $\Delta X = X \oplus X'$ and $\Delta V = V \oplus V'$ respectively, where X and X' are two-bit input vectors, and V and V' are one-bit controlled vectors. Then we get the following equations.

a) If $\Delta X = 10$(or 01) and $\Delta V = 0$ then the corresponding output difference of $P_{2/1}$-box is $\Delta Y = 10$(or 01) with probability 2^{-1} and $\Delta Y = 01$(or 10) with probability 2^{-1}.

b) If $\Delta X = 00$ and $\Delta V = 1$ then the corresponding output difference of $P_{2/1}$-box is $\Delta Y = 00$ with probability 2^{-1} and $\Delta Y = 11$ with probability 2^{-1}.

The above properties are also expanded into the following properties.

Property 2. [12,13] Let V and V' be m-bit control vectors for $P_{n/m}$-box such that $V \oplus V' = e_i$ ($1 \leq i \leq m$). Then $P_{n/m(V)}(X) = P_{n/m(V')}(X)$ with probability 2^{-1} where $X \in \{0,1\}^n$. It also holds in $P^{-1}_{n/m}$-box.

Property 3. [12,13] Let X and X' be n-bit inputs for $P_{n/m}$-box such that $X \oplus X' = e_i$ ($1 \leq i \leq n$). Then $P_{n/m(V)}(X) \oplus P_{n/m(V)}(X') = e_j$ for some j ($1 \leq j \leq n$).

Property 4. Let $P_{n/m(V)}(X) \oplus P_{n/m(V)}(X \oplus e_i) = e_j$ for some i and j.

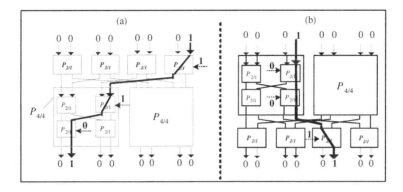

Fig. 3. An example of the difference routes when the input and output differences of $P_{8/12}$ and $P_{8/12}^{-1}$ are fixed

a) If $n = 8$, $m = 12$ then the exact one difference route from i to j via three $P_{2/1}$-boxes is fixed. It also holds in $P_{8/12}^{-1}$-box.

b) If $n = 32$, $m = 96$ then the exact two difference routes from i to j via six $P_{2/1}$-boxes are fixed. It also holds in $P_{32/96}^{-1}$-box.

c) If $n = 64$, $m = 192$ then the exact one difference route from i to j via six $P_{2/1}$-boxes is fixed. It also holds in $P_{64/192}^{-1}$-box.

For example, consider $i = 8$ and $j = 2$ in the *Property* 4-a). Then, we can exactly know the 3 bits of control vectors $(1,1,0)$ corresponding to three elements $P_{2/1}$-boxes of $P_{8/12}$-box with probability 1. See Fig. 3. In Fig. 3, the bold line denotes the possible difference route when the input and output differences of $P_{8/12}$ and $P_{8/12}^{-1}$ are fixed.

3   Cobra-H64 and Cobra-H128

In this section, we briefly describe two block ciphers, Cobra-H64, Cobra-H128 [17] and introduce their properties used in our attacks. These ciphers use same iterative structure and are composed of the initial transformation (IT), e-dependent round function $Crypt^{(e)}$, and the final transformation (FT) where $e = 0$ ($e = 1$) denotes encryption (decryption) as follow:

1. An input data block is divided into two subblocks L and R.
2. Perform initial transformation :
   $L_0 = L \oplus O_3$ and $R_0 = R \oplus O_4$, where O_3 and O_4 are subkeys;
3. For $j = 1$ to $r - 1$ do :
   ○ $(L_j, R_j) := Crypt^{(e)}(L_{j-1}, R_{j-1}, Q_j^{(e)})$, where $Q_j^{(e)}$ is the j-th round key;
   ○ Swap the data subblocks : $T = R_j$, $R_j = L_j$, $L_j = T$;
4. $j = r$ do :
   ○ $(L_r, R_r) := Crypt^{(e)}(L_{r-1}, R_{r-1}, Q_r^{(e)})$;

5. Perform final transformation :
   $C_L = L_r \oplus O_1$ and $C_R = R_r \oplus O_2$, where O_1 and O_2 are subkeys;
6. Return the ciphertext block $C = (C_L, C_R)$.

3.1 A Description of Cobra-H64

Cobra-H64 encrypts 64-bit data blocks with a 128-bit key by iterating a round function 10 times. The $Crypt^{(e)}$ used in Cobra-H64 consists of an extension box E, a switchable fixed permutation $\pi^{(e)}$, a permutational involution I, a nonlinear operation G, and two CP-boxes $P_{32/96}$, $P_{32/96}^{-1}$. See Fig. 4. The extension box E provides the following relation between its input $L = (l_1, \cdots, l_{32})$ and output $V = (V_1, \cdots, V_6)$:

$$V_1 = L_l, \quad V_2 = L_l^{\lll 6}, \quad V_3 = L_l^{\lll 12}, \quad V_4 = L_h, \quad V_5 = L_h^{\lll 6}, \quad V_6 = L_r^{\lll 12}$$

where $L_l = (l_1, \cdots, l_{16})$, $L_h = (l_{17}, \cdots, l_{32})$, $|l_i| = 1$ $(1 \leq i \leq 32)$ and $|V_i| = 16$ $(1 \leq i \leq 6)$.

The switchable fixed permutation $\pi^{(e)}$ performs permutation $\pi^{(0)}$ when enciphering, and $\pi^{(1)}$ when deciphering. Both of them contain two cycles. The first cycle corresponds to identical permutation of the least significant input bit x_{32}. The second cycle is described by the following equations:

$$\pi^{(0)}(x_1, x_2, \cdots, x_{31}) = (x_1, x_2, \cdots, x_{31})^{\lll 5}, \quad \pi^{(1)}(x_1, x_2, \cdots, x_{31}) = (x_1, x_2, \cdots, x_{31})^{\lll 26}$$

The permutational involution I which is used to strengthen the avalanche effect is performed as follows:

$$I = \ (1,17)(2,21)(3,25)(4,29)(5,18)(6,22)(7,26)(8,30)(9,19)(10,23)$$
$$(11,27)(12,31)(13,20)(14,24)(15,28)(16,32).$$

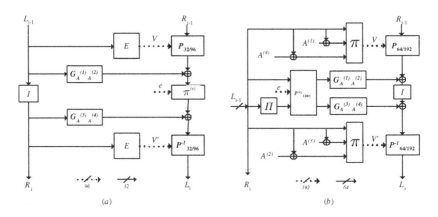

Fig. 4. (a) $Crypt^{(e)}$ of Cobra-H64, (b) $Crypt^{(e)}$ of Cobra-H128

Table 2. Key schedule of Cobra-H64

j	1	2	3	4	5	6	7	8	9	10
$A^{(1)}_j$	O_1	O_4	O_3	O_2	O_1	O_1	O_2	O_3	O_4	O_1
$A^{(2)}_j$	O_2	O_1	O_4	O_3	O_4	O_4	O_3	O_4	O_1	O_2
$A^{(3)}_j$	O_3	O_2	O_1	O_4	O_3	O_3	O_4	O_1	O_2	O_3
$A^{(4)}_j$	O_4	O_3	O_2	O_1	O_2	O_2	O_1	O_2	O_3	O_4

The operation $G_{A'A''}(L)$, which is only nonlinear part in Cobra-H64, is described by the following expression ((A', A'') can be round keys $(A^{(1)}, A^{(2)})$ or $(A^{(3)}, A^{(4)})$):

$$W = L_0 \oplus A'_0 \oplus (L_2 \cap L_3) \oplus (L_1 \cap L_2) \oplus (L_1 \cap L_3) \oplus (L_2 \cap A''_1) \oplus (A'_1 \cap L_3) \oplus (A''_0 \cap L_1 \cap L_2)$$

where binary vectors L_j, A'_j, and A''_j are expressed as follows:

$$L_0 = L = (l_1, l_2, \cdots, l_{32}), L_1 = (1, l_1, l_2, \cdots, l_{31}), L_2 = (1, 1, l_1, \cdots, l_{30}),$$
$$L_3 = (1, 1, 1, l_1, \cdots, l_{29}), A'_0 = A' = (a'_1, a'_2, \cdots, a'_{32}), A'_1 = (1, a'_1, a'_2, \cdots, a'_{31})$$
$$A''_0 = A'' = (a''_1, a''_2, \cdots, a''_{32}), A''_1 = (1, a''_1, a''_2, \cdots, a''_{31}), A''_2 = (1, 1, a''_1, \cdots, a''_{30})$$

The key schedule of Cobra-H64 is very simple. An 128-bit master key K is split into four 32-bit blocks, i.e., $K = (K_1, K_2, K_3, K_4)$. Then, in order to generate 10 e-dependent round keys $Q^{(e)}_j = (A^{(1)}_j, A^{(2)}_j, A^{(3)}_j, A^{(4)}_j)$ ($1 \leq j \leq 10$), K_1, K_2, K_3 and K_4 are rearranged as specified in Table 2 in which $O_i = K_i$ if $e = 0$, $O_1 = K_3$, $O_2 = K_4$, $O_3 = K_1$, $O_4 = K_2$ if $e = 1$.

3.2   A Description of Cobra-H128

Cobra-H128 is a 128-bit block cipher with a 256-bit key and the number of 12 rounds. The $Crypt^{(e)}$ of Cobra-H128 uses two fixed permutations π, Π, a permutational involution I, a nonlinear operation G, and two CP-boxes $P^{(V)}_{64/192}$, $(P^{-1}_{64/192})^{(V')}$ (See Fig. 4). These components are a little bit different from those of Cobra-H64.

1. The permutation Π contains four cycles of the length 16 represented as follows;

   (1,50,9,42,17,34,25,26,33,18,41,10,49,2,57,58)(3,64,43,24,19,48,59,8,35,32,11,56,51,16,27,40)
   (4,7,28,47,52,23,12,63,36,39,60,15,20,55,44,31)(5,14,13,6,21,62,29,54,37,46,45,38,53,30,61,22).

2. π forms the control vectors V and V' using three 64-bit input values for the $P_{64/192}$ and $P^{-1}_{64/192}$-box respectively. For example, let us consider formation of the vector $V = (V_1, V_2, V_3, V_4, V_5, V_6) = \pi(L, A_1, A_4)$ where $V_i \in \{0, 1\}^{32}$

Table 3. 192 bits control vector V and the corresponding positions for $P_{64/192}$-box

V	$P_{64/192}$																															
V_1	31	32	3	4	5	6	7	8	9	10	11	12	13	14	15	16	17	18	19	20	21	22	23	24	25	26	27	28	29	30	1	2
V_2	10'	24'	25'	26'	29'	13'	27'	16'	1'	2'	31'	32'	3'	4'	19'	6'	7'	8'	9'	23'	11'	12'	28'	15'	14'	30'	17'	18'	5'	20'	21'	22'
V_3	13"	14"	15"	16"	17"	18"	19"	20"	21"	22"	23"	24"	25"	26"	27"	28"	29"	30"	31"	32"	1"	2"	3"	4"	5"	6"	7"	8"	12"	10"	11"	9"
V_4	33	34	35	36	37	38	39	40	41	42	43	44	45	46	47	48	49	50	51	52	53	54	55	56	57	58	59	60	61	62	63	64
V_5	55'	56'	57'	58'	59'	60'	61'	62'	63'	64'	33'	34'	35'	36'	37'	38'	39'	40'	41'	42'	43'	44'	45'	46'	47'	48'	49'	50'	51'	52'	53'	54'
V_6	45"	46"	47"	48"	49"	50"	51"	52"	53"	54"	55"	56"	57"	58"	59"	60"	61"	62"	63"	64"	33"	34"	35"	36"	37"	38"	39"	40"	41"	42"	43"	44"

and $L,A_1,A_4 \in \{0,1\}^{64}$ $(1 \leq i \leq 32)$. Table 3 depicts the distribution of the 192 controlling bits in $P_{64/192}$-box.

In Table 3, (V_1,V_4), (V_2,V_5) and (V_3,V_6) are represented as respective rearrangement of bits of $L=(l_1,l_2,\cdots,l_{64})$, $L \oplus A_1=(l_1 \oplus a_1^1, l_2 \oplus a_2^1, \cdots, l_{64} \oplus a_{64}^1)$ and $L \oplus A_4=(l_1 \oplus a_1^4, l_2 \oplus a_2^4, \cdots, l_{64} \oplus a_{64}^4)$, i.e., $i=l_i$, $j'=l_j \oplus a_j^1$ and $k''=l_k \oplus a_k^4$ where $l_i, a_j^1, a_k^4 \in \{0,1\}$ and $1 \leq i,j,k \leq 64$.

3. I is a permutational involution. It is described as follows:
$Y = (Y_1, Y_2, \cdots, Y_8) = I(X_1, X_2, \cdots, X_8)$, where $Y_1 = X_6^{\lll 4}, Y_2 = X_5^{\lll 4}$, $Y_3 = X_4^{\lll 4}$, $Y_4 = X_3^{\lll 4}, Y_5 = X_2^{\lll 4}, Y_6 = X_1^{\lll 4}, Y_7 = X_8^{\lll 4}, Y_8 = X_7^{\lll 4}$ $(1 \leq i \leq 8)$.

4. G is the only non-linear part of $Crypt^{(e)}$. If $L=(l_1,\ldots,l_{64})$ is a 64-bit input value, and $A'=(a_1',\ldots,a_{64}')$ and $A''=(a_1'',\ldots,a_{64}'')$ are 64-bit subkeys of G then the output value $W=G(L, A', A'')=G_{(A',A'')}(L)$ of G is computed as follows;

$$W = L_0 \oplus A_0' \oplus (L_1 \cap A_0'') \oplus (L_2 \cap L_5) \oplus (L_6 \cap A_1') \oplus (A_1'' \cap A_2') \oplus (L_4 \cap L_3) \oplus (L_1 \cap L_6 \cap L_4) \oplus (L_2 \cap L_6 \cap A_1'') \oplus (L_1 \cap A_1'' \cap L_2 \cap L_4),$$

where $\forall i \in \{0,1,2\}$, $\forall j \in \{0,1,...,6\}$, the binary vectors L_j and A_i are defined as : $L_j = L^{\lll 64-j}$, $A_0 = A$, $A_1 = (1, a_1, ..., a_{63})$, $A_2 = (1, 1, a_1, ..., a_{62})$, $(A=A'$ or $A'')$.

The key schedule of Cobra-H128 is also very simple and uses Table 4 as a rearrangement of the master key sequences $(K_1, K_2, K_3 \ K_4)$ where $|K_i|=64$.

Table 4. Key schedule of Cobra-H128

j	1	2	3	4	5	6	7	8	9	10	11	12
$A^{(1)}_j$	O_1	O_4	O_3	O_2	O_1	O_3	O_3	O_1	O_2	O_3	O_4	O_1
$A^{(2)}_j$	O_2	O_3	O_4	O_1	O_2	O_4	O_4	O_2	O_1	O_4	O_3	O_2
$A^{(3)}_j$	O_3	O_2	O_1	O_4	O_3	O_1	O_1	O_3	O_4	O_1	O_2	O_3
$A^{(4)}_j$	O_4	O_1	O_2	O_3	O_4	O_2	O_2	O_4	O_3	O_2	O_1	O_4

3.3   Properties of Cobra-H64 and Cobra-H128

In this subsection, we describe some properties for components of $Crypt^{(e)}$ of Cobra-H64 and Cobra-H128, which allow us to construct strong related key differential characteristics.

Property 5. This is a property for components of $Crypt^{(e)}$ of Cobra-H64.

a) If L is a random input and A', A'' are two random round keys then $G_{A'A''}(L)$ $\oplus G_{A'\oplus e_{32}A''\oplus e_{32}}(L) = 0$ with probability $1/4$ (i.e., it holds only when $(l_{30}, l_{31}) = (1, 1)$) and $G_{A'A''}(L) \oplus G_{A'\oplus e_{32}A''\oplus e_{32}}(L) = e_{32}$ with probability $3/4$ (i.e., it holds only when $(l_{30}, l_{31}) = (0, 0), (0, 1)$ or $(1, 0)$).

b) For any fixed i, j ($1 \leq i, j \leq 32$) $\Delta P_{32/96(\Delta V = 0)}(\Delta X = e_i) = e_j$ with probability 2^{-5}. (For any fixed i, j there can be two difference routes in $P_{32/96(V)}(X)$, and each route occurs with probability 2^{-6}.) Similarly, it also holds in $P_{32/96}^{-1}$.

Property 6. This is a property for components of $Crypt^{(e)}$ of Cobra-H128.

a) For the control vector V of $P_{64/192}$-box, $\pi(L, A', A'') \oplus \pi(L, A' \oplus e_{64}, A'')=$ e_{138} and $\pi(L, A', A'') \oplus \pi(L, A', A'' \oplus e_{64})=e_{180}$. For the control vector V' of $P_{64/192}^{-1}$-box, $\pi(L, A', A'') \oplus \pi(L, A' \oplus e_{64}, A'')=e_{42}$ and $\pi(L, A', A'') \oplus \pi(L, A', A'' \oplus e_{64})=e_{20}$. where $L, A', A''\in\{0, 1\}^{64}$ and $V, V' \in \{0, 1\}^{192}$.

b) If L is a random input and A', A'' are two random round keys then $G_{A'A''}(L)$ $\oplus G_{A'\oplus e_{64}A''\oplus e_{64}}(L) = 0$ with probability $1/2$ (i.e., it holds only when $l_{63} = 1$) and $G_{A'A''}(L) \oplus G_{A'\oplus e_{64}A''\oplus e_{64}}(L) = e_{64}$ with probability $1/2$ (i.e., it holds only when $l_{63} = 0$).

c) For any fixed i, j ($1 \leq i, j \leq 64$) $\Delta P_{64/192(\Delta V=0)}(\Delta X = e_i) = e_j$ with probability 2^{-6}. (For any fixed i, j there can be one difference route in $P_{64/192(V)}(X)$, and this route occurs with probability 2^{-6}.) Similarly, it also holds in $P_{64/192}^{-1}$.

4   Related-Key Differential Characteristics on Cobra-H64 and Cobra-H128

In this section, we construct related-key differential characteristics for Cobra-H64 and Cobra-H128 using the properties mentioned in the previous subsection.

4.1   Related-Key Differential Characteristic on Cobra-H64

As stated earlier, the key schedule of the Cobra-H64 is very simple, i.e., the round keys are only 32-bit parts of the 128-bit master key, and there are many useful properties of $P_{32/96}$ and $P_{32/96}^{-1}$ which allow us to construct useful related-key differential characteristics.

In this subsection, we show how to construct full-round (10 rounds) related-key differential characteristics with a high probability. We consider the situation that we encrypt plaintexts $P = (P_L, P_R)$ and $P' = (P'_L, P'_R)$ under an unknown key $K=(K_1, K_2, K_3, K_4)$ and an unknown related-key $K' = (K'_1, K'_2, K'_3, K'_4)$ such that $P \oplus P' = (e_{32}, e_{32})$ and $K \oplus K' = (e_{32}, e_{32}, e_{32}, e_{32})$, respectively. Then we can obtain 32 desired full-round related-key differential characteristics $\alpha \to \beta_j$ with the same probability of $2^{-12.5}$, where $\alpha = (e_{32}, e_{32})$ and $\beta_j = (e_{32}, e_{j,32})$ for each j ($1 \le j \le 32$) as depicted in Table 5.

The related-key differential characteristics described in Table 5 exploits one round iterative differential characteristic whose input and output differences are $(0, 0)$ and key difference is $(e_{32}, e_{32}, e_{32}, e_{32})$. This one round iterative differential characteristic holds with probability $10/16$, which can be obtained as follows. Since the $\pi^{(e)}$ function does not affect the least significant bit (i.e., 32-th bit), if the output differences of the first and second G functions are both 0 or e_{32} then the one round iterative differential characteristic should be satisfied. According to *Property* 5-a), the output differences of the first and second G functions are both 0 with probability $1/16(= 1/4 \cdot 1/4)$ and the output differences of the first and second G functions are both e_{32} with probability $9/16(= 3/4 \cdot 3/4)$ and thus the one round iterative differential characteristic holds with probability $10/16$.

In order to make a key recovery attack of Cobra-H64 easily, we use another differential characteristic in the last round whose input difference is $(0, 0)$, output difference is $(0, e_j)$ and key difference is $(e_{32}, e_{32}, e_{32}, e_{32})$. This one round differential characteristic holds with probability $(3/8) \cdot 2^{-5}$, which can be obtained as follows. For getting the desired output difference it should be satisfied that one of output differences of the first and second G functions is e_{32} and

Table 5. Related-Key Differential Characteristic of Cobra-H64

Round (i)	ΔRI^i	ΔRK^i	Prob.
IT	(e_{32}, e_{32})	(e_{32}, e_{32})	1
1	$(0,0)$	$(e_{32}, e_{32}, e_{32}, e_{32})$	$10/16$
2	$(0,0)$	$(e_{32}, e_{32}, e_{32}, e_{32})$	$10/16$
3	$(0,0)$	$(e_{32}, e_{32}, e_{32}, e_{32})$	$10/16$
4	$(0,0)$	$(e_{32}, e_{32}, e_{32}, e_{32})$	$10/16$
5	$(0,0)$	$(e_{32}, e_{32}, e_{32}, e_{32})$	$10/16$
6	$(0,0)$	$(e_{32}, e_{32}, e_{32}, e_{32})$	$10/16$
7	$(0,0)$	$(e_{32}, e_{32}, e_{32}, e_{32})$	$10/16$
8	$(0,0)$	$(e_{32}, e_{32}, e_{32}, e_{32})$	$10/16$
9	$(0,0)$	$(e_{32}, e_{32}, e_{32}, e_{32})$	$10/16$
10	$(0,0)$	$(e_{32}, e_{32}, e_{32}, e_{32})$	$(3/8) \cdot 2^{-5}$
FT	$(0, e_j)$	(e_{32}, e_{32})	1
Output	$(e_{32}, e_{j,32})$	.	.
Total	.	.	$2^{-12.5}$

$1 \le j \le 32$: fixed value, if $j = 32$, $e_{j,32} = 0$

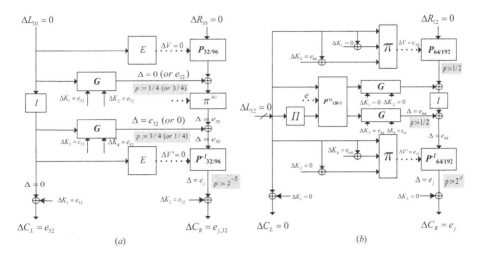

Fig. 5. Propagation of the difference in the last round

the other is 0. According to *Property* 5-a), this event occurs with probability $3/8(= 2 \cdot (1/4) \cdot (3/4))$. Since for any fixed j $(1 \leq j \leq 32)$ $\Delta P_{32/96(\Delta V=0)}^{-1}(\Delta X = e_{32}) = e_j$ with probability 2^{-5} (refer to *Property* 5-b)), the last round differential characteristic holds with probability $(3/8) \cdot 2^{-5}$.

In order to verify these results we performed a series of simulations with randomly chosen 2^{14} plaintexts and randomly chosen 3000 related key pairs, respectively. As a result, we checked that there exist more than 3 pairs on average satisfying each of ciphertext differences in Table 5. Our simulation result is higher than our expectation $2^{1.5}(= 2^{14} \times 2^{-12.5})$. This difference is due to the fact that our estimation only considers one differential characteristic rather than a differential.

4.2   Related-Key Differential Characteristic on Cobra-H128

Using the same method presented in the previous subsection, we construct full-round related-key differential characteristics for Cobra-H128. We consider the situation that we encrypt plaintexts P and P' under an unknown key K and an unknown related-key K' such that $P \oplus P' = (e_{64}, e_{64})$ and $K \oplus K' = (0, 0, e_{64}, e_{64})$, respectively. Then we can obtain 64 desired full-round related-key differential characteristics $(e_{64}, e_{64}) \rightarrow (0, e_j)$ $(1 \leq j \leq 64)$ with the same probability of 2^{-42}, as depicted in Table 6.

Since we consider a related-key pair (K, K') satisfying $K \oplus K' = (0, 0, e_{64}, e_{64})$, we know the difference form of each round key is satisfied with $RK = (0, 0, e_{64}, e_{64})$ or $RK = (e_{64}, e_{64}, 0, 0)$ (See Table. 4). Now, according to the condition of RK, we describe one round differential characteristic of $Crypt^{(e)}$ used in our attack.

Table 6. Related-Key Differential Characteristic of Cobra-H128

Round (i)	ΔRI^i	ΔRK^i	P1/P2/P3	Prob.	Case
IT	(e_{64}, e_{64})	(e_{64}, e_{64})	$\cdot$	1	$\cdot$
1	$(0,0)$	$(0,0,e_{64},e_{64})$	$2^{-1}/2^{-1}/2^{-1}$	2^{-3}	$C1$
2	$(0,0)$	$(e_{64},e_{64},0,0)$	$2^{-1}/2^{-1}/2^{-1}$	2^{-3}	$C2$
3	$(0,0)$	$(e_{64},e_{64},0,0)$	$2^{-1}/2^{-1}/2^{-1}$	2^{-3}	$C2$
4	$(0,0)$	$(0,0,e_{64},e_{64})$	$2^{-1}/2^{-1}/2^{-1}$	2^{-3}	$C1$
5	$(0,0)$	$(0,0,e_{64},e_{64})$	$2^{-1}/2^{-1}/2^{-1}$	2^{-3}	$C1$
6	$(0,0)$	$(e_{64},e_{64},0,0)$	$2^{-1}/2^{-1}/2^{-1}$	2^{-3}	$C2$
7	$(0,0)$	$(e_{64},e_{64},0,0)$	$2^{-1}/2^{-1}/2^{-1}$	2^{-3}	$C2$
8	$(0,0)$	$(0,0,e_{64},e_{64})$	$2^{-1}/2^{-1}/2^{-1}$	2^{-3}	$C1$
9	$(0,0)$	$(0,0,e_{64},e_{64})$	$2^{-1}/2^{-1}/2^{-1}$	2^{-3}	$C1$
10	$(0,0)$	$(e_{64},e_{64},0,0)$	$2^{-1}/2^{-1}/2^{-1}$	2^{-3}	$C2$
11	$(0,0)$	$(e_{64},e_{64},0,0)$	$2^{-1}/2^{-1}/2^{-1}$	2^{-3}	$C2$
12	$(0,0)$	$(0,0,e_{64},e_{64})$	$2^{-1}/2^{-1}/2^{-7}$	2^{-9}	$C1'$
FT	$(0,e_j)$	$(0,0)$	$\cdot$	1	$\cdot$
Output	$(0,e_j)$	$\cdot$	$\cdot$	$\cdot$	$\cdot$
Total	$\cdot$	$\cdot$	$\cdot$	2^{-42}	$\cdot$

$1 \leq j \leq 64$: fixed value

$C1$: $RK = (0,0,e_{64},e_{64})$

If the input difference of $Crypt^{(e)}$ is zero then, by *Property 6-a*), the output difference of the first π is (e_{180}) with probability 1. Thus, by *Property 2*, the output difference of $P_{64/192}$ is zero with probability $P1 = 2^{-1}$ because the input and controlled vector differences are 0 and e_{180}, respectively. Since the input and round key differences of the second G are 0 and (e_{64},e_{64}), respectively, the corresponding output difference of the second G is 0 with probability $P2 = 2^{-1}$. Similarly, the output difference of the second π is e_{42} and the output difference of $P_{64/192}^{-1}$ is 0 with probability $P3 = 2^{-1}$. Hence if the input difference of $Crypt^{(e)}$ is 0 under $RK = (0,0,e_{64},e_{64})$ then the corresponding output difference of $Crypt^{(e)}$ is 0 with probability 2^{-3}.

$C2$: $RK = (e_{64},e_{64},0,0)$

If the input difference of $Crypt^{(e)}$ is zero then, by *Property 6-a*), the output difference of the first π is (e_{138}) with probability 1. Thus, by *Property 2*, the output difference of $P_{64/192}$ is zero with probability $P1 = 2^{-1}$ because the input and controlled vector differences are 0 and e_{138}, respectively. Since the input and round key differences of the first G are 0 and (e_{64},e_{64}), respectively, the corresponding output difference of the second G is 0 with probability $P2 = 2^{-1}$. Similarly, the output difference of the second π is e_{20} and the output difference of $P_{64/192}^{-1}$ is 0 with probability $P3 = 2^{-1}$. Hence if the input difference of $Crypt^{(e)}$ is 0 under $RK = (e_{64},e_{64},0,0)$ then the corresponding output difference of $Crypt^{(e)}$ is 0 with probability 2^{-3}.

We alternatively use $C1$ and $C2$ to construct the first 11 rounds of our differential characteristics (See Table 6). In the last round, however, we use a little bit different characteristic from $C1$ and $C2$ for our key recovery attack. In Table 6, the case of $C1'$ means that the output difference of the second G in the last round is not zero but e_{64} with probability 2^{-1}, and then by *Property* 2 and *Property* 6-c) we have $P3 = 2^{-7}$ in the last round. See Fig. 5.

5 Key Recovery Attacks on Cobra-H64 and Cobra-H128

We now present key recovery attacks on Cobra-H64 and Cobra-H128 using our related-key differential characteristics.

5.1 Attack Procedure on Cobra-H64

To begin with, we encrypt $2^{14.5}$ plaintext pairs $P = (P_L, P_R)$ and $P' = (P_L \oplus e_{32}, P_R \oplus e_{32})$ under an unknown key $K = (K_1, K_2, K_3, K_4)$ and an unknown related-key $K' = (K_1 \oplus e_{32}, K_2 \oplus e_{32}, K_3 \oplus e_{32}, K_4 \oplus e_{32})$, respectively, and then get the $2^{14.5}$ corresponding ciphertext pairs $C = (C_L, C_R)$ and $C' = (C'_L, C'_R)$, i.e., $E_K(P) = C$ and $E_{K'}(P') = C'$, where E is the block cipher Cobra-H64. Since our full-round related-key differential characteristic of Cobra-H64 has a probability of $2^{-12.5}$, we expect about four ciphertext pair (C, C') such that $C \oplus C' = (e_{32}, e_{j,32})$ for each j $(1 \leq j \leq 32)$. According to our differential trail described in Table 5, we can deduce that the j-th one-bit difference in such (C, C') is derived from the output difference of $P_{2/1}^{(V_6'^{16})}$ in $P_{32/96}^{-1}$ of the last round (Refer to Fig 6). That is, we can expect that there are two differential routes: one is from $P_{2/1}^{(V_6'^{16})}$ and $P_{2/1}^{(V_5'^{14})}$, and the other is from $P_{2/1}^{(V_6'^{16})}$ and $P_{2/1}^{(V_5'^{16})}$ (the second one is described in Fig. 6). From each of these two routes, we can extract 6 bits of control vectors by using Property 3. However, since in the attack procedure one route is a right route and the other is a wrong route, one of the two extracted 6 bits may not be correct.

For example, assume that output difference is e_{27}. Then we have the following two 6 bits of control vectors (Refer to Fig. 6 and Table 7, 8).

$$- \ v_{96} = C_L^{28} \oplus K_1^{28} = 0, \ v_{80} = C_L^{22} \oplus K_1^{22} = 0, \ v_{64} = C_L^{32} \oplus K_1^{32} = 0,$$
$$v_{48} = C_L^{12} \oplus K_1^{12} = 1, \ v_{31} = C_L^5 \oplus K_1^5 = 0, \ v_{14} = C_L^{14} \oplus K_1^{14} = 1$$
$$- \ v_{96} = C_L^{28} \oplus K_1^{28} = 1, \ v_{78} = C_L^{20} \oplus K_1^{20} = 0, \ v_{62} = C_L^{30} \oplus K_1^{30} = 0,$$
$$v_{46} = C_L^{10} \oplus K_1^{10} = 1, \ v_{29} = C_L^3 \oplus K_1^3 = 0, \ v_{14} = C_L^{14} \oplus K_1^{14} = 0$$

From this procedure we can increase counters of extracted keys. If we use enough plaintext pairs to follow the above procedure, we can distinguish the right key from wrong keys by the maximum likelihood method. Based on this idea we can devise a related-key differential attack on full-round Cobra-H64.

1. Prepare $2^{14.5}$ plaintext pairs (P_i, P'_i), $i = 1, \cdots, 2^{14.5}$, which have the (e_{32}, e_{32}) difference. All P_i are encrypted using a master key K and all P'_i are

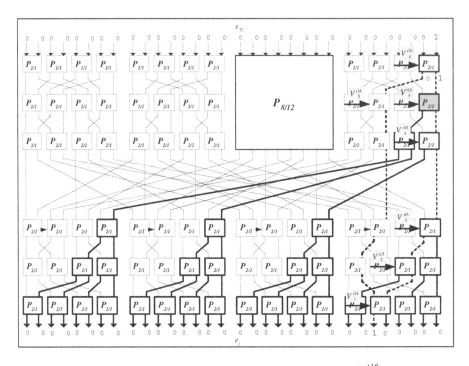

Fig. 6. The possible routes of the non-zero output difference of $P_{2/1}^{(V'_6{}^{16})}$-box in $P_{32/96}^{-1}$

encrypted using a master key K' where K and K' have the $(e_{32}, e_{32}, e_{32}, e_{32})$ difference. Encrypt each plaintext pair (P_i, P'_i) to get the corresponding ciphertext pair (C_i, C'_i).

2. Check that $C_i \oplus C'_i = (e_{32}, e_{j,32})$ for each i and j. We call the bit position of j whose values are 1 OBP(One Bit Position).

3. For each ciphertext pair (C_i, C'_i) passing Step 2, extract two 6 bits of control vectors by chasing two difference routes between the OBP and the position of the second input bit in $P_{2/1}^{(V_6{}'^{16})}$. Compute candidates of the corresponding bits of K_1 and K'_1 by using Tables 7, 8. Output each 6-bit subkey pair with maximal number of hits (here, each of 6-bit subkey pairs corresponds to one of difference routes).

The data complexity of this attack is $2^{15.5}$ related-key chosen plaintexts. The time complexity of Step 1 is $2^{15.5}$ full-round Cobra-H64 encryptions and the time complexity of Steps 2 and 3 is much less than that of Step 1. By our related-key differential characteristic each ciphertext pair can pass Step 2 with probability at least $2^{-12.5}$ and thus the expectation of ciphertext pairs with the $(e_{32}, e_{j,32})$ differences for each j that pass this test is at least 4. This means that the expected number of hits for each 6-bit right key is 4 (Note that the expected number of hits for each 6-bit right key is not 8, since one of two associated difference routes is wrong). On the other hands, the expected number of hits for each 6-bit wrong key is $8 \cdot 2^{-6}$.

Hence we can retrieve 16 bits of keys in the lower layer of $P^{-1}_{32/96}$ and 7 bits of keys in the upper layer of $P^{-1}_{32/96}$ with a data and a time complexity of $2^{15.5}$. Moreover, this attack can be simply extended to retrieve the whole of master key pair (K, K') by performing an exhaustive search for the remaining keys.

5.2   Attack Procedure on Cobra-H128

Unlike the above attack procedure, a related-key attack on Cobra-H128 directly finds some bits of keys by using difference routes.

1. Prepare 2^{43} plaintext pairs (P_i, P'_i), $i = 1, \cdots, 2^{43}$, which have the (e_{64}, e_{64}) difference. All P_i are encrypted using a master key K and all P'_i are encrypted using a master key K' where K and K' have the $(0, 0, e_{64}, e_{64})$ difference. Encrypt each plaintext pair (P_i, P'_i) to get the corresponding ciphertext pair (C_i, C'_i).
2. Check that $C_i \oplus C'_i = (0, e_j)$ for each i and j $(1 \leq j \leq 64)$.
3. For each ciphertext pair (C_i, C'_i) passing Step 2, extract some bits of control vector by chasing a difference route between this OBP and the position of the 64-th input bit in $P^{-1}_{64/192}$ (See Fig. 7). Then find the corresponding bits of K_1, $K_1 \oplus K_2$, and $K_1 \oplus K_3$. Note that the controlled vector V' of $P^{-1}_{64/192}$ in the last round is formatted with $C_L \oplus K_1$, $C_L \oplus K_1 \oplus K_2$, and $C_L \oplus K_1 \oplus K_3$.

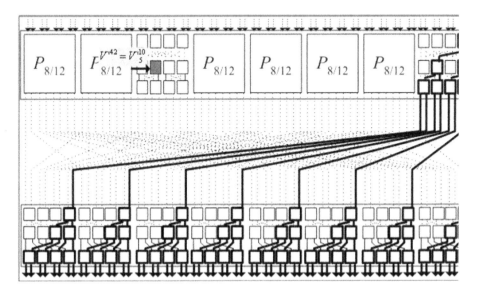

Fig. 7. The possible routes of the 64-th difference of $P^{-1}_{64/192}$

The data complexity of this attack is 2^{44} related-key chosen plaintexts. The time complexity of Step 1 is 2^{44} full-round Cobra-H64 encryptions and the time complexity of Steps 2 and 3 is much less than that of Step 1. By our related-key differential characteristics each ciphertext pair can pass Step 2 with probability at least 2^{-42} and thus the expectation of ciphertext pairs with the $(0, e_j)$ difference that pass this test is at least 2. So we have at least one ciphertext pairs with the $(0, e_j)$ difference for each $1 \leq j \leq 64$. Thus we can retrieve 56 bits of information of keys in the lower layer of $P_{64/192}^{-1}$ and 7 bits of information of keys in the upper layer of $P_{64/192}^{-1}$ with a data and a time complexity of 2^{44}. Similarly, this attack can be simply extended to retrieve the whole of master key pair (K, K') by performing an exhaustive search for the remaining keys.

6   Conclusion

We presented related-key attacks on Cobra-H64 and Cobra-H128. These ciphers are designed suitable for wireless communications networks which require high-speed, but they have a weak diffusion, a weak non-linear operation, and a simple key schedule. So they are vulnerable to related-key differential attacks. According to our results full-round Cobra-H64 can be broken by a complexity of $2^{15.5}$ and full-round Cobra-H128 by a complexity of 2^{44}.

Acknowledgments

We would like to thank the anonymous referees and Jesang Lee for helpful comments about this work. This research was supported by the MIC(Ministry of Information and Communication), Korea, under the ITRC(Information Technology Research Center) support program supervised by the IITA(Institute of Information Technology Assessment). Furthermore, the second author was financed by Ph.D. grants of the Katholieke Universiteit Leuven and of CIST, Korea University and supported by the Concerted Research Action (GOA) Ambiorics 2005/11 of the Flemish Government and by the European Commission through the IST Programme under Contract IST2002507932 ECRYPT.

References

1. E. Biham and A. Shamir, "Differential Cryptanalysis of the Data Encryption Standard", Springer-Verlag, 1993.
2. N. D. Goots, B. V. Izotov, A. A. Moldovyan, and N. A. Moldovyan, "Modern cryptography: Protect Your Data with Fast Block Ciphers", Wayne, A-LIST Publish., 2003.
3. N. D. Goots, B. V. Izotov, A. A. Moldovyan, and N. A. Moldovyan, "Fast Ciphers for Cheap Hardware : Differential Analysis of SPECTR-H64", *MMM-ACNS'03*, LNCS 2776, Springer-Verlag, 2003, pp. 449-452.

4. N. D. Goots, N. A. Moldovyan, P. A. Moldovyanu and D. H. Summerville, "Fast DDP-Based Ciphers: From Hardware to Software", *46th IEEE Midwest International Symposium on Circuits and Systems*, 2003.

5. N. D. Goots, A. A. Moldovyan, N. A. Moldovyan, "Fast Encryption Algorithm Spectr-H64", *MMM-ACNS'01*, LNCS 2052, Springer-Verlag, 2001, pp. 275-286.

6. S. Kavut and M. D. Yücel, "Slide Attack on Spectr-H64", *INDOCRYPT'02*, LNCS 2551, Springer-Verlag, 2002, pp. 34-47.

7. J. Kelsey, B. Schneier, and D. Wagner, "Key Schedule Cryptanalysis of IDEA, G-DES, GOST, SAFER, and Triple-DES", *Advances in Cryptology - CRYPTO '96*, LNCS 1109, Springer-Verlag, 1996, pp. 237-251.

8. J. Kelsey, B. Schneier, and D. Wagner, "Related-Key Cryptanalysis of 3-WAY, Biham-DES, CAST, DES-X, NewDES, RC2, and TEA", *ICICS'97*, LNCS 1334, Springer-Verlag, 1997, pp. 233-246.

9. J. Kim, G. Kim, S. Hong, S. Lee and D. Hong, "The Related-Key Rectangle Attack - Application to SHACAL-1", *ACISP 2004*, LNCS 3108, Springer-Verlag, 2004, pp. 123-136.

10. J. Kim, G. Kim, S. Lee, J. Lim and J. Song, "Related-Key Attacks on Reduced Rounds of SHACAL-2", *INDOCRYPT 2004*, LNCS 3348, Springer-Verlag, 2004, pp. 175-190.

11. Y. Ko, D. Hong, S. Hong, S. Lee, and J. Lim, "Linear Cryptanalysis on SPECTR-H64 with Higher Order Differential Property", *MMM-ACNS03*, LNCS 2776, Springer-Verlag, 2003, pp. 298-307.

12. Y. Ko, C. Lee, S. Hong and S. Lee, "Related Key Differential Cryptanalysis of Full-Round SPECTR-H64 and CIKS-1 ", *ACISP 2004*, LNCS 3108, 2004, pp. 137-148.

13. Y. Ko, C. Lee, S. Hong, J. Sung and S. Lee, "Related-Key Attacks on DDP based Ciphers: CIKS-128 and CIKS-128H ", *Indocrypt 2004*, LNCS 3348, Springer-Verlag, 2004, pp. 191-205.

14. C. Lee, D. Hong, S. Lee, S. Lee, H. Yang, and J. Lim, "A Chosen Plaintext Linear Attack on Block Cipher CIKS-1", *ICICS 2002*, LNCS 2513, Springer-Verlag, 2002, pp. 456-468.

15. C. Lee, J. Kim, S. Hong, J. Sung, and Sangjin Lee, "Related Key Differential Attacks on Cobra-S128, Cobra-F64a, and Cobra-F64b", *MYCRYPT 2005*, LNCS 3715, Springer-Verlag, 2005, pp. 245-263.

16. A. A. Moldovyan and N. A. Moldovyan, "A cipher Based on Data-Dependent Permutations", *Journal of Cryptology*, volume 15, no. 1 (2002), pp. 61-72

17. N. Sklavos, N. A. Moldovyan, and O. Koufopavlou, "High Speed Networking Security: Design and Implementation of Two New DDP-Based Ciphers", *Mobile Networks and Applications-MONET*, Kluwer Academic Publishers, Vol. 25, Issue 1-2, pp. 219-231, 2005.

18. R. C.-W Phan and H. Handschuh, "On Related-Key and Collision Attacks: The case for the IBM 4758 Cryptoprocessor", *ISC 2004*, LNCS 3225, Springer-Verlag, 2004, pp. 111-122.

A   Classes of the Key Bits Corresponding to the Possible Routes

The following two tables represent classes of the key bits corresponding to the possible routes when the non-zero input difference of $P_{2/1}^{(V'^{16}_6)}$ and output difference e_i in $P_{32/96}^{-1}$-box are fixed

Table 7. Classes of the controlled vectors and key bits corresponding to the possible routes when the non-zero input difference of $P_{2/1}^{(V'^{16}_5)}$ and output difference e_i in $P_{32/96}^{-1}$-box are fixed

Class	e_i	Controlled vectors	Key bits
CL_1	e_1	$v_{96} = C_L^{28} \oplus K_1^{28} = 0(0),\ v_{80} = C_L^{22} \oplus K_1^{22} = 1(1),\ v_{63} = C_L^{31} \oplus K_1^{31} = 1(1)$	K_1^1, K_1^9, K_1^{16}
	(e_2)	$v_{36} = C_L^{16} \oplus K_1^{16} = 1(1),\ v_{19} = C_L^9 \oplus K_1^9 = 1(1),\ v_1 = C_L^1 \oplus K_1^1 = 1(0)$	$K_1^{22}, K_1^{28}, K_1^{31}$
CL_2	e_3	$v_{96} = C_L^{28} \oplus K_1^{28} = 0(0),\ v_{80} = C_L^{22} \oplus K_1^{22} = 1(1),\ v_{63} = C_L^{31} \oplus K_1^{31} = 1(1)$	K_1^1, K_1^9, K_1^{16}
	(e_4)	$v_{36} = C_L^{16} \oplus K_1^{16} = 1(1),\ v_{19} = C_L^9 \oplus K_1^9 = 0(0),\ v_2 = C_L^2 \oplus K_1^2 = 1(0)$	$K_1^{22}, K_1^{28}, K_1^{31}$
CL_3	e_5	$v_{96} = C_L^{28} \oplus K_1^{28} = 0(0),\ v_{80} = C_L^{22} \oplus K_1^{22} = 1(1),\ v_{63} = C_L^{31} \oplus K_1^{31} = 1(1)$	$K_1^3, K_1^{10}, K_1^{16}$
	(e_6)	$v_{36} = C_L^{16} \oplus K_1^{16} = 0(0),\ v_{20} = C_L^{10} \oplus K_1^{10} = 1(1),\ v_3 = C_L^3 \oplus K_1^3 = 1(0)$	K_1^{22}, K_1^{31}
CL_4	e_7	$v_{96} = C_L^{28} \oplus K_1^{28} = 0(0),\ v_{80} = C_L^{22} \oplus K_1^{22} = 1(1),\ v_{63} = C_L^{31} \oplus K_1^{31} = 1(1)$	$K_1^4, K_1^{11}, K_1^{16}$
	(e_8)	$v_{36} = C_L^{16} \oplus K_1^{16} = 0(0),\ v_{21} = C_L^{11} \oplus K_1^{11} = 0(0),\ v_4 = C_L^4 \oplus K_1^4 = 1(0)$	$K_1^{22}, K_1^{28}, K_1^{31}$
CL_5	e_9	$v_{96} = C_L^{28} \oplus K_1^{28} = 0(0),\ v_{80} = C_L^{22} \oplus K_1^{22} = 1(1),\ v_{63} = C_L^{31} \oplus K_1^{31} = 0(0)$	K_1^4, K_1^5, K_1^{13}
	(e_{10})	$v_{40} = C_L^4 \oplus K_1^4 = 1(1),\ v_{23} = C_L^{13} \oplus K_1^{13} = 1(1),\ v_5 = C_L^5 \oplus K_1^5 = 1(0)$	$K_1^{22}, K_1^{28}, K_1^{31}$
CL_6	e_{11}	$v_{96} = C_L^{28} \oplus K_1^{28} = 0(0),\ v_{80} = C_L^{22} \oplus K_1^{22} = 1(1),\ v_{63} = C_L^{31} \oplus K_1^{31} = 0(0)$	K_1^4, K_1^6, K_1^{13}
	(e_{12})	$v_{40} = C_L^4 \oplus K_1^4 = 1(1),\ v_{23} = C_L^{13} \oplus K_1^{13} = 0(0),\ v_6 = C_L^6 \oplus K_1^6 = 1(0)$	$K_1^{22}, K_1^{28}, K_1^{31}$
CL_7	e_{13}	$v_{96} = C_L^{28} \oplus K_1^{28} = 0(0),\ v_{80} = C_L^{22} \oplus K_1^{22} = 1(1),\ v_{63} = C_L^{31} \oplus K_1^{31} = 0(0)$	K_1^4, K_1^7, K_1^{14}
	(e_{14})	$v_{40} = C_L^4 \oplus K_1^4 = 0(0),\ v_{24} = C_L^{14} \oplus K_1^{14} = 1(1),\ v_7 = C_L^7 \oplus K_1^7 = 1(0)$	$K_1^{22}, K_1^{28}, K_1^{31}$
CL_8	e_{15}	$v_{96} = C_L^{28} \oplus K_1^{28} = 0(0),\ v_{80} = C_L^{22} \oplus K_1^{22} = 1(1),\ v_{63} = C_L^{31} \oplus K_1^{31} = 0(0)$	K_1^4, K_1^8, K_1^{14}
	(e_{16})	$v_{40} = C_L^4 \oplus K_1^4 = 0(0),\ v_{24} = C_L^{14} \oplus K_1^{14} = 0(0),\ v_8 = C_L^8 \oplus K_1^8 = 1(0)$	$K_1^{22}, K_1^{28}, K_1^{31}$
CL_9	e_{17}	$v_{96} = C_L^{28} \oplus K_1^{28} = 0(0),\ v_{80} = C_L^{22} \oplus K_1^{22} = 0(0),\ v_{64} = C_L^{32} \oplus K_1^{32} = 1(1)$	K_1^1, K_1^8, K_1^9
	(e_{18})	$v_{44} = C_L^8 \oplus K_1^8 = 1(1),\ v_{27} = C_L^1 \oplus K_1^1 = 1(1),\ v_9 = C_L^9 \oplus K_1^9 = 1(0)$	$K_1^{22}, K_1^{28}, K_1^{32}$
CL_{10}	e_{19}	$v_{96} = C_L^{28} \oplus K_1^{28} = 0(0),\ v_{80} = C_L^{22} \oplus K_1^{22} = 0(0),\ v_{64} = C_L^{32} \oplus K_1^{32} = 1(1)$	K_1^1, K_1^8, K_1^{10}
	(e_{20})	$v_{44} = C_L^8 \oplus K_1^8 = 1(1),\ v_{27} = C_L^1 \oplus K_1^1 = 0(0),\ v_{10} = C_L^{10} \oplus K_1^{10} = 1(0)$	$K_1^{22}, K_1^{28}, K_1^{32}$
CL_{11}	e_{21}	$v_{96} = C_L^{28} \oplus K_1^{28} = 0(0),\ v_{80} = C_L^{22} \oplus K_1^{22} = 0(0),\ v_{64} = C_L^{32} \oplus K_1^{32} = 1(1)$	K_1^2, K_1^8, K_1^{11}
	(e_{22})	$v_{44} = C_L^8 \oplus K_1^8 = 0(0),\ v_{28} = C_L^2 \oplus K_1^2 = 1(1),\ v_{11} = C_L^{11} \oplus K_1^{11} = 1(0)$	$K_1^{22}, K_1^{28}, K_1^{32}$
CL_{12}	e_{23}	$v_{96} = C_L^{28} \oplus K_1^{28} = 0(0),\ v_{80} = C_L^{22} \oplus K_1^{22} = 0(0),\ v_{64} = C_L^{32} \oplus K_1^{32} = 1(1)$	K_1^2, K_1^8, K_1^{12}
	(e_{24})	$v_{44} = C_L^8 \oplus K_1^8 = 0(0),\ v_{28} = C_L^2 \oplus K_1^2 = 0(0),\ v_{12} = C_L^{12} \oplus K_1^{12} = 1(0)$	$K_1^{22}, K_1^{28}, K_1^{32}$
CL_{13}	e_{25}	$v_{96} = C_L^{28} \oplus K_1^{28} = 0(0),\ v_{80} = C_L^{22} \oplus K_1^{22} = 0(0),\ v_{64} = C_L^{32} \oplus K_1^{32} = 0(0)$	$K_1^5, K_1^{12}, K_1^{13}$
	(e_{26})	$v_{48} = C_L^{12} \oplus K_1^{12} = 1(1),\ v_{31} = C_L^5 \oplus K_1^5 = 1(1),\ v_{13} = C_L^{13} \oplus K_1^{13} = 1(0)$	$K_1^{22}, K_1^{28}, K_1^{32}$
CL_{14}	e_{27}	$v_{96} = C_L^{28} \oplus K_1^{28} = 0(0),\ v_{80} = C_L^{22} \oplus K_1^{22} = 0(0),\ v_{64} = C_L^{32} \oplus K_1^{32} = 0(0)$	$K_1^5, K_1^{12}, K_1^{14}$
	(e_{28})	$v_{48} = C_L^{12} \oplus K_1^{12} = 1(1),\ v_{31} = C_L^5 \oplus K_1^5 = 0(0),\ v_{14} = C_L^{14} \oplus K_1^{14} = 1(0)$	$K_1^{22}, K_1^{28}, K_1^{32}$
CL_{15}	e_{29}	$v_{96} = C_L^{28} \oplus K_1^{28} = 0(0),\ v_{80} = C_L^{22} \oplus K_1^{22} = 0(0),\ v_{64} = C_L^{32} \oplus K_1^{32} = 0(0)$	$K_1^6, K_1^{12}, K_1^{15}$
	(e_{30})	$v_{48} = C_L^{12} \oplus K_1^{12} = 0(0),\ v_{32} = C_L^6 \oplus K_1^6 = 1(1),\ v_{15} = C_L^{15} \oplus K_1^{15} = 1(0)$	$K_1^{22}, K_1^{28}, K_1^{32}$
CL_{16}	e_{31}	$v_{96} = C_L^{28} \oplus K_1^{28} = 0(0),\ v_{80} = C_L^{22} \oplus K_1^{22} = 0(0),\ v_{64} = C_L^{32} \oplus K_1^{32} = 0(0)$	$K_1^6, K_1^{12}, K_1^{16}$
	(e_{32})	$v_{48} = C_L^{12} \oplus K_1^{12} = 0(0),\ v_{32} = C_L^6 \oplus K_1^6 = 0(0),\ v_{16} = C_L^{16} \oplus K_1^{16} = 1(0)$	$K_1^{22}, K_1^{28}, K_1^{32}$

Table 8. Classes of the controlled vectors and key bits corresponding to the possible routes when the non-zero input difference of $P_{2/1}^{(V'^{14}_5)}$ and output difference e_i in $P_{32/96}^{-1}$-box are fixed

Class	e_i	Controlled vectors	Key bits
$\mathcal{CL}_{1'}$	e_1	$v_{96} = C_L^{28} \oplus K_1^{28} = 1(1),\ v_{78} = C_L^{20} \oplus K_1^{20} = 1(1),\ v_{61} = C_L^{29} \oplus K_1^{29} = 1(1)$	K_1^1, K_1^7, K_1^{14}
	(e_2)	$v_{34} = C_L^{14} \oplus K_1^{14} = 1(1),\ v_{17} = C_L^7 \oplus K_1^7 = 1(1),\ v_1 = C_L^1 \oplus K_1^1 = 0(1)$	$K_1^{20}, K_1^{28}, K_1^{29}$
$\mathcal{CL}_{2'}$	e_3	$v_{96} = C_L^{28} \oplus K_1^{28} = 1(1),\ v_{78} = C_L^{20} \oplus K_1^{20} = 1(1),\ v_{61} = C_L^{29} \oplus K_1^{29} = 1(1)$	K_1^2, K_1^7, K_1^{14}
	(e_4)	$v_{34} = C_L^{14} \oplus K_1^{14} = 1(1),\ v_{17} = C_L^7 \oplus K_1^7 = 0(0),\ v_2 = C_L^2 \oplus K_1^2 = 0(1)$	$K_1^{20}, K_1^{28}, K_1^{29}$
$\mathcal{CL}_{3'}$	e_5	$v_{96} = C_L^{28} \oplus K_1^{28} = 1(1),\ v_{78} = C_L^{20} \oplus K_1^{20} = 1(1),\ v_{61} = C_L^{29} \oplus K_1^{29} = 1(1)$	K_1^3, K_1^8, K_1^{14}
	(e_6)	$v_{34} = C_L^{14} \oplus K_1^{14} = 0(0),\ v_{18} = C_L^8 \oplus K_1^8 = 1(1),\ v_3 = C_L^3 \oplus K_1^3 = 0(1)$	$K_1^{20}, K_1^{28}, K_1^{29}$
$\mathcal{CL}_{4'}$	e_7	$v_{96} = C_L^{28} \oplus K_1^{28} = 1(1),\ v_{78} = C_L^{20} \oplus K_1^{20} = 1(1),\ v_{61} = C_L^{29} \oplus K_1^{29} = 1(1)$	K_1^4, K_1^8, K_1^{14}
	(e_8)	$v_{34} = C_L^{14} \oplus K_1^{14} = 0(0),\ v_{18} = C_L^8 \oplus K_1^8 = 0(0),\ v_4 = C_L^4 \oplus K_1^4 = 0(1)$	$K_1^{20}, K_1^{28}, K_1^{29}$
$\mathcal{CL}_{5'}$	e_9	$v_{96} = C_L^{28} \oplus K_1^{28} = 1(1),\ v_{78} = C_L^{20} \oplus K_1^{20} = 1(1),\ v_{61} = C_L^{29} \oplus K_1^{29} = 0(0)$	K_1^1, K_1^5, K_1^{11}
	(e_{10})	$v_{38} = C_L^1 \oplus K_1^1 = 1(1),\ v_{21} = C_L^{11} \oplus K_1^{11} = 1(1),\ v_5 = C_L^5 \oplus K_1^5 = 0(1)$	$K_1^{20}, K_1^{28}, K_1^{29}$
$\mathcal{CL}_{6'}$	e_{11}	$v_{96} = C_L^{28} \oplus K_1^{28} = 1(1),\ v_{78} = C_L^{20} \oplus K_1^{20} = 1(1),\ v_{61} = C_L^{29} \oplus K_1^{29} = 0(0)$	K_1^1, K_1^6, K_1^{11}
	(e_{12})	$v_{38} = C_L^1 \oplus K_1^1 = 1(1),\ v_{21} = C_L^{11} \oplus K_1^{11} = 0(0),\ v_6 = C_L^6 \oplus K_1^6 = 0(1)$	$K_1^{20}, K_1^{28}, K_1^{29}$
$\mathcal{CL}_{7'}$	e_{13}	$v_{96} = C_L^{28} \oplus K_1^{28} = 1(1),\ v_{78} = C_L^{20} \oplus K_1^{20} = 1(1),\ v_{61} = C_L^{29} \oplus K_1^{29} = 0(0)$	K_1^1, K_1^7, K_1^{12}
	(e_{14})	$v_{38} = C_L^1 \oplus K_1^1 = 0(0),\ v_{22} = C_L^{12} \oplus K_1^{12} = 1(1),\ v_7 = C_L^7 \oplus K_1^7 = 0(1)$	$K_1^{20}, K_1^{28}, K_1^{29}$
$\mathcal{CL}_{8'}$	e_{15}	$v_{96} = C_L^{28} \oplus K_1^{28} = 1(1),\ v_{78} = C_L^{20} \oplus K_1^{20} = 1(1),\ v_{61} = C_L^{29} \oplus K_1^{29} = 0(0)$	K_1^1, K_1^8, K_1^{12}
	(e_{16})	$v_{38} = C_L^1 \oplus K_1^1 = 0(0),\ v_{22} = C_L^{12} \oplus K_1^{12} = 0(0),\ v_8 = C_L^8 \oplus K_1^8 = 0(1)$	$K_1^{20}, K_1^{28}, K_1^{29}$
$\mathcal{CL}_{9'}$	e_{17}	$v_{96} = C_L^{28} \oplus K_1^{28} = 1(1),\ v_{78} = C_L^{20} \oplus K_1^{20} = 0(0),\ v_{62} = C_L^{30} \oplus K_1^{30} = 1(1)$	K_1^6, K_1^9, K_1^{15}
	(e_{18})	$v_{42} = C_L^6 \oplus K_1^6 = 1(1),\ v_{25} = C_L^{15} \oplus K_1^{15} = 1(1),\ v_9 = C_L^9 \oplus K_1^9 = 0(1)$	$K_1^{20}, K_1^{28}, K_1^{30}$
$\mathcal{CL}_{10'}$	e_{19}	$v_{96} = C_L^{28} \oplus K_1^{28} = 1(1),\ v_{78} = C_L^{20} \oplus K_1^{20} = 0(0),\ v_{62} = C_L^{30} \oplus K_1^{30} = 1(1)$	$K_1^6, K_1^{10}, K_1^{15}$
	(e_{20})	$v_{42} = C_L^6 \oplus K_1^6 = 1(1),\ v_{25} = C_L^{15} \oplus K_1^{15} = 0(0),\ v_{10} = C_L^{10} \oplus K_1^{10} = 0(1)$	$K_1^{20}, K_1^{28}, K_1^{30}$
$\mathcal{CL}_{11'}$	e_{21}	$v_{96} = C_L^{28} \oplus K_1^{28} = 1(1),\ v_{78} = C_L^{20} \oplus K_1^{20} = 0(0),\ v_{62} = C_L^{30} \oplus K_1^{30} = 1(1)$	$K_1^6, K_1^{11}, K_1^{16}$
	(e_{22})	$v_{42} = C_L^6 \oplus K_1^6 = 0(0),\ v_{26} = C_L^{16} \oplus K_1^{16} = 1(1),\ v_{11} = C_L^{11} \oplus K_1^{11} = 0(1)$	$K_1^{20}, K_1^{28}, K_1^{30}$
$\mathcal{CL}_{12'}$	e_{23}	$v_{96} = C_L^{28} \oplus K_1^{28} = 1(1),\ v_{78} = C_L^{20} \oplus K_1^{20} = 0(0),\ v_{62} = C_L^{30} \oplus K_1^{30} = 1(1)$	$K_1^6, K_1^{12}, K_1^{16}$
	(e_{24})	$v_{42} = C_L^6 \oplus K_1^6 = 0(0),\ v_{26} = C_L^{16} \oplus K_1^{16} = 0(0),\ v_{12} = C_L^{12} \oplus K_1^{12} = 0(1)$	$K_1^{20}, K_1^{28}, K_1^{30}$
$\mathcal{CL}_{13'}$	e_{25}	$v_{96} = C_L^{28} \oplus K_1^{28} = 1(1),\ v_{78} = C_L^{20} \oplus K_1^{20} = 0(0),\ v_{62} = C_L^{30} \oplus K_1^{30} = 0(0)$	$K_1^3, K_1^{10}, K_1^{13}$
	(e_{26})	$v_{46} = C_L^{10} \oplus K_1^{10} = 1(1),\ v_{29} = C_L^3 \oplus K_1^3 = 1(1),\ v_{13} = C_L^{13} \oplus K_1^{13} = 0(1)$	$K_1^{20}, K_1^{28}, K_1^{30}$
$\mathcal{CL}_{14'}$	e_{27}	$v_{96} = C_L^{28} \oplus K_1^{28} = 1(1),\ v_{78} = C_L^{20} \oplus K_1^{20} = 0(0),\ v_{62} = C_L^{30} \oplus K_1^{30} = 0(0)$	$K_1^3, K_1^{10}, K_1^{14}$
	(e_{28})	$v_{46} = C_L^{10} \oplus K_1^{10} = 1(1),\ v_{29} = C_L^3 \oplus K_1^3 = 0(0),\ v_{14} = C_L^{14} \oplus K_1^{14} = 0(1)$	$K_1^{20}, K_1^{28}, K_1^{30}$
$\mathcal{CL}_{15'}$	e_{29}	$v_{96} = C_L^{28} \oplus K_1^{28} = 1(1),\ v_{78} = C_L^{20} \oplus K_1^{20} = 0(0),\ v_{62} = C_L^{30} \oplus K_1^{30} = 0(0)$	$K_1^4, K_1^{10}, K_1^{15}$
	(e_{30})	$v_{46} = C_L^{10} \oplus K_1^{10} = 0(0),\ v_{30} = C_L^4 \oplus K_1^4 = 1(1),\ v_{15} = C_L^{15} \oplus K_1^{15} = 0(1)$	$K_1^{20}, K_1^{28}, K_1^{30}$
$\mathcal{CL}_{16'}$	e_{31}	$v_{96} = C_L^{28} \oplus K_1^{28} = 1(1),\ v_{78} = C_L^{20} \oplus K_1^{20} = 0(0),\ v_{62} = C_L^{30} \oplus K_1^{30} = 0(0)$	$K_1^4, K_1^{10}, K_1^{16}$
	(e_{32})	$v_{46} = C_L^{10} \oplus K_1^{10} = 0(0),\ v_{30} = C_L^4 \oplus K_1^4 = 0(0),\ v_{16} = C_L^{16} \oplus K_1^{16} = 0(1)$	$K_1^{20}, K_1^{28}, K_1^{30}$

The Physically Observable Security of Signature Schemes

Alexander W. Dent[1] and John Malone-Lee[2]

[1] Information Security Group,
Royal Holloway, University of London,
Egham, Surrey, TW20 0EX, UK
a.dent@rhul.ac.uk
http://www.isg.rhul.ac.uk/~alex
[2] Department of Computer Science,
University of Bristol,
Merchant Venturers Building, Woodland Road,
Bristol, BS8 1UB, UK
malone@cs.bris.ac.uk
http://www.cs.bris.ac.uk/~malone

Abstract. In recent years much research has been devoted to producing formal models of security for cryptographic primitives and to designing schemes that can be proved secure in such models. This line of research typically assumes that an adversary is given black-box access to a cryptographic mechanism that uses some secret key. One then proves that this black-box access does not help the adversary to achieve its task.

An increasingly popular environment for cryptographic implementation is the smart-card. In such an environment a definition of security that provides an adversary with only black-box access to the cryptography under attack may be unrealistic. This is illustrated by attacks such as the power-analysis methods proposed by Kocher and others.

In this paper we attempt to formally define a set of necessary conditions on an implementation of a cryptosystem so that security against an adversary with black-box access is preserved in a more hostile environment such as the smart-card. Unlike the previous work in this area we concentrate on high-level primitives. The particular example that we take is the digital signature scheme. [1]

1   Introduction

The idea of formally modelling cryptographic security originates in *Probabilistic Encryption* [11], the seminal work of Goldwasser and Micali. Since the publication of that paper, a huge body of research has been devoted to designing

---

[1] The work described in this paper has been supported in part by the European Commission through the IST Programme under Contract IST-2002-507932 ECRYPT. The information in this document reflects only the author's views, is provided as is and no guarantee or warranty is given that the information is fit for any particular purpose. The user thereof uses the information at its sole risk and liability.

N.P. Smart (Ed.): Cryptography and Coding 2005, LNCS 3796, pp. 220–232, 2005.

schemes that provably meet some definition of security. The basic tenet of all this work is as follows. Start with an assumption about some atomic primitive (for example, the assumption that a particular function is one-way); design a scheme in such a way that an adversary cannot break the scheme without violating the assumption about the atomic primitive. This property is demonstrated using a complexity-theoretic reduction: one shows that, if an adversary of the scheme exists, then this adversary could be used as a subroutine in an algorithm to violate the assumption about the atomic primitive. Following this procedure one concludes that, if the assumption about the atomic primitive is correct, the scheme satisfies the chosen security definition.

Until recently, the idea of showing that the security of a cryptosystem relies only on the properties of a set of critical atomic primitives has been applied in a "black-box" manner. That is to say, an adversary is given black-box access to an instance of the cryptosystem with a randomly generated secret key. It may have complete control over all inputs, and see all outputs, but it has no knowledge of the internal state of the black-box implementing the cryptosystem. This approach has been used in both the symmetric-key [2] and asymmetric-key [3] settings.

Such models may be appropriate for applications in which all potential adversaries are remote from the legitimate user; however, increasingly cryptography is used in, for example, smart card applications where this is not the case. An attack in this setting was discovered by Kocher *et al.* [14]. They showed how, by monitoring the power consumption of a smart card running DES [9], the secret key could be recovered. These attacks have since become known as *side-channel attacks*; the side channel is the information leaked by the physical implementation of an algorithm – the power consumption in the example above.

There is no avoiding the fact that a physical device performing sensitive operations may leak information. Countermeasures preventing specific attacks can often be designed but they may be expensive. It is therefore imperative that security models are developed that are capable of explicitly isolating the security-critical operations in the implementation of a cryptosystem. This will allow appropriate countermeasures to be focused exactly where they are necessary.

Recently, the first steps have been taken towards formally defining security models for environments where an adversary is able to mount side-channel attacks. One such approach, proposed by Micali and Reyzin [15], is known as *physically observable cryptography*. In this model, every cryptographic operation gives off physical *observables*. For example, these observables could be related to the power consumption [14] or the electro-magnetic radiation emitted by a device during computation [1]. The model does not deal with attacks such as those proposed in [5,6] in which an adversary actively attempts to alter the operation of a cryptosystem.

The original paper of Micali and Reyzin [15], very properly, concentrates on physically observable cryptography on a "micro" scale: it examines the impact that physical observables have on proofs of security for fundamental primitives such as one-way functions and permutations. They left open the question of how their model could be applied on a more "macro" scale, to primitives such as encryption and signatures – for which there already exist schemes with black-box security proofs. In this paper we start to address this question. More specifically, we will describe how a set of necessary conditions can be established on the implementations of components of certain cryptosystems so that a black-box security proof holds in the setting where an adversary is given access to the implementation of the cryptosystem.

2   Physically Observable Cryptography

In this section we begin by reviewing the model proposed by Micali and Reyzin [15]. Once we have done so we describe how to extend the model to deal with higher-level primitives than those considered to-date.

2.1   Informal Axioms

The model of Micali and Reyzin [15] requires several "informal axioms". These axioms are assumed to apply to any computational device used to implement the primitives under consideration. We state these axioms briefly below and elaborate on them where necessary. Further details may be found in [15].

Axiom 1. Computation, and only computation, leaks information. Hence, un-accessed memory is totally secure.

Axiom 2. The same computation leaks different information on different devices.

Axiom 3. Information leakage depends upon the chosen measurement.

Axiom 4. Information measurement is local: the information leaked by a component function is independent of the computations made by any of the other component functions.

Axiom 5. All leaked information is efficiently computable from the device's internal configuration. In particular, this means that the leakage is efficiently simulatable if you know all the inputs to a component function.

We note that these axioms cannot be applied indiscriminately to all devices implementing cryptography. For example the cache-based cryptanalysis proposed by Page [16] and developed by Tsunoo *et al.* [18] exploits certain implementations of DES where Axiom 4 above fails to hold. In particular these attacks exploit the fact that, when implemented on a piece of hardware with cache memory, the time taken to access S-box data may vary according to previous S-box access. This time-based side-channel can be exploited to recover the key.

2.2   Computational Model

In the traditional Turing machine (TM) model the tape of the machine is accessed sequentially. This is not consistent with Axiom 1 in Section 2.1: to move from one cell to another the machine may need to scan many intermediate cells thereby leaking information about data that is not involved directly in computation. To overcome this problem Micali and Reyzin augment the model with random access memory so that each bit can be addressed and accessed independently of all other bits.

As we noted at the end of Section 2.1, Axiom 4 requires us to work in a model where the leakage of a given device is independent of all the computation that precedes it. Micali and Reyzin point out that this means we cannot work in a single TM model [15]. To provide modularity for physically observable cryptography, the model of computation consists of multiple machines. These machines may call one another as subroutines. A second requirement of the model in order to preserve independence of leakage is that each machine has its own memory space that only it can see. To implement this requirement the model is augmented with a *virtual memory manager*.

We now proceed to formalise these concepts following the ideas of Micali and Reyzin [15]. We comment on the differences between our version and the original as and when they occur.

Abstract Virtual-Memory Computers. An *abstract virtual-memory computer* (abstract VMC or simply VMC for short) consists of a collection of special Turing machines. We call these machines *abstract virtual-memory Turing machines* (abstract VTMs or simply VTMs for short). We write $A = (T_1, \ldots, T_n)$ to denote the fact that abstract virtual-memory computer A consists of abstract VTMs $T_1, \ldots, T_n$, where T_1 is distinguished: it is invoked first and its inputs and outputs coincide with those of A.

The specialisation of this model that we will use will have the following features. We will assume that T_1 calls each of $T_2, \ldots, T_n$ in turn; that none of these is called more than once; and that T_i does not call T_j if $i \neq 1$. We will demonstrate these properties with a concrete example in Section 3.2.

Virtual-Memory Management. In addition to the standard input, output, work and random tapes of a probabilistic TM, a VTM has random access to its own *virtual address space* (VAS). There is in fact a single *physical address space* (PAS) for the entire VMC. A *virtual-memory manager* takes care of mapping the PAS to individual VASs. Complete details of how the virtual-memory manager works may be found in [15]. We only mention here the properties that we require.

Each VTM has a special VAS-access tape. To obtain a value from its VAS it simply writes the location of the data that it wishes to read on its VAS-access tape. The required data then appears on the VTMs VAS-access tape. The mechanics of writing to VAS are equally simple: a VTM simply writes the data and the location that it wishes it to be stored at on its VAS-access tape.

The only special requirement that we have in this context is that the virtual-memory manager should only remap memory addresses, but never access the data.

Input and Output. The implementations of the cryptosystems that we will consider in this paper will have the following form. The only VTM that will take any external input or produce any external output is T_1. So, if $A = (T_1, \ldots, T_n)$, the external input and external output of A are exactly those of T_1.

At the start of the computation the input is on the start of T_1's VAS. At the end of the computation the output occupies a portion of T_1's VAS. Further details of this may be found in [15].

Calling VTMs as Subroutines. As we mentioned above, for the implementations of the cryptosystems that fit our version of the model, the only VTM that will call any other VTM is T_1. This VTM has a special *subroutine-call tape*. To call a subroutine T_i, T_1 specifies where the input for T_i is located on its – T_1s – VAS. It also specifies where it wants the output. The virtual memory manager takes care of mapping address locations.

2.3   Physical Security Model

The physical implementation of a cryptosystem, itself modelled as a virtual-memory computer $T = (T_1, \ldots T_n)$, will be modelled as a *physical virtual-memory computer* (physical VMC). This is a collection of *physical virtual Turing machines* (physical VTMs) $\mathcal{P} = (P_1, P_2, \ldots, P_n)$. Each physical VTM P_i consists of a pair $P_i = (L_i, T_i)$ where T_i is a VTM and L_i is a *leakage function* associated with the physical implementation of T_i.

A leakage function is used to model the information leaked to an adversary by a particular implementation. As defined by Micali and Reyzin [15], it has three inputs: (1) the current configuration of the physical VTM under consideration; (2) the setting of the measuring apparatus used by the adversary; and (3) a random string to model the randomness of the measurement. Further details may be found in [15].

We say that an adversary *observes* a physical VTM if it has access to the output of the leakage function for the VTM and can decide upon the second input: the measuring apparatus to use. As in [15], we denote the event that an adversary A outputs y_A after being run on input x_A and observing a physical VTM P_i being executed on an input $x_{\mathcal{P}}$ and producing an output $y_{\mathcal{P}}$, by

$$y_{\mathcal{P}} \leftarrow \mathcal{P}(x_{\mathcal{P}}) \rightsquigarrow A(x_A) \rightarrow y_A.$$

3   A Definition of Security for Physical Virtual-Memory Computers

Recall from Section 2.2 that we are interested in implementations of cryptosystems with the following form. The cryptosystem must be susceptible to being modelled as a VMC $T = (T_1, \ldots, T_n)$ such that the input and output of T correspond exactly to the input and output of T_1, and T_1 calls each T_i once and once only. We will also require in our model that T_1 is not responsible for any computation itself. It simply maps addresses in its VAS to the VAS-access tapes of the VTMs $T_2, \ldots, T_n$. It therefore follows from Axiom 1 in Section 2.1 that T_1 does not leak any side-channel information to an adversary.

The final point that we should make about T_1 is that it has the secret key of the cryptosystem concerned hard-coded into its VAS before any adversary is given access to the implementation. We will also assume that the secret key for the cryptosystem is of the form $sk = (sk_2, \ldots, sk_n)$ where the sk_i is the secret key material used by T_i. At present our model can only deal with the case where the sk_i are distinct and generated independently of one another by the key generation algorithm for the scheme (modulo some common parameter such as a group or the bit-length of an RSA modulus). A good example of a cryptosystem with such a property is the CS1a scheme of Cramer and Shoup [8]. We will comment on why this property is necessary at the appropriate point in our proof of security.

At this point we will provide an example to illustrate the concepts that we are introducing. The example that we will use is a version of the PSS variant [4] of the RSA signature scheme [17]. We also introduce this example because we will prove our result specifically for the case of digital signature schemes.

Before going into details of the specific scheme we remind ourselves of the definition of a signature scheme and the definition of security for signature schemes.

3.1   Signature Schemes and Their Security

A signature scheme SIG consists of three algorithms $KeyGen, Sig$ and Ver. These have the following properties.

- The *key generation algorithm KeyGen* is a probabilistic algorithm that takes as input a security parameter 1^k and produces a public/secret key pair (pk, sk).
- The *signing algorithm Sig* takes as input the secret key sk and a message m; it outputs a signature s. The signing algorithm may be probabilistic or deterministic.
- The *verification algorithm Ver* takes as input a message m, the public key pk and a purported signature s; it outputs 1 if s is a valid signature on m under pk, otherwise it outputs 0.

Let us also recall the standard definition of (black-box) security for a signature scheme: *existential unforgeability under adaptive chosen message attack* [12]. This notion is described using the experiment below involving a signature scheme $SIG = (KeyGen, Sig, Ver)$, an adversary $\mathcal{A}$ and a security parameter k.

Exp$_{SIG,\mathcal{A}}^{\mathrm{euacma}}(k)$

Stage 1. The key generation algorithm $KeyGen$ for the signature scheme in question is run on input of a security parameter 1^k. The resulting public key is given to adversary $\mathcal{A}$.

Stage 2. Adversary $\mathcal{A}$ makes a polynomial number (in the security parameter) of queries to a signing oracle. This oracle produces signatures for $\mathcal{A}$ on messages of its choice. The oracle produces these signatures using the secret key generated in Stage 1 and the algorithm Sig.

Stage 3. Adversary $\mathcal{A}$ attempts to output a message and a valid signature such that the message was never a query to the signing oracle in Stage 2. If it succeeds in doing this we say that $\mathcal{A}$ *wins* and we output 1, otherwise we output 0.

The adversary $\mathcal{A}$'s advantage is the probability that it wins in the above. We say

$$\mathbf{Adv}_{SIG,\mathcal{A}}^{\text{euacma}}(k) = \Pr[\mathbf{Exp}_{SIG,\mathcal{A}}^{\text{euacma}}(k) = 1]. \qquad (1)$$

If, for all probabilistic polynomial time $\mathcal{A}$, (1) is a negligible function k then SIG is said to be *existentially unforgeable under adaptive chosen message attack*.

Having defined the black-box version of the definition of security for a signature scheme, it is a straightforward manner to define the physically observable analogue. To do this we simply replace the black-box queries that $\mathcal{A}$ has in Stage 2 of the above definition with *physically observable queries* as defined in Section 2.3. In other words, we give $\mathcal{A}$ access to an oracle for the leakage function of the system. The adversary supplies the measurement information that is input to the leakage function, and the oracle uses the machines current state and randomness as the other inputs. We denote the experiment where $\mathcal{A}$ has access to a leakage oracle $\mathbf{Exp}_{T(SIG),\mathcal{A}}^{\text{eupoacma}}(k)$ and define the advantage of an adversary $\mathcal{A}$ in this game by

$$\mathbf{Adv}_{T(SIG),\mathcal{A}}^{\text{eupoacma}}(k) = \Pr[\mathbf{Exp}_{T(SIG),\mathcal{A}}^{\text{eupoacma}}(k) = 1]. \qquad (2)$$

In the above $T(SIG)$ denotes the fact that we are concerned with the actual implementation T of the scheme SIG rather than SIG itself.

3.2   The RSA-PSS Signature Scheme

In order to make the concepts we are discussing more concrete, we give an example: the RSA-PSS signature scheme [4,17].

The signing algorithm for RSA-PSS involves formatting the message and then performing modular exponentiation using a secret exponent. Here we describe how this process can be decomposed into its various subroutines in our model.

Suppose that signing of n-bit messages is performed using a k-bit RSA modulus N and a secret exponent d. This requires two hash functions

$$H : \{0,1\}^{n+k_0} \to \{0,1\}^{k_1} \text{ and } G : \{0,1\}^{k_1} \to \{0,1\}^{n+k_0} \qquad (3)$$

where $k = n + k_0 + k_1 + 1$.

The signing procedure will be modelled as a VMC $T = (T_1, \ldots, T_5)$. We describe the roles of the various VTMs below. The message to be signed is m.

- T_2 requires no input. It simply generates a k_0 bit random number r and writes r to the appropriate location in its VAS.

- T_3 requires m and r as input. The addresses of these are provided by T_1 and the virtual memory manager does the appropriate address mapping. Having recovered m and r from its VAS, T_3 computes $u = H(m||r)$ and writes u to the appropriate location in its VAS.
- T_4 requires m, r and u as input. The addresses of these are provided by T_1 and the virtual memory manager does the appropriate address mapping. Having recovered m, r and u from its VAS, T_4 computes $v = G(u) \oplus (m||r)$ and writes v to the appropriate location in its VAS.
- T_5 requires u, v, d and N as input. The addresses of these are provided by T_1 and the virtual memory manager does the appropriate address mapping. Having recovered u, v, d and N from its VAS, T_5 converts the bit-string $0||u||v$ into an integer x and computes $s = x^d \bmod N$. It then writes s to the appropriate location in its VAS.
- T_1 takes external input m and has d and N hard-coded into its VAS. Its role is simply to write the appropriate addresses from its VAS to its subroutine-call tape and invoke $T_2, \ldots, T_5$ in turn (the appropriate addresses are implicit in the descriptions of $T_2, \ldots, T_5$ above). The final job of T_1 is to output the data from the portion of its VAS where s is located after T_5 has been called.

Note that in the description of T_5 we include the 0 in the string $0||u||v$ to insure that, once the string is converted into an integer, that integer is less than N.

3.3   Definition of Security for Implementations

In this paper we are starting with the assumption that a cryptosystem satisfying the constraints that we outlined above is secure in a black-box setting. For a signature scheme such as RSA-PSS this means existential unforgeability under adaptive chosen message attack. Our aim is to provide sufficient conditions on the various components of the implementation such that security in the black-box setting translates into security of the physical implementation.

Let us consider an implementation $T = (T_1, \ldots, T_n)$ of some cryptosystem with public key pk and secret key $sk = (sk_2, \ldots, sk_n)$. For $i = 2, \ldots, n$ we let

$$\underline{x}_i \leftarrow T|^i(m, pk, sk_i)$$

denote the action of executing T and halting after T_i has been run. We denote by $\underline{x}_i$ the vector of outputs from $T_i, \ldots, T_2$.

Also, for $i = 3, \ldots, n$ we let

$$s \leftarrow T|_i(m, \underline{x}_{i-1}, pk, sk_i)$$

denote the action of executing T from the point of T_i onwards where $\underline{x}_{i-1}$ denotes the vector of outputs produced by $T_{i-1}, \ldots, T_2$. Note that in a complete execution of T, T_1 would know the locations of $\underline{x}_{i-1}$ VAS. Using these it would be able to provide the necessary input for $T_i, \ldots, T_n$.

We say that T_i is secure if there exists a polynomial-time simulator S_i such that no adversary $\mathcal{A}$ can win the following game ($\mathbf{Exp}^{\text{lor}}_{\mathcal{A}, T_i}(k)$) with probability significantly greater than $1/2$. Note that, since S_i produces no output that is used by any later process, it can be thought of as either a VTM with a leakage function or a function that simulates the leakage function of the VTM T_i. In the description below, the symbol q represents an upper bound on the number of *queries* that $\mathcal{A}$ is able to make to the implementation that it is attacking. In the description of $\mathbf{Exp}^{lor}_{\mathcal{A},T_i}(k)$ below lor is an acronym for left-or-right; either $\mathcal{A}$ ends up being run in the experiment on the left or in the experiment on the right. This idea has been used extensively in the literature, see [2] for example.

> **Experiment $\mathbf{Exp}^{lor}_{\mathcal{A},T_i}(k)$**
> Run the key generation algorithm for the scheme
>   on input 1^k to produce a key-pair (pk, sk)
> Prepare implementation of T_i using sk_i (where $sk = (sk_2, \ldots, sk_n)$)
> Choose b at random from $\{0, 1\}$
> If $b = 1$ run $\mathbf{Exp}^{real}_{\mathcal{A},T_i}(k)$
> If $b = 0$ run $\mathbf{Exp}^{sim}_{\mathcal{A},T_i}(k)$
> If $b' = b$ return 1, otherwise return 0

Experiment $\mathbf{Exp}^{real}_{\mathcal{A},T_i}(k)$
$state \leftarrow (pk, \{sk_l\}_{l \neq i})$
for$(j = 0, j < q, j = j + 1)$
$\{$
$\quad m_j \leftarrow \mathcal{A}(state)$
$\quad \underline{x}_{i-1} \leftarrow T|^{i-1}(m_j, pk, sk_i)$
$\quad x_i \leftarrow T_i(\underline{x}_{i-1}, pk, sk_i)$
$\quad \leadsto \mathcal{A}(state) \rightarrow state$
$\quad \underline{x}_i = (x_i, \underline{x}_{i-1})$
$\quad s_j \leftarrow T|_{i+1}(m_j, \underline{x}_i, pk, sk_i)$
$\quad state \leftarrow \mathcal{A}(state, s_j)$
$\}$
$b' \leftarrow \mathcal{A}(state)$

Experiment $\mathbf{Exp}^{sim}_{\mathcal{A},T_i}(k)$
$state \leftarrow (pk, \{sk_l\}_{l \neq i})$
for$(j = 0, j < q, j = j + 1)$
$\{$
$\quad m_j \leftarrow \mathcal{A}(state)$
$\quad \underline{x}_i \leftarrow T|^i(m_j, pk, sk_i)$
$\quad Null \leftarrow S_i(m_j, pk)$
$\quad \leadsto \mathcal{A}(state) \rightarrow state$
$\quad s_j \leftarrow T|_{i+1}(m_j, \underline{x}_i, pk, sk_i)$
$\quad state \leftarrow \mathcal{A}(state, s_j)$
$\}$
$b' \leftarrow \mathcal{A}(state)$

Note that we assume that $\mathcal{A}$ is given all the secret-key material for all the T_l with $l \neq i$. This is crucial for our security proof.

We define

$$\mathbf{Adv}^{lor}_{T_i,\mathcal{A}}(k) = |2 \cdot \Pr[\mathbf{Exp}^{lor}_{T_i,\mathcal{A}}(k) = 1] - 1|.$$

We say that it is possible to implement T_i securely if there exists and S_i such that, for all probabilistic polynomial-time $\mathcal{A}$, $\mathbf{Adv}^{lor}_{T_i,\mathcal{A}}(k)$ is a negligible function of k. Henceforth we refer to S_i as the *physical simulator* for T_i.

3.4   Result

In this section we present our result for an implementation $T = (T_1, \ldots, T_n)$ of a signature scheme *SIG*. Our result holds in the model described in Section 2. We state it formally below.

Theorem 1. Suppose that $\mathcal{A}$ is an adversary that succeeds in forging a *SIG* signature by using a physical adaptive chosen message attack on the implementation $T = (T_1, \ldots, T_n)$. We show that there are adversaries $\mathcal{A}_2, \ldots, \mathcal{A}_n, \mathcal{A}'$ such that

$$\mathbf{Adv}_{T(SIG),\mathcal{A}}^{\text{eupoacma}}(k) \leq \mathbf{Adv}_{T_2,\mathcal{A}_2}^{\text{lor}}(k) + \ldots + \mathbf{Adv}_{T_n,\mathcal{A}_n}^{\text{lor}}(k) + \mathbf{Adv}_{SIG,\mathcal{A}'}^{\text{euacma}}(k). \quad (4)$$

The execution times of $\mathcal{A}_2, \ldots, \mathcal{A}_n$ and $\mathcal{A}'$ are all essentially the same as that of $\mathcal{A}$ and the number of oracle calls made by each one is the same as the number make by $\mathcal{A}$.

Proof. To prove our result we define a sequence $\mathbf{G}_0, \ldots, \mathbf{G}_{n-1}$ of modified attack games. The only difference between games is how the environment responds to $\mathcal{A}$'s oracle queries. For any $0 \leq i \leq n - 1$, we let W_i be the event that $\mathcal{A}$ succeeds in producing a valid forged signature in game $\mathbf{G}_i$. This probability is taken over the random choices of $\mathcal{A}$ and those of $\mathcal{A}$'s oracles.

The first game $\mathbf{G}_0$ is the real attack game in which $\mathcal{A}$ my physically observe the execution of $T = (T_1, \ldots, T_n)$ on chosen input m. It may do this q times and may choose its inputs adaptively based on information gleaned from previous queries. From the definition of $\mathbf{Adv}_{T(SIG),\mathcal{A}}^{\text{eupoacma}}(k)$ it follows that

$$\Pr[W_0] = \mathbf{Adv}_{T(SIG),\mathcal{A}}^{\text{eupoacma}}(k). \quad (5)$$

In the second game $\mathbf{G}_1$ we replace the implementation of T_n with which $\mathcal{A}$ interacts with S_i - the physical simulator for T_n. We claim that there exists a polynomial time adversary $\mathcal{A}_n$, whose execution time is essentially the same as that of $\mathcal{A}$, such that

$$|\Pr[W_0] - \Pr[W_1]| \leq \mathbf{Adv}_{T_n,\mathcal{A}_n}^{\text{lor}}(k). \quad (6)$$

It is easy to construct such an adversary $\mathcal{A}_n$. According to the definition of $\mathbf{Exp}_{T_i,\mathcal{A}_n}^{\text{lor}}(k)$, $\mathcal{A}_n$ is given as input $pk, (sk_2, \ldots, sk_{n-1})$. Now, to construct $\mathcal{A}_n$ we simply prepare implementations of $T_2, \ldots, T_{n-1}$ which we can use to simulate $\mathcal{A}$'s view in its attack on T. Now, in the case where the bit hidden form $\mathcal{A}_n$ is 1, $\mathcal{A}$ is run by $\mathcal{A}_n$ in exactly the same way that the former would be run in game $\mathbf{G}_0$. Also, in the case where the bit hidden from $\mathcal{A}_n$ is 0, $\mathcal{A}$ is run by $\mathcal{A}_n$ in exactly the same way that the former would be run in game $\mathbf{G}_1$. It follows that any perceptible difference in $\mathcal{A}$'s performance in the transition from game $\mathbf{G}_0$ to game $\mathbf{G}_1$ would provide us with an adversary $\mathcal{A}_n$ of the implementation of T_i.

We repeat this procedure, replacing T_{n-1} with S_{n-1} and so on, until we have replaced T_2 with S_2. For $j = 1, \ldots, n - 2$ this gives us

$$|\Pr[W_j] - \Pr[W_{j+1}]| \leq \mathbf{Adv}_{T_{n-j},\mathcal{A}_{n-j}}^{\text{lor}}(k), \quad (7)$$

where $\mathcal{A}_{n-j}$ is an adversary of the implementation of T_{n-j} whose execution time is essentially the same as that of $\mathcal{A}$.

Finally, once this process has been completed, $\mathcal{A}$ does not have access to any genuine physically observable components of T. We may therefore consider that in game $\mathbf{G}_{n-1}$, $\mathcal{A}$ is an adversary of the scheme in the black-box setting. We conclude that there exists some adversary $\mathcal{A}'$ such that

$$\Pr[W_{n-1}] \le \mathbf{Adv}_{SIG,\mathcal{A}'}^{\text{euacma}}(k). \tag{8}$$

The result now follows from (5), (6), (7) and (8).

4   Conclusion

We have provided a set of sufficient conditions for the implementation of a cryptosystem to be no less secure than the abstract cryptosystem itself. The sufficient conditions come in two parts. Firstly we assume that the implementation of the cryptosystem fits a computational model based on that proposed by Micali and Reyzin [15]. This is the model that we described in Section 2. Secondly, in Section 3.3, we gave an indistinguishability-based definition of security that should be satisfied by the subroutines used by the implementation of the cryptosystem. In Theorem 1 we proved that, in this model, if the subroutines satisfy our definition then the implementation is no less secure than the cryptosystem itself.

The model that we have considered here is designed to cope with attacks that are in some sense passive; the adversary is assumed not to actually tamper with the internal workings of the implementation. This means that the model does not say anything about attacks such as the fault attacks of Biham *et al.* [5]. However, there has been some preliminary research into the possibility of a theoretical model to treat such cases [10].

Although we believe that a formal security treatment for side-channel environments may provide valuable insight, we recognise that it may be very difficult to prove indistinguishability results about implementations in a complexity-theoretic sense. An orthogonal line of research is to develop concrete tests that can be applied to implementations in order to assess and compare their security. For example, Coron *et al.* have proposed a set of statistical tests for the detection of leaked secret information [7]. A second interesting approach is the hidden-markov model technique proposed by Karlof and Wagner [13]. The aim of this technique is to infer information about a secret key based on side-channel information.

While our theoretical approach provides us with sufficient conditions for secure implementations, the statistical techniques above provides us with necessary conditions. By working on the problem from both these ends (so to speak), one hopes to over time converge on a realistic, well-defined set of conditions that a secure implementation of a cryptosystem should satisfy.

References

1. Dakshi Agrawal, Bruce Archambeault, Josyula R. Rao, and Pankaj Rohatgi. The EM side-channel(s). In Burton S. Kaliski Jr., Çetin Kaya Koç, and Christof Paar, editors, *Chryptographic Hardware and Embedded Systems – CHES 2002, 4th International Workshop*, volume 2523 of *Lecture Notes in Computer Science*, pages 29–45, Redwood Shores, CA, USA, August 13–15 2003. Springer-Verlag, Berlin, Germany.

2. Mihir Bellare, Anand Desai, Eric Jokipii, and Phillip Rogaway. A concrete security treatment of symmetric encryption. In *38th Annual Symposium on Foundations of Computer Science*, pages 394–403, Miami Beach, Florida, October 19–22, 1997. IEEE Computer Society Press.

3. Mihir Bellare, Anand Desai, David Pointcheval, and Phillip Rogaway. Relations among notions of security for public-key encryption schemes. In Hugo Krawczyk, editor, *Advances in Cryptology – CRYPTO '98*, volume 1462 of *Lecture Notes in Computer Science*, pages 26–45, Santa Barbara, CA, USA, August 23–27, 1998. Springer-Verlag, Berlin, Germany.

4. Mihir Bellare and Phillip Rogaway. The exact security of digital signatures: How to sign with RSA and rabin. In Ueli M. Maurer, editor, *Advances in Cryptology – EUROCRYPT '96*, volume 1070 of *Lecture Notes in Computer Science*, Saragossa, Spain, May 12–16, 1996. Springer-Verlag, Berlin, Germany.

5. Eli Biham and Adi Shamir. Differential fault analysis of secret key cryptosystems. In Walter Fumy, editor, *Advances in Cryptology – EUROCRYPT '97*, volume 1233 of *Lecture Notes in Computer Science*, pages 513–525, Konstanz, Germany, May 11–15, 1997. Springer-Verlag, Berlin, Germany.

6. Dan Boneh, Richard A. DeMillo, and Richard J. Lipton. On the importance of checking cryptographic protocols for faults. In Walter Fumy, editor, *Advances in Cryptology – EUROCRYPT '97*, volume 1233 of *Lecture Notes in Computer Science*, pages 37–51, Konstanz, Germany, May 11–15, 1997. Springer-Verlag, Berlin, Germany.

7. Jean-Sébastien Coron, David Naccache, and Paul Kocher. Statistics and secret leakage. *ACM SIGOPS Operating Systems Review*, 3(3):492–508, August 2004.

8. Ronald Cramer and Victor Shoup. Design and analysis of practical public-key encryption schemes secure against adaptive chosen ciphertext attack. *SIAM Journal on Computing*, 33(1):167–226, 2003.

9. Federal Information Processing Standards Publication 46-3 (FIPS PUB 46-3): Data Encryption Standard, October 1999.

10. Rosario Gennaro, Anna Lysyanskaya, Tal Malkin, Silvio Micali, and Tal Rabin. Algorithmic tamper-proof (ATP) security: Theoretical foundations for security against hardware tampering. In Moni Naor, editor, *TCC 2004: 1st Theory of Cryptography Conference*, volume 2951 of *Lecture Notes in Computer Science*, pages 258–277, Cambridge, MA, USA, February 19–21, 2004. Springer-Verlag, Berlin, Germany.

11. Shafi Goldwasser and Silvio Micali. Probabilistic encryption. *Journal of Computer and System Sciences*, 28:270–299, 1984.

12. Shafi Goldwasser, Silvio Micali, and Ronald L. Rivest. A digital signature scheme secure against adaptive chosen-message attacks. *SIAM Journal on Computing*, 17(2):281–308, April 1988.

13. Chris Karlof and David Wagner. Hidden markov model cryptanalysis. In Colin D. Walter, Çetin Kaya Koç, and Christof Paar, editors, *Chryptographic Hardware and Embedded Systems – CHES 2003, 5th International Workshop*, volume 2779 of *Lecture Notes in Computer Science*, page 2, Cologne, Germany, September 8–10 2003. Springer-Verlag, Berlin, Germany.

14. Paul Kocher, Joshua Jaffe, and Benjamin Jun. Differential power analysis. In Michael J. Wiener, editor, *Advances in Cryptology – CRYPTO '99*, volume 1666 of *Lecture Notes in Computer Science*, pages 388–397, Santa Barbara, CA, USA, 1999.

15. Silvio Micali and Leonid Reyzin. Physically observable cryptography (extended abstract). In Moni Naor, editor, *TCC 2004: 1st Theory of Cryptography Conference*, volume 2951 of *Lecture Notes in Computer Science*, pages 278–296, Cambridge, MA, USA, February 19–21, 2004. Springer-Verlag, Berlin, Germany. Full version available at http://eprint.iacr.org2003/120.

16. Dan Page. Theoretical use of cache memory as a cryptanalytic side-channel. Technical Report CSTR-02-003, University of Bristol Department of Computer Science, June 2002.

17. Ronald L. Rivest, Adi Shamir, and Leonard M. Adleman. A method for obtaining digital signature and public-key cryptosystems. *Communications of the Association for Computing Machinery*, 21(2):120–126, 1978.

18. Yukiyasu Tsunoo, Teruo Saito, Tomoyasu Suzaki, Maki Shigeri, and Hiroshi Miyauchi. Cryptanalysis of DES implemented on computers with cache. In Burton S. Kaliski Jr., Çetin Kaya Koç, and Christof Paar, editors, *Chryptographic Hardware and Embedded Systems – CHES 2002, 4th International Workshop*, volume 2523 of *Lecture Notes in Computer Science*, pages 62–ŋ76, Redwood Shores, CA, USA, August 13–15 2003. Springer-Verlag, Berlin, Germany.

On the Automatic Construction of Indistinguishable Operations

M. Barbosa[1,*] and D. Page[2]

[1] Departamento de Informática, Universidade do Minho,
Campus de Gualtar, 4710-057 Braga, Portugal
mbb@di.uminho.pt
[2] Department of Computer Science, University of Bristol,
Merchant Venturers Building, Woodland Road,
Bristol, BS8 1UB, United Kingdom
page@cs.bris.ac.uk

Abstract. An increasingly important design constraint for software running on ubiquitous computing devices is security, particularly against physical methods such as side-channel attack. One well studied methodology for defending against such attacks is the concept of indistinguishable functions which leak no information about program control flow since all execution paths are computationally identical. However, constructing such functions by hand becomes laborious and error prone as their complexity increases. We investigate techniques for automating this process and find that effective solutions can be constructed with only minor amounts of computational effort.

Keywords: Side-channel Cryptanalysis, Simple Power Analysis, Countermeasures, Indistinguishable Operations.

1 Introduction

As computing devices become increasingly ubiquitous, the task of writing software for them has presented programmers with a number of problems. Firstly, devices like smart-cards are highly constrained in both their computational and storage capacity; due to their low unit cost and small size, such devices are significantly less powerful than PDA or desktop class computers. This demands selection and implementation of algorithms which are sensitive to the demands of the platform. Coupled with these issues of efficiency, which are also prevalent in normal software development, constrained devices present new problems for the programmer. For example, one typically needs to consider the power characteristics and communication frequency of any operation since both eat into the valuable battery life of the device.

Perhaps the most challenging aspect of writing software for ubiquitous computers is the issue of security. Performing computation in a hostile, adversarial

---

* Funded by scholarship SFRH/BPD/20528/2004, awarded by the Fundação para a Ciência e Tecnologia, Ministério da Ciência e do Ensino Superior, Portugal.

N.P. Smart (Ed.): Cryptography and Coding 2005, LNCS 3796, pp. 233–247, 2005.
© Springer-Verlag Berlin Heidelberg 2005

environment demands that software is robust enough to repel attackers who hope to retrieve data stored on the device. Although cryptography provides a number of tools to aid in protecting the data, the advent of physical attacks such as side-channel analysis and fault injection mean one needs to consider security of the software implementation as well as the mathematics it implements. By passive monitoring of execution features such as timing variations [15], power consumption [16] or electromagnetic emission [1,2] attackers can remotely re-cover secret information from a device with little fear of detection. Typically attacks consist of a collection phase which provides the attacker with profiles of execution, and an analysis phase which recovers the secret information from the profiles. Considering power consumption as the collection medium from here on, attack methods can be split into two main classes. Simple power analysis (SPA) is where the attacker is given only one profile and is required to recover the secret information by focusing mainly on the operation being executed. In contrast, differential power analysis (DPA) uses statistical methods to form a correlation between a number of profiles and the secret information by focusing mainly on the data items being processed.

As attack methods have become better understood, so have the related de-fence methods. Although new vulnerabilities are regularly uncovered, one can now deploy techniques in hardware and software which will vastly reduce the effectiveness of most side-channel attacks and do so with fairly minor overhead. Very roughly, defence methods fall into one of two main categories:

Randomisation. One method of reducing the chance of leaking secret informa-tion is to introduce a confusion or randomisation element into the algorithm being executed. This is particularly effective in defending against DPA-style attacks but may also be useful in the SPA-style case. Essentially, randomisa-tion ensures the execution sequence and intermediate results are different for every invocation and hence reduces the correlation of a given profile with the secret information. This method exists in many different forms, for example the addition of blinding factors to exponents; dynamically randomising the parameters or control flow in exponentiation algorithms; and using redun-dant representations.

Indistinguishability. To prevent leakage of secret information to an SPA-style attack by revealing the algorithm control flow, this approach aims to modify operations sequences so that every execution path is uniform. Again, there are several ways in which this can be achieved. One way is to work directly on the mathematical formulae that define the operations and modify them so that the resulting implementations have identical structure. Another method is to work directly on the code, rearranging it and inserting dummy opera-tions, to obtain the same effect.

A key difference between issues of efficiency and security is that the program-mer is assisted by a compiler in the former case but not in the later. That is, the programmer is entirely responsible for constructing defence methods against side-channel analysis. Although the general technique of creating indistinguish-able functions to foil SPA style attack is well understood, the general barrier

---

Algorithm 1. The double-and-add method for ECC point multiplication

---

Input: point P, integer d
Output: point $Q = d \cdot P$
  1: $Q \leftarrow \mathcal{O}$
  2: **for** $i = |d| - 1$ **downto** 0 **do**
  3:    $Q \leftarrow 2 \cdot Q$
  4:    **if** $d_i = 1$ **then**
  5:       $Q \leftarrow Q + P$
  6:    **end if**
  7: **end for**
  8: **return** Q

---

to implementation is how labour intensive and error prone the process is. This is especially true when operation sequences in the functions are more complex than in the stock example of elliptic curve cryptography (ECC), for example systems like XTR or hyperelliptic curve cryptography (HECC). However, the task is ideally suited to automation; to this end our focus in this paper is the realisation of such automation to assist the development of secure software. To conclude this introduction we introduce the concept and use of indistinguishable functions in more detail and present an overview of related work. Then, in Section 2 we describe the construction of such functions as an optimisation problem and offer an algorithm to produce solutions in Section 3. Finally, we present some example results in Section 4 and concluding remarks in Section 5.

1.1   Using Indistinguishable Functions

One of the most basic forms of side-channel attack is that of simple power analysis (SPA): the attacker is presented with a single profile from the collection phase and tasked with recovering the secret information. Such an attack can succeed if one can reconstruct the program control flow by observing the operations performed in an algorithm. If decisions in the control flow are based on secret information, it is leaked to the attacker. We focus here on point multiplication as used in ECC [4] and described by Algorithm 1.

Restricting ourselves to working over the field $K = \mathbb{F}_p$, where p is a large prime, our elliptic curve is defined by:

$$E(K) : y^2 = x^3 + Ax + B$$

for some parameters A and B. The set of rational points $P = (x, y)$ on this curve, together with the identity element $\mathcal{O}$, form an additive group. ECC based public key cryptography typically derives security by presenting an intractable discrete logarithm problem over this curve group. That is, one constructs a secret integer d and performs the operation $Q = d \cdot P$ for some public point P. Since reversing this operation is believed to be hard, one can then transmit Q without revealing the value of d.

Point addition and doubling on an elliptic curve are often distinguishable from each other in a profile of power consumption because one is composed from a different sequence of operations than the other. Denoting addition by A and doubling by D, the collection phase of an SPA attack presents the attacker with a profile detailing the operations performed during execution of the algorithm. For example, by monitoring execution of using the multiplier $d = 1001_2 = 9_{10}$, one obtains the profile:

$$DADDDA$$

Given this single profile, the analysis phase can recover the secret value of d simply by spotting where the point additions occur. If the sequence DA occurs during iteration i we have that $d_i = 1$ whereas if the sequence D occurs then $d_i = 0$.

One way to avoid this problem is to employ a double-and-add-always method, due to Coron [8], whereby a dummy addition is executed if the real one is not. Although the cases where $d_i = 0$ and $d_i = 1$ are now indistinguishable, this method significantly reduces the performance of the algorithm since many more additions are performed.

However, the ECC group law is very flexible in terms of how the point addition and doubling operations can be implemented through different curve parameterisations, point representations and so on. We can utilise this flexibility to force indistinguishability by manipulating the functions for point addition and doubling so that they are no longer different. This is generally achieved by splitting the more expensive point addition into two parts, each of which is identical in terms of the operations it performs to a point doubling. Put more simply, instead of recovering the profile above from the SPA collection phase, an attacker gets:

$$DDDDDDDD$$

from which they can get no useful information. Note that although we present the use of indistinguishable functions solely for point multiplication or exponentiation, the technique is more generally useful and can be applied in many other contexts.

1.2   Related Work

Gebotys and Gebotys [12] analyse the SPA resistance of a DSP-based implementation of ECC point multiplication using projective coordinates on curves over $\mathbb{F}_p$. They show that by hand-modifying the doubling and adding implementation code, simply by inserting dummy operations, it is possible to obtain significant improvements. Likewise, Trichina and Bellezza [21] analyse the overhead associated with the same approach using mixed coordinates on curves over $\mathbb{F}_{2^n}$, and again find an efficient hand-constructed solution. Brier and Joye [6] present unified addition and doubling functions by observing that operations for calculating slope can be shared between the two cases. Joye and Quisquater [14] and Liardet and Smart [18] take a different approach by finding different curve parameterisations that offer naturally indistinguishable formulae; they utilise Hessian and

Jacobi form elliptic curves respectively. In other contexts than ECC, Page and Stam [20] present hand-constructed indistinguishable operations for XTR.

Chevallier-Mames et al. [7] propose a generalised formulation for constructing indistinguishable functions and apply it to processor-level sequences of instructions. SPA attacks typically exploit conditional instructions that depend on secret information: the solution is to make the sequences of instructions (processes) associated with both branches indistinguishable. The authors introduce the concept of side channel atomicity. All processes are transformed, simply by padding them with dummy operations, so that they execute as a repetition of a small instruction sequence (a pattern) called a side-channel atomic block. This idea is closely related to our work.

2   Indistinguishable Functions

In this section we enunciate the problem of building indistinguishable functions as an optimisation problem. We begin by defining a problem instance.

Definition 1. *Let F be a list of N functions $F = F_1, F_2, ..., F_N$ where each function F_i is itself a list of instructions from a finite instruction set L:*

$$F_i = F_i[1], F_i[2], ..., F_i[|F_i|]$$

where $|F_i|$ denotes the length of function F_i, and $F_i[j] \in L$ denotes instruction j of function F_i, with $1 \leq j \leq |F_i|$. Also, let $F_i[k..j]$ denote instructions k to j in function F_i, with $1 \leq k \leq j \leq |F_i|$.

For concreteness one should think of the simple case of two functions F_1 and F_2 as performing ECC point addition and doubling. Further, the instruction set L is formed from three-address style operations [19] on elements in the base field, for example addition and multiplication, and the functions are straight-line in that they contain no internal control flow structure.

We aim to manipulate the original functions into new versions F_i' such that the execution trace of all of them is some multiple of the execution trace of a shorter sequence. We term this shorted sequence Π, the fixed pattern of operations which is repeated to compose the larger functions. Clearly we might need to add some dummy instructions to the original functions as well as reordering their instructions so that the pattern is followed. To allow for instruction reordering, we extend our problem definition to include information about the data dependencies between instructions within each function. We represent these dependencies as directed graphs.

Definition 2. *Given a set F as in Definition 1, let P be the list of pairs*

$$P = (F_1, G_1), (F_2, G_2), ..., (F_N, G_N)$$

where $G_i = (V_i, E_i)$ is a directed graph in which V_i and E_i are the associated sets of nodes and edges, respectively. Let $|V_i| = |F_i|$ and, to each instruction

$F_i[j]$, associate node $v_j \in V_i$. Let E_i contain an edge from node v_j to node v_k if and only if executing instruction $F_i[j]$ before instruction $F_i[k]$ disrupts the normal data flow inside the function. We say that instruction $F_i[j]$ depends on instruction $F_i[k]$.

In general terms, given a straight-line function F_i described using three-address operations from our instruction set L, the pair of function and graph (F_i, G_i) can be constructed as follows:

1. Add $|F_i|$ nodes to V_i so that each instruction in the function is represented by a node in the graph.
2. For every instruction $F_i[j]$ add an edge (v_j, v_k) to E_i if and only if $F_i[j]$ uses a result directly modified by some instruction $F_i[k]$. Note that we assume that symbols for intermediate results are not reused. That is, the function is in single-static-assignment (SSA) form [19]. If reuse is permitted, additional edges must be inserted in the dependency graph to prevent overwriting intermediate results.
3. Calculate (V_i, E_i'), the transitive closure of the graph (V_i, E_i), and take $G_i = (V_i, E_i')$.

We use the dependency graphs from Definition 2 to guarantee that the transformations we perform on the functions F_i are sound. That is, as long as we respect the dependencies, the program is functionally correct even though the instructions are reordered. Definition 3 captures this notion.

Definition 3. *A function F_i' is a valid transformation of a function F_i (written $F_i' \Leftarrow F_i$) if given the dependency graph G_i, F_i' can be generated by modifying F_i as follows:*

1. *Reorder the instructions in F_i, respecting the dependency graph G_i i.e. if there is an edge $(v_j, v_k) \in E_i$ then instruction $F_i[j]$ must occur after instruction $F_i[k]$ in F_i'.*
2. *Insert a finite number of dummy instructions.*

The goal is to find Π and matching F_i' whose processing overhead compared to the original programs is minimised. Hence, our problem definition must also include the concept of computational cost. For the sake of generality, we assign to each basic instruction in set L an integer weight value that provides a relative measure of it's computational weight.

Definition 4. *Let $\omega : L \to \mathbb{N}$ be a weight function that, for each basic instruction $l \in L$, provides a relative value $\omega(l)$ for the computational load associated with instruction l.*

Given this cost function, we are now in a position to provide a formulation of the problem of building indistinguishable functions as an optimisation problem.

Definition 5. *Given a pair (P, ω) as in Definitions 1, 2 and 4, find a pattern Π and a list of functions $F' = F_1', F_2', ..., F_N'$ such that*

$$\begin{cases} \Pi = \Pi[1], \Pi[2], ..., \Pi[|\Pi|] & \Pi[k] \in L, 1 \le k \le |\Pi| \\ F_i' \Leftarrow F_i & 1 \le i \le N \\ |F_i'| = 0 \pmod{|\Pi|} & 1 \le i \le N \\ F_i'[j] = \Pi[(j \bmod |\Pi|) + 1] & 1 \le i \le N, 1 \le j \le |F_i'| \end{cases}$$

and that

$$\sum_{i=1}^{N} \sum_{j=1}^{|F_i'|} \omega(F_i'[j])$$

is minimal.

To reiterate, from this definition we have that each function must be composed of a number of instances of the pattern which constrains the type of each instruction. As a consequence, each instruction within each function matches the same instruction, modulo the pattern size, of every other function. Two functions are hence indistinguishable since one cannot identify their boundaries within a larger sequence of such patterns. In context, the only leaked information is potentially the Hamming weight and length of d: this is undesirable but unavoidable given the scope of our work.

Intuition on the hardness of satisfying these constraints comes from noticing similarities with well-known NP-complete optimisation problems such as the Minimum Bin Packing, Longest Common Subsequence and Nearest Codeword problems [9].

2.1 A Small Example

Recalling our definition of the elliptic curve $E(K)$ in Section 1.1, Algorithm 2 details two functions for affine point addition and doubling on such a curve. Denoting the addition and doubling as functions F_1 and F_2 respectively, we find $|F_1| = 10$ while $|F_2| = 13$. From these functions, we also find our instruction set is $L = \{x + y, x - y, x^2, x \times y, 1/x\}$ with all operations over the base field $K = \mathbb{F}_p$. Thus, we setup our costs as $\omega(x + y) = 1$, $\omega(x - y) = 1$, $\omega(x^2) = 10$, $\omega(x \times y) = 20$ and $\omega(1/x) = 100$.

Notice the role of dependencies in the functions: operation three in F_1 depends on operation two but not on operation one. In fact, we can relocate operation one after operation three to form a valid function F_1' since it respects the data dependencies that exist.

The graphs in Figure 1 represent the direct dependencies between the instructions in the addition method (top) and the doubling method (bottom). Complete dependency graphs as specified in Definition 2 can be obtained by calculating the transitive closure over the graphs in Figure 1.

Algorithm 2. Methods for ECC affine point addition (left) and doubling (right)

Input: $P = (x_1, y_1), Q = (x_2, y_2)$	**Input:** $P = (x_1, y_1)$
Output: $R = (x_3, y_3) = P + Q$	**Output:** $R = (x_3, y_3) = 2 \cdot P$
1: $\lambda_1 \leftarrow y_2 - y_1$	1: $\lambda_1 \leftarrow x_1^2$
2: $\lambda_2 \leftarrow x_2 - x_1$	2: $\lambda_2 \leftarrow \lambda_1 + \lambda_1$
3: $\lambda_3 \leftarrow \lambda_2^{-1}$	3: $\lambda_3 \leftarrow \lambda_2 + \lambda_1$
4: $\lambda_4 \leftarrow \lambda_1 \cdot \lambda_3$	4: $\lambda_4 \leftarrow \lambda_3 + A$
5: $\lambda_5 \leftarrow \lambda_4^2$	5: $\lambda_5 \leftarrow y_1 + y_1$
6: $\lambda_6 \leftarrow \lambda_5 - x_1$	6: $\lambda_6 \leftarrow \lambda_5^{-1}$
7: $x_3 \leftarrow \lambda_6 - x_2$	7: $\lambda_7 \leftarrow \lambda_4 \cdot \lambda_6$
8: $\lambda_7 \leftarrow x_1 - x_3$	8: $\lambda_8 \leftarrow \lambda_7^2$
9: $\lambda_8 \leftarrow \lambda_4 \cdot \lambda_7$	9: $\lambda_9 \leftarrow x_1 + x_1$
10: $y_3 \leftarrow \lambda_8 - y_1$	10: $x_3 \leftarrow \lambda_8 - \lambda_9$
	11: $\lambda_{10} \leftarrow x_1 - x_3$
	12: $\lambda_{11} \leftarrow \lambda_{10} \cdot \lambda_7$
	13: $y_3 \leftarrow \lambda_{11} - y_1$

Algorithm 3 shows a solution for this instance of the optimisation problem. The cost of the solution is 12 since we add an extra square and two extra additions both denoted by the use of λ_d as their arguments. It is easy to see that it is actually an absolute minimal value. To clarify the criteria specified in Definition 5, let us see how they apply to this case.

The pattern Π is given by the operation sequence of the doubling method, and we have $|\Pi| = 13$. To ensure both $|F_1'| = 0 \pmod{|\Pi|}$ and $|F_2'| = 0 \pmod{|\Pi|}$ we need to add three dummy instructions to F_1. The solution presents no mismatches between the instruction sequences of either function and the pattern Π, so the restriction $F_i'[j] = \Pi[(j \bmod |\Pi|) + 1]$ holds for all valid i and j values. Finally, it is easy to see that both F_i' are valid transformations of the original F_i. Instruction reordering occurs only once in F_1' (instructions 3 and 5), and these are independent in F_1. F_2' is identical to F_2.

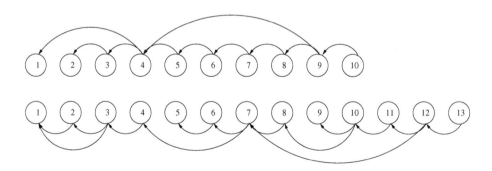

Fig. 1. Dependency graphs for the methods in Algorithm 2

Algorithm 3. Indistinguishable versions of the methods in Algorithm 2

Input: $P = (x_1, y_1), Q = (x_2, y_2)$	**Input:** $P = (x_1, y_1)$
Output: $R = (x_3, y_3) = P + Q$	**Output:** $R = (x_3, y_3) = 2 \cdot P$
1: $\lambda_d \leftarrow \lambda_d^2$	1: $\lambda_1 \leftarrow x_1^2$
2: $\lambda_d \leftarrow \lambda_d + \lambda_d$	2: $\lambda_2 \leftarrow \lambda_1 + \lambda_1$
3: $\lambda_2 \leftarrow x_2 - x_1$	3: $\lambda_3 \leftarrow \lambda_2 + \lambda_1$
4: $\lambda_d \leftarrow \lambda_d + \lambda_d$	4: $\lambda_4 \leftarrow \lambda_3 + A$
5: $\lambda_1 \leftarrow y_2 - y_1$	5: $\lambda_5 \leftarrow y_1 + y_1$
6: $\lambda_3 \leftarrow \lambda_2^{-1}$	6: $\lambda_6 \leftarrow \lambda_5^{-1}$
7: $\lambda_4 \leftarrow \lambda_1 \cdot \lambda_3$	7: $\lambda_7 \leftarrow \lambda_4 \cdot \lambda_6$
8: $\lambda_5 \leftarrow \lambda_4^2$	8: $\lambda_8 \leftarrow \lambda_7^2$
9: $\lambda_6 \leftarrow \lambda_5 - x_1$	9: $\lambda_9 \leftarrow x_1 + x_1$
10: $x_3 \leftarrow \lambda_6 - x_2$	10: $x_3 \leftarrow \lambda_8 - \lambda_9$
11: $\lambda_7 \leftarrow x_1 - x_3$	11: $\lambda_{10} \leftarrow x_1 - x_3$
12: $\lambda_8 \leftarrow \lambda_4 \cdot \lambda_7$	12: $\lambda_{11} \leftarrow \lambda_{10} \cdot \lambda_7$
13: $y_3 \leftarrow \lambda_8 - y_1$	13: $y_3 \leftarrow \lambda_{11} - y_1$

3   An Optimisation Algorithm

Our approach to solving the problem as described above is detailed in Algorithm 4. The algorithm represents an adaptation of Threshold Accepting [10], a generic optimisation algorithm. Threshold Accepting is a close relative of simulated annealing, where candidate solutions are deterministically accepted or rejected according to a predefined threshold schedule: a proposed solution is accepted as long as it does not increase the current cost by more than the current threshold value. Note that we are not aiming to find the optimal solution, but to find a good enough approximation of it that can be used in practical applications.

Algorithm 4 makes S attempts to find an optimal pattern size, which is selected randomly in each s-iteration (line 2). In each of these attempts, the original functions are taken as the starting solution, with the minor change that they are padded with dummy instructions, so as to make their size multiple of the pattern size (lines 3 to 6).

The inner loop (lines 10 to 18), which runs T times, uses a set of randomised heuristics to obtain a neighbour solution. This solution is accepted if it does not represent a relative cost increase greater than the current threshold value. The threshold varies with t, starting at a larger value for low values of t and gradually decreasing. The number of iterations S and T must be adjusted according to the size of the problem.

The quality of a solution is measured using a cost function that operates as follows:

- The complete set of instructions in a solution $\mathbf{x}$ is seen as a matrix with $|\Pi|$ columns and $(\sum_{i=1}^{N} |F_i'|)/|\Pi|$ rows (see Figure 2), in which each function occupies $|F_i'|/|\Pi|$ consecutive rows.

Algorithm 4. An optimisation algorithm for indistinguishable operations.

Input: (P, ω)
Output: (Π, F'), a quasi-optimal solution to the problem in Definition 5
 1: **for** $s = 1$ to S **do**
 2:     $|\Pi| \leftarrow$ random pattern size
 3:     $\mathbf{x} \leftarrow \{F_i, 1 \leq i \leq N\}$
 4:     **for all** $F_i' \in \mathbf{x}$ **do**
 5:         Append $|\Pi| - (|F_i'| \pmod{|\Pi|})$ dummy instructions to F_i'
 6:     **end for**
 7:     $cost \leftarrow cost(\mathbf{x})$
 8:     $result \leftarrow \mathbf{x}$
 9:     $best \leftarrow cost$
10:     **for** $t = 1$ to T **do**
11:         $\mathbf{x}' \leftarrow neighbour(\mathbf{x})$
12:         $thresh \leftarrow threshold(t, T)$
13:         $cost' \leftarrow cost(\mathbf{x}')$
14:         **if** $(cost'/cost) - 1 < thresh$ **then**
15:             $\mathbf{x} \leftarrow \mathbf{x}'$
16:             $cost \leftarrow cost'$
17:         **end if**
18:     **end for**
19:     **if** $cost < best$ **then**
20:         $result \leftarrow \mathbf{x}$
21:         $best \leftarrow cost$
22:     **end if**
23: **end for**
24: **return** $result$

- Throughout the algorithm, the pattern Π is adjusted to each solution by taking $\Pi[k]$ as the most common basic instruction in column k of the matrix.
- A dummy instruction is always taken to be of the same type as the pattern instruction for its particular column, so dummy instructions are ignored when adjusting Π to a particular solution.
- The overall cost of a solution has two components: c and d. The former is the cost associated with deviations from the pattern and it is evaluated as the sum, taken over all non-matching instructions in the matrix, of the weight difference relative to the corresponding pattern instruction. The latter is the cost associated with dummy operations and it is evaluated as the accumulated weight of all the dummy instructions in the matrix.
- The relative importance of these components can be tuned to put more or less emphasis on indistinguishability. This affects the trade-off between indistinguishability and processing overhead.

Throughout its execution, the algorithm keeps track of the best solution it has been able to reach (lines 19 to 22); this is returned when the algorithm terminates (line 24).

$$\Pi[1] \qquad\qquad \Pi[2] \qquad \ldots \quad \Pi[|\Pi|-1] \qquad \Pi[|\Pi|]$$

$$\begin{bmatrix}
F_1'[1] & F_1'[2] & \ldots & F_1'[|\Pi|-1] & F_1'[|\Pi|] \\
F_1'[|\Pi|+1] & F_1'[2] & \ldots & F_1'[2|\Pi|-1] & F_1'[2|\Pi|] \\
\vdots & \vdots & \vdots & \vdots & \vdots \\
F_1'[|F_1'|-|\Pi|+1] & F_1'[|F_1'|-|\Pi|+2] & \ldots & F_1'[|F_1'|-1] & F_1'[|F_1'|] \\
F_2'[1] & F_2'[2] & \ldots & F_2'[|\Pi|-1] & F_2'[|\Pi|] \\
\vdots & \vdots & \vdots & \vdots & \vdots \\
F_N'[|F_N'|-|\Pi|+1] & F_N'[|F_N'|-|\Pi|+2] & \ldots & F_N'[|F_N'|-1] & F_N'[|F_N'|]
\end{bmatrix}$$

Fig. 2. A solution as a matrix of instructions

A neighbour solution is derived from the current solution by randomly selecting one of the following heuristics:

Tilt function left. A function F_i' is selected randomly, and its instructions are all moved as far to the left as possible, filling spaces previously occupied by dummy instructions or just freed by moving other instructions. The order of the instructions is preserved, and an instruction is only moved if it matches the pattern instruction for the target column.

Tilt function right. Same as previous, only that instructions are shifted to the right.

Move instruction left. A function F_i' is selected randomly, and an instruction $F_i'[j]$ is randomly picked within it. This instruction is then moved as far to the left as possible. An instruction is only moved if this does not violate inter-instruction dependencies, and it matches the pattern instruction for the target column.

Move instruction right. Same as previous, only that the instruction is shifted to the right.

After application of the selection heuristic, the neighbour solution is optimised by removing rows or columns containing only dummy instructions. If the final solution produced by Algorithm 4 includes deviations from the chosen pattern, these can be eliminated in an optional post-processing phase. In this phase we increase the pattern size to cover the mismatches and introduce extra dummy operations to retain indestinguishability. If the number of mismatches is large, this produces a degenerate solution which is discarded due to the high related cost.

Our experimental results indicate the following rules of thumb that should be considered when parameterising Algorithm 4:

- S should be at least half of the length of the longest function.
- T should be a small multiple of the total number of instructions.
- An overall cost function calculated as $c^2 + d$ leads to a good compromise between indistinguishability and processing overhead.

- The threshold should decrease quadratically, starting at 70% for $t = 1$ and reaching 10% when $t = T$.
- The neighbour generation heuristics should be selected with equal probability, or with a small bias favouring moving over tilting mutations.

4   Results

Using Algorithm 4 we have been able to produce results equivalent to various hand-made solutions published in the literature for small sized problems, and to construct indistinguishable versions of the much larger functions for point addition and doubling in genus 2 hyper-elliptic curves over finite fields. To save space, we refer to Appendices within the full version of this paper for a complete set of results [3].

Appendix 1 contains the results produced by Algorithm 4 when fed with the instruction sequences for vanilla EC affine point addition over $\mathbb{F}_p$ using projective coordinates, as presented by Gebotys and Gebotys in (Figure 1,[12]). This result has exactly the same overhead as the version presented in the same reference. Appendix 2 contains the instruction sequences corresponding to formulae for finite field arithmetic in a specific degree six extension as used in a number of torus based constructions [13,11], together with the results obtained using Algorithm 4 for this problem instance.

Table 1 shows the number of dummy field operations added to each of the functions in Appendix 2. Also shown in Table 1 is the estimated overhead for an exponentiation. We assume the average case in which the number of squarings is twice the number of multiplications. This is roughly equivalent to the best hand-made solution that we were able to produce in reasonable time, even if the number of dummy multiplications is slightly larger in the automated version.

Appendix 3 includes indistinguishable versions of hyper-elliptic curve point adding and doubling functions. These implementations correspond to the general case of the explicit formulae for genus 2 hyperelliptic curves over finite fields using affine coordinates. provided by Lange in [17]. A pseudo-code implementation of these formulae is also included in Appendix 3. In our analysis, we made no assumptions as to the values of curve parameters because our objective was to work over a relatively large problem instance. Operations involving curve parameters were treated as generic operations.

In this example, the group operations themselves contain branching instructions. To accommodate this, we had to first create indistinguishable versions of the smaller branches inside the addition and doubling functions, separately.

Table 1. Overheads for the indistinguishable functions in Appendix 2

Operations	*Square*	*Multiply*	Overhead
Add	9	11	24%
Multiply	4	2	19%
Shift	0	12	400%

The process was then applied globally to the two main branches of the addition function and to the doubling function as a whole, which meant processing three functions of considerable size, simultaneously.

Table 2 (left) shows the number of dummy field operations added to each of the functions in Appendix 3. Note that functions $Add2'$ and $Double'$ correspond to cases that are overwhelmingly more likely to occur. The overhead, in these cases, is within reasonable limits. Also shown in Table 2 (right) is the estimated overhead for a point multiplication. We assume the average case in which the number of doublings is twice the number of additions, and consider only the most likely execution sequences for these operations ($Add2'$ and $Double'$).

Table 2. Overheads for the indistinguishable functions in Appendix 3

Operations	$Add1$	$Add2'$	$Add2''$	$Double'$	$Double''$	Overhead
Add	35	20	37	4	16	19%
Multiply	28	14	27	5	16	25%
Square	7	5	6	2	4	60%
Invert	1	1	1	1	1	100%

5 Conclusion

Defence against side-channel attacks is a vital part of engineering software that is to be executed on constrained devices. Since such devices are used within an adversarial environment, sound and efficient techniques are of value even if they are hard to implement. To this end we have investigated an automated approach to constructing indistinguishable functions, a general method of defence against certain classes of side-channel attack which are notoriously difficult to implement as the functions grow more complex. Our results show that efficient versions of such functions, which are competitive with hand-constructed versions, can be generated with only minor computational effort.

This work is pitched in the context of cryptography-aware compilation: the idea that programmers should be assisted in describing secure software just like they are offered support to optimise software. We have embedded our algorithm in such a compiler which can now automatically manipulate a source program so the result is more secure. As such, interesting further work includes addressing the relationship between register allocation and construction of indistinguishable functions. Ideally, the number of registers used is minimised using, for example, a graph colouring allocator. Manipulation of functions can alter the effectiveness of this process, a fact that requires some further investigation. Equally, the relationship between the presented work and side-channel atomicity might provide an avenue for further work. One would expect a similar method to the one presented here to be suitable for automatic construction of side-channel atomic patterns, and that aggressive inlining within our compiler could present the opportunity to deploy such a method.

Acknowledgements

The authors would like to thank Nigel Smart and Richard Noad for their input during this work, and the anonymous reviewers whose comments helped to improve it.

References

1. D. Agrawal, B. Archambeault, J.R. Rao and P. Rohatgi. The EM Side-Channel(s). In *Cryptographic Hardware and Embedded Systems (CHES)*, Springer-Verlag LNCS 2523, 29–45, 2002.
2. D. Agrawal, J.R. Rao and P. Rohatgi. Multi-channel Attacks. In *Cryptographic Hardware and Embedded Systems (CHES)*, Springer-Verlag LNCS 2779, 2–16, 2003.
3. M. Barbosa and D. Page. On the Automatic Construction of Indistinguishable Operations. In *Cryptology ePrint Archive*, Report 2005/174, 2005.
4. I.F. Blake, G. Seroussi and N.P. Smart. *Elliptic Curves in Cryptography*. Cambridge University Press, 1999.
5. I.F. Blake, G. Seroussi and N.P. Smart. *Advances in Elliptic Curve Cryptography*. Cambridge University Press, 2004.
6. É. Brier and M. Joye. Weierstraß Elliptic Curves and Side-channel Attacks. In *Public Key Cryptography (PKC)*, Springer-Verlag LNCS 2274, 335–345, 2002.
7. B. Chevallier-Mames, M. Ciet and M. Joye. Low-Cost Solutions for Preventing Simple Side-Channel Analysis: Side-Channel Atomicity. In *IEEE Transactions on Computers*, 53(6), 760–768, 2004.
8. J-S. Coron. Resistance against Differential Power Analysis for Elliptic Curve Cryptosystems. In *Cryptographic Hardware and Embedded Systems (CHES)*, Springer-Verlag LNCS 1717, 292–302, 1999.
9. P. Crescenzi and V. Kann. A Compendium of NP Optimization Problems. Available at: http://www.nada.kth.se/$\sim$viggo/problemlist/.
10. G. Dueck and T. Scheuer. Threshold Accepting: A General Purpose Optimization Algorithm Appearing Superior to Simulated Annealing. In *Journal of Computational Physics*, 90(1), 161–175, 1990.
11. M. van Dijk, R. Granger, D. Page, K. Rubin, A. Silverberg, M. Stam and D. Woodruff. Practical Cryptography in High Dimensional Tori. *Advances in Cryptology (EUROCRYPT)*, Springer-Verlag LNCS 3494, 234–250, 2005.
12. C.H. Gebotys and R.J. Gebotys. Secure Elliptic Curve Implementations: An Analysis of Resistance to Power-Attacks in a DSP Processor. In *Cryptographic Hardware and Embedded Systems (CHES)*, Springer-Verlag LNCS 2523, 114–128, 2002.
13. R. Granger, D. Page and M. Stam. A Comparison of CEILIDH and XTR. In *Algorithmic Number Theory Symposium (ANTS)*, Springer-Verlag LNCS 3076, 235–249, 2004.
14. M. Joye and J-J. Quisquater. Hessian Elliptic Curves and Side-Channel Attacks. In *Cryptographic Hardware and Embedded Systems (CHES)*, Springer-Verlag LNCS 2162, 402–410, 2001.
15. P.C. Kocher. Timing Attacks on Implementations of Diffie-Hellman, RSA, DSS, and Other Systems. In *Advances in Cryptology (CRYPTO)*, Springer-Verlag LNCS 1109, 104–113, 1996.
16. P.C. Kocher, J. Jaffe and B. Jun. Differential Power Analysis. In *Advances in Cryptology (CRYPTO)*, Springer-Verlag LNCS 1666, 388–397, 1999.

17. T. Lange. Efficient Arithmetic on Genus 2 Hyperelliptic Curves over Finite Fields via Explicit Formulae. In *Cryptology ePrint Archive*, Report 2002/121, 2002.
18. P-Y. Liardet and N.P. Smart. Preventing SPA/DPA in ECC Systems Using the Jacobi Form. In *Cryptographic Hardware and Embedded Systems (CHES)*, Springer-Verlag LNCS 2162, 391–401, 2001.
19. S.S. Muchnick. *Advanced Compiler Design and Implementation*, Morgan Kaufmann, 1997.
20. D. Page and M. Stam. On XTR and Side-Channel Analysis. In *Selected Areas in Cryptography (SAC)*, Springer-Verlag LNCS 3357, 54–68, 2004.
21. E. Trichina and A. Bellezza. Implementation of Elliptic Curve Cryptography with Built-In Counter Measures against Side Channel Attacks. In *Cryptographic Hardware and Embedded Systems (CHES)*, Springer-Verlag LNCS 2523, 98–113, 2002.

Efficient Countermeasures for Thwarting the SCA Attacks on the Frobenius Based Methods

Mustapha Hedabou

INSA de Toulouse, LESIA,
135, avenue de Rangueil, 31077 Toulouse cedex 4 France
`hedabou@insa-toulouse.fr`

Abstract. The Frobenius endomorphism τ is known to be useful for efficient scalar multiplication on elliptic curves defined over a field with small characteristic $(E(\mathbb{F}_{q^m}))$. However, on devices with small resources, scalar multiplication algorithms using Frobenius are, as the usual double-and-add algorithms, vulnerable to Side Channel Attacks (SCA). The more successful countermeasure for thwarting the SCA attacks on the Frobenius-based $\tau - adic$ method seems to be the multiplier randomization technique introduced by Joye and Tymen. This technique increases the computational time by about 25%. In this paper, we propose two efficient countermeasures against SCA attacks, including the powerful RPA and ZPA attacks. First, we propose to adapt the Randomized Initial Point technique (RIP) to the $\tau - adic$ method for Koblitz curves with trace 1 by using a small precomputed table (only 3 points stored). We present also an efficient fixed base $\tau - adic$ method SCA-resistant based on the Lim and Lee technique. For this purpose we modify the $\tau - NAF$ representation of the secret scalar in order to obtain a new sequence of non-zero bit-strings. This, combined with the use of Randomized Linearly-transformed coordinates (RLC), will prevent the SCA attacks on the fixed base $\tau - adic$ method, including RPA and ZPA. Furthermore, our algorithm optimizes both the size of the precomputed table and the computation time. Indeed, we only store 2^{w-1} points instead of $\frac{3^w - 1}{2}$ for the fixed-base $\tau - adic$ method, with a more advantageous running time.

Keywords: Elliptic curve, scalar multiplication, Frobenius map, $\tau - adic$ method, Side Channel Attacks, precomputed table.

1 Introduction

Since they provide the same level of security as other systems for keys with shorter length, Elliptic Curve Cryptosystems (ECC) are of great interest for cryptographic applications on devices with small resources. However, they are vulnerable to Side Channel Attacks (SCA), introduced first by Kocher et al. [Koc96, KJJ99], which have become an important threat for cryptosystems on such devices. Particularly, the improved and sophisticated RPA [Gou03] and ZPA [AT03] attacks, recently introduced, are effective on ECC, as the randomization countermeasures efficient against other SCA [JT01, Cor99] are not sufficient against them.

N.P. Smart (Ed.): Cryptography and Coding 2005, LNCS 3796, pp. 248–261, 2005.

This communication will focus on the countermeasures thwarting SCA attacks on the scalar multiplication method based on the use of the Frobenius map, which speeds up the scalar multiplication on certain categories of elliptic curves (Koblitz curves). It will present two new efficient countermeasures, the first one adapting the Randomized Initial Point (RIP) technique to the $\tau - adic$ method by using a small precomputed table (3 stored points are needed), and the second one is based on on a new $\tau - adic$ representation of the secret scalar k that allows to render the fixed base $\tau - adic$ method SCA-resistant.

The paper is organized as follows: section 2 briefly reviews the properties of the Frobenius map in the setting of ECC and describes the Frobenius-based scalar multiplication. In section 3, we introduce the Side Channel Attacks and their countermeasures. In section 4, we propose our SCA-resistant schemes: in subsection 4.1, the one based on the RIP technique to protect the $\tau - adic$ method, and in subsection 4.2 the fixed base $\tau - adic$ method SCA-resistant , which is based on a new $\tau - adic$ representation of the secret scalar.

2   Elliptic Curves Cryptosystems and the Frobenius Map

An elliptic curve is the set of the solutions of a Weierstrass equation over a field with a formal point $\mathcal{O}$, called the point at infinity . For the finite field $\mathbb{F}_{2^m}$, the standard Weierstrass equation is:

$$y^2 + xy = x^3 + ax^2 + b \quad \text{with } a, b \in \mathbb{F}_{2^m} \text{ and } b \neq 0.$$

Koblitz curves [Kob91] are defined over $\mathbb{F}_2$ by the following equations $y^2 + xy = x^3 + x^2 + 1$ and $y^2 + xy = x^3 + 1$.

2.1   The Frobenius Map

In this section we introduce briefly the Frobenius map. The reader can refer to [Men92] for details. Let $E(\mathbb{F}_q)$ an elliptic curve defined over a finite field $\mathbb{F}_q$.

We define the q-th power Frobenius map τ on $E(\mathbb{F}_q)$ as follows

$$\tau : (x, y) \leftarrow (x^q, y^q).$$

The Frobenius map satisfies the equation $\tau^2 - t\tau + q = 0$, where t is the trace of the curve $E(\mathbb{F}_q)$ ($\#E(\mathbb{F}_q) = q + 1 - t$). For Koblitz curves the characteristic equation of the Frobenius map is $\tau^2 - (-1)^{1-a}\tau + 2 = 0$

Let $E(\mathbb{F}_{q^m})$ be an elliptic curve defined over an extension $\mathbb{F}_{q^m}$ of $\mathbb{F}_q$. By the Weil theorem on elliptic curves, we know that

$$\#E(\mathbb{F}_{q^m}) = q^m + 1 - t_m$$

where t_i is the sequence satisfying

$$t_0 = 2, t_1 = t \text{ and } t_i = t_1 t_{i-1} - q t_{i-2} \text{ for } i \geq 2.$$

Since $\tau^m(x,y) = (x^{q^m}, y^{q^m}) = (x,y)$ for all $(x,y) \in \mathbf{F}_{q^m} \times \mathbf{F}_{q^m}$, it is clear that the Frobenius map on $E(\mathbb{F}_q)$ verifies

$$\tau^m(R) = R \text{ for all points } R \in E(\mathbb{F}_{q^m}).$$

2.2   The Frobenius-Based Scalar Multiplication

Let τ denote the Frobenius endomorphism of $\mathbb{F}_2$. In this section, we will describe the Frobenius-based $\tau - adic$ method. The results about the $\tau - adic$ representation in $\mathbb{Z}[\tau]$ are presented without proof; more details can be found in [Kob91, Mül98, MS92].

In [Kob91], Koblitz showed that the use of the Frobenius map τ can speed up the multiplication of a point P of the curve by a scalar k on certain categories of elliptic curves defined over fields with a characteristic $q = 2$ (Koblitz curves), as k may be written in the form $k = \sum_{i=0}^{l-1} k_i \tau^i$ with $k_i \in \{-1, 0, 1\}$, after what the computation of kP may be performed by applying the usual left-to-right point multiplication scheme. The representation $(k_{l-1}, \cdots, k_0)$ such that $k = \sum_{i=0}^{l-1} k_i \tau^i$ with $k_i \in \{-1, 0, 1\}$ is called the $\tau - adic$ representation of k.

For special curves with $q = 2$ and $t = 1$, the length of the $\tau - adic$ representation given by [MS92] can be reduced to $\min\{m-1, 2\log_2(m)+1\}$ by a suitable reduction modulo $\#E(\mathbb{F}_{2^m})$ [Mül98], i.e $k = \sum_{i=0}^{m-1} k_i \tau^i$, $k_i \in \{-1, 0, 1\}$. The $\tau - adic$ method computes the scalar multiplication kP on Koblitz curves with trace 1 as follows.

Algorithm 1 : $\tau - adic$ method
Input : an integer k, and a point $P \in E(\mathbb{F}_{2^m})$.
Output : kP.
1. Computation of the $\tau - adic$ representation: $k = \sum_{i=0}^{m-1} k_i \tau^i$, with $k_i \in \{-1, 0, 1\}$.
2. $Q \leftarrow P$.
3. for $i = m - 2$ down to 0 do
3.1 $Q \leftarrow \tau(Q)$.
3.2 if $k_i = 1$ then $Q \leftarrow Q + P$.
3.3 if $k_i = -1$ then $Q \leftarrow Q - P$.
4. Return Q.

Remark 1: *in the same manner as the NAF representation of the secret scalar k gives improvement over the binary representation in the case of the integers, we can reduce the number of the non-zero digits in the scalar k by using the $\tau - NAF$ representation of k (algorithm 4 in [Sol97]).*

3   The SCA Attacks and Their Countermeasures

Side Channel Attacks exploit some data leaking information such as power consumption and computing time to detect a part or the whole of the bits of the secret key. We can distinguish two types of SCA attacks:

- The Simple Power Analysis (SPA) attacks which analyzes the information leaking from a single execution of the algorithm. The $\tau - adic$ method computes a Frobenius map and an adding of points if $k_i \neq 0$, and only a Frobenius map if $k_i = 0$. By observing the power consumption, an SPA attacker can detect whether the secret digits k_i are zero or not. To prevent SPA attacks, many countermeasures have been proposed; the standard approach is to use fixed pattern algorithms [Cor99, Mon87].
- The Differential Power Analysis (DPA) attacks which collect informations from several executions of the algorithm and interpret them with statistical tools. To prevent DPA attacks, randomization of parameters seems to be an efficient technique [Cor99, JT01]. The usual approach is to randomize the base point P. Coron proposes to transform the affine point $P = (x, y)$ into randomized Jacobian projective coordinates $P = (r^2x, r^3y, r)$ for a random non-zero integer r. Joye and Tymen use a random curve belonging to the isomorphism class of the elliptic curve. A point $P = (x, y)$ of an elliptic curve E is transformed into $P' = (r^2x, r^3y)$ which is a point of the corresponding isomorphic curve E' of E.
- The RPA attack proposed by Goubin [Gou03] belongs to a new generation of DPA attacks that use special points to deduce the bits of the secret key. The fundamental remark of Goubin is that randomizing points with a 0-coordinate $((x, 0)$ or $(0, y))$ yields points that possess still a 0-coordinate. Supposing that the bits $k_{l-1}, \cdots, k_{j+1}$ of the secret scalar k are known by the attacker, and that he wants to guess the value of the next bit k_j, he just needs to choose the point $P = (c^{-1}mod\#E)(0, y)$ with $c = 2^j + \sum_{i=j+1}^{l-1} 2^i k_i$. If, in the process of the computation of kP, the scalar multiplication computes the point $cP = (0, y)$, the power consumption for the next step is significantly distinct. Thus, the attacker can know whether cP has been computed or not, and hence if k_j was 1 or 0. Iterating this process, all bits of the secret key can be determined. Akishita-Takagi [AT03] generalize Goubin's idea to elliptic curves without points with a 0-coordinate. Their attack focuses on the auxiliary registers which might contain a zero value, when the adding and doubling operations a re performed by the scalar multiplication. The ZPA attack is in particular efficient on several standard curves with no 0-coordinate point. To prevent the RPA and ZPA attacks, the authors in [IIT04, MMM04] have proposed the Randomized Linearly-transformed coordinates (RLC) technique.
- To prevent the SCA attacks on the $\tau - adic$ method, Joye and Tymen [JT01] have proposed to randomize the secret scalar k. The scalar k is reduced modulo ρ $(\tau^m - 1)$, where ρ is a random element of $\mathbb{Z}[\tau]$. For the same purpose, Hasan [Has00] proposed previously three countermeasures. In the Key Masking with Localized Operations (KMLO) technique, the symbols of the $\tau - adic$ representation can be replaced in more than one way on a window of three and more symbols, since we have $2 = \tau - \tau^2 = -\tau^3 - \tau$ which is derived from the equation $\tau^2 - \tau + 2 = 0$ (assuming that $t = 1$). The Random Rotation of Key (RRK) technique proposes to compute the scalar multiplication kP as $k'P'$ where $P' = \tau^r P$, and r is a random integer such as $r \leq m - 1$.

Finally, the Random Insertion of Redundant Symbols (RIRS) technique proposes to insert in the $\tau - adic$ representation of the secret scalar k a number of redundant symbols such as they collectively neutralize their own effects. Another countermeasure was proposed by Smith [Smi02]; it consists in decomposing the $\tau - adic$ representation of k into r groups of g coefficients, r being a random element such as $r \leq m$ and $g = \lceil \frac{m}{r} \rceil$. The point multiplication between each group and the base point P is performed in a random order. The countermeasure of Joye and Tymen seems to be more efficient in thwarting the SCA attacks, since it randomizes the entire digits of the secret scalar k.

In the following section, we propose two new efficient countermeasures for preventing SCA attacks on the Frobenius-based scalar multiplication algorithms.

4   Proposed Countermeasures

As explained in the previous section, the $\tau - adic$ method for scalar multiplication is vulnerable to SPA attacks. It is also not secure against RPA and ZPA attacks, even if the countermeasures against SPA and the usual randomization techniques are used. In this section, we propose two efficient countermeasures that aim to protect the Frobenius-based algorithms against SCA attacks.

First, we propose to adapt the Randomized Initial Point (RIP) technique proposed in [IIT04, MMM04] to the $\tau - adic$ method in order to protect the always-add-and-double method against SCA attacks. The initial point is changed into a random point R. This will randomize all intermediate informations. As a second countermeasure, we propose to convert the fixed-base $\tau - adic$ method, which is based on the Lim and Lee technique, into a SPA-resistant scheme by changing the $\tau - NAF$ representation of the scalar k; the use of Randomized Linearly-transformed Coordinates [IIT04] with the obtained scheme achieves the security against SCA attacks

4.1   The $\tau - adic$ Method with a Randomized Initial Point (RIP)

Let $E(\mathbb{F}_{2^m})$ be an elliptic curve with $t = 1$, and let τ the Frobenius map on $E(\mathbb{F}_2)$, i.e $\tau((x, y)) = (x^2, y^2)$.

It is clear that τ verifies $\tau^m = 1$ in End_E. Thus

$$\tau^m - 1 = (\tau - 1) \sum_{i=0}^{m-1} \tau^i = 0.$$

Then for all points $R \in E(\mathbb{F}_{2^m}) \setminus E(\mathbb{F}_2)$ (i.e $\tau(R) \neq R$) we have

$$\sum_{i=0}^{m-1} \tau^i(R) = 0. \quad (1)$$

The basic idea of our countermeasure is to exploit the equation (1) by introducing a random point $R \in E(\mathbb{F}_{2^m}) \setminus E(\mathbb{F}_2)$ in such a way that the $\tau - adic$ method computes $kP + \sum_{i=0}^{m-1} \tau^i(R)$, which is equal to kP, by using the equation (1). For this purpose, we store the points $iP + R$ for $i \in \{-1, 0, 1\}$, and run the $\tau - adic$ method as a Window method [BSS99]. The following algorithm implements in detail the proposed countermeasure.

Algorithm 2 : Secure $\tau - adic$ method
Input : an integer k, and a point $P \in E(\mathbb{F}_{2^m})$.
Output : kP.
1. $R \leftarrow Randompoint()$ $(R \in E(\mathbb{F}_{2^m}) \setminus E(\mathbb{F}_2))$.
2. Precomputation. Compute and store $P_i = iP + R$ for $i = -1, 0, 1$.
2. Computation of the $\tau - adic$ representation. $k = \sum_{i=0}^{m-1} k_i \tau^i$, with $k_i \in \{-1, 0, 1\}$.
3. $Q \leftarrow P_{k_{m-1}}$.
4. for $i = m - 2$ to 0 do
4.1 $Q \leftarrow \tau(Q)$.
4.2 $Q \leftarrow Q + P_{k_i}$.
5. Return Q.

$Randompoint()$ is a function that generates a random point R on $E(\mathbb{F}_{2^m}) \setminus E(\mathbb{F}_2)$. The simplest way to obtain a random point R is to generate a random x-coordinate belonging to $\mathbb{F}_{2^m} \setminus \mathbb{F}_2$ and to compute the corresponding y-coordinate if it exists, but this process is probabilistic and may require many computations. The optimized way is to randomize a fixed stored point $Q \in E(\mathbb{F}_{2^m}) \setminus E(\mathbb{F}_2)$ by using Randomized Projective Coordinates [Cor99].

In our modified $\tau - adic$ method, the adding point operation $Q + \mathcal{O}$ corresponding to a zero digit in the $\tau - adic$ representation of k is replaced by the adding point operation $Q + R$, which implies that the algorithm performs the scalar multiplication with a uniform behaviour, computing exactly a Frobenius map and an adding point at each step. Consequently, the execution of a SPA attack can not reveal any information on the digits of k. Since R is chosen randomly by some way mentioned above, all intermediate values will be randomized and thus, the algorithm will also resist to DPA attacks, and to the more powerful RPA and ZPA attacks.

Now we will estimate the cost of the proposed countermeasure. The cost of generating the random point R in the optimized way explained above is nearly free ($5M$, where M denotes a field multiplication). In the precomputation phase, we perform two more point additions. Thus the total cost of this countermeasure is only $2A + 5M$, where A denotes an adding point operation. The major disadvantage of the RIP technique is that it imposes the computation of a Frobenius map and an addition of points at each step of the algorithm 2 even if the digit k_i is zero, which makes the cost of the algorithm 2 much higher than that of the $\tau - NAF$ method Since the average density of the non-zero terms among $\tau - NAF$ representation of the scalar is only about $1/3$.

4.2   Changing the Representation of k Combined with Lim-Lee Technique

If the the base point P is fixed and some storage is available, the Lim-Lee technique allows to speed up the scalar multiplication on elliptic curve. In the following we describe the fixed-base $\tau - adic$ method based on the Lim and Lee [LL94] technique.

Let $(k_{l-1}, \cdots, k_1, k_0)$ be the $\tau - adic$ representation of an integer k, i.e $k = \sum_{i=0}^{i=l-1} k_i \tau^i$, with $k_i \in \{-1, 0, 1\}$, and let w be an integer such as $w \geq 2$; we set $d = \lceil \frac{l}{w} \rceil$. P being an elliptic curve point, for all $(b_{w-1}, \cdots, b_1, b_0) \in \mathbb{Z}_2{}^w$, we define $[b_0, b_1, \cdots, b_{w-1}]P = b_0 P + b_1 \tau^d(P) + b_2 \tau^{2d}(P) + \cdots + b_{w-1} \tau^{(w-1)d}(P)$. The comb method considers that k is represented by a matrix of w rows and d columns, and processes k columnwise.

Algorithm 3 : Fixed-base $\tau - adic$ method
Input: a $\tau - adic$ representation $(k_{l-1}, \cdots, k_1, k_0)$ of a positive integer k, an elliptic   curve point P and a window width w such as $w \geq 2$.
Output: kP.
1. $d = \lceil \frac{l}{w} \rceil$.
2. Precomputation: compute $[b_{w-1}, \cdots, b_1, b_0]P$ for all $(b_{w-1}, \cdots, b_1, b_0) \in \mathbb{Z}_2{}^w$.
3. By padding the $\tau - adic$ representation $(k_{l-1}, \cdots, k_1, k_0)$ on the left with 0's if necessary, write $k = K^{w-1} \| \cdots \| K^1 \| K^0$ where each K^j is a bit-strings of length d. Let K_i^j denote the i-th bit of K^j.
4. $Q \leftarrow [K_{d-1}^{w-1}, \cdots, K_{d-1}^1, K_{d-1}^0]P$.
5. For i from $d - 2$ down to 0 do
     5.1 $Q \leftarrow \tau(Q)$
     5.2 $Q \leftarrow Q + [K_i^{w-1}, \cdots, K_i^1, K_i^0]P$.
6. Return Q.

The execution of a SPA attack on the fixed-base $\tau - adic$ method can allow to deduce the bits of the secret scalar. This is because the fixed base $\tau - adic$ method performs only a Frobenius map operation if the bit-string $[K_i^{w-1}, \cdots, K_i^1, K_i^0]$ is equal to zero, and an adding and Frobenius map operation in the other case; thus, the analysis of power consumption during the execution of the algorithm can reveal whether the bit-string $[K_i^{w-1}, \cdots, K_i^1, K_i^0]$ is zero or not. Since the probability to have a zero bit-string ($[K_i^{w-1}, \cdots, K_i^1, K_i^0] = (0, \cdots, 0)$) is less important than the probability to get a single zero bit ($k_i = 0$), the fixed base $\tau - adic$ method offers a better resistance against SPA attacks than the $\tau - adic$ method, but it is not totally secure against them.

Geiselmann and Steinwandt's attack will be efficient against the fixed-base $\tau - adic$ method, since it uses a precomputed table, even if the usual randomization techniques are used. The fixed base $\tau - adic$ method is also not secure against the more powerful RPA and ZPA attacks.

In this section, our first aim is to generate a new representation of k as a sequence of bit-strings different from zero, so as to thwart the SPA attack. For this purpose, we modify the $\tau - NAF$ representation of k by eliminating all its zero digits and using only digits equal to 1 or -1. We then combine the obtained SPA-resistant algorithm with Randomized Linearly-transformed Coordinates [IIT04] in order to prevent the DPA, RPA and ZPA attacks.

4.2.1   A New Representation for k

Let $\mathbb{E}(\mathbb{F}_2)$ be a Koblitz curve and $\tau^2 - t\tau + 2 = 0$ be the equation verified by the Frobenius map ($t = 1$ or -1). The curve $\mathbb{E}(\mathbb{F}_{2^m})$ denote the curve regarded over the extension $\mathbb{F}_{2^m}$ of $\mathbb{F}_2$.

Let $k = \sum_{i=0}^{l-1} k_i \tau^i$, where $k_i \in \{-1, 0, 1\}$, be the $\tau - NAF$ representation of k, i.e $k_j k_{j+1} = 0$ for $j = 0, \cdots, l - 2$, and suppose that the Frobenius equation is $\tau^2 - \tau + 2 = 0$ (the trace t of the curve is 1).

Our algorithm proposes to replace every zero digit k_i by 1 or -1, depending of its neighbour bits. Assuming that the first digit k_0 of the $\tau - NAF$ representation of k is different from zero, let k_i be the first bit equal to 0. We then set $k_i \leftarrow k_i + k_{i-1}$, $k_{i+1} \leftarrow k_{i+1} - k_{i-1}$ and $k_{i-1} \leftarrow k_{i-1} - 2k_{i-1} = -k_{i-1}$. After this modification we are sure that $k_{i-1}, k_i \in \{-1, 1\}$ and $k_{i+1} \in \{-2, -1, 0, 1, 2\}$. To eliminate all the zero digits k_i of the $\tau - NAF$ representation of k we propose to proceed as follows

$$\begin{cases} k_i \leftarrow k_i + k_{i-1} \\ k_{i+1} \leftarrow k_{i+1} - k_{i-1} \\ k_{i-1} \leftarrow k_{i-1} - 2k_{i-1} = -k_{i-1} \end{cases}$$

if k_i is even and keep the digits k_{i-1}, k_i, k_{i+1} unchanged otherwise.

Remark 2: *If we denote k'_{i-1}, k'_i, k'_{i+1} the obtained digits, then we have*

$k'_{i-1}\tau^{i-1} + k'_i \tau^i + k'_{i+1}\tau^{i+1} = k_{i-1}\tau^{i-1} + k_i\tau^i + k_{i+1}\tau^{i+1} - k_{i-1}\tau^{i+1} + k_{i-1}\tau^i - 2k_{i-1}\tau^{i-1}$

$$\begin{aligned} &= k_{i-1}\tau^{i-1} + k_i\tau^i + k_{i+1}\tau^{i+1} - k_{i-1}\tau^{i-1}(\tau^2 - t\tau + 2) \\ &= k_{i-1}\tau^{i-1} + k_i\tau^i + k_{i+1}\tau^{i+1} + 0 \\ &= k_{i-1}\tau^{i-1} + k_i\tau^i + k_{i+1}\tau^{i+1} \end{aligned}$$

Thus we can conclude that this process does not change the value of k.

Example. Let $(1, 0, -1, 0, 1)$ be the $\tau - NAF$ representation of an integer k.

$(1, 0, -1, 0, 1) \xrightarrow{k_1=0} (1, 0, -1 - 1, 0 + 1, 1 - 2) = (1, 0, -2, 1, -1)$
$(1, 0, -2, 1, -1) \xrightarrow{k_2=-2} (1, 0 - 1, -2 + 1, 1 - 2, -1) = (1, -1, -1, -1, -1)$.

The new $\tau - adic$ representation of k is $(1, -1, -1, -1, -1)$

But we have also to make the generation of our new Frobenius representation SPA-resistant. In the present state, the digits k_{i-1}, k_i, k_{i+1} are either touched if k_i is a zero digit or kept unchanged otherwise; hence, a SPA attack on this algorithm can occur. To deal with this threat, we modify our method to ensure that each digit is touched, independently of its value.

If the first digit k_0 of the Frobenius representation of k is zero, we make $k_1 = k + 1$, and we compute $k_1 P$. The result of the scalar multiplication kP is then recovered by performing the substraction $k_1 P - P$. This too might give way to a SPA attack, due to the difference in the treatment of scalars k with $k_0 = 0$ and those with $k_0 = \pm 1$. So we convert as well the scalars k with $k_0 = \pm 1$ to $k_1 = k + 2$ if $k_0 = -1$ and $k_1 = k - 2$ if $k_0 = 1$, and recover in this case kP by performing the substraction $k_1 P - 2P$ or $k_1 P + 2P$. Finally, we arrive at the following algorithm:

Algorithm 4 : Modified Frobenius representation
Input: a $\tau - NAF$ representation $k = (k_{l-1}, \cdots, k_1, k_0)_2$ of an integer k.
    ($k_i \in \{-1, 0, 1\}$ for $i = 0, \cdots, l - 1$ and $k_i = 0$ for $i \geq l$).
Output: modified $\tau - adic$ representation $(k_{l-1}, ..., k_0)$ of k, $k_i \in \{-1, 1\}$.
1. If $k_0 \mod 2 = 0$ then $k \leftarrow k + 1$ else $k \leftarrow k - 2k_0$.
2. For $i = 1$ to $l - 1$ do
2.1 $b[0] \leftarrow k_i$, $b[1] \leftarrow k_i + k_{i-1}$, $c[0] \leftarrow k_{i-1}$, $c[1] \leftarrow -k_{i-1}$, $d[0] \leftarrow k_{i+1}$
$d[1] \leftarrow k_{i+1} - k_{i-1}$
2.2 $k_i \leftarrow b[1 - | k_i \mod 2 |]$, $k_{i+1} \leftarrow d[1 - | k_i \mod 2 |]$, $k_{i-1} \leftarrow c[1 - | k_i \mod 2 |]$.
3. If $k_l = 0$ return $(k_{l-1}, ..., k_0)$ else $(k_l, ..., k_0)$.

The length of the obtained new representation of k is at most one digit longer. Indeed it is exactly l if the modified k_{l-1} is an odd integer and $l + 1$ otherwise. Furthermore, it is clear that we can extend the length of the modified $\tau - adic$ representation to $l + 2j$ where $j = 1, 2, \cdots$ if the modified k_{l-1} is an odd integer or to $l + 2j + 1$ where $j = 1, 2, \cdots$ otherwise by replacing $l - 1$ in step 2 by $l - 1 + j$ for $j = 1, 2, \cdots$.

Now we will implement an SPA-resistant fixed-base $\tau - adic$ method (algorithm 3). Let w be the window width and suppose the length of the modified $\tau - NAF$ representation of k is $l + 1$ (when the length is l we proceed in the same way) .

If the $l + 1$ is divisible by w, it is clear that that all the obtained bit-strings do not contain any zero digit and thus they are all different from zero. Consequently the algorithm 3 implemented with the modified $\tau - NAF$ representation of the scalar k is SPA-resistant. On the other hand, if $l + 1$ is not divisible by w, we extend the length of the modified $\tau - NAF$ of the k to $l + (2j_0 + 1)$ for some j_0 such as w divide $l + (2j_0 + 1) - 1$.

The length of the new representation of k is $l + (2j_0 + 1)$, thus we can write

$$kP = \sum_{i=0}^{i=l+2j_0} k_i \tau^i(P) = k_0 P + \sum_{i=1}^{i=l+2j_0} k_i \tau^i(P)$$
$$= k_0 P + \tau(\sum_{i=1}^{i=l+2j_0} k_i \tau^{i-1}(P)) = k_0 P + \tau(\sum_{i=0}^{i=l+2j_0-1} k_{i+1} \tau^i(P))$$

If we set $k_2 = \sum_{i=0}^{i=l+2j_0-1} k_{i+1} \tau^i$ the length of the $\tau - adic$ representation of k_2 is $l + 2j_0$ which is divisible by w. Thus, to make the algorithm 3 SPA-resistant we propose to compute the scalar multiplication $k_2 P$ via the algorithm 3 and we recover kP by performing $\tau(k_2 P) + k_0 P$.

It is clear from this process that the length of the new representation used by the algorithm 3 is the same as the $\tau - NAF$ representation of k. Since the length of the $\tau - NAF$ and $\tau - adic$ representations of the scalar k are about m [Sol97, MS92], we can conclude that the number of the bit-strings required by the proposed method is the same as that of the fixed base $\tau - adic$ method. Consequently the proposed scheme needs only a more adding and Frobenius map operation than the fixed base $\tau - adic$ method.

In section 1 of appendix we prove the following theorem which ensures that algorithm 4 outputs a new $\tau - adic$ expansion that represents correctly k. The algorithm that describes how we can obtain the modified $\tau - adic$ representation of k for curves with trace -1 is given in section 2.

Theorem 1: *Algorithm 3, when given a positif integer k, outputs a new sequence of digits $(k_l, ..., k_0)$ such as $k = \sum_{i=0}^{i=l} k_i \tau^i$, with $k_i \in \{-1, 1\}$ for $i = 0, \cdots, l$.*

4.2.2   The Size of the Precomputed Table

The fixed-base $\tau - adic$ method precomputes the points $[b_{w-1}, \cdots, b_1, b_0]P$ for all $(b_{w-1}, \cdots, b_1, b_0) \in \{-1, 0, 1\}^w$, which may be represented as $\{-Q, Q\}$, where $Q = \{[b_{w-1}, \cdots, b_1, b_0]P$, with $b_0 = 1$. Thus, the number of points stored in the precomputed table is $\frac{3^w - 1}{2}$.

On the other hand, the set of points stored in the precomputed table of the proposed scheme is $\{[b_{w-1}, \cdots, b_1, b_0]P$, for all $b_i = \pm 1\}$, which is symmetric set. Thus We need only to store in the precomputed table the points $[b_{w-1}, \cdots, b_1, 1]P$ with $b_i = \pm 1\}$. Consequently, the number of the points stored in the precomputation phase of the proposed scheme is 2^{w-1}, which is at most the half of what is required by the fixed-base $\tau - adic$ method.

4.2.3   Security Considerations

This section discusses the security of the proposed scheme against the SPA, DPA and second-order DPA, Geiselmann and Steinwandt, RPA and ZPA attacks.

• SPA: as explained before, our method builds a new sequence of bit-strings which form the new scalar's representation, in which all the bit-strings are different from zero. At each step, the main phase of the multiplication algorithm performs then exactly an adding and Frobenius map operation, and the elliptic curve scalar multiplication behaves in a fixed pattern. Consequently, the execution of a SPA attack can not reveal any information on the bits of the secret scalar.

• DPA and second-order DPA: the use of projective randomization methods, such as randomized projective coordinates[Cor99] or random isomorphic curves [JT01], prevents DPA attacks. Okeya and Sakurai's second-order DPA attack [OS02] may be applied against the proposed algorithm, since it uses a precomputed table. For each bit-string, we access the table to get a point $[K_i^{w-1}, \cdots,$

$K_i^1, K_i^0]P$ to be added to Q. An attacker could thus manage to detect whether or not a bit-string $[K_i^{w-1}, \cdots, K_i^1, K_i^0]$ is equal to $[K_j^{w-1}, \cdots, K_j^1, K_j^0]$, by monitoring some information related to the power consumption. To prevent this atack, we propose to change the randomization of each precomputed point after getting it from the table. Thus, even if we have got the same point for different bit-strings, the new point randomization implies that we load a different data.

• Geiselmann and Steinwandt, RPA and ZPA attacks: to prevent these attacks, we propose to use the Randomized Linearly-transformed Coordinates (RLC) introduced by Itoh and al [IIT04]. This technique converts a point (x, y, z) into a randomized point (x', y', z') such as

$$x' = \lambda_x(\lambda)(x - \mu_x) + \mu_x, \ y' = \lambda_y(\lambda)(y - \mu_y) + \mu_y, \ z' = \lambda_z(\lambda)(z - \mu_z) + \mu_z$$

where $\lambda_x, \lambda_y, \lambda_z$ are functions of λ and μ_x, μ_y, μ_z.

The RLC technique with $\mu_x, \mu_y \neq 0$ allows to randomize also the points with a 0-coordinate and all the intermediate values, and thus makes the proposed algorithm secure against RPA, ZPA and Geiselmann-Steinwandt's attacks

4.2.4   Computation Cost

The proposed multiplication algorithm performs an adding (A) and Frobenius map (D) operation at each step and a more adding and Frobenius map operation for recovering the scalar kP when the length of the modified $\tau - adic$ representation of the scalar is not divisible by w, so the cost of the main computation phase is $d(A + D)$.

Now, we evaluate the cost of the precomputation phase. In this phase, we generate the sequence of points $[b_{w-1}, \cdots, b_1, 1]P$, for all $(b_{w-1}, \cdots, b_1, 1) \in \{-1, 1\}^{w-1}$, such as

$$[b_{w-1}, \cdots, b_1, 1]P = b_{w-1}2^{(w-1)d}P + \cdots + b_2 2^{2d}P + b_1 2^d + P.$$

To perform the precomputing phase, we first compute $2^d P, \ 2^{2d} P, \cdots, 2^{(w-1)d} P$, which will cost $((w-1)d)$ Frobenius map operation. The second step consists in computing all possible combinations $b_{w-1}2^{(w-1)d}P + \cdots + b_2 2^{2d}P + b_1 2^d P + P$. where $b_i \in \{-1, 1\}$, for $i = 2, \cdots, w - 1$. The cost of this second step is at most $2^w - w$ adding operations for $w = 2, \ 3, \ 4, \ 5$, which are the optimum choices for w in elliptic curve cryptosystems. The total cost of the precomputing phase is then

$$[(w-1)d]D + [2^w - w]A$$

Thus the total cost of the proposed method including efforts for preventing SPA attacks is

$$[wd]D + [2^w - w + d]A.$$

4.2.5 Efficiency Comparison

In this section, we will compare the efficiency (cost, size of the table) of the proposed algorithm with the fixed-base $\tau - NAF$ method. We recall that precomputation phase of the fixed-base $\tau - adic$ method computes $[b_{w-1}, \cdots, b_1, b_0] P$ for all $(b_{w-1}, \cdots, b_1, 1)$ where $b_i \in \{-1, 0, 1\}$ for $i = 0, \cdots, w - 1$, thus the cost of this phase is $[\frac{3^w - 1}{2} - w]A + (w - 1) * dD$. Since it's main phase computes an adding and Frobenius map of an elliptic curve point at each step, the total cost of fixed-base $\tau - NAF$ method is

$$[(w - 1) * d + d - 1]D + [\frac{3^w - 1}{2} - w + d - 1]A$$

The following table compares the efficiency of the proposed method with that of the fixed-base $\tau - adic$ method, including only efforts for preventing SPA attacks and using randomized Jacobian coordinates. S will denote the number of points stored in the precomputation phase, and T the number of field multiplications; k is a scalar with length 163 ($log_2(k) = 163$).

Method	$w = 2$		$w = 3$		$w = 4$		$w = 5$	
	S	T	S	T	S	T	S	T
Fixed-base $\tau - adic$ method	4	1986	13	1608	40	1846	121	3288
Proposed method	2	2208	4	1530	8	1388	16	1530

5   Conclusion

In this paper, we have presented two efficient countermeasures for preventing SCA attacks on the Frobenius based methods. The first countermeasure is an adaptation of the Randomized Initial point (RIP) technique for the $\tau - adic$ method on Koblitz curves. This method needs a precomputed table with small size (only 3 points stored).

The second proposed method converts the fixed base $\tau - adic$ method to an SPA-resistant scheme by changing the $\tau - NAF$ representation of the scalar. The obtained scheme is combined with Randomized Linearly-transformed Coordinates to achieve resistance against SCA attacks, with optimized performances. Indeed, the proposed scheme requires to store only 2^{w-1} points in a precomputed table instead of $\frac{3^w - 1}{2}$ for the fixed-base $\tau - adic$ method, with a more advantageous running time.

References

[AT03]   T. AKISHITA, T. TAKAGI. *Zero-value point attacks on elliptic curve crytosystems*. In: Information Security Conference-ISC'2003, vol. 2851, Lecture Notes in Computer Science (LNCS), pp. 218-233, Springer-Verlag, 2003.

[BSS99]   I. BLAKE, G. SEROUSSI, N. SMART. *Elliptic curves in cryptography.* Cambridge University Press, 1999.

[Cor99]   J.S. CORON. *Resistance against differential power analysis for elliptic curve cryptosystems.* In: Cryptography Hardware and Embedded Systems-CHES'99, C.K. Koç and C.Paar, editors, vol. 1717, LNCS, pp. 292-302, 1999.

[Gou03]   L. GOUBIN. *A refined power-analysis attack on elliptic curve cryptosystems.* In: Public Key Cryptography International Workshop-PKC'2003, vol. 2567, LNCS, pp. 199-210, 2003.

[Has00]   M. ANWAR HASAN. *Power analysis attacks and algorithmic approaches to their countermeasures for Koblitz curve cryptosystems.* In: Cryptography Hardware and Embedded Systems-CHES'00, vol. 1965, LNCS, pp. 93-108, 2000.

[IIT04]   K. ITOH, T. IZU, M. TAKENAKA. *Efficient countermeasures against power analysis for elliptic curve cryptosystems.* In: Proceedings of CARDIS-WCC 2004.

[JT01]   M. JOYE, C. TYMEN. *Protections against differential analysis for elliptic curve cryptography: an algebraic approach.* In: Cryptography Hardware and Embedded Systems-CHES'01, C. Koç, D. Naccache and C. Paar, editors, vol. 2162, LNCS, pp. 386-400, 2001.

[KJJ99]   P. KOCHER, J. JAFFE, B. JUN. *Differential power analysis.* In: Advances in Cryptology-CRYPTO'99, M. Wiener, editor, vol. 1666, LNCS, pp. 388-397, 1999.

[Kob91]   N. KOBLITZ. *CM-curves with good cryptographic properties.* In: Advances in Cryptology-CRYPTO'91, J. Feigenbaum, editor, vol. 576, LNCS, pp. 279-287, 1991.

[Koc96]   P. KOCHER. *Timing attacks on implementations of Diffie-Hellman, RSA, DSA and other systems.* In: Advances in Cryptology-CRYPTO'96, N. Koblitz, editor, vol. 1109, LNCS, pp. 104-113, 1996.

[LL94]   C. LIM, P. LEE. *More flexible exponentiation with precomputation.* In: Advances in Cryptology-CRYPTO'94, vol 839, LNCS, pp. 95-107, 1994.

[Men92]   A. MENEZES. *Elliptic curve public key cryptosystems.* The Kluwer Academic publishers, vol. 234, 1993. pp. 333-344, 1992.

[MMM04]   H. MAMIYA, A. MIYAJI, H. MORIMOTO. *Efficient Countermeasures against RPA, DPA, and SPA.* In: Cryptography Hardware and Embedded Systems-CHES'04, M. Joye, J.J. Quisquater, editors, vol. 3156, LNCS, pp. 343-356, 2004.

[Mon87]   P.L. MONTGOMERY. *Speeding up the Pollard and elliptic curve methods of factorization.* Mathematics of Computation, 48(177), pp. 243-264, January 1987.

[MS92]   W. MEIER, O. STAFFELBACH. *Efficient multiplication on certain non-supersingular elliptic curves.* In: Advances in Cryptology-CRYPTO'92, vol.740,LNCS, pp. 333-344, 1992.

[Mül98]   V. MÜLLER. *Fast multiplication on elliptic curves over small fields of characteristic two.* Journal of Cryptology (1998)11, pp. 219-234, January 1998.

[OS02]   K. OKEYA, K. SAKURAI. *A Second-Order DPA attacks breaks a window-method based countermeasure against side channel attacks.* In: Information Security Conference-ISC'2002, LNCS 2433, pp. 389-401, 2002.

[Smi02]   E. W. SMITH. *The implementation and analysis of the ECDSA on the Motorola StarCore SC140 DSP primarily targeting portable devices.* Master thesis, University of Waterloo, Ontario, Canada, 2002.

[Sol97]   J. A. SOLINAS. *An improved algorithm for arithmetic on a family of elliptic curves*. In: Advances in Cryptology-CRYPTO'97, vol. 1294, LNCS, pp. 357-371, 1997.

Appendix

A-1. Proof of the Theorem 1

Proof. It is clear from the remark 1 that $k = \sum_{i=0}^{i=l} k_i \tau^i$. It remains to prove that all the digits k_i belong to $\{-1, 1\}$.

As mentioned above, after the first modification, the digits k_{i-1}, k_i may take only the values 1 or -1, but the digit k_{i+1} may be equal to $-2, -1, 0, 1$ or 2.

Suppose that $k_{i+1} = -2$ and $k_i = 1$ (or $k_{i+1} = 2$ and $k_i = -1$), then we have after modification $k_i = k_i - 2k_i = -1$, $k_{i+1} = k_{i+1} + k_1$ and $k_{i+2} = k_{i+2} - k_i = 1$ (or $k_i = k_1 - 2k_i = 1$, $k_{i+1} = k_{i+1} - k_i$ and $k_{i+2} = k_{i+2} - k_i = 1$). Since the $\tau - NAF$ representation of the scalar k is used (at least one of two consecutive digits is zero) it is clear that $k_{i+2} = 0$ and thus $k_{i+2} = 1$ or -1.

On the other hands if $k_{i+1} = 2$ and $k_i = 1$ (or $k_{i+1} = -2$ and $k_i = -1$) the new obtained digits are $k_i = k_i - 2k_i = -1$, $k_{i+1} = k_{i+1} + k_i = 3$ and $k_{i+2} = k_{i+2} - 1$ (or $k_i = k_1 - 2k_i = 1$, $k_{i+1} = k_{i+1} - 1 = -3$ and $k_{i+2} = k_{i+2} + 1$).

Thus to achieve the proof of the theorem 1 we have cheek that the two last cases do not occur. Suppose that after modification, we get $k_{j+1} = k_{j+1} - k_{j-1} = 2$ and $k_j = k_j + k_{j-1} = 1$ for some $j \leq m$, this means that before modification $k_{j-1} = 1$, $k_j = 0$ and $k_{j+1} = 3$, which is impossible since all the not modified digits of the $\tau - NAF$ representation of k belong to $\{-1, 0, 1\}$.

By the same way we prove that the case $k_{i+1} = -2$ and $k_i = -1$ do not occur, which complete the proof.

A-1. Algorithm for Curves with Trace -1.

For curves with trace -1 we proceed as follows

$$\begin{cases} k_i \leftarrow k_i - k_{i-1} \\ k_{i+1} \leftarrow k_{i+1} - k_{i-1} \\ k_{i-1} \leftarrow k_{i-1} - 2k_{i-1} = -k_{i-1} \end{cases}$$

if k_i is even and keep the digits k_{i-1}, k_i, k_{i+1} unchanged otherwise.

Example. Let $(1, 0, -1, 0, 1)$ the $\tau - NAF$ representation of an integer k.

$(1, 0, -1, 0, 1) \xrightarrow{k_1 = 0} (1, 0, -1 - 1, 0 - 1, 1 - 2) = (1, 0, -2, -1, -1)$
$(1, 0, -2, -1, -1) \xrightarrow{k_2 = -2} (1, 0 + 1, -2 + 1, -1 + 2, -1) = (1, 1, -1, 1, -1)$.

The new $\tau - adic$ representation of k is $(1, 1, -1, 1, -1)$.

Complexity Estimates for the F_4 Attack on the Perturbed Matsumoto-Imai Cryptosystem

J. Ding[1], J.E. Gower[1], D. Schmidt[2], C. Wolf [3], and Z. Yin[1]

[1] Department of Mathematical Sciences,
University of Cincinnati, Cincinnati,
OH 45211-0025, USA
{ding, gowerj, yinzhi}@math.uc.edu
[2] Department of Electrical & Computer Engineering and Computer Science,
University of Cincinnati, Cincinnati,
OH 45211-0030, USA
dieter.schmidt@uc.edu
[3] K.U. Leuven ESAT-COSIC,
Kasteelpark Arenberg 10,
B-3001 Leuven-Heverlee, Belgium
Christopher.Wolf@esat.kuleuven.ac.be or chris@Christopher-Wolf.de

Abstract. Though the Perturbed Matsumoto-Imai (PMI) cryptosystem is considered insecure due to the recent differential attack of Fouque, Granboulan, and Stern, even more recently Ding and Gower showed that PMI can be repaired with the Plus (+) method of externally adding as few as 10 randomly chosen quadratic polynomials. Since relatively few extra polynomials are added, the attack complexity of a Gröbner basis attack on PMI+ will be roughly equal to that of PMI. Using Magma's implementation of the F_4 Gröbner basis algorithm, we attack PMI with parameters $q = 2$, $0 \leq r \leq 10$, and $14 \leq n \leq 59$. Here, q is the number of field elements, n the number of equations/variables, and r the perturbation dimension. Based on our experimental results, we give estimates for the running time for such an attack. We use these estimates to judge the security of some proposed schemes, and we suggest more efficient schemes. In particular, we estimate that an attack using F_4 against the parameters $q = 2, r = 5, n = 96$ (suggested in [7]) has a time complexity of less than 2^{50} 3-DES computations, which would be considered insecure for practical applications.

Keywords: public-key, multivariate, quadratic polynomials, perturbation, Gröbner basis.

1 Introduction

1.1 Multivariate Quadratic Cryptosystems and Perturbation

Multivariate Quadratic ($\mathcal{MQ}$) public key cryptosystems, first introduced in [6], have become a serious alternative to number theory based cryptosystems such as RSA, especially for small devices with limited computing resources. Since

N.P. Smart (Ed.): Cryptography and Coding 2005, LNCS 3796, pp. 262–277, 2005.

solving a set of multivariate polynomial equations over a finite field appears to be difficult (analogous to integer factorization, though it is unknown precisely how difficult either problem is), it seems reasonable to expect that we can build secure multivariate public key cryptosystems and signature schemes. In the last ten years, there has been significant effort put into realizing practical implementations of this idea, and many schemes have been proposed: Matsumoto-Imai, HFE, HFEv, Sflash, Oil & Vinegar, Quartz, TTM, and TTS, to name but a few.

At this stage, we seem to be more successful in building multivariate signature schemes than encryption schemes. For example, Sflash^{v2} [1] has been recommended by the New European Schemes for Signatures, Integrity, and Encryption (NESSIE, [17]) as a signature scheme for constrained environments. For encryption schemes, the best choice is probably HFE [19]. However, for a secure system, one must choose parameters which lead to a rather inefficient scheme.

Internal perturbation [7] was introduced as a general method to improve the security of multivariate public key cryptosystems. Roughly speaking, the idea is to "perturb" the system in a controlled way so that the resulting system is invertible, efficient, and much more difficult to break. The first application of this method was to the Matsumoto-Imai (MI) cryptosystem, a system that is otherwise vulnerable to the linearization attack [18]. The resulting system, called the perturbed Matsumoto-Imai cryptosystem (PMI), is slower as one needs to go through a search process on the perturbation space. However, we believe that for realistic choices of parameters, PMI is still much faster than HFE and provides superior security against all known attacks, except the recent differential attack of Fouque, Granboulan, and Stern [13]. Fortunately PMI is easily repaired with the Plus (+) [20] method of externally adding relatively few random quadratic polynomials. In fact, in the most general case of PMI, as few as 10 polynomials will be sufficient to protect PMI from the differential attack. Since so few extra polynomials are needed to create a secure Perturbed Matsumoto-Imai-Plus (PMI+) scheme, there is no significant difference between the two schemes regarding the Gröbner bases attack complexity [5,26]. Therefore, for simplicity we will consider Gröbner bases attacks on PMI.

1.2   Attacks Against Perturbed Multivariate Cryptosystems

In [2] it is shown that the XL algorithm will always need more time and space than either the F_4 or F_5 version of the Gröbner basis algorithm. Hence, it suffices to consider only Gröbner basis attacks. Both algorithms are quite similar in that they use the original Buchberger algorithm to compute a Gröbner basis for a given ideal of polynomials, and so for practical reasons we use only the F_4 version. Therefore, in this paper we analyse the security of PMI against Gröbner basis attacks as it depends on the parameter r, the perturbation dimension, and n, the message length. Specifically we give estimates for the time complexity of the F_4 Gröbner basis attack on PMI. Based on our experimental results, we give formulæ for these estimates that can be used to evaluate the security of proposed PMI systems against such attacks, and suggest parameters that may give better performance while providing sufficient security. These results can then be used to infer similar statements regarding the security of PMI+. Since [8] shows that differential analysis cannot be effectively used against PMI+, it is sufficient to consider Gröbner attacks against PMI+ to determine its security. Hence, the most successful attack against PMI+ can be found in [12] while the most successful one against PMI is [13].

1.3   Outline

The remainder of this paper is organised as follows. After introducing the MI and PMI cryptosystems in Section 2, we describe our experimental evaluation of the security of PMI in Section 3. We then interpret the data and make some suggestions for improvement and give some predictions for the security of proposed instances of PMI in Section 4. We present our conclusions in Section 5.

2   The Perturbed Matsumoto-Imai Cryptosystem

2.1   The Original Matsumoto-Imai Cryptosystem

Let k be a finite field of size q and characteristic 2, and fix an irreducible polynomial of $g(x) \in k[x]$ of degree n. Then $K = k[x]/g(x)$ is an extension of degree n over k, and we have an isomorphism $\phi : K \longrightarrow k^n$ defined by $\phi(a_0 + \cdots + a_{n-1}x^{n-1}) = (a_0, \ldots, a_{n-1})$.

Fix θ so that $\gcd(1 + q^\theta, q^n - 1) = 1$ and define $F : K \longrightarrow K$ by $F(X) = X^{1+q^\theta}$. Then F is invertible and $F^{-1}(X) = X^t$, where $t(1+q^\theta) \equiv 1 \bmod q^n - 1$. Define the map $\tilde{F} : k^n \longrightarrow k^n$ by $\tilde{F}(x_1, \ldots, x_n) = \phi \circ F \circ \phi^{-1}(x_1, \ldots, x_n) = (\tilde{F}_1, \ldots, \tilde{F}_n)$. In this case, the $\tilde{F}_i(x_1, \ldots, x_n)$ are quadratic polynomials in the variables $x_1, \ldots, x_n$. Finally, let L_1 and L_2 be two randomly chosen invertible affine linear maps over k^n and define $\overline{F} : k^n \longrightarrow k^n$ by $\overline{F}(x_1, \ldots, x_n) = L_1 \circ \tilde{F} \circ L_2(x_1, \ldots, x_n) = (\overline{F}_1, \ldots, \overline{F}_n)$. The public key of the Matsumoto-Imai cryptosystem (MI or C*) consists of the polynomials $\overline{F}_i(x_1, \ldots, x_n)$. See [16] for more details.

2.2   The Perturbed Matsumoto-Imai Cryptosystem

Fix a small integer r and randomly choose r invertible affine linear functions $z_1, \ldots, z_n$, written $z_j(x_1, \ldots, x_n) = \sum_{i=1}^{n} \alpha_{ij}x_i + \beta_j$, for $j = 1, \ldots, r$. This defines a map $Z : k^n \longrightarrow k^r$ by $Z(x_1, \ldots, x_n) = (z_1, \ldots, z_r)$. Now randomly choose n quadratic polynomials $f_1, \ldots, f_n$ in the variables $z_1, \ldots, z_r$. The f_i define a map $f : k^r \longrightarrow k^n$ by $f(z_1, \ldots, z_r) = (f_1, \ldots, f_n)$. Define $\tilde{f} : k^n \longrightarrow k^n$ by $\tilde{f} = f \circ Z$, and $\overline{\overline{F}} : k^n \longrightarrow k^n$ by $\overline{\overline{F}} = \tilde{F} + \tilde{f}$. The map $\overline{\overline{F}}$ is called the perturbation of $\tilde{F}$ by $\tilde{f}$, and as with MI, its components are quadratic polynomials in the variables $x_1, \ldots, x_n$. Finally, define the map $\hat{F} : k^n \longrightarrow k^n$ by $\hat{F}(x_1, \ldots, x_n) = L_1 \circ \overline{\overline{F}} \circ L_2(x_1, \ldots, x_n) = (y_1, \ldots, y_n)$. The public key of the perturbed Matsumoto-Imai (PMI) cryptosystem consists of the components y_i of $\hat{F}$. See Fig. 1 for an illustration of this idea, and [7] for more details.

Note that for MI there is a bijective correspondence between plaintext and ciphertext. However, PMI does not enjoy this property. Indeed, for a given ciphertext $c \in k^n$, $\hat{F}^{-1}(c)$ may have as many as q^r elements, though we may use the technique suggested for HFE to distinguish the plaintext from the other preimages. It has been proposed [7] that we can choose the parameters of PMI ($q = 2, r = 6, n = 136$) so that the resulting system is faster than HFE, and also claiming a very high level of security.

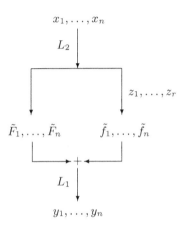

Fig. 1. Structure of PMI

2.3   Known Attacks Against MI and PMI

The most successful attack against MI is that of Patarin [18]. At present it is not clear whether this approach can be generalised to attack PMI, the main difficulty being that PMI mixes the operations in the extension field K from MI with the operations in the ground field k from the perturbation of MI. Another approach might involve ideas from the cryptanalysis of Sflash [14,15], though it is not immediately clear how this might work.

The differential attack [13] has rendered PMI insecure. However, it is easily repaired [8] using the Plus method of externally adding relatively very few Plus (+) polynomials [20]. The resulting scheme is call the Perturbed Matsumoto-Imai-Plus (PMI+) cryptosystem. Since the number of polynomials in PMI+ exceeds the number of unknowns by such a small amount, the attack complexity of a Gröbner basis attack very close to that of the same attack mounted against PMI. As a result, for simplicity we henceforth consider only PMI. These extra polynomials are added between the two linear transformations L_1, L_2. In particular, this means that we have $L_1 : k^{n+a} \to k^{n+a}$ now with $a \in \mathcal{N}$ added polynomials. These polynomials have the form

$$f_{n+1}(x_1, \ldots, x_n) := \gamma'_{n+1,1,2} x_1 x_2 + \ldots + \gamma'_{n+1,n-1,n} x_{n-1} x_n +$$
$$\beta'_{n+1,1} x_1 + \ldots \beta'_{n+1,n} x_n + \alpha'_{n+1}$$

$$\vdots$$

$$f_{n+a}(x_1, \ldots, x_n) := \gamma'_{n+a,1,2} x_1 x_2 + \ldots + \gamma'_{n+a,n-1,n} x_{n-1} x_n +$$
$$\beta'_{n+a,1} x_1 + \ldots \beta'_{n+a,n} x_n + \alpha'_{n+a}$$

for $\gamma', \beta', \alpha' \in_R k$ being random coefficients.

In [4] and [12], Gröbner bases have been used to break instances of HFE. By exploiting the underlying algebraic structure, they are able to break HFE in a far shorter time than it would take to solve a system of random equations [11,12]. For a fixed number of monomials in HFE, it can be shown that the running time will be polynomial. This result applies to MI as it uses only one monomial. The running time of this attack applied to PMI is not known.

3   Experiments with the F_4 Gröbner Basis Algorithm

3.1   Methodology

We attempted to experimentally determine the running time and memory requirements for a Gröbner basis attack on PMI. To this end we generated several instances of PMI. For each resulting set of polynomials $y_1, \ldots, y_n$ we chose several $(y'_1, \ldots, y'_n) \in k^n$ and timed how long it takes to find a Gröbner basis for the ideal $(y_1 - y'_1, \ldots, y_n - y'_n)$. Such a basis allows us to swiftly determine all $(x'_1, \ldots, x'_n) \in k^n$ such that $\hat{F}(x'_1, \ldots, x'_n) = (y'_1, \ldots, y'_n)$.

More specifically, we randomly generated 101 instances of PMI in Magma [3] for several values of n and r with $q = 2$, $14 \le n \le 59$, and $0 \le r \le 10$. In addition we randomly generated 101 elements in k^n and applied the F_4 version of the Gröbner basis implementation in Magma to each instance/element pair. In both cases, we used a uniform distribution on the private key/element from k^n. For all runs, we measured the memory and time needed until the algorithm found a solution. It did happen that some elements had no preimage under PMI, which is the same as with random systems of multivariate quadratic equations, hence we kept these timings in the sample. This decision was made as we were mainly interested in understanding the security of a signature scheme. For an encryption scheme, a more obvious choice would have been to encrypt random vectors $x \in k^n$ and then solving the corresponding equations, i.e., $\hat{F}(x) = y$ for given $\hat{F}$ and y and "unknown" x.

We note that in theory it would be best to measure the maximal degree of the equations generated during a run of the F_4 algorithm. Unfortunately, Magma does not provide this feature, and so we had to use the indirect measurements of time and memory. It should also be noted that the F_5 algorithm [10] is said to be faster than the previous algorithm F_4 [9]. However, recent experiments by Steel show that the Magma implementation of F_4 is superior in the case of

HFE systems [23]. In particular, Steel was able to solve HFE Challenge 1 in less operations than Faugère with his own implementation of F_5. This is a rather surprising fact as F_5 should be faster than F_4 from a theoretical perspective in *all* cases. Still, Magma's implementation of F_4 achieves better timings than Faugère's implementation of F_5 when applied to HFE Challenge 1. For our experiments, we decided to use the F_4 implementation of Magma as it is the fastest, publicly available implementation of Gröbner base algorithms. We benchmarked its performance by solving random instances of PMI for various parameters n, r for the finite field $k = GF(2)$. Although other ground fields with characteristic 2 are possible, we avoid them since solving PMI for a given private key takes an additional workload of $O(q^r)$. Also, the running time of Gröbner algorithms is very sensitive to the ground field k. Hence, it is difficult to obtain enough data for the cases $q = 4, 8$, and 16.

To ensure the accuracy and reliability of the data, we conducted the experiments with two independent teams, Team Q and Team Ω. Team Q used a cluster of identical AMD Athlon XP 2000+ with 900 MB of memory each, and Team Ω used an UltraSPARC-III+ 1.2 GHz dual processor with 8.0 GB of main memory. Because of these hardware and software differences, we expected to see differences in our measurements. However both data sets point to the same asymptotic behaviour. For brevity, we include only Team Q's data.

3.2   Empirical Data

It is clear that the case of $r = 0$ corresponds to MI, while the case of $r = n$ corresponds to a system of n randomly selected polynomials in n variables. Thus we expected the Gröbner basis attack on a system with $r = 0$ to be polynomial in time [12], while the same attack on a system with $r = n$ is expected to be exponential in time. Using our data, we wanted to answer two questions. First, for a fixed n we wanted to find the so-called "optimal perturbation dimension," *i.e.*, the minimal value of r for which a PMI system with parameters n and r is indistinguishable from a set of random polynomials. We also sought to obtain formulæ which would allow us to predict the running time behaviour of F_4 applied to PMI for any n and r.

The number of steps involved in attacking PMI with F_4 can be found in Table 5, while the memory requirements are shown in Table 6. Since no $\theta \in \mathbb{N}$ exists such that $\gcd(1 + 2^\theta, 2^n - 1) = 1$ for $n = 16, 32$, there is no corresponding instance of PMI and we hence have no data for these two cases. Each entry in these two tables is the median of 101 computations. This relatively small sample size was justified by additional experiments to determine the variation in larger (1001 computations) data sets. We found that the ratio of the maximum value to minimum value was always less than 2 in these larger sets. Data sets with median time below 0.05 seconds, or with memory requirements greater than 900 MB of memory were excluded from consideration in the final analysis on the grounds that they were either too noisy or suffered from the effects of extensive memory swapping. However, we actually performed many more experiments than are listed in Tables 5 and 6. Moreover, all experiments that terminated prematurely were due to memory shortage and not time constraints.

4   Interpretation

From the point of view of cryptanalysis, most agree that it is the computational complexity that essentially determines the security of a cryptosystem. In our experiments we notice that the time and memory tables are closely correlated. The explanation for this can easily be seen from the structure of the F_4 algorithm. Therefore we believe it suffices to analyse the timing data, and hence we omit a detailed analysis of the memory usage. However, from our experiments we notice that the memory usage is on the same scale as that of the time complexity. Since memory is a much more critical constraint than time, in the end we believe it will be memory that will determine how far F_4 can go.

4.1   Polynomial and Exponential Models

It is known that the case $r = 0$ is precisely MI. Hence the attack from Faugère and Joux [12] using Gröbner bases should be polynomial. Thus we first consider the hypothesis that the data is well approximated by a polynomial model. Let $t(n, r)$ be the time to attack PMI with parameters n and r. We assume that our computer can perform $2 \cdot 10^9$ steps per second, and define the number of steps, $s(n, r) = 2 \cdot 10^9 t(n, r)$. We use $s(n, r)$ instead of $t(n, r)$ for all calculations. The polynomial model predicts the existence of constants $\alpha = \alpha(r)$ and $\beta = \beta(r)$ such that $s(n, r)$ is well approximated by αn^β. Table 1 shows the fitting obtained from applying the method of least squares for a fixed r on the data $\{\log_2 n, \log_2 s(n, r)\}$, where ε is the error sum of squares for this data set. We note that for $r = 0$, the exponent $\beta = 7.16$ is greater than that predicted in [12], though we speculate that the difference may be due to the fact that F_5 is used instead of F_4.

Table 1. Polynomial fittings

r	0	1	2	3	4	5	6	7	8	9	10
$\log_2 \alpha$	-4.81	-3.33	-12.64	-7.71	-13.74	-10.87	-29.38	-29.30	-29.84	-31.09	-30.27
β	7.16	7.12	9.50	9.22	10.81	10.18	14.90	15.02	15.17	15.48	15.28
ε	4.20	5.28	6.97	2.17	1.51	3.29	6.51	2.32	0.91	1.10	0.80

It should also be observed that there is some sort of "phase transition" that occurs in the fitting behaviour as r increase from 5 to 6. This is most obviously seen by looking at the values of $\log_2 \alpha$, which should not be either unusually small or unusually large, as this quantity represents the expected complexity for small n. Our data shows that as r increases from 5 to 6, α decreases from

Table 2. Exponential fittings

r	0	1	2	3	4	5	6	7	8	9	10
$\log_2 \alpha$	21.88	22.37	19.83	20.62	18.02	19.10	11.58	11.65	11.05	11.16	10.94
$\log_2 \beta$	0.27	0.30	0.45	0.55	0.72	0.68	1.15	1.18	1.22	1.21	1.23
ϵ	6.73	11.44	2.57	5.22	2.95	6.34	2.54	1.91	0.82	1.10	0.89

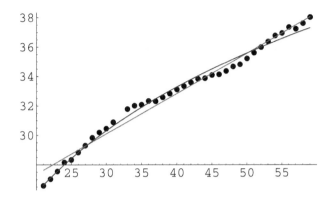

Fig. 2. Graph of $\log_2 s(n,0)$

the reasonable scale of approximately 2^{-10} to 2^{-30}, which would seem to be unreasonable. Therefore, we suspect that it is at $r = 6$ where the transition from polynomial to exponential behaviour occurs.

To examine this possibility, we also consider the hypothesis that the data is well approximated by an exponential model. As before, let $t(n, r)$ be the time to attack PMI with parameters n and r. The exponential model predicts the existence of constants $\alpha = \alpha(r)$ and $\beta = \beta(r)$ such that $s(n, r)$ is well approximated by $\alpha\beta^n$. Table 2 shows the fitting obtained from applying the method of least squares for a fixed r on the data $\{n, \log_2 s(n, r)\}$, where again ε is the error sum of squares. To illustrate the fittings we present Table 2 and Fig. 2–5.

Observe that ε does not help to decide which fitting is more appropriate, so we must study the other parameters of the fitting. In particular, from Table 2 we note that again there is a transition happening with $\log_2 \alpha$ between $r = 5$ and $r = 6$. Once again, the important feature is the transition in $\log_2 \alpha$, which again happens between $r = 5$ and 6. Our reasoning is as before; *i.e.*, we do not believe α should either be too large or too small. In the case of the exponential model, α is too large for $r < 6$. Hence, we find the exponential model much more convincing for the case of $r \geq 6$, and the polynomial model a better fit for $r < 6$.

In summary, from this data we observe that the complexity makes a transition between two distinct regions, where the first region represents polynomial behaviour such as that of MI, and the second represents the exponential behaviour of a system defined by a random set of polynomials. We call the point at

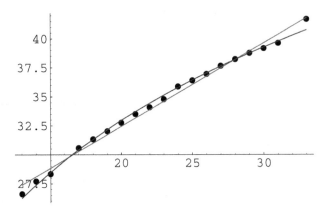

Fig. 3. Graph of $\log_2 s(n, 4)$

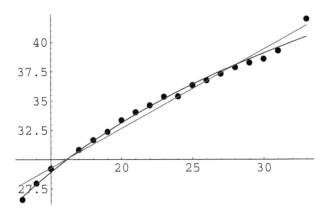

Fig. 4. Graph of $\log_2 s(n, 5)$

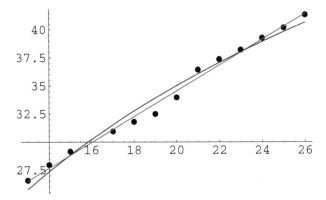

Fig. 5. Graph of $\log_2 s(n, 6)$

which this transition occurs the *phase transition point*, which we believe is $r = 6$. In our data, we did not find any other transition point. In particular, this behaviour fits well with the corresponding theory: as soon as the number of linearly independent monomials reaches a certain threshold, Gröbner base algorithms like F_4 or F_5 cannot make use of the structure of the private key anymore.

4.2   Optimal Perturbation Dimension

The analysis in the previous two sections assumed a fixed r and variable n. We now consider fixing n and varying r in order to study how the complexity achieves its maximum as r increases from zero (polynomial) to n (exponential). Fig. 6 illustrates the typical features of such a transition process.

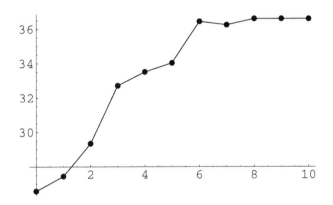

Fig. 6. Graph of $\log_2 s(21, r)$

It is important not to confuse the phase transition point with the minimal value of r (for a fixed n) for which the maximal complexity is achieved and no further advantage is gained by increasing r. We call such an r the *optimal perturbation dimension* for a given n. Based on the experiments above, we empirically determined this dimension and summarise our findings in Table 3.

Table 3. Optimal perturbation dimension

n	14...15	17...21	22...24
r	4	7	10

This data agrees very well with the theoretical explanation of the behaviour of Gröbner bases algorithms for solving HFE. There, the maximal degree of the polynomial equations derived during one computation is also discrete. It seems that we get similar behaviour here. As was already pointed out, we could not make this observation directly as Magma does not provide the maximal degree. In order to confirm this behaviour and find formulæ that can be used to predict the location of the optimal perturbation dimension, we would need more data for larger values of n. However, this is not possible at present as the memory requirements for F_4 are quite severe, cf Table 6.

4.3   Practical Security

We now evaluate the security of some implementations of PMI. As was already pointed out, PMI by itself is insecure under the differential attack. Therefore the following analysis assumes that the Plus method has been applied. We first use the fittings to evaluate the security of the practical system proposed in [7]. We then use our fittings to propose some new optimised practical systems.

Evaluation of Practical Examples. In [7] the practical example suggested is the case of $q = 2, r = 6$, and $n = 136$. First we evaluate this example using the exponential model. The predicted security in this model is greater than 2^{160}. Assuming the validity of this model, the proposed system is very secure against Gröbner based attacks. In particular, according to the exponential model, since $\log_2 \beta > 1$ for $r \geq 6$, the running time of F_4 increases faster than exhaustive key search for $q = 2$. Therefore, these instances of PMI should be secure against these types of attacks assuming the validity of the exponential model. In fact, a practical and secure instance of PMI can use the parameters $q = 2, r = 6$, and $n = 83$ to meet the NESSIE requirements of 2^{80} 3-DES computations, if our model is valid. In particular, the security of the exponential model suggests a strength of 2^{100} 3-DES computations. However, for $q = 2, n < 80$, a brute-force search would take less than the required 2^{80} computations in 3-DES. Therefore, we decided to chose the first prime above 80, to rule out subfield-attacks as suggested in [22].

To be on the safe side, we also evaluate the system with $q = 2, r = 6$, and $n = 136$ using the polynomial model. The predicted security for this model is roughly 2^{70} 3-DES computations. According to this model, n must be greater than 227 to achieve the required NESSIE security level. However, if we consider the memory requirements needed to attack such a system, breaking these systems will be practically impossible with currently available resources. In particular, it is not clear at present how such an attack could be distributed over different machines, e.g., using a distributed network of machines all agreeing to collaborate or possibly being captured by Trojan horses.

It is speculated in [7] that $q = 2$, $r = 5$, and $n = 96$ may also be secure. To evaluate this claim it is more appropriate to assume the polynomial model. According to this fitting, the security level is less than 2^{50} 3-DES computations, which is much less than the security level as requested in NESSIE. Therefore we conclude that this speculation may be overly optimistic. But again, we point out the severe memory requirements. Based on our experiments, we expect a number well above 2^{60} bytes.

The PMI scheme was originally proposed for use as an encryption scheme. One can easily modify the system for signature purposes; for example, one can use the Feistel-Patarin-Network, as in the signature scheme Quartz [21]. The scheme with the parameters $q = 2$, $r = 6$, $n = 83$ suggested above can be used in this way to build a secure signature scheme with only 83 bits.

Practical Perturbation Dimension. For PMI cryptosystems, one of the fundamental questions one should answer is how to choose r appropriately for a fixed n. From the behaviour of the complexity, a natural choice would be the optimal perturbation dimension, where the system becomes indistinguishable from a system defined by a random set of polynomials. However, we notice that this number may be too large for practical purposes, so we must choose some smaller value for r such that the system is practical and secure. Since the phase transition point is the minimal r that provides exponential behaviour we suggest this value for the perturbation dimension in practice.

4.4 Further Research

While the security of MI and also the behaviour of Gröbner base algorithm is well understood, this is not the case for PMI. Hence, it would be nice to have empirical data about the behaviour of other algorithms, e.g., F_5 in the case of PMI. This proves difficult at present as F_5 is not available in a public implementation.

Using the maximal degree of the polynomials generated during a run on F_4 would have been more telling than our time or memory measurements. From [12], we expect this degree to be far more stable than the time or memory requirements. However, as Magma is also closed source, we could not modify the code to obtain this information. Hence, an open source implementation of F_4 would be preferable. In any case, we believe that this information would help us find the optimal perturbation dimension for a fixed n, the determination of which is important to completely understand PMI.

Apart from this, PMI seems secure against Gröbner attacks, and so we conclude that PMI+ is secure against both differential and Gröbner basis attacks.

5   Conclusions

In this paper we presented a security analysis of the PMI cryptosystem against Gröbner basis attacks. From this analysis we saw that for reasonable choices of parameters, PMI is secure against such attacks. Since PMI can be protected from the differential attack by externally adding as few as 10 random Plus polynomials, we conclude that this security analysis extends to that of PMI+.

Running various experiments with random instances of PMI, we established that PMI with small security parameter r can likely be solved in polynomial time. However, the rather large memory requirements will prevent such an attack from being practical with currently available resources. On the other hand, for

Table 4. Comparison between Quartz and PMI

	Quartz-7m	PMI
ground field k	GF(2)	
variables n	107	83
equations m	100	83
Signature Size [bits]	128	83
Public Key Size [kByte]	71	35

$r \geq 6$, we saw that the attacks using Gröbner bases become less efficient than exhaustive key search. Hence, we conclude that PMI is secure against these type of attacks. In particular, we suggest $q = 2, r = 6$, and $n = 83$ as a secure instance of PMI.

These results suggest that we can obtain a signature scheme from PMI that allows shorter signatures than other multivariate schemes, and in particular Quartz (128 bit, cf Table 4). For our comparison, we use the tweaked version Quartz-7m from [24–Sect. 4.3]. Moreover, this scheme would be the only known instance that gives a security level of 2^{80} 3-DES computations and still allows efficient decryption of messages in a multivariate public key scheme.

Acknowledgements

Christopher Wolf was supported in part by the Concerted Research Action (GOA) Mefisto-2000/06 and GOA Ambiorix 2005/11 of the Flemish Government and the European Commission through the IST Programme under Contract IST-2002-507932 ECRYPT.

References

1. M.-L. Akkar, N. T. Courtois, R. Duteuil and L. Goubin. A Fast and Secure Implementation of Sflash. In *PKC 2003*, LNCS 2567:267–278.
2. G. Ars, J.-C. Faugère, H. Imai, M. Kawazoe and M. Sugita. Comparison Between XL and Gröbner Basis Algorithms. In *Asiacrypt 2004*, LNCS 3329:338–353.
3. University of Sydney Computational Algebra Group. The MAGMA computational algebra system for algebra, number theory and geometry. http://magma.maths.usyd.edu.au/magma.
4. N. Courtois, M. Daum and P. Felke. On the Security of HFE, HFEv- and Quartz. In *PKC 2003*, LNCS 2567:337–350.
5. N. Courtois, A. Klimov, J. Patarin, and A. Shamir. Efficient Algorithms for Solving Overdefined Systems of Multivariate Polynomial Equations. In *Eurocrypt 2000*, LNCS 1807:392–407.
6. W. Diffie and M. Hellman. New Directions in Cryptography. In *IEEE Transactions on Information Theory*, 22(6):644–654, 1976.
7. J. Ding. A New Variant of the Matsumoto-Imai Cryptosystem Through Perturbation. In *PKC 2004*, LNCS 2947:305–318.

8. J. Ding and Jason E. Gower. Inoculating Multivariate Schemes Against Differential Attacks. Pre-print, 12 pages. http://math.uc.edu/~aac/pub/pmi+.pdf

9. J.-C. Faugère. A New Efficient Algorithm for Computing Gröbner Bases (F_4). In *Journal of Applied and Pure Algebra*, 139:61–88, June 1999.

10. J.-C. Faugère. A New Efficient Algorithm for Computing Gröbner Bases Without Reduction to Zero (F_5). In *ISSAC 2002*, pp. 75-83, ACM Press.

11. J.-C. Faugère. Algebraic Cryptanalysis of (HFE) Using Gröbner Bases. Technical report, Institut National de Recherche en Informatique et en Automatique, February 2003. http://www.inria.fr/rrrt/rr-4738.html, 19 pages.

12. J.-C. Faugère and A. Joux. Algebraic Cryptanalysis of Hidden Field Equation (HFE) Cryptosystems Using Gröbner Bases. In *Crypto 2003*, LNCS 2729:44-60.

13. P.-A. Fouque, L. Granboulan, and J. Stern. Differential Cryptanalysis for Multivariate Schemes. In *Eurocrypt 2005*, LNCS 3494:341–353.

14. W. Geiselmann, R. Steinwandt and T. Beth. Attacking the Affine Parts of SFlash. In *Cryptography and Coding – 8th IMA International Conference*, LNCS 2260:355–359, 2001.

15. H. Gilbert and M. Minie. Cryptanalysis of SFLASH. In *Eurocrypt 2002*, LNCS 2332:288–298.

16. T. Matsumoto and H. Imai. Public Quadratic Polynomial-Tuples for Efficient Signature-Verification and Message-Encryption. In *Eurocrypt 1988*, LNCS 330:419–453.

17. NESSIE. European project IST-1999-12324 on New European Schemes for Signature, Integrity and Encryption. http://www.cryptonessie.org.

18. J. Patarin. Cryptanalysis of the Matsumoto and Imai Public Key Scheme of Eurocrypt'88. In *Crypto 1995*, LNCS 963:248–261.

19. J. Patarin. Hidden Fields Equations (HFE) and Isomorphisms of Polynomials (IP): Two New Families of Asymmetric Algorithms. In *Eurocrypt 1996*, LNCS 1070:33–48. Extended version: http://www.minrank.org/hfe.pdf.

20. J. Patarin, L. Goubin and N. Courtois. C^*_{-+} and HM: Variations Around Two Schemes of T. Matsumoto and H. Imai. In *Asiacrypt 1998*, LNCS 1514:35–50.

21. J. Patarin, L. Goubin and N. Courtois. QUARTZ, 128-Bit Long Digital Signatures. In *CT-RSA 2001*, LNCS 2020:298–307.

22. A. V. Sidorenko and E. M. Gabidulin. The Weak Keys for HFE. In *Proceedings of ISCTA 2003*, 6 pages.

23. A. Steel. Allan Steel's Gröbner Basis Timings Page. http://magma.maths.usyd.edu.au/users/allan/gb.

24. C. Wolf and B. Preneel. Asymmetric Cryptography: Hidden Field Equations. In *European Congress on Computational Methods in Applied Sciences and Engineering 2004*, 20 pages. Extended version: http://eprint.iacr.org/2004/072.

25. Christopher Wolf and Bart Preneel. Taxonomy of public key schemes based on the problem of multivariate quadratic equations. Cryptology ePrint Archive, Report 2005/077, 12th of May 2005. http://eprint.iacr.org/2005/077/, 64 pages.

26. B.-Y. Yang, J.-M. Chen, and N. Courtois. On Asymptotic Security Estimates in XL and Gröbner Bases-Related Algebraic Cryptanalysis. In *ICICS 2004*, LNCS 3269:410–423

Appendix

Table 5. Steps $s(n,r)$ in $\log_2$ for solving instances of PMI

n \ r	0	1	2	3	4	5	6	7	8	9	10
14	24.10	24.10	25.18	27.56	27.70	27.96	27.96	27.96	27.96	27.96	27.96
15	24.10	25.18	25.84	28.25	28.34	29.21	29.16	29.16	29.17	29.16	29.16
16											
17	25.18	25.79	26.84	29.78	30.58	30.81	30.94	32.35	32.35	32.35	32.35
18	25.32	25.84	27.83	31.13	31.33	31.66	31.79	33.37	33.37	33.37	33.37
19	25.84	26.55	28.34	31.71	32.02	32.37	32.50	34.33	34.32	34.31	34.32
20	26.22	27.58	28.83	32.16	32.75	33.35	33.97	35.32	35.35	35.34	35.34
21	26.58	27.44	29.34	32.71	33.51	34.03	36.45	36.25	36.61	36.61	36.61
22	27.04	28.51	30.14	33.12	34.09	34.62	37.38	37.16	37.46	38.28	38.29
23	27.56	28.42	30.16	33.68	34.79	35.30	38.21	37.97	38.98	39.14	38.97
24	28.17	29.57	30.66	33.84	35.88	35.39	39.26	40.16	40.28	39.82	40.02
25	28.34	29.21	31.03	34.84	36.41	36.34	40.19	41.31	41.40	41.28	
26	28.84	30.25	31.51	35.20	36.96	36.75	41.34				
27	29.30	30.73	31.97	35.84	37.66	37.31					
28	29.84	31.27	32.40	36.47	38.22	37.85					
29	30.21	31.53	32.69	36.75	38.76	38.27					
30	30.46	31.88	33.10	37.28	39.18	38.59					
31	30.90	32.30	33.57	38.08	39.63	39.29					
32											
33	31.79	33.01	35.08	39.28	41.68	41.99					
34	32.02	33.18	35.58	39.46							
35	32.08	33.30	36.01	39.75							
36	32.33	33.85	36.33	40.11							
37	32.31	33.80	37.09	40.55							
38	32.58	33.99	37.24	40.87							
39	32.84	34.18	37.46								
40	33.11	35.14	38.00								
41	33.34	34.48	38.63								
42	33.60	34.84	38.74								
43	33.83	34.94	39.06								
44	33.88	35.76	39.35								
45	34.09	35.41	40.07								
46	34.13	35.68	40.23								
47	34.36	35.67	40.36								
48	34.67	37.50	41.28								
49	34.82	36.01									
50	35.22	36.59									
51	35.61	36.75									
52	36.00	37.90									
53	36.37										
54	36.78										
55	36.95										
56	37.35										
57	37.23										
58	37.61										
59	38.02										

Table 6. Memory requirements in $\log_2$ for solving instances of PMI

n \ r	0	1	2	3	4	5	6	7	8	9	10
14	21.69	21.73	21.82	22.19	22.03	22.33	22.32	22.33	22.32	22.33	22.32
15	21.83	21.90	22.00	22.46	22.67	22.78	22.66	22.67	22.66	22.67	22.66
16											
17	22.34	22.33	22.44	23.12	23.52	23.59	23.53	24.74	24.73	24.74	24.73
18	23.01	23.01	22.97	23.55	24.06	24.07	24.47	25.41	25.41	25.41	25.41
19	23.81	23.81	23.47	24.08	24.60	24.63	24.71	26.08	26.09	26.09	26.09
20	24.46	24.46	24.22	24.72	25.17	25.20	25.77	26.71	26.71	26.71	26.71
21	24.46	24.46	24.22	24.72	25.17	25.20	28.00	27.10	27.10	27.10	27.10
22	24.46	24.46	24.22	24.85	25.40	25.64	28.35	27.61	27.61	28.56	28.57
23	24.46	24.46	23.47	25.25	25.75	26.36	28.94	28.12	29.18	29.29	29.05
24	24.46	24.46	24.22	25.50	26.24	26.83	29.61	29.90	29.99	29.72	29.78
25	24.46	24.46	24.22	26.11	26.44	27.35	29.90	30.41	30.55	30.43	
26	24.46	24.46	24.22	26.51	26.80	27.91	30.33				
27	24.46	23.36	24.22	26.87	27.85	28.30					
28	24.46	24.46	24.22	27.24	27.90	28.64					
29	24.46	24.46	24.35	27.55	28.37	29.05					
30	24.46	24.46	24.73	27.93	29.02	29.45					
31	24.46	24.46	24.79	28.16	29.40	29.79					
32											
33	24.46	24.64	25.70	28.82	30.39	30.69					
34	24.46	24.64	26.14	29.11							
35	24.46	24.64	26.57	29.40							
36	24.46	25.34	26.86	29.68							
37	24.46	25.34	27.01	29.89							
38	24.85	25.34	27.60	30.16							
39	24.85	26.05	27.87								
40	25.29	26.00	28.01								
41	25.30	26.06	28.21								
42	25.30	26.65	28.77								
43	25.30	26.77	28.71								
44	25.31	26.77	28.96								
45	25.51	27.30	29.51								
46	25.61	27.30	29.78								
47	25.72	27.30	29.68								
48	25.92	27.97	30.13								
49	25.98	27.93									
50	26.09	28.40									
51	26.28	28.41									
52	26.33	28.41									
53	26.43										
54	26.65										
55	26.67										
56	26.77										
57	27.03										
58	27.03										
59	27.11										

An Algebraic Framework for Cipher Embeddings

C. Cid[1,*], S. Murphy[1], and M.J.B. Robshaw[2]

[1] Information Security Group,
Royal Holloway, University of London,
Egham, Surrey, TW20 0EX, U.K.
[2] France Télécom Research and Development,
38–40 rue du Général-Leclerc,
92794 Issy les Moulineaux, France

Abstract. In this paper we discuss the idea of block cipher embeddings and consider a natural algebraic framework for such constructions. In this approach we regard block cipher state spaces as algebras and study some properties of cipher extensions on larger algebras. We apply this framework to some well-known examples of AES embeddings.

1 Introduction

Cryptosystems are designed to hinder the efforts of the cryptanalyst. However it is good cryptographic practice to ensure that the cryptosystem is presented in a clear and natural manner to facilitate independent scrutiny. Since there is rarely one single viewpoint for looking at a cipher, one approach for the cryptanalyst is to consider alternative presentations. At first sight it may appear that there is little to be gained by studying the same cryptosystem in a different way. However, new perspectives may well:

- reveal mathematical structure that was hidden;
- permit calculations that were previously considered intractable;
- encourage the development of ideas about the analysis of the cryptosystem;
- provide implementation insights.

In fact, finding different presentations is fundamentally the only technique available for the analysis of many asymmetric cryptosystems, as the following two examples of widely used asymmetric cryptosystems demonstrate.

- **RSA.** The security of the RSA cryptosystem [16] with modulus n, where $n = pq$ (two unknown primes), is believed to fundamentally depend on the inability of an attacker to factor the integer n. This is equivalent to the inability of an attacker to find a presentation based on the ring "factoring" isomorphism $\mathbb{Z}_n \to \mathbb{Z}_p \times \mathbb{Z}_q$.

* This author was supported by EPSRC Grant GR/S42637.

N.P. Smart (Ed.): Cryptography and Coding 2005, LNCS 3796, pp. 278–289, 2005.
© Springer-Verlag Berlin Heidelberg 2005

- **Cryptosystems Based on Discrete Logarithms.** There are many cryptosystems, such as ElGamal [5], the Digital Signature Standard [15] or Elliptic Curves [10,7], whose security is believed to fundamentally depend on the inability of an attacker to calculate discrete logarithms in certain finite cyclic groups. For such a group G of order p, this is equivalent to the inability of an attacker to find a presentation based on the isomorphism $G \to \mathbb{Z}_p^+$, where $\mathbb{Z}_p^+$ is the additive cyclic group of integers modulo p.

For an asymmetric cryptosystem, it is often fairly obvious which presentation of the cryptosystem might be of greatest use to the analyst. The difficulty for an alternative presentation of an asymmetric cryptosystem is "merely" one of calculating this presentation.

For a symmetric cryptosystem it is unlikely to be so obvious. One standard technique for analysing a mathematical structure is to *embed* it as a sub-structure within a larger one. In this way the original mathematical structure is presented within the context of a larger structure and such an approach has yielded great insights in many areas of mathematics. In this paper, therefore, we consider a framework for the embeddings of block ciphers.

While embeddings are usually constructed in the hope that they can provide a further insight to the cryptanalyst, they can also be useful when considering some implementation issues: an alternative presentation could provide a more efficient implementation of the cipher, or might be used to protect against some forms of side-channel attacks, such as timing or power analysis.

Our discussion is conducted with a view to providing a basic framework for block cipher embeddings. The question of whether a particular block cipher embedding yields any new insights might be, to some extent, a subjective judgement. However we observe that there are embeddings that are in some sense "trivial" and they cannot possibly offer extra insight into the cipher. In this paper we seek to provide a framework to provide some initial discrimination between embeddings of different types.

We begin the development of this framework by considering the natural mathematical structure for a block cipher state space, namely the algebra. The embedding of one block cipher into a larger one is then discussed in terms of the embedding function between the two state space algebras. This leads to a natural mathematical derivation of the extended cryptographic functions of the larger block cipher. To illustrate our approach, we discuss some well-known examples of AES embeddings [1,11,12].

2   State Space Algebras

The state space of a block cipher is usually composed of a number of identical components. For example, the state space of the Data Encryption Standard (DES) [13] consists of 64 bits, whereas the state space of the Advanced Encryption Standard (AES) [14,3,4] is usually thought of as consisting of 16 bytes. For many block ciphers, these components are viewed as elements of some algebraic structure, and internal block cipher computations often depend on using operations based on this structure. Thus, it is natural to regard a component of the DES state space as an element of the finite field $GF(2)$, and a component of the AES state space as an element of the field $GF(2^8)$.

For a block cipher, in which a component of the state space is naturally regarded as an element of a field K, the entire state space is given by the set K^n, where n is the number of components ($n = 64$ for the DES and $n = 16$ for the AES). The set K^n has a natural ring structure as the direct sum of n copies of the field K, as well as a natural vector space structure as a vector space of dimension n over K. A set with such structure is known as an *algebra* [8]. More formally, we have the following definition.

Definition 1. *Let K be a field. An associative K-algebra (or simply algebra) is a vector space A over K equipped with an associative K-bilinear multiplication operation.*

Informally, we can regard an algebra as a vector space which is also a ring. Algebras can be also generalised to the case when K is a commutative ring (in which case A is a K-module rather than a vector space). The dimension of the algebra A is the dimension of A as a vector space. The set $A' \subset A$ is a subalgebra of A if A' is an algebra in its own right, and A' is an ideal subalgebra if it is also an ideal of the ring A. We can also classify mappings between two algebras in the obvious way, so an algebra homomorphism is a mapping that is both a ring homomorphism and a vector space homomorphism (linear transformation).

Considering block ciphers, the algebra of most interest cryptographically is formed by the set K^n. This is an algebra of dimension n over K, where "scalar" multiplication by the field element $\lambda \in K$ is identified with multiplication by the ring element $(\lambda, \ldots, \lambda) \in K^n$. This algebra is the natural algebraic structure for most block cipher state spaces and we term the algebra K^n the *state space algebra*. We note that even in cases where K is not a field (for example, a component of the state space of the SAFER family of block ciphers [9] is most naturally thought of as an element of the ring $\mathbb{Z}_{2^8}$), the K-algebra K^n still remains the most interesting structure for our analysis, and most of the discussion following can be suitably modified.

The algebraic transformations of a state space algebra, that is transformations that preserve most of the structure of the algebra, are necessarily based either on a linear transformation of the state space or on a ring-based transformation of the state space. However, a secure design often requires some non-algebraic block cipher transformations; for example in each round there is often a transformation using a substitution or look-up table. There are cases however (most notably the AES) where the round transformations are dominated by algebraic operations and, in such cases, it may be interesting to study the cipher by means of an embedding of the state space algebra in a larger algebra. An embedding may be defined so that all transformations of the embedded state space are also algebraic transformations with respect to the larger algebra. The hope is that this new representation may offer new insights on the essential structure of the original cipher.

3   Block Cipher Embeddings

Suppose that $A = K^n$ is the state space algebra of dimension n over K for some block cipher, and that the encryption process consists of a family of (key-dependent) functions $f : A \rightarrow A$. A block cipher embedding is constructed from an injective mapping $\eta : A \rightarrow B$ from the algebra A into some (possibly larger) algebra B and suitably extended versions of the functions f defined on B. We now consider different methods of embedding block ciphers.

3.1   Identity Embeddings

The are clearly many ways of embedding A in an algebra B of higher dimension. One obviously unproductive way to construct a cipher embedding is to embed the algebra A and the cryptographic function f into the algebra B by means of the *identity* mapping.

Suppose that the algebra B can be written as the direct sum

$$B = A \oplus A',$$

with the embedding mapping given by $\eta : a \mapsto (a, 0)$. The functions f can easily be extended in a trivial manner, so that $(a, 0) \mapsto (f(a), 0)$. This provides a direct mirror of the cipher within B, and the A'-component of the embedding (and the value of the extended function beyond $\eta(A)$) is irrelevant to the definition of the original cipher in its embedded form. Clearly, this idea can also be extended to any embedding mapping of the form $\eta : a \mapsto (a, ?)$ and any extension of f to a function $(a, ?) \mapsto (f(a), ?)$. For example, the cryptographic function $f : A \rightarrow A$ could be extended to a cryptographic function $\widehat{f} : B \rightarrow B$ given by $(a, a') \mapsto (f(a), g(a'))$ for some function $g : A' \rightarrow A'$.

Knudsen essentially gives an example of such an embedding where f is the Data Encryption Standard (DES) [13] encryption function and g is the RSA [16] encryption function, with the same key used (in very different ways) for each of these encryption functions [6]. Based on this embedding function, Knudsen makes statements about the security of DES in terms of the security of RSA and vice versa [6]. The readers of [6] are left to draw their own conclusions about a security statement made about one cipher but which is based on the analysis of a different arbitrary cipher.

We term such an embedding an *identity-reducible* embedding. Apart from possibly providing another presentation of the cipher, identity-reducible embeddings are of little interest mathematically or cryptographically.

3.2   Induced Embeddings

The starting point for a cipher embedding is an injective function $\eta : A \rightarrow B$. We denote by $B_A = \eta(A) \subset B$ the image of this mapping. We now discuss the natural method of extending the cipher functions f to functions on B using η.

We first consider how to define the induced embedded cryptographic functions $f_\eta : B_A \to B_A$. These functions need to mirror the action of f on A, but within B_A. They must therefore be given by

$$f_\eta(b) = \eta \left(f \left(\eta^{-1}(b) \right) \right) \text{ for } b \in B_A,$$

which is illustrated in the diagram below (Figure 1). To illustrate this induced

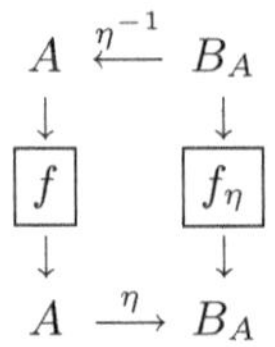

$$
\begin{array}{ccc}
A & \xleftarrow{\eta^{-1}} & B_A \\
\downarrow & & \downarrow \\
\boxed{f} & & \boxed{f_\eta} \\
\downarrow & & \downarrow \\
A & \xrightarrow{\eta} & B_A
\end{array}
$$

Fig. 1. Induced Function given by the Embedding η

function, consider the usual case where subkeys are introduced into a block cipher by addition, that is by the function $f(a) = a + k$. Suppose we choose an additive embedding transformation η, whose definition can be extended to subkeys. In this case, the corresponding embedded method of introducing subkeys can be naturally defined by addition, as

$$f_\eta \left(\eta(a) \right) = \eta \left(f \left(\eta^{-1} \left(\eta(a) \right) \right) \right) = \eta \left(f(a) \right) = \eta(a + k) = \eta(a) + \eta(k).$$

3.3   Natural Extensions of Embeddings

The reason for considering an embedding is to analyse a cipher within a (possibly) larger cipher in the hope of gaining a better insight into the original cipher. In general however, $B_A = \eta(A)$ is not a subalgebra of B, but it exists within the context of the algebra B and operations on B_A are defined by the algebra B. It is thus far more natural to work within the algebra B and define B_A mathematically within B, than it is to consider B_A in isolation. For example, while B_A may not be a vector space, the induced encryption functions may be defined in terms of a matrix multiplication of the elements of B_A. It is clearly mathematically appropriate in such an example to consider the vector space on which the linear transformation is defined instead of just one subset (e.g. B_A). It is thus desirable in an embedding to *naturally* extend the function f_η to B, so that the extension retains the main algebraic properties of f_η.

The most appropriate structure to consider in this case would be the set $\overline{B}_A$, the (algebraic) *closure* of B_A. This is the minimal algebra containing B_A, and is generated (as an algebra) by the elements of B_A. The set B_A can now be considered algebraically entirely within the context of the closure $\overline{B}_A$, and the operations on B_A are defined within the algebra $\overline{B}_A$.

It is clear how this notion of closure can give the appropriate extension of the induced functions $f_\eta : B_A \to B_A$ to an extended induced functions $\overline{f}_\eta : \overline{B}_A \to$

$\overline{B}_A$. In particular, such an extension $\overline{f}_\eta$ preserves algebraic relationships between the input and output of the functions f_η. It should also be clear that $\overline{B}_A$ is the absolute extent of the cipher within B. No elements outside $\overline{B}_A$ (that is $B \setminus \overline{B}_A$) can be generated by the embedded versions of elements of the state space algebra A. Thus the extension of any function beyond $\overline{B}_A$ is not determined algebraically by the original cipher function, and can thus be considered arbitrary. There seems to be no need (or point) in considering anything beyond the closure of the embedding.

We have thus described a natural three-step process for embedding a cipher within another cipher with a larger state space algebra:

1. Define an injective embedding function from the original state space algebra to the larger state space algebra.
2. Based on the embedding function, define the induced cryptographic functions on the embedded image of the original state space algebra.
3. Extend these induced cryptographic functions in a *natural* manner to the larger state space algebra by algebraic closure.

This general approach seems to be an appropriate framework for considering cipher embeddings, particularly for ciphers with a highly algebraic structure (note that key schedules can usually be similarly embedded). However it is clear that each embedding should be considered on its own merits. Furthermore, not every property of the embedded cipher is of immediate relevance to the original cipher. Indeed, an example of a weakness of the larger algebraically embedded cipher that does not translate to the original cipher was given in [12]. However our framework allows us to immediately identify some embeddings that inevitably have little cryptanalytical value.

4   Embeddings of the AES

The AES [14] is a cipher with a highly algebraic structure, and it is a suitable cipher on which to apply and analyse different embedding methods. We look at three different approaches that have been proposed in the literature and consider their merits in terms of the framework given in Section 3.

The AES encryption process is typically described using operations on an array of bytes, which we can regard as an element of the field $\mathbb{F} = GF(2^8)$. Without loss of generality, we consider the version of the AES with 16-byte message and key spaces, and 10 encryption rounds. The state space algebra of the AES is thus the algebra $\mathbb{F}^{16}$, which we denote by $\mathbf{A}$.

4.1   Dual Ciphers of the AES

In [1] Barkan and Biham construct a number of alternative representations of the AES, which they call *dual ciphers* of Rijndael. These distinct representations are derived from the automorphisms of the finite field $\mathbb{F} = GF(2^8)$ (based on the Fröbenius map $a \mapsto a^2$) and the different representations of the field itself (via the explicit isomorphisms between fields of order 2^8). Each representation can clearly be seen as a form of embedding; the embedding functions are isomorphisms and therefore $B \cong A$. The AES cryptographic functions are extended to B according to these isomorphisms. These embeddings are essentially mirrors of the AES, although the different representations may permit us to gain a better insight of algebraic structure of the cipher, such as the importance of some of the choices made in the design of the AES. For instance, by analysing different representations, it is concluded that a change of the "Rijndael polynomial" (used to represent the finite field $GF(2^8)$ within the cipher) should not affect the strength of the cipher [1]. Such alternative representations can also be useful in providing additional insights into efficient and secure implementation practices.

4.2   The BES Extension of the AES

The embedding of the AES in a larger cipher called Big Encryption System (BES) was introduced in [12]. The main goal of this construction was to represent the AES within a framework where the cipher could be expressed through simple operations (inversion and affine transformation) in the field $\mathbb{F} = GF(2^8)$.

The BES operates on 128-byte blocks with 128-byte keys and has a very simple algebraic structure. The state space algebra of the AES is the algebra $\mathbf{A} = \mathbb{F}^{16}$, while the state space algebra of the BES is the algebra $\mathbb{F}^{128}$ (denoted by $\mathbf{B}$). The embedding function for the BES embedding is based on the vector conjugate mapping $\phi : \mathbb{F} \to \mathbb{F}^8$ [12], which maps an element of $\mathbb{F}$ to a vector of its eight conjugates. Thus ϕ is an injective ring homomorphism given by

$$\phi(a) = \left(a^{2^0}, a^{2^1}, a^{2^2}, a^{2^3}, a^{2^4}, a^{2^5}, a^{2^6}, a^{2^7} \right).$$

This definition can be extended in the obvious way to an embedding function $\phi : \mathbf{A} \to \mathbf{B}$ given by $\phi(\mathbf{a}) = \phi(a_0, \ldots, a_{15}) = (\phi(a_0), \ldots, \phi(a_{15}))$, which is an injective ring homomorphism. We note that the image of this ring homomorphism, $\mathbf{B}_A = Im(\phi)$, is a subring of $\mathbf{B}$, but not a subalgebra. However, it contains a basis for $\mathbf{B}$ as a vector space, and so $\mathbf{B}$ is the closure of $\mathbf{B}_A$. Thus ϕ is not an identity-reducible embedding.

The three-step process of Section 3 shows how this embedding gives an embedded cipher on $\mathbf{B}$. Based on the embedding function ϕ, an encryption function $f : \mathbf{A} \to \mathbf{A}$ of the AES induces an embedded encryption function $f_\phi : \mathbf{B}_A \to \mathbf{B}_A$. This can be naturally extended by closure to a function $\overline{f}_\phi : \mathbf{B} \to \mathbf{B}$. This extension to $\mathbf{B}$ can be expressed by simple operations over $GF(2^8)$, namely inversion and affine transformation. These are natural extensions of the algebraic operations of the AES to the larger algebra $\mathbf{B}$, based on the embedding function.

The AES embedding in the BES is an example of a cipher embedding which yields insights into the cipher that are not apparent from the original description. This is demonstrated by the multivariate quadratic equation system for the AES that is based on the BES embedding [12,2], which is a much simpler multivariate quadratic equation system than can be obtained directly from the AES. More generally, it is clear that the AES embedding in the BES offers a more natural environment in which to study the algebraic properties of the AES.

4.3   AES Extensions of Monnerat and Vaudenay

Monnerat and Vaudenay recently considered extensions of the AES and the BES, namely the CES and the Big-BES [11]. The authors showed that these were *weak* extensions in which cryptanalytic attacks could be easily mounted. They observed however that the weaknesses in the larger ciphers did not translate to weaknesses in the AES and BES, and were therefore of no consequence to the security of the AES. Within the framework established in Section 3 it is now very easy to see why the extensions given in [11] are inevitably divorced from the original cipher.

The extensions of the AES to CES and the Big-BES are similar, so we only consider the extension of the AES to the CES in this paper. A component of the state space for the CES can be considered as an element of the set $\mathbf{R} = \mathbb{F} \times \mathbb{F}$. The set $\mathbf{R}$ is given a ring structure $(\mathbf{R}, \oplus, \otimes)$ with binary operations defined by:

$$\begin{aligned}
\text{Addition} \qquad & (x_1, y_1) \oplus (x_2, y_2) = (x_1 + x_2 ,\ y_1 + y_2), \\
\text{Multiplication} \quad & (x_1, y_1) \otimes (x_2, y_2) = (x_1 x_2 ,\ x_1 y_2 + x_2 y_1).
\end{aligned}$$

The state space algebra for the CES is the algebra $\mathbf{C} = \mathbf{R}^{16}$, which is an algebra of dimension 32 over $\mathbb{F}$, with scalar multiplication by the field element $\lambda \in \mathbb{F}$ being identified with multiplication by the ring element $(\lambda, 0, \lambda, 0, \ldots, \lambda, 0) \in \mathbf{C}$.

The embedding of the AES in the CES is based on the injective algebra homomorphism $\theta : \mathbb{F} \to \mathbf{R}$ given by $\theta(a) = (a, 0)$. This definition can be extended in the obvious way to the injective algebra homomorphism $\theta : \mathbf{A} \to \mathbf{C}$

$$\theta(\mathbf{a}) = \theta(a_0, \ldots, a_{15}) = (\theta(a_0), \ldots, \theta(a_{15})) = ((a_0, 0), \ldots, (a_{15}, 0)).$$

The AES cryptographic functions were then induced based on this embedding map, and extended to the entire state space $\mathbf{C}$ to define the cipher CES.

There are several reasons why the cryptographic and algebraic relevance of such an embedding would be immediately questionable. Firstly, the definition of the function on the embedded image does not appear to be appropriate since some important algebraic properties are not retained within the CES. For instance, the AES "inversion" function satisfies $x^{(-1)} = x^{254}$, but this algebraic relationship is not satisfied by the CES "inversion" function. Secondly, the algebra $\mathbf{C}$ can be expressed as the direct sum of $\mathbf{C}_A = Im(\theta)$ and some other ideal subalgebra $\mathbf{C}'$. Thus, in our terminology, θ is a identity-reducible embedding. As shown earlier, this means that the way the embedded encryption function $f_\theta : \mathbf{C}_A \to \mathbf{C}_A$ is extended beyond $\mathbf{C}_A$ is irrelevant and has no consequences in the analysis of the AES. However, the cryptanalysis of the CES given in [11] is based on the properties of this arbitrary $\mathbf{C}'$-component of the CES. The fundamental reason for this separation into two components is clearly seen using the framework presented in this paper. The other embedding mappings proposed in [11] (based on $a \mapsto (a, \lambda a)$) are also identity-reducible and so at a fundamental level they are bound to have the same ineffectiveness in tying together the properties of the underlying cipher and the extension cipher.

5   Regular Representations of State Space Algebras

A very powerful and widely used technique in the study of algebras is to embed an algebra in a matrix algebra. Such an embedding of an algebra is known as a *representation* of the algebra. Thus a representation of a state space algebra gives an embedding of a cipher in a matrix algebra. In this section, we consider how a cipher state space algebra may be represented as matrix algebra, and how such a matrix representation can highlight properties of the cipher and its state space.

A *representation* of an n-dimensional algebra A is formally defined as an algebra homomorphism from A to a subalgebra of $M_l(K)$ [8], where $M_l(K)$ denotes the set of $l \times l$ matrices over the field K. Thus a representation of the algebra A identifies A with an n-dimensional subalgebra of the $l \times l$ matrices. If the algebra homomorphism is an isomorphism, then we may identify A with this n-dimensional subalgebra of the $l \times l$ matrices. Clearly, there are many ways in to define a representation. One standard technique is the *regular* representation, which is the algebra homomorphism $\nu : A \to M_n(K)$ that maps $a \in A$ to the matrix corresponding to the linear transformation $z \mapsto az$ (z a K-vector of length n) [8].

An illustration of a regular representation is given by the complex numbers, which form a 2-dimensional algebra over the real numbers. The complex number $x + iy$ can be identified with its regular representation as a matrix, which is given by

$$\nu(x + iy) = \begin{pmatrix} x & y \\ -y & x \end{pmatrix}.$$

The set of all such matrices forms a 2-dimensional algebra over the real numbers and can be identified with the complex numbers.

5.1   Regular Representation of the AES and the BES

The regular representations of the AES and the BES state spaces are algebra homomorphisms to diagonal matrix algebras. Thus we identify elements of these state spaces with the obvious diagonal matrix. An element of the AES state space **A** has a regular representation as a 16×16 diagonal matrix over $\mathbb{F}$, so **A** can be thought of as the 16×16 diagonal matrices. An embedded element of the AES state space in the BES has a regular representation as a diagonal 128×128 matrix over $\mathbb{F}$ in which the diagonal consists of octets of conjugates. The closure under matrix (algebra) operations of such embedded elements is clearly the algebra of all 128×128 diagonal matrices, which is the regular representation of the state space algebra **B**. The BES, and hence the AES, can thus be defined in terms of standard matrix operations in the regular representation of **B**. Suppose B is the diagonal 128×128 matrix that is the regular representation of some $\mathbf{b} \in \mathbf{B}$, then these BES transformations are given in matrix terms below.

- **Inversion.** For diagonal matrix B, this is the mapping $B \mapsto B^{(-1)} = B^{254}$. For an invertible diagonal matrix B, this is matrix inversion.

- **Linear Diffusion.** For diagonal matrix B, there exist diagonal matrices D_i and permutation matrices P_i $(i = 0, \ldots, 31)$ such that this linear transformation can be defined by

$$B \mapsto \sum_{i=0}^{31} D_i P_i B P_i^T.$$

- **Subkey Addition.** For diagonal matrix B and round subkey diagonal matrix K, this is the mapping $B \mapsto B + K$.

Thus the BES can be defined in matrix terms through the regular representation of the algebra **B** as the subalgebra of diagonal matrices, with the operations of the BES being represented by algebraic operations on these matrices.

The natural algebraic method of generalising operations on diagonal matrices is to extend these operations by some method to a larger algebra of matrices that contain the diagonal matrices. Thus we could define a "Matrix-AES" or "Matrix-BES" defined on some algebra of matrices that coincides with the AES or the BES for diagonal matrices. As we discuss below, this is in fact the approach taken in [11] to give the definition of the CES and the Big-BES. However, the functional "inversion" operation $M \mapsto M^{254}$ is not an invertible mapping on any subalgebra containing non-diagonal matrices. Thus there is no algebraic extension of the AES or BES state spaces beyond the diagonal matrices. In any case, the regular representation of the AES and the BES state spaces as diagonal matrices illustrates very well the point made in Section 3. From the viewpoint of the AES or the BES, all extensions beyond diagonal matrices are arbitrary and algebraically indistinguishable.

5.2   Regular Representations of Monnerat–Vaudenay Embeddings

We now consider the regular representation corresponding to the Monnerat and Vaudenay embedding. The algebra R has dimension 2 over $\mathbb{F}$, so its regular representation is given by a 2-dimensional subalgebra of the 2×2 matrices over $\mathbb{F}$. For an element $(x, y) \in \mathbf{R}$, the regular representation (with right matrix multiplication) is given by

$$\nu\left((x, y)\right) = \begin{pmatrix} x & y \\ 0 & x \end{pmatrix}.$$

Thus the regular representation of R is as the algebra of triangular matrices with constant diagonals, with the subalgebra corresponding to the embedding of $\mathbb{F}$ in **R** being the 1-dimensional subalgebra of 2×2 diagonal matrices with constant diagonals. It is clear that in any matrix operation related to the AES, the value of the diagonal elements never depends on any off-diagonal element. The regular representation of the CES state space (**C**) is a subalgebra of the 32×32 matrices over $\mathbb{F}$ given by the 32-dimensional subalgebra of 2×2 block diagonal matrices of the form given above. The regular representation of the AES subalgebra of the CES is given by the 16-dimensional subalgebra of diagonal matrices with pairs of constant terms. As noted above, the off-diagonal elements never have any effect on the diagonal elements and are entirely arbitrary. However, from the algebraic viewpoint of the AES, this subalgebra of diagonal matrices is the only subalgebra with any relevance. We note that this subalgebra of diagonal matrices is a representation in the 32×32 matrices of the algebra $\mathbf{A}$, and is clearly algebra isomorphic to the subalgebra of diagonal 16×16 matrices, which is the regular representation of the AES state space $\mathbf{A}$.

All the regular representations of the AES subset of the various state spaces considered consist of diagonal matrices. Those diagonal matrices given by the regular representation of Monnerat and Vaudenay embeddings merely use diagonal matrices of twice the size with diagonal entries repeated. Every cipher considered (AES, BES, CES and Big-BES) can all be defined solely in matrix terms within the subalgebra of diagonal matrices. Any extension of the block cipher definitions beyond diagonal matrices is arbitrary. The use of other embeddings based on similar algebraic structures is also suggested in [11]. However, it can be seen that the regular representations of the state space algebras of such embeddings merely correspond to other matrix subalgebras containing the diagonal matrix subalgebra. Thus such other embeddings also have the same cryptographic relevance as the original embeddings of Monnerat and Vaudenay [11]. Any conclusions drawn about diagonal matrices (AES embeddings) by considering the effect of these arbitrary block ciphers on non-diagonal matrices is arbitrary.

6   Conclusions

In this paper, we have presented a natural framework for the analysis of block cipher embeddings. This has been done in terms of the algebra of their state spaces, but takes into consideration the construction of the embedding function, how to "naturally" induce the cryptographic function on the embedded image, and how to (possibly) extend this image to the algebraic closure.

In this way we have shown that different approaches to embeddings in the literature are not algebraically equivalent. By way of example we have looked at three embedding strategies that have been discussed in the context of the AES. It is clear that while some embeddings might bring benefits such as cryptanalytic or implementation insights, it is possible to define other embeddings that, by their very construction, cannot possibly offer additional insights into the cipher.

References

1. E. Barkan and E. Biham. In How Many Ways Can You Write Rijndael?. *ASIACRYPT 2002*, LNCS vol. 2501, pp 160–175, Springer, 2002.
2. C. Cid, S. Murphy, and M.J.B. Robshaw. Computational and Algebraic Aspects of the Advanced Encryption Standard. In V. Ganzha *et al.*, editors, Proceedings of the *Seventh International Workshop on Computer Algebra in Scientific Computing - CASC 2004*, St. Petersburg, Russia, pages 93–103, Technische Universität München". 2004.

3. J. Daemen and V. Rijmen. AES Proposal: Rijndael (Version 2). NIST AES website csrc.nist.gov/encryption/aes, 1999.

4. J. Daemen and V. Rijmen. *The Design of Rijndael: AES—The Advanced Encryption Standard*. Springer–Verlag, 2002.

5. T. ElGamal. A Public Key Cryptosystem and a Signature Scheme based on Discrete Logarithms. *IEEE Transactions on Information Theory*, vol. 31, pp. 469–472, 1985.

6. L.R. Knudsen. New Directions in Cryptography (Volume II). *Journal of Craptology*, available at http://www2.mat.dtu.dk/people/Lars.R.Knudsen/crap.html, Vol. 1 No. 0, December 2000.

7. N. Koblitz. Elliptic Curve Cryptosystems. *Mathematics of Computation*, Vol. 48, pp 321-348, 1987.

8. A.I. Kostrikin. *Introduction to Algebra*. Springer-Verlag, 1981.

9. J.L. Massey. SAFER K-64: A byte-oriented block-ciphering algorithm. *Fast Software Encryption 1993*, LNCS vol. 809, pp. 1–17, Springer-Verlag, 1994.

10. V.S. Miller. Uses of Elliptic Curves in Cryptography. *CRYPTO '85*, LNCS vol. 218, pp. 417-426, Springer-Verlag, 1986.

11. J. Monnerat and S. Vaudenay. On some Weak Extensions of AES and BES. *Sixth International Conference on Information and Communications Security 2004*, LNCS vol. 3269, pp414–426, Springer, 2004.

12. S. Murphy and M.J.B. Robshaw. Essential Algebraic Structure within the AES. *CRYPTO 2002*, LNCS vol. 2442, pp. 1–16, Springer, 2002.

13. National Bureau of Standards. Data Encryption Standard. FIPS 46. 1977.

14. National Institute of Standards and Technology. Advanced Encryption Standard. FIPS 197. 26 November 2001.

15. National Institute of Standards and Technology. Digital Signature Standard. FIPS 186. 1994.

16. R.L. Rivest, A. Shamir and L.M. Adleman. A Method for Obtaining Digital Signatures and Public-Key Cryptosystems, *Communications of the ACM*, vol. 21, pp. 120-126, 1978.

Probabilistic Algebraic Attacks

An Braeken and Bart Preneel

Katholieke Universiteit Leuven,
Dept. Elect. Eng.-ESAT/SCD-COSIC,
Kasteelpark Arenberg 10, 3001 Heverlee, Belgium
{an.braeken, bart.preneel}@esat.kuleuven.be

Abstract. This paper investigates a probabilistic algebraic attack on
LFSR-based stream ciphers. We consider two scenarios (S3a and S3b)
proposed by Meier et al. at Eurocrypt 2004. In order to derive the probability in this new algebraic attack, we quantify the distance between
a Boolean function and a function with annihilator of a certain degree.
We show that in some cases the approximations can improve the algebraic attacks. Moreover, this distance leads to other theoretical results
such as the weights of the subfunctions and the distance to normal functions; it also provides information on the Walsh spectrum of the function.

Keywords: Algebraic attack, algebraic immunity, annihilator, Walsh
spectrum, combination and filter generator.

1 Introduction

Filter and combination generators are two well-known models of stream ciphers
based on Linear Feedback Shift Registers (LFSRs); both use a linear state update
function that guarantees a large period and a highly nonlinear Boolean function
that provides nonlinearity and a high linear complexity. Filter generators are key
stream generators that consist of only one LFSR to which a nonlinear filtering
function is applied. Combination generators use a nonlinear function to combine
several linear feedback shift registers. Until 2002, (fast) correlation attacks (see
for instance [21, 16, 13, 3]) were the most important and strongest attacks on
stream ciphers and in particular on the filter and combination generators. Fast
correlation attacks search for the solution of a highly noisy system of linear
equations. Their strength mainly depends on the nonlinearity and correlation-
immunity of the combining or filtering function, which defines the noise of the
system.

Recently, algebraic attacks have received a lot of attention. Algebraic attacks
try to solve a system of nonlinear equations. For stream ciphers this system is
always highly overdetermined. The complexity of the methods for solving systems of nonlinear equations is mainly determined by the degree of the equations.
The first type of probabilistic attack was considered in [6], where functions were
approximated by low degree functions. For instance, this attack was very successful for the stream cipher Toyocrypt [17] since its filter function contains only

N.P. Smart (Ed.): Cryptography and Coding 2005, LNCS 3796, pp. 290–303, 2005.
© Springer-Verlag Berlin Heidelberg 2005

one term of high degree. A further improvement in the area of algebraic attacks was the method to lower the degree of the equations in such a way that they still hold with probability one as shown in [5]. Therefore, the existence of low degree annihilators of the function is exploited, which is related to the concepts of algebraic immunity and annihilator function. For instance, for functions of dimension n, annihilators of degree less or equal than $\lceil n/2 \rceil$ always exist.

In this paper, we investigate the probabilistic version of the algebraic attack which uses the low degree annihilators. We call this attack the extended probabilistic algebraic attack. For this purpose, we first derive the distance between a given function and a function with low algebraic immunity by means of the Walsh spectrum of the function. From this distance we derive the probability that the equations used in the probabilistic attack hold. We show an example where this probabilistic algebraic attack outperforms the algebraic attack.

The paper is organized as follows. We first explain, in Sect. 2, the framework of the algebraic attack of the second type. In Sect. 3, we briefly describe some properties of Boolean functions. Section 4 contains the derivation of theoretical results concerning the lower bound on the distance to functions with low degree annihilators. In Sect. 5, we show how the formula for the distance can be used to derive information on the on the Walsh spectrum of the function, the weights of the subfunctions, and the distance to normal functions. Section 6 concludes the paper.

2   Extended Probabilistic Algebraic Attack

Consider a filter or combining generator with corresponding filtering or combining function f on $\mathbb{F}_2^n$. The secret key K is the initial state of the LFSR(s). The i-th output of the generator is given by $f(\mathcal{L}^i(K))$, where $\mathcal{L}$ is the linear transformation describing the LFSR(s). Denote the length of the LFSR by L. In [15], Meier et al. reduced the three attack scenarios for algebraic attacks proposed by Courtois and Meier in [5] to only two scenarios.

S3a. There exists a nonzero function h of low degree such that $fh = h$.
S3b. There exists a nonzero function g of low degree such that $fg = 0$.

The attack works as follows. If the output bit $f(x) = 0$, we are in Scenario S3a and conclude that $h(x) = 0$. If the output bit $f(x) = 1$, we use the low degree annihilator g (Scenario S3b) and thus $g(x) = 0$. Note that h is a low degree annihilator of $f \oplus 1$. Consequently, we only need to search for low degree annihilators of f and $f \oplus 1$ (see the definition of algebraic immunity in Sect. 3).

This paper investigates the probabilistic versions of these scenarios. They can be seen as a second type of probabilistic algebraic attack. Consider the sets $S_0^f = \{x \in \mathbb{F}_2^n : f(x) = 0\}$ and $S_1^f = \{x \in \mathbb{F}_2^n : f(x) = 1\}$ with respect to the Boolean function f on $\mathbb{F}_2^n$.

S4a. There exists a nonzero function h of low degree such that $fh = h$ on S_0^f with high probability p.

S4b. There exists a nonzero function g of low degree such that $fg = 0$ on S_1^f with high probability p.

Note that the probabilistic version of criterion S3a is also described in [5] but not further used.

In order to compute the probability for applying scenarios S4a and S4b, we will search for the best approximation to a function with low degree annihilator. We derive in Sect. 4 a formula for measuring this distance X. The corresponding probability for attack scenario S4a is then equal to $p = 1 - \frac{X}{2^n - \text{wt}(f)}$ since we need to restrict to the set S_0^f, and equal to $p = 1 - \frac{X}{\text{wt}(f)}$ for scenario S4b.

Consequently, an overdetermined system of nonlinear equations of degree k is obtained which hold with probability p. These equations are derived from the approximation function which has multiple of degree k. We will show in Sect. 4 that $p \geq 1 - 2^{-k}$ for a balanced function. There are several methods to solve a system of nonlinear equations of degree k such as linearization, XL-algorithm [6], and Gröbner bases algorithms such as F4 and F5 [11]. However, note that we consider a *probabilistic* nonlinear system of equations. Let w be the exponent of the Gaussian reduction ($w \approx 2.807$ [1]). We can take the following approaches:

- Solve the system with F5 (F4 and XL are proven to be worse than F5 [10]). We need L equations which are satisfied with probability p^{-L}. However, solving systems of nonlinear equations is believed to be hard, hence the complexity is high.
- Use the linearization algorithm which has reasonable complexity $\binom{L}{k}^w$, which is reasonable. However, in this algorithm we use $\binom{L}{k}$ equations which hold with a very small probability $p^{-\binom{L}{k}}$.

We conclude that both approaches are not very satisfactory. The first approach leads to a reasonable complexity related to the probability since we need a small number of equations, but it requires a high complexity for solving the system. The second approach has a low complexity for solving the system but has a high complexity corresponding to the probability since we need a large number of equations.

It is an interesting open problem if there exist more efficient methods for solving this type of equations. For instance, is it possible to find the solution by means of a least square solution? Or, can we combine somehow the methods used in the correlation attack (where the probability is much lower than in the probabilistic algebraic attack) together with the linearization method? As we show in Sect. 4, the probability for applying the probabilistic algebraic attacks with respect to a function with affine annihilators corresponds to the probability of a linear approximation. In some sense, this type of attack can also be considered as an attack in between a correlation and algebraic attack.

We want to stress that the goal of this paper is to measure the distance X in order to obtain equations with low degree which hold with high probability. We do not investigate methods for solving system of nonlinear equations in this paper.

3   Background

Let $f(x)$ be a Boolean function on $\mathbb{F}_2^n$. Denote $S_1^f = \{x \in \mathbb{F}_2^n : f(x) = 1\}$ (also called the support of f) and $S_0^f = \{x \in \mathbb{F}_2^n : f(x) = 0\}$. The weight $\text{wt}(w)$ of a vector $w \in \mathbb{F}_2^n$ is defined by the number of nonzero positions in w. The truth table of a Boolean function f is a vector consisting of the function values $(f(0), \ldots, f(2^n - 1))$. Any Boolean function f can be uniquely expressed in the algebraic normal form (ANF):

$$f(x) = \bigoplus_{(a_1,\ldots,a_n)\in\mathbb{F}_2^n} h(a_1,\ldots,a_n)x_1^{a_1} \cdots x_n^{a_n},$$

with h a function on $\mathbb{F}_2^n$, defined by $h(a) = \sum_{x \leq a} f(x)$ for any $a \in \mathbb{F}_2^n$, where $x \leq a$ means that $x_i \leq a_i$ for all $i \in \{1, \ldots, n\}$. The algebraic degree of f, denoted by $\deg(f)$, is defined as the number of variables in the highest term $x_1^{a_1} \cdots x_n^{a_n}$ in the ANF of f, for which $h(a_1, \ldots, a_n) \neq 0$.

With respect to the algebraic attacks, Meier et al. introduced the concepts of algebraic immunity (AI) and annihilator [15].

Definition 1. *Let f be a Boolean function from $\mathbb{F}_2^n$ into $\mathbb{F}_2$. The lowest degree of the function g from $\mathbb{F}_2^n$ into $\mathbb{F}_2$ for which $f \cdot g = 0$ or $(f \oplus 1) \cdot g = 0$ is called the algebraic immunity (AI) of the function f. The function g for which $f \cdot g = 0$ is called an annihilator function of f.*

As proven in [15–Theorem 1], the set of annihilators of f is an ideal and consists of the functions $g(f \oplus 1)$ where g is an arbitrary Boolean function on $\mathbb{F}_2^n$.

An important tool in the study of Boolean functions is the *Walsh transform*. The Walsh transform of $f(x)$ is a real-valued function over $\mathbb{F}_2^n$ that is defined for all $v \in \mathbb{F}_2^n$ as

$$W_f(v) = \sum_{x\in\mathbb{F}_2^n} (-1)^{f(x)+v\cdot x} = 2^n - 2\,\text{wt}(f(x) \oplus v \cdot x). \qquad (1)$$

The Walsh spectrum can be used to express the nonlinearity N_f of the function f, which is the minimum distance between f and the set of all affine functions:

$$N_f = 2^{n-1} - \frac{1}{2} \max_{w\in\mathbb{F}_2^n} |W_f(w)|.$$

Functions for which $|W_f(v)| = 2^t$ or $W_f(v) = 0$ with $\left\lceil \frac{n}{2} \right\rceil < t \leq n$ for all $v \in \mathbb{F}_2^n$ are called plateaued functions with amplitude 2^t [23]. If n is even and $t = \frac{n}{2}$, then the function is said to be maximum nonlinear or bent [8].

Another property related to the Walsh spectrum is resiliency [20]. A Boolean function is said to be t-resilient if and only if $W_f(w) = 0$ for all $w \in \mathbb{F}_2^n$ with $\text{wt}(w) \leq t$. A 0-resilient function is also called a balanced function, i.e., a function for which the output is equally distributed.

4   Approximation by Functions with Low Degree Annihilators

4.1   How to See Annihilator Functions

Let k be an integer such that $1 \leq k \leq \deg(f)$. Suppose that $f : \mathbb{F}_2^n \to \mathbb{F}_2$ can be written as $f(x) = l_{d_1}(x)g_1(x) \oplus \cdots \oplus l_{d_k}(x)g_k(x)$ where $l_{d_i}(x)$ are functions of degree d_i for $1 \leq i \leq k$ with $d_1 + d_2 + \cdots + d_k = d$ such that $S_1^{l_{d_1}} \cup \cdots \cup S_1^{l_{d_k}} \neq \mathbb{F}_2^n$ and $g_1(x), \ldots, g_k(x)$ are arbitrary functions. Then, the corresponding annihilator of degree d is equal to $(l_{d_1}(x) \oplus 1) \cdots (l_{d_k}(x) \oplus 1)$, which is a nonzero function by the property that $S_1^{l_{d_1}} \cup \cdots \cup S_1^{l_{d_k}} \neq \mathbb{F}_2^n$.

In general, if for an arbitrary function f on $\mathbb{F}_2^n$, there exist functions $l_{d_1}, \ldots, l_{d_k}$ such that $S_1^{l_{d_1}} \cup \cdots \cup S_1^{l_{d_k}} \neq \mathbb{F}_2^n$ and $S_1^f \subseteq (S_1^{l_{d_1}} \cup \cdots \cup S_1^{l_{d_k}})$, where $\deg(l_{d_1}) + \cdots + \deg(l_{d_k}) = d$, then f has an annihilator of degree d. Therefore, we obtain

Theorem 1. *[15] If a balanced function has an annihilator of degree equal to one, then the function is affine.*

If $X = |S_1^f| - |S_1^f \cap (S_1^{l_{d_1}} \cup \cdots \cup S_1^{l_{d_k}})|$, then f is at distance X from a function f' with annihilator of degree d. The function f' is equal to $f'(x) = l_{d_1}(x)g_1(x) \oplus \cdots \oplus l_{d_k}(x)g_k(x)$, where $g_1, \ldots, g_k$ are arbitrary functions that satisfy $l_{d_1}(x)g_1(x) \oplus \cdots \oplus l_{d_k}(x)g_k(x) = 1$ for all x such that $f(x) = 1$. Moreover, the probability corresponding to scenario S4b is equal to $p = 1 - \frac{X}{\text{wt}(f)}$.

A similar property holds for scenario S4a, because we now need to search for annihilators of $f \oplus 1$. If $S_1^{f \oplus 1} = S_0^f \subseteq (S_1^{l_{d_1}} \cup \cdots \cup S_1^{l_{d_k}})$, where $S_1^{l_{d_1}} \cup \cdots \cup S_1^{l_{d_k}}$ and $\deg(l_{d_1}) + \cdots + \deg(l_{d_k}) = d$, then $f \oplus 1$ has a nonzero annihilator of degree d. If $X = |S_0^f| - |S_0^f \cap (S_1^{l_{d_1}} \cup \cdots \cup S_1^{l_{d_k}})|$, then $f \oplus 1$ is at distance X from a function with annihilator of degree d. The probability corresponding to scenario S4a is equal to $p = 1 - \frac{X}{2^n - \text{wt}(f)}$. It is easy to see that if $S_1^f \subseteq (S_1^{l_{d_1}} \cup \cdots \cup S_1^{l_{d_k}})$ holds then $S_0^f \subseteq (S_0^{l_{d_1}} \cup \cdots \cup S_0^{l_{d_k}})$ if and only if $\text{wt}((f \oplus 1)l_{d_1} \cdots l_{d_k}) = 0$ (i.e. there does not exist an $x \in \mathbb{F}_2^n$ such that $f(x) = 0$ and $l_{d_1}(x) = \cdots = l_{d_k}(x) = 1$).

In the following, we will mainly concentrate on the annihilators of f, corresponding to scenario S4b, since both cases are similar.

4.2   Computing the Distance to Functions with Low Degree Annihilators

We now derive a formula for determining the distance X between a given function and a function with low algebraic immunity by means of the Walsh spectrum of the function. Therefore, we first assume the general case where the annihilator consists of the product of functions of arbitrary degree. We then explain that for balanced functions the best approximations in general are obtained by functions having annihilators that consist of affine factors.

Theorem 2. *Let $l_{d_1}(x), \cdots, l_{d_k}(x)$ be k different functions on $\mathbb{F}_2^n$, such that $S_1^{l_{d_1}} \cup \cdots \cup S_1^{l_{d_k}} \neq \mathbb{F}_2^n$. The distance X between a Boolean function f on $\mathbb{F}_2^n$*

and a function with annihilator $g(x) = (l_{d_1}(x) \oplus 1) \cdot \ldots \cdot (l_{d_k}(x) \oplus 1)$ is equal to $\mathrm{wt}(f \cdot (l_{d_1}(x) \oplus 1) \cdot \ldots \cdot (l_{d_k}(x) \oplus 1))$.

Proof. Let f' be the function equal to $f'(x) = l_{d_1}(x)g_1(x) \oplus \cdots \oplus l_{d_k}(x)g_k(x)$ where $g_1 \ldots, g_k$ are arbitrary Boolean functions on $\mathbb{F}_2^n$ which satisfy the property that if $l_{i_1}(x) = \cdots = l_{i_l}(x) = 1$ and $f(x) = 1$ then $g_{i_1}(x) \oplus \cdots \oplus g_{i_l}(x) = 1$ for $i_1, \ldots, i_l \in \{1, \ldots, k\}$. It is clear that f' has the function g as annihilator and is closest to f. More precisely, f' differs from f only in the points where $l_{d_1}(x) = \cdots = l_{d_k}(x) = 0$ and $f(x) = 1$. $\qquad\square$

In order to compute the weight of the product of functions, we make use of the following lemma.

Lemma 1. *Let* $l_{d_1}, \ldots, l_{d_k}, f$ *be Boolean functions on* $\mathbb{F}_2^n$, *then*

$$W_{l_{d_1}(x) \cdot \ldots \cdot l_{d_k}(x) \cdot f(x)}(w)$$
$$= \frac{1}{2^k}((2^k - 1)2^n \delta(w) + W_{l_{d_1}}(w) + \cdots + W_{l_{d_k}}(w) + W_f(w)$$
$$- W_{l_{d_1} \oplus l_{d_2}}(w) - \cdots - W_{l_{d_k} \oplus f}(w)$$
$$+ W_{l_{d_1} \oplus l_{d_2} \oplus l_{d_3}}(w) + \cdots + W_{l_{d_{k-1}} \oplus l_{d_k} \oplus f}(w)$$
$$+ \cdots + (-1)^k W_{l_{d_1} \oplus \cdots \oplus l_{d_k} \oplus f}(w)), \qquad (2)$$

where $\delta(w)$ *represents the Kronecker* δ *function* $(\delta(w) = 1$ *if and only if* $w = 0)$.

Proof. One can easily check by induction that

$$(-1)^{l_{d_1}(x) \cdot \ldots \cdot l_{d_k}(x) \cdot f(x)}$$
$$= \frac{1}{2^k}(2^k - 1 + (-1)^{l_{d_1}(x)} + \cdots (-1)^{l_{d_k}(x)} + (-1)^{f(x)}$$
$$- (-1)^{l_{d_1}(x) \oplus l_{d_2}(x)} - \cdots - (-1)^{l_{d_k}(x) \oplus f(x)}$$
$$+ (-1)^{l_{d_1}(x) \oplus l_{d_2}(x) \oplus l_{d_3}(x)} + \cdots + (-1)^{l_{d_{k-1}}(x) \oplus l_{d_k}(x) \oplus f(x)}$$
$$+ \cdots + (-1)^k (-1)^{l_{d_1}(x) \oplus \cdots \oplus l_{d_k} \oplus f(x)}).$$

Multiplying both sides with $\sum_{x \in \mathbb{F}_2^n} (-1)^{w \cdot x}$ leads to Eqn. (2). $\qquad\square$

Remark 1. We note that Eqn. (2) was already proven by Daemen for $k = 1$ [7]. Applying this formula leads to

$$W_{l_{d_1} \cdot f}(w) = \frac{1}{2}(2^n \delta(w) + W_{l_{d_1}}(w) + W_f(w) - W_{l_{d_1} \oplus f}(w)).$$

Consequently, l_{d_1} is annihilator of f if and only if

$$W_{l_{d_1}}(w) + W_f(w) = W_{l_{d_1} \oplus f}(w), \quad \text{for all } w \neq 0$$
$$W_{l_{d_1}}(0) + W_f(0) - 2^n = W_{l_{d_1} \oplus f}(0).$$

We conclude that only for the product of a function with its annihilating function, we obtain a kind of linearity property in the Walsh spectrum.

Equation (1) shows the relation between Walsh spectrum and weight. Consequently, we can immediately derive from (2) a formula for the distance X.

Corollary 1. Let $l_{d_1}(x), \cdots, l_{d_k}(x)$ be k different functions on $\mathbb{F}_2^n$, such that $S_1^{l_{d_1}} \cup \cdots \cup S_1^{l_{d_k}} \neq \mathbb{F}_2^n$ and f be a Boolean function on $\mathbb{F}_2^n$. The distance X between a Boolean function f on $\mathbb{F}_2^n$ and a function with annihlator $g(x) = (l_{d_1}(x) \oplus 1) \cdot \ldots \cdot (l_{d_k}(x) \oplus 1)$ is equal to

$$X = 2^{n-(k+1)} - \frac{1}{2^{k+1}}(W_{l_{d_1} \oplus 1}(0) + \cdots + W_{l_{d_k} \oplus 1}(0) + W_f(0)$$
$$-W_{l_{d_1} \oplus l_{d_2}}(0) + \cdots + (-1)W_{l_{d_1} \oplus \cdots \oplus l_{d_k} \oplus f}(0)).$$

In particular, let $l_{d_i} = w_i \cdot x \oplus w_i'$ where $w_i \in \mathbb{F}_2^n, w_i' \in \mathbb{F}_2$ for all $1 \le i \le k$, then

$$X = 2^{n-(k+1)} - \frac{1}{2^{k+1}} \sum_{(i_1,\ldots,i_k) \in \mathbb{F}_2^k} (-1)^{w_1' i_1 + \cdots + w_k' i_k} W_f(w_1^{i_1} \oplus \cdots \oplus w_k^{i_k}),$$

$$(3)$$

where $w_j^{i_j}$ is equal to w_j if $i_j = 1$ and 0 else for $1 \le j \le k$.

Remark 2. Note that the support of the product of k affine functions $l_{d_1}(x) \cdots l_{d_k}(x)$ corresponds to an affine subspace V of dimension $n - k$ determined by $l_{d_1}(x) = 1, \ldots, l_{d_k}(x) = 1$. Therefore, Equation (3) can also be obtained from the weight of the restriction of f to a subspace of dimension $n - k$ (see e.g. [2–Proposition 1]).

For computing the distance between $f \oplus 1$ and a function with low degree annihilator, we obtain more or less the same formula. Only the terms $W_{f \oplus h}(w)$ for all linear combinations h of $l_{d_1}, \ldots, l_{d_k}$ switch sign since $W_{f \oplus 1}(w) = -W_f(w)$ for all $w \in \mathbb{F}_2^n$. In particular, for the affine functions $l_{w_i}(x) = w_i \cdot x \oplus w_i'$ for $1 \le i \le k$, we obtain:

Corollary 2. Let f be a function on $\mathbb{F}_2^n$. The distance between $f \oplus 1$ and a function with annihilator g of degree k, i.e. $g(x) = (l_{w_1}(x) \oplus 1) \cdots (l_{w_k}(x) \oplus 1)$ is equal to

$$X = 2^{n-(k+1)} + \frac{1}{2^{k+1}} \sum_{(i_1,\ldots,i_k) \in \mathbb{F}_2^k} (-1)^{w_1' i_1 + \cdots + w_k' i_k} W_f(w_1^{i_1} \oplus \cdots \oplus w_k^{i_k}). \quad (4)$$

We expect to obtain in general very good results for the approximation by using the covering of only affine functions (annihilators which consist of the product of affine functions), corresponding with equations (3) and (4). To confirm this, we make use of the following theorem.

Theorem 3. Let l_k be a Boolean function of degree k. Then l_k has the highest possible weight equal to $2^n - 2^{n-k}$ in the set of all functions with degree k if and only if $l_k = l_{w_1} \vee \cdots \vee l_{w_k}$ where $l_{w_i}(x) = w_i \cdot x \oplus 1$ for $1 \le i \le k$ and $\{w_1, \ldots, w_k\}$ linearly independent.

Proof. By induction, we derive that $l_k(x) = (w_1 \cdot x) \cdots (w_k \cdot x) \oplus 1$. Since the minimum distance of $RM(k,n)$ is equal to 2^{n-k}, this function has highest possible weight in the set of functions with degree k.   □

Consequently, the support of the or-sum of k linearly independent affine functions is always greater or equal than the support of all other or-sums of j functions for which the sum of their degrees is equal to k. Moreover, since the functions l_{w_i} for $1 \leq i \leq k$ are affine functions, the vectors in $S_1^{l_{w_1} \vee \cdots \vee l_{w_k}}$ are more or less random and have more chance to belong to the support of a random balanced function f than the vectors of $S_1^{l_k}$ where l_k is a function obtained by the or-sum of functions for which the sum of the degrees is equal to k.

For the best approximation with respect to a function with annihilators of degree 1, we obtain that $X = 2^{n-2} - \frac{1}{4}\max_{w \in \mathbb{F}_2^n} |W_f(0) + W_f(w)|$. If f is balanced, then $X = \frac{N_f}{2}$ and the corresponding function can be written as $g(x)l_w(x)$, where $g(x) = f(x)$ if $f(x) = 1$. However, the probability used in attack criterion S4b is equal to $\frac{1}{2} - \frac{1}{2^{n+1}}W_f(w)$, which is exactly the probability that the function f can be approximated by the linear function l_w. Consequently, in the affine case, the probabilistic algebraic attack of the first and second type coincide. Moreover, we do not find better approximations for the (fast) correlation attacks.

4.3   How to Find the Approximation Functions

As we explained in the previous section, we expect to find the best approximations for functions having annihilators consisting of affine factors. Therefore, we can restrict ourself to equations (3) and (4), which determine the distance with respect to a function which has low degree annihilators of f resp. $f \oplus 1$. From these equations, we deduce the following.

Theorem 4. *If we find a subspace of dimension k in the Walsh spectrum, determined by $W = <w_1, \ldots, w_k>$ and k elements $w'_1, \ldots, w'_k \in \mathbb{F}_2$, for which the sum of the Walsh values $(-1)^{w'_1 i_1 + \cdots + w'_k i_k} W_f(w_1^{i_1} \oplus \cdots \oplus w_k^{i_k})$ for all $(i_1, \ldots, i_k) \in \mathbb{F}_2^k$ is high, then the function is close to a function with annihilating function $(l_{w_1} \oplus 1) \cdot \ldots \cdot (l_{w_k} \oplus 1)$.*

As a subcase, consider the situation where $w'_i = 0$ for all $1 \leq i \leq k$. Then, if the sum of the Walsh values $W_f(w)$ for all $w \in W$ is high, we obtain a low distance to a function which has an annihilator of degree k of f. Similar, if the sum of the Walsh values $-W_f(w)$ for all $w \in W$ is small, we obtain a low distance to a function which has an annihilator of degree k of $f \oplus 1$.

Since we can always find a subspace for which the corresponding sum of the Walsh values belonging to the subspace is greater or equal than zero, we can conclude that $X \geq 2^{n-(k+1)}$ for the distance of a balanced function with respect to a function with AI equal to k. Therefore, the probability for the probabilistic algebraic attack is greater or equal than $1 - 2^{-k} = 2^{-1} + 2^{-2} + \cdots + 2^{-k}$ when the approximation with respect to a function with AI equal to k is used. For instance, for $k = 2$, the probability is greater or equal than 0.75.

4.4   Example

We present a toy example to illustrate the extended probabilistic algebraic attack. Consider the balanced Boolean function f of degree 4 on $\mathbb{F}_2^6$ with ANF representation $f(x_1, x_2, x_3, x_4, x_5, x_6) = x_1 x_3 x_4 x_5 \oplus x_1 x_4 x_5 \oplus x_2 x_4 x_5 \oplus x_2 x_3 \oplus x_4 x_5 \oplus x_1 \oplus x_6$. One can check that the function f has AI = 3. Since $W_f(35) = 24, W_f(37) = 32$ and $W_f(0) = W_f(6) = 0$, we have that $X = 2^3 - \frac{1}{8}(0 + 0 + 24 + 32) = 1$ corresponding to Eqn. (3). This also means that f is at distance 1 to the set of functions g which have $(l_{d_1} \oplus 1)(l_{d_2} \oplus 1)$ as annihilator, where l_{d_1} is the linear function $x_1 \oplus x_2 \oplus x_6$ and l_{d_2} the linear function $x_1 \oplus x_3 \oplus x_6$. These functions g are equal to $l_{d_1}(x)g_1(x) \oplus l_{d_2}(x)g_2(x)$, where g_1, g_2 are arbitrary functions on $\mathbb{F}_2^6$ such that if $l_{d_1}(x) = l_{d_2}(x) = f(x) = 1$ then $g_1(x) \oplus g_2(x) = 1$, and if $l_{d_i}(x) = f(x) = 1$ then $g_i(x) = 1$ for $i = \{1, 2\}$. Consequently, the corresponding extended probabilistic algebraic attack can be performed with probability $p = 1 - \frac{1}{32} = 0.96875$. We now compare the complexity of the extended probabilistic algebraic attack with respect to equations of degree 2 with probability p, and the algebraic attack with equations of degree 3 and probability 1 when using linearization for solving the system of equations in both cases. This means that we need to compare the complexity $p^{-\binom{L}{2}} \binom{L}{2}^{2.807}$ for the extended probabilistic algebraic attack with the complexity of the algebraic attack $\binom{L}{3}^{2.807}$. One can check that for lenghts of the LFSR L strictly less than 18 the probabilistic attack will outperform the deterministic one. In general, the probabilistic attack will of course be faster when p is close to one.

We note that for the stream cipher LILI-128, the probabilistic attack does not outperform the usual algebraic attack which has a complexity of 2^{57} CPU clocks [5] by using annihilators of degree 4, if we use the linearization algorithm. Nevertheless, we note the relatively small distances between the Boolean filtering function, a function on $\mathbb{F}_2^{10}$ which has degree 6 and is 2-resilient, and a function with annihilator 2 resp. 3. The maximum absolute Walsh value is equal to 64. We refer to [22] for more details on the description of LILI-128. In order to obtain the distance with respect to annihilators of degree 2, we use Eqn. (3) for X from Theorem 1; the result is $X = 104$. Therefore, we use the linear functions

$$l_{w_1}(x) = x_1 \oplus x_2 \oplus x_4 \oplus x_5 \qquad (w_1 = 27)$$
$$l_{w_2}(x) = x_1 \oplus x_3 \oplus x_4 \oplus x_6 \oplus x_7. \quad (w_2 = 109).$$

Since $W_f(27) = W_f(109) = W_f(118) = 64$, this is the best possible value we can obtain for X with respect to second order functions. The corresponding probability for the S4b attack is equal to 0.79687.

The distance becomes very small when we look at functions with annihilators of degree 3. We now consider the supports of three linear functions:

$$l_{w_1}(x) = x_1 \oplus x_2 \oplus x_4 \oplus x_5 \qquad\qquad (w_1 = 27)$$
$$l_{w_2}(x) = x_1 \oplus x_3 \oplus x_4 \oplus x_6 \oplus x_7 \qquad (w_2 = 109)$$
$$l_{w_3}(x) = x_1 \oplus x_2 \oplus x_3 \oplus x_5 \oplus x_6 \oplus x_8 \quad (w_3 = 183).$$

By applying Eqn. (3) for X, we obtain $X = 40$ since 6 of the 7 vectors (obtained by the nonzero linear combinations of w_1, w_2 and w_3) have Walsh value 64 and one has Walsh value 0. The corresponding probability for the S4b attack is equal to 0.921875. Unfortunately, these probabilities are still far too low for applying fast probabilistic algebraic attacks.

5   Other Interpretations

We show that the distance X (corresponding with Eqn. (3) from Corollary 1) is related to the weights of the subfunctions and to the distance to normal functions. We also deduce general information on the structure of the Walsh spectrum of the function. Note that here we will only consider the special case where the functions $l_{d_i}(x)$ are the affine functions $l_{w_i}(x) = w_i \cdot x \oplus w'_i$ for $1 \leq i \leq k$.

Weight of Subfunctions

Corollary 3. *The weight of a Boolean function f on $\mathbb{F}_2^n$ when restricted to the $(n - k)$-dimensional subspace, determined by $< l_{w_1}(x) = 0, \ldots, l_{w_k}(x) = 0 >$, is equal to X.*

Remark 3. Note that a general formula for the weight of a function when restricted to a subspace is also derived in [2–Proposition 1].

Let wt_i denote the minimum possible weight of the subfunction of a Boolean function f on $\mathbb{F}_2^n$ obtained by restriction to a subspace of dimension $n - i$ for $1 \leq i \leq n$. Then, we have the following inequalities:

$$2^{n-2} - 2^{\lceil \frac{n}{2} \rceil - 4} \geq \mathrm{wt}_1 \geq \cdots \geq \mathrm{wt}_i \geq 2\,\mathrm{wt}_{i+1} \geq \cdots \geq \mathrm{wt}_n = 0.$$

The first inequality follows from the property of bent functions. The last inequality is satisfied by the all-one function. The inequality in the middle is obtained by the fact that at each step, we can cover at least half of the nonzero elements, i.e., $\mathrm{wt}_{i+1} \leq \mathrm{wt}_i - \frac{\mathrm{wt}_i}{2}$. We know that any function has AI $\leq \lceil \frac{n}{2} \rceil$. However, this does not imply that $\mathrm{wt}_{\lceil \frac{n}{2} \rceil} = 0$ in general. Only for $n \leq 7$ or quadratic functions, we have that $\mathrm{wt}_{\lceil \frac{n}{2} \rceil} = 0$ [9].

For t-resilient functions, we already know that the subfunctions obtained by fixing at most $n - t$ coordinates are balanced. This can be generalized by saying that all subfunctions obtained by restriction to the subspace $W = < l_{w_1}(x) = 0, \ldots, l_{w_k}(x) = 0 >$ where $\mathrm{wt}(w) \leq t$ for all $w \in W$ and $k \leq t$ are balanced. Moreover, Eqn. (3) also gives information on the other subfunctions.

We also derive a relation between the ANF coefficients and the weight of the subfunction.

Theorem 5. *The weight of the subfunction with respect to the subspace W of dimension l given by $< x_{i_1} = \cdots = x_{i_{n-l}} = 0 >$, where $1 \leq i_1 \leq \cdots \leq i_{n-l} \leq n$ is even if and only if the ANF coefficient $h(v_1 \oplus 1, \cdots, v_n \oplus 1)$ is equal to zero where $\mathrm{wt}(v) = l$ and $v_{i_1} = \cdots = v_{i_{n-l}} = 0$.*

Proof. Suppose the weight of the subfunction with respect to W is even and is equal to $2k$ for $k \in \mathbb{Z}$. From Eqn. (3), we obtain that

$$\sum_{w \in W} W_f(w) = 2^n - 2^{l+2}k. \tag{5}$$

Moreover, [12] describes the following relation between the ANF coefficients and Walsh values:

$$h(v_1 \oplus 1, \cdots, v_n \oplus 1) = 2^{n-\text{wt}(v)-1} - 2^{-\text{wt}(v)-1} \sum_{w \preceq v} W_f(w) \mod 2$$

$$= 2^{n-l-1} - 2^{-l-1} \sum_{w \in W} W_f(w) \mod 2. \tag{6}$$

Substituting (5) in (6) shows the equivalence. □

Remark 4. Proposition 2.8 in [14] shows that $S_m(f) + D_{n-m}(f) = \binom{n}{m}$, where $S_m(f)$ denotes the number of subfunctions with even weight obtained by fixing m variables of f with zero values, and $D_{n-m}(f)$ is the number of terms of degree $(n - m)$. It is clear that the previous theorem can be seen as a generalization of this result.

Remark 5. For t-resilient functions f on $\mathbb{F}_2^n$, Siegenthaler's inequality states that $\deg(f) \leq n-t-1$. Moreover, the condition on the degree $d \leq n-t-1$ for functions which have Walsh values divisible by 2^t is well-known [4]. Both results follow from the previous theorem.

Distance to Normal Functions

The distance X is also of independent interest in the study of normality of Boolean functions. By the equivalence relation between k-normality and algebraic immunity of degree $n-k$ as shown in [15], we obtain the following theorem.

Corollary 4. *The value X determines the distance between a Boolean function and an $(n - k)$-normal function. The value*

$$X' = 2^{n-(k+1)} - \frac{1}{2^{k+1}} \sum_{(i_1, \ldots, i_k) \in \mathbb{F}_2^k} (-1)^{w_1' i_1 + \cdots + w_k' i_k} W_f(a \oplus w_1^{i_1} \oplus \cdots \oplus w_k^{i_k})$$

determines the distance between the function $f \oplus l_a(x)$ and an $(n - k)$-normal function. As a consequence, X' measures the distance between the function f and an $(n-k)$-weakly normal function. The corresponding $(n-k)$-normal function is given by $f'(x) = l_{w_1}(x)g_1(x) \oplus \cdots \oplus l_{w_k}(x)g_k(x)$, where $g_1, \ldots, g_k$ are arbitrary functions that satisfy $l_{w_1}(x)g_1(x) \oplus \cdots \oplus l_{w_k}(x)g_k(x) = 1$ for all x such that $f(x) = 1$.

For instance, a function is $(n-1)$-normal if and only if $|W_f(0) + W_f(w)| = 2^n$ for a certain $w \in \mathbb{F}_2^n$. If there exist $a, w \in \mathbb{F}_2^n$ such that $|W_f(a) + W_f(a \oplus w)| = 2^n$, then f is $(n-1)$-weakly normal. The smallest distance to $(n-1)$-normal functions is equal to $2^{n-2} - \frac{1}{4}\max_{w\in\mathbb{F}_2^n}|W_f(0) + W_f(w)|$, and to $(n-1)$-weakly normal functions is equal to $2^{n-2} - \frac{1}{4}\max_{a,w\in\mathbb{F}_2^n}|W_f(a) + W_f(a\oplus w)|$. Consequently, any Boolean function can be approximated by an $(n-1)$-normal function with probability greater or equal than 0.75. More generally, any Boolean function can be approximated by an $(n-k)$-normal function with probability greater or equal than $1 - 2^{-(k+1)} = 2^{-1} + 2^{-2} + \cdots + 2^{-(k+1)}$.

Information on Walsh Spectrum

Equation (3) leads to several interesting results on the Walsh spectrum of a function. Since X represents a real number greater or equal than zero, we can conclude that the sum of the Walsh values of vectors belonging to a flat of dimension k is always divisible by 2^{k+1}. This can be seen as a generalization of Theorem 5 from [18].

Moreover, from the definition of the Walsh spectrum, we can conclude that the sum

$$\sum_{(i_1,\dots,i_k)\in\mathbb{F}_2^k} (-1)^{w_1'i_1 + \cdots + w_k'i_k} W_f(w_1^{i_1} \oplus \cdots \oplus w_k^{i_k})$$

determines the Walsh spectrum of the function obtained by restriction with respect to the subspace $< l_{w_1} = 0,\dots,l_{w_k} = 0 >$. For instance, as already proven in [24], the Walsh spectrum of subfunctions of bent functions with respect to a subspace of dimension $n-1$ is equal to $\{0, \pm 2^{n/2+1}\}$ (plateaued function with amplitude $2^{n/2+1}$). More general, the Walsh spectrum of subfunctions of bent functions with respect to a subspace of dimension $n-k$ is equal to $\{0, \pm 2^{n/2+l}\}$ where $1 \le l \le k$.

By the property that $\sum_{w\in\mathbb{F}_2^n} W_f(w) = \pm 2^n$, we can conclude that if the support of the Walsh spectrum is contained in a subspace of dimension k then f admits an annihilator of degree k. Unfortunately, this observation will not lead to better results, since $\mathrm{wt}(W_f) \ge 2^{\deg(f)}$ as proven in [19].

6   Conclusion

In this paper we have discussed probabilistic versions of the algebraic attack on stream ciphers. By means of the Walsh spectrum, we have derived an exact formula for computing the distance to a function with annihilators of a certain degree k. From the distance, we have computed the probability in this new algebraic attack. In future work, we will concentrate on more efficient algorithms to solve a system of probabilistic nonlinear equations.

Acknowledgement

We thank Frederik Armknecht, Joseph Lano, Matthew Parker, and the anonymous referees for their useful comments and suggestions. This work was supported in part by the Concerted Research Action (GOA) Ambiorics 2005/11 of the Flemish Government and by the European Commission through the IST Programme under Contract IST2002507932 ECRYPT. An Braeken is an F.W.O. Research Assistant, sponsored by the Fund for Scientific Research - Flanders (Belgium).

References

1. D.H. Bailey, K. Lee, and H.D. Simon. Using Strassens algorithm to accelerate the solution of linear systems. Journal of Supercomputing, 4:357371, 1990.
2. A. Canteaut and P. Charpin. Decomposing bent functions. IEEE Transactions on Information Theory, IT-49(8):20042019, 2003.
3. A. Canteaut and M. Trabbia. Improved fast correlation attacks using parity-check equations of weight 4 and 5. In Advances in Cryptology EUROCRYPT 2000, volume 1807 of Lecture Notes in Computer Science, pages 573588. Bart Preneel, editor, Springer, 2000.
4. A. Canteaut and M. Videau. Degree of composition of highly nonlinear functions and applications to higher order differential cryptanalysis. In Advances in Cryptology EUROCRYPT 2002, volume 2332 of Lecture Notes in Computer Science, pages 518533. Lars R. Knudsen, editor, Springer, 2002.
5. N. Courtois and W. Meier. Algebraic attacks on stream ciphers with linear feedback. In Advances in Cryptology EUROCRYPT 2003, volume 2656 of Lecture Notes in Computer Science, pages 345359. Eli Biham, editor, Springer, 2003.
6. N.T. Courtois. Higher order correlation attacks, XL algorithm, and cryptanalysis of Toyocrypt. In International Conference on Information Security and Cryptology ICISC 2002, volume 2587 of Lecture Notes in Computer Science, pages 182199. Pil Joong Lee and Chae Hoon Lim, editors, Springer, 2002. ISBN 3-540-00716-4.
7. J. Daemen. Cipher and Hash Function Design. PhD thesis, Katholieke Universiteit Leuven, 1995.
8. J. Dillon. A survey of bent functions. Technical report, NSA Technical Journal, 1972. pp. 191215, unclassified.
9. Sylvie Dubuc. Etude des proprietes de degenerescene et de normalite des fonctions booleennes et construction des fonctions q-aires parfaitement non-lineaires. PhD thesis, Universite de Caen, 2001.
10. J.-C. Faug'ere, M. Sugita, M. Kawazoe, and H. Imai. Comparison between XL and Grobner basis algorithms. In Advances in Cryptology ASIACRYPT 2004, volume 3329 of Lecture Notes in Computer Science. Pil Jong Lee, editor, Springer, 2004. ISBN 3-540-00171-9.
11. Jean-Charles Faug'ere. A new efficient algorithm for computing Grobner bases without reduction to zero (F5). In International Symposium on Symbolic and Algebraic Computation ISSAC 2002, pages 7583. ACM Press, 2002.
12. X. Guo-Zhen and J. Massey. A spectral characterization of correlation-immune combining functions. IEEE Transactions on Information Theory, IT-34(3):596 571, 1988. 16
13. T. Johansson and F. Jonsson. Fast correlation attacks based on Turbo Code techniques. In Advances in Cryptology CRYPTO 1999, volume 1666 of Lecture Notes in Computer Science, pages 181197. Michael Wiener, editor, Springer, 1999.
14. A. Koholosha. Investigation in the Design and Analysis of Key-Stream Generators. PhD thesis, Technische Universiteit Eindhoven, 2003.

15. W. Meier, E. Pasalic, and C. Carlet. Algebraic attacks and decomposition of Boolean functions. In Advances in Cryptology EUROCRYPT 2004, volume 3027 of Lecture Notes in Computer Science, pages 474491. Christian Cachin and Jan Camenisch, editors, Springer, 2004.

16. W. Meier and O. Staffelbach. Fast correlation attacks on certain stream ciphers. Journal of Cryptology, 1(3):6786, 1992.

17. M.J. Mihaljevic and H. Imai. Cryptanalysis of Toyocrypt-HS1 stream cipher. IEICE Transactions on Fundamentals, E85-A:6673, 2002.

18. B. Preneel, W. Van Leekwijck, L. Van Linden, R. Govaerts, and J. Vandewalle. Propagation characteristics of Boolean functions. In Advances in Cryptology EUROCRYPT 1990, volume 473 of Lecture Notes in Computer Science, pages 161173. I.B. Damgard, editor, Springer, 1990.

19. M. Quisquater. Applications of Character Theory and The Moobius Inversion Principle to the Study of Cryptographic Properties of Boolean Functions. PhD thesis, Katholieke Universiteit Leuven, 2004.

20. T. Siegenthaler. Correlation-immunity of nonlinear combining functions for cryptographic applications. IEEE Transactions on Information Theory, IT-30(5):776 780, 1984.

21. T. Siegenthaler. Decrypting a class of stream ciphers using ciphertext only. IEEE Transactions on Computers, C-34(1):8185, January 1985.

22. L. Simpson, E. Dawson, J. Golic, and W. Millan. LILI keystream generator. In Selected Areas in Cryptography SAC 2000, volume 2012 of Lecture Notes in Computer Science, pages 248261. D.R. Stinson and S.E. Tavares, editors, Springer, 2001.

23. Y. Zheng and X.M. Zhang. Plateaud functions. In International Conference on Information Communication Security ICICS 1999, volume 1726 of Lecture Notes in Computer Science, pages 284300. Vijay Varadharajan and Yi Mu, editors, Springer, 1999.

24. Y. Zheng and X.M. Zhang. Relationships between bent functions and complementary plateaud functions. In International Conference on Information Security and Cryptology ICISC 1999, volume 1787 of Lecture Notes in Computer Science, pages 6075. JooSeok Song, editor, Springer, 1999.

Unconditionally Secure Information Authentication in Presence of Erasures

Goce Jakimoski[*]

Computer Science Department, 253 Love Bldg,
Florida State University, Tallahassee, FL 32306-4530, USA
`jakimosk@cs.fsu.edu`

Abstract. The traditional authentication model assumes that the data loss on the communication channel between the sender and the receiver is handled by mechanisms that should be studied separately. In this paper, we consider a more general setting where both unconditionally secure information authentication and loss-resilience are achieved at the same time via erasure-tolerant authentication codes (or η-codes), and we address some fundamental questions concerning erasure-tolerant authentication codes. Namely, we adapt several lower bounds on the probability of successful deception derived for the traditional authentication model to the setting that we consider here. We also analyze the distance properties of the η-codes and the security properties of the η-codes constructed by means of concatenation and composition. One interesting class of η-codes is the class of η-codes with minimal probabilities of successful impersonation and substitution. We show that all members of this class can be represented as a composition of an authentication code with minimal impersonation and substitution probabilities and an erasure-resilient code. Finally, we present some examples of η-code constructions.

Keywords: message authentication, authentication codes, erasure-resilient codes, erasure-tolerant authentication.

1 Introduction

1.1 Erasure-Tolerant Information Authentication

Authentication codes (or A-codes) are cryptographic primitives that are used to achieve one of the basic information security goals, information authentication, and they have been extensively studied in the past. Some of the results relevant to our discussion include the following. Simmons [18,19] developed the theory of unconditional authentication and derived some lower bounds on the deception probability. Stinson [20,21,22] (see also [9]) studied the properties of authentication codes that have minimum possible deception probabilities and minimum number of encoding rules. He characterized the existence of authentication codes

---

[*] This work has been suported in part by the National Science Foundation under grants CCR-0209092 and CCR-008588.

N.P. Smart (Ed.): Cryptography and Coding 2005, LNCS 3796, pp. 304–321, 2005.

that have the aforementioned desirable properties in terms of existence of orthogonal arrays and balanced incomplete block designs (BIBD). The problem of unconditionally secure multicast message authentication has been studied by Desmedt et al [7]. Carter and Wegman [5,24] introduced the notion of universal classes of hash function and proposed their use in authentication. The use of universal hashing to construct unconditionally secure authentication codes without secrecy has also been studied by Stinson [23] and by Bierbrauer et al [4]. Rogaway [17] showed that a complexity-theoretic secure MAC can be obtained from the unconditionally secure if a pseudo random function is used for key generation. Afanassiev et al [1] proposed an efficient procedure for polynomial evaluation that can be used to authenticate messages in about 7-13 instructions per word.

In the authentication model that is used to derive the aforementioned results, it is assumed that the messages arrive intact. That is, there is no data loss on the communication channel between the sender and the receiver. The mechanisms that provide such communication channel (e.g., retransmission, erasure-resilient codes [10,11,15], etc.). are considered an issue that should be studied separately. We consider a more general setting where the problems of authentication and data loss are studied together. Particularly, we allow parts of the message to be erased. In that case, the receiver should be able to verify the authenticity of a part (possibly the whole) of the plaintext. We refer to the codes that allow such verification as erasure-tolerant authentication codes (or η-codes).

1.2   Motivating Application

While most of the network applications are based on the client-server paradigm and make use of point-to-point packet delivery, many emerging applications are based on the group communications model. In particular, a packet delivery from one or more authorized sender(s) to a possibly large number of authorized receivers is required. One such class of application is the class of multicast stream applications like digital audio and video, data feeds (e.g., news feeds, stock market quotes), etc.

Gennaro and Rohatgi [8] have proposed a stream signing scheme based on a chain of one-time signatures. Several schemes have been proposed subsequently [2,6,14,3]. Most of the aforementioned schemes are founded on the following paradigm. The sender first commits to some key K_i by sending a commitment $H(K_i)$, then he sends the chunk of the stream M_i authenticated using the key K_i, and finally, after all recipients have received the authenticated chunk, he sends the key K_i.

When incorporated in the existing multicast stream authentication schemes, erasure-resilient authentication codes can be used both to achieve loss-resilient stream authentication and to improve the security and/or performance of the schemes that already handle packet loss due to the following:

– The main advantage of using computationally secure message authentication schemes over unconditionally secure message authentication is that a single key can be used to authenticate many messages. Since the keys in the multicast stream authentication paradigm described above are used only once, the use of η-codes will not reduce the efficiency of the schemes. On the other hand, η-codes provide unconditional security.

– If the commitment $H(K_i)$ is lost or the key K_i is lost, then the receiver cannot verify the authenticity of the chunk M_i. Since η-codes provide erasure tolerance, the schemes will work even in the case of lossy channels.

1.3 Our Contribution

We address several fundamental issues concerning the erasure-tolerant information authentication setting: lower bounds on the deception probability, distance properties, the structure of the η-codes that have the desirable properties of minimum impersonation and substitution probabilities, relations to authentication codes and erasure-resilient codes, and construction of η-codes.

2   The Setting

The authentication model that we are going to analyze is based on the model described in [18,19,12]. As it is case in the usual model, there are three participants: a *transmitter*, a *receiver* and an *adversary*. The transmitter wants to communicate the *source state* (or *plaintext* [1]) to the receiver using a public communication channel. We assume that all plaintexts are strings of length k whose letters are from some q-set Q. First, the plaintext is encrypted using some *encoding rule* (or *key*) into a q-ary string of length n. The derived *message* (or *ciphertext*) is sent through the public channel. The transmitter has a key source from which he obtains a key. Prior to any message being sent, the key is communicated to the receiver through a secure channel. The receiver uses the key to verify the validity of the received message. If at most t $(t < n)$ letters are missing from the original intact valid message and the position of the missing letters within the message is known, then the received message is still considered valid. In this case, the receiver accepts a plaintext that is derived from the original plaintext by erasing at most r $(r < k)$ letters. When r is zero (i.e., we can recover and verify the authenticity of the complete plaintext), we say that the code is *full-recovery*. If the received message is not derived from some intact valid message by erasing at most t letters, then the receiver does not accept a plaintext.

We denote by $\mathcal{S}$ the set of plaintexts (source states). Let $\mathcal{M}_0$ be the set of all possible messages derived by applying each encoding rule (key) to the source states and let $\mathcal{M}_i$ $(0 < i \leq t)$ be the set of all possible messages derived

---

[1] In general, η-codes can provide secrecy (privacy). Hence, we adhere to the terminology used by Simmons and use the terms plaintext (or source state), ciphertext (or message) and encryption. This can be slightly confusing, since most of the current works on MAC schemes use the term message for the source state, the term message authentication code only for codes without secrecy, and the term authenticated encryption for schemes that provide both authenticity and privacy.

by erasing i letters from the messages in $\mathcal{M}_0$. The set of all possible messages is $\mathcal{M} = \bigcup_{i=0}^{t} \mathcal{M}_i$. Finally, we will use $\mathcal{E}$ to denote the set of encoding rules, and $X, Y_0, \ldots, Y_t, Y$ and Z to denote random variables that take values from $\mathcal{S}, \mathcal{M}_0, \ldots \mathcal{M}_t, \mathcal{M}$ and $\mathcal{E}$ correspondingly. Note that the probability distribution of Y_0 is uniquely determined by the probability distributions of X, Z and the randomness used by the code. However, there are infinitely many possible probability distributions for each of the random variables $Y_1, \ldots, Y_t$ depending on how we erase the letters of the messages.

An η-code can be represented by an $|\mathcal{E}| \times |\mathcal{M}|$ *encoding matrix* A. The rows of the matrix are indexed by the encoding rules $e_z \in \mathcal{E}$ and the columns of the matrix are indexed by the messages $y \in \mathcal{M}$. The entries of the matrix are either empty or contain a string derived from a source state by erasing at most r (possibly 0) letters. We use the characteristic functions $\chi(y, z)$ to indicate whether an entry is empty or not. In particular, the function $\chi(y, z)$ is one if $A[e_z, y]$ is not empty (i.e., $y \in \mathcal{M}$ is a valid message when the key in use is z), and $\chi(y, z)$ is zero if $A[e_z, y]$ is empty (i.e., $y \in \mathcal{M}$ is not a valid message when the key in use is z). We will also use the characteristic function $\phi(y_1, y_2, z)$, which is one if both y_1 and y_2 are valid when encoding rule e_z is used, and zero otherwise.

The following restrictions are imposed on the encoding matrix in order to capture the important aspects of the setting that we have discussed above. Let the message $y \in \mathcal{M}_0$ be valid under the encoding rule e_z. Then, there is a plaintext $x \in \mathcal{S}$ such that $y = e_z(x)$. In that case, the entry $A[e_z, y]$ in the encoding matrix should be x. Furthermore, the transmitter should be able to send any plaintext to the receiver regardless of the encoding rule in use. Therefore, for each encoding rule e_z and each source state x, there is at least one message $y \in \mathcal{M}_0$ such that $y = e_z(x)$. If there is exactly one such message regardless of the plaintext and the encoding rule, then the code is deterministic. Otherwise, the code is randomized. If the message $y \in \mathcal{M}_0$ is valid under the encoding rule e_z, then any message y' derived from y by erasing at most t letters is also valid. The entry $A[e_z, y']$ should be derivable from the entry $A[e_z, y]$ by erasing at most r letters. Note that if we discard from the encoding matrix each column that does not correspond to an element in $\mathcal{M}_0$, then the resulting encoding matrix defines an authentication code. We will refer to this code as the *underlying authentication code*.

We assume that the secret key will be used only once (although this can be relaxed as in [1,24]) and we are considering only two types of threats: *impersonation* and *substitution*. In an impersonation attack, the adversary, based only on his knowledge of the authentication scheme, can send a fraudulent message to the receiver when in fact no message has yet been sent by the transmitter. In a substitution attack, the adversary can intercept one valid message and replace it with his fraudulent message. The *probability of successful impersonation, P_I*, is defined as the probability of success when the adversary employs optimum impersonation strategy. The *probability of successful substitution, P_S*, is defined as the probability of success when the adversary employs optimal substitution strategy. Finally, the adversary may be able to select whether to employ an impersonation or a substitution attack (a *deception attack*). The *probability of successful deception*, P_d, is the probability of success when an optimum deception strategy is employed.

Simmons [18] used a game-theoretic approach in which the transmitter can choose the encoding strategy (key statistics) to foil the type of attack that the adversary may apply. In this case, one can only assert that $P_d \geq max(P_I, P_S)$. Later, due to the Kerckhoff's assumption, authors (e.g., [12,22]) assume that the encoding strategy is fixed and known to all participants. In this case, which is assumed hereafter, it holds that $P_d = max(P_I, P_S)$.

A simple example of an encoding matrix for a binary η-code is given in Table 1. We use * to denote the missing letters. Assuming that the keys are equiprobable, the code has the following parameters: $k = 1, n = 3, t = 1, r = 0, S = \{0, 1\}, \mathcal{M}_0 = \{000, 011, 101\}, \mathcal{E} = \{e_1, e_2, e_3\}, P_S = 1/2, P_d = P_I = 2/3$. Note that the code is full recovery ($r = 0$), and that it is with secrecy since the adversary cannot determine the plaintext given a ciphertext. For example, assume that the adversary has intercepted the message 000. Since he doesn't know the key that is in use, he can not tell whether the plaintext is 0 or 1.

Table 1. An example of a binary η-code with secrecy

	000	00*	0*0	*00	011	01*	0*1	*11	101	10*	1*1	*01
e_1	0	0	0	0	1	1	1	1				
e_2	1	1	1	1					0	0	0	0
e_3					0	0	0	0	1	1	1	1

3   Lower Bounds on the Deception Probability

In this section, we are going to address the question of how much security we can achieve given some parameters of the η-code. The following theorem gives some lower bounds on the probability of successful deception for the model described in Section2.

Theorem 1 (Lower Bounds). *The following inequalities hold for the probability of successful deception:*

1. $P_d \geq \dfrac{\min\limits_{\mathcal{E}} |\mathcal{M}(e_z)|}{|\mathcal{M}|}$
2. $P_d \geq \dfrac{|\mathcal{S}|}{|\mathcal{M}_0|}$
3. $P_d \geq 2^{- \inf_{Y_t} I(Y_t, Z)} \geq \ldots \geq 2^{- \inf_{Y_1} I(Y_1, Z)} \geq 2^{-I(Y_0, Z)}$
4. $P_d \geq \dfrac{2^{\frac{1}{2}(I(Y_0, Z) - \inf_{Y_t} I(Y_t, Z))}}{\sqrt{|\mathcal{E}|}} \geq \dfrac{1}{\sqrt{|\mathcal{E}|}}$

where $|\mathcal{M}(e_z)|$ is the number of messages in $\mathcal{M}$ that are valid when the key z is used, and $\inf_{Y_i} I(Y_i, Z)$ denotes the infimum of the mutual information $I(Y_i, Z)$ over all possible probability distributions of Y_i.

Assume that the sender and the receiver have already agreed that they are going to use the encoding rule e_z. One possible impersonation strategy is to select uniformly at random a message from $\mathcal{M}$. The attack will succeed if the selected message is one of the messages that are valid under that key z. Since there are $|\mathcal{M}(e_z)|$ messages that are valid under the key z and the fraudulent message was selected uniformly at random, the probability of success given a key z is $|\mathcal{M}(e_z)|/|\mathcal{M}|$. Clearly, the probability of successful deception will be greater or equal than $\frac{\min_{\mathcal{E}} |\mathcal{M}(e_z)|}{|\mathcal{M}|}$, which is our first lower bound.

In Section 2, we mentioned that if we discard from the encoding matrix all columns that are not indexed by the messages in $\mathcal{M}_0$, then we will get an authentication code. It is obvious that an attack on the underlying authentication code is also an attack on the η-code. Hence, any lower bound on the probability of successful deception for authentication codes can be translated into a lower bound on the probability of success in a deception attack for η-codes. One such bound is given in Theorem 1(2) and it follows from Corollary 1 [18]. Corollary 1 [18] trivially follows from Theorem 1 [18] since the number of valid messages for a given key is at least $|\mathcal{S}|$ in the case of authentication codes. However, that is not the case for their counterparts. Namely, the bound of Theorem 1(2) is not a consequence of Theorem 1(1), that is $\frac{\min_{\mathcal{E}} |\mathcal{M}(e_z)|}{|\mathcal{M}|}$ is not always greater or equal than $\frac{|\mathcal{S}|}{|\mathcal{M}_0|}$. A counterexample is given in Table 2. The set of source states is $\mathcal{S} = \{00, 01, 11\}$, the set of encoding rules is $\mathcal{E} = \{e_1, e_2\}$, and the set of all possible intact messages is $\mathcal{M}_0 = \{000, 001, 011, 110\}$. It is not hard to verify that $\frac{\min_{\mathcal{E}} |\mathcal{M}(e_z)|}{|\mathcal{M}|} = \frac{10}{14} < \frac{|\mathcal{S}|}{|\mathcal{M}_0|} = \frac{3}{4}$.

Table 2. Counterexample for $\frac{\min_{\mathcal{E}} |\mathcal{M}(e_z)|}{|\mathcal{M}|} \geq \frac{|\mathcal{S}|}{|\mathcal{M}_0|}$

	000	00*	0*0	*00	001	0*1	*01	011	01*	*11	110	11*	1*0	*10
e_1	00	0*	00	00	01	*1	01	11	11	11				
e_2	00	00	00	00			01	01	01	11	11	11	11	

The third bound (Theorem 1(3)) is an erasure-tolerant analogue to the authentication channel capacity bound [18], and the final lower bound (Theorem 1(4)) is a bound on the security that we can achieve for a given key length (see Appendix A for proofs).

4 Some Properties and Constructions

In this section, we investigate some additional properties of η-codes.

4.1 Distance Properties

The Hamming distance[2] between the messages is not important in the traditional authentication model because it is assumed that the messages arrive intact. How-

---

[2] The number of letters that need to be changed in one message to obtain the other.

ever, the distances[3] between the messages of an η-code play a crucial role since they determine the erasure-tolerant aspects of the code. The following theorem describes several distance properties of the erasure-tolerant authentication codes.

Theorem 2 (Distance properties).

1. *If the distance $d(x_1, x_2)$ between two plaintexts x_1 and x_2 is greater than r, then the distance $d(y_1, y_2)$ between the corresponding ciphertexts $y_1 = e_z(x_1)$ and $y_2 = e_z(x_2)$ is greater than t.*
2. *If there is a code (set of codewords) $\zeta \subseteq S$ whose distance $d(\zeta)$ is greater than r, then, for any encoding rule e_z, there is a code $\varsigma \subseteq \mathcal{M}_0(e_z)$ such that $|\varsigma| \geq |\zeta|$ and $d(\varsigma) > t$, where $\mathcal{M}_0(e_z)$ is the set of valid messages in $\mathcal{M}_0$ when the encoding rule e_z is in use.*
3. *For each encoding rule e_z of a full-recovery η-code, there is a code $\varsigma \subseteq \mathcal{M}_0(e_z)$ such that $|\varsigma| \geq |S|$ and $d(\varsigma) > t$.*
4. *For each encoding rule e_z of a deterministic full-recovery η-code, the distance of the code $\mathcal{M}_0(e_z)$ is greater than t.*
5. *Let $p_s = \min_{y_1, y_2 \in \mathcal{M}_0} P(y_2 \text{ valid } | y_1 \text{ valid})$ be greater than zero and let the erasure-tolerant code be deterministic and full-recovery. Then, the distance between any two elements of $\mathcal{M}_0$ is greater than t.*

The proof is given in Appendix B.

4.2 Concatenation and Composition of η-Codes

In this subsection, we are going to consider two common construction techniques: concatenation and composition.

In the case of concatenation, the set of source states S of the new erasure-tolerant authentication code consists of all possible concatenations $x_1||x_2$, where $x_1 \in S'_0$ is a source state of the first code and $x_2 \in S''_0$ is a source state of the second code. To encode a source state $x = x_1||x_2$, we select two encoding rules $e_{z_1} \in \mathcal{E}'$ and $e_{z_2} \in \mathcal{E}''$, and compute the message y as $y = e_{z_1}(x_1)||e_{z_2}(x_2)$. It is not hard to verify that if we erase $t = t_1 + t_2 + 1$ letters from a message y, then we can "lose" at most $r = \max(r_1 + k_2, r_2 + k_1)$ letters of the corresponding source state x. Note that we allow more than t_1 (resp., t_2) letters to be erased from $e_{z_1}(x_1)$ (resp., $e_{z_2}(x_2)$). In that case, we check the validity of the second (resp., first) part of the message and discard the letters of the first (resp., second) part of the source state. For the composition construction, the set of source states S'' of the second code is equal to the set $\mathcal{M}'_0$ of all possible intact messages of the first code. The set of source states of the new code S is equal to the set of source states S' of the first code, and the set $\mathcal{M}_0$ of all possible intact messages is equal to the set $\mathcal{M}''_0$ of all intact messages of the second code. The message y corresponding to a given source state x is computed as $y = e_{z_2}(e_{z_1}(x))$, where the encoding rules $e_{z_1} \in \mathcal{E}'$ and $e_{z_2} \in \mathcal{E}''$ are chosen independently according to the corresponding probability distributions. We require r_2 to be less or equal

---

[3] Hereafter, when we say distance we mean Hamming distance.

than t_1, and if we erase at most $t = t_2$ letters from a message y, then we can "lose" at most $r = r_1$ letters of the corresponding source state x. Some relations between the impersonation and substitution probabilities of the new code and the impersonation and substitution probabilities of the component codes are provided below.

Theorem 3. *The following relations hold for the probability of successful deception of a concatenation and composition of η-codes:*

1. Concatenation: $P_d = \max(P_d', P_d'')$
2. Composition: $P_I \leq P_I' P_I''$, $P_S \leq \tilde{P}_S' P_S''$, *where*

$$\tilde{P}_S' = \frac{\max_{y',y'' \in \mathcal{M}'} P((y', y'') \text{ valid})}{\min_{y \in \mathcal{M}'} P(y \text{ valid})}.$$

The proof is given in Appendix C.

Note that authentication codes and erasure-resilient codes can also be used as component codes. Namely, authentication codes form a class of η-codes whose members provide authenticity, but no erasure-resilience ($t = 0$). The erasure-resilient codes on the other hand form a class of η-codes whose members provide erasure-resilience, but no authentication ($P_d = 1$).

Finally, we will consider the case when the probabilities $P((y', y'') \text{ valid})$ and $P(y \text{ valid})$ are uniformly distributed. In this case, the deception probability is characterized by the following corollary (see Appendix C for proof).

Corollary 1. *If the probabilities $P((y', y'') \text{ valid})$ and $P(y \text{ valid})$ $(y, y', y'' \in \mathcal{M}')$, are uniformly distributed, then the approximation $\tilde{P}_S'$ is equal to P_S', and we have*

$$P_d \leq P_d' P_d''.$$

4.3 Erasure-Tolerant Authentication Codes with Minimal Impersonation and Substitution Probabilities

Not all η-codes can be represented as a composition of an authentication code and an erasure-resilient code. However, the members of one interesting class of erasure-tolerant authentication codes can always be represented as a composition of an authentication code and an erasure-resilient code since the messages in $\mathcal{M}_0$ form a code whose distance is greater than t. This is the class of η-codes whose probabilities of successful impersonation and substitution are minimal.

Theorem 4. *An η-code without secrecy (resp., with secrecy) has probability of successful impersonation $P_I = \frac{|\mathcal{S}|}{|\mathcal{M}_0|} < 1$ and probability of successful substitution $P_S = \frac{|\mathcal{S}|}{|\mathcal{M}_0|}$ (resp., $P_S = \frac{|\mathcal{S}|-1}{|\mathcal{M}_0|-1}$) if and only if*

1. *$d(\mathcal{M}_0) > t$ and*
2. *the underlying authentication code is an authentication code without secrecy (resp., with secrecy) such that $P_{uI} = \frac{|\mathcal{S}|}{|\mathcal{M}_0|}$ and $P_{uS} = \frac{|\mathcal{S}|}{|\mathcal{M}_0|}$ (resp., $P_{uS} = \frac{|\mathcal{S}|-1}{|\mathcal{M}_0|-1}$).*

The proof is given in Appendix D. As we mentioned in the introduction, Stinson [20,21,22] has characterized the existence of authentication codes that have minimal impersonation and substitution probabilities in terms of existence of orthogonal arrays and balanced incomplete block designs (BIBD). Using these characterizations and Theorem 4, one can easily derive a relation between the existence of orthogonal arrays and BIBDs and the existence of η-codes with minimal P_I and P_S.

5   Examples

In this section, we give some examples of erasure-tolerant authentication codes.

5.1 η-Codes from Set Systems

A set system is a pair $(X, \mathcal{B})$ of a set $X = \{a_1, a_2, \ldots, a_k\}$ and a multiset $\mathcal{B}$ whose elements are subsets (or blocks) of X.

We can construct an η-code from a set system as follows. The set X will consist of the letters of the source state x. Then, we use an authentication code to compute an authentication tag for each block $B_i \in \mathcal{B}$. The message y is constructed by appending the authentication tags of the blocks to the source state x. Now, if some letter of the message is erased, and the erased letter does not belong to a block B_i or to the authentication tag of the block B_i, then we can still check the authenticity of the letters of the plaintext that belong to B_i.

One possible construction is from a complementary design of a covering. The set system $(X, \mathcal{B})$ is a (k, m, t)-covering design if all blocks are m-subsets of X, and any t-subset of X is contained in at least one block. Some efficient constructions of coverings can be found in [13,16]. The complementary set system of a set system $(X, \mathcal{B})$ is the set system $(X, \mathcal{B}^c)$, where $\mathcal{B}^c = \{X \backslash B_i | B_i \in \mathcal{B}\}$. It is not hard to prove the following property of the complementary design of a covering.

Lemma 1. Let $(X, \mathcal{B})$ be a (k, m, t)-covering design. Then, for the complementary design, we have

1. $|B_i| = k - m$ for all $B_i \in \mathcal{B}^c$
2. For any subset $F \subset X$ such that $|F| \le t$, there is a block $B_i \in \mathcal{B}^c$ such that $F \bigcap B_i = \emptyset$.

The following proposition trivially follows from the previous lemma.

Proposition 1. If at most t letters are erased from a message of an η-code derived from a complementary design of a (k, m, t)-covering, then we can verify the authenticity of at least $k - m$ letters of the plaintext ($r \le m$).

Now, we are going to consider a specific example. Assume that the source state is a sequence of v^2 packets arranged in a square matrix

$$P_{0,0}, \ldots, P_{0,v-1}, \ldots, P_{v-1,0}, \ldots, P_{v-1,v-1},$$

where each packet $P_{i,j}$ is a sequence of l letters. We divide the set of blocks $\mathcal{B}$ into two disjoint subsets $\mathcal{R}$ and $\mathcal{D}$. The blocks in $\mathcal{R}$ are constructed from the rows of the matrix

$$\mathcal{R} = \{R_i | R_i = \{P_{i,0}, \dots, P_{i,v-1}\}, 0 \leq i < v\}.$$

The blocks in $\mathcal{D}$ are "parallel" to the main diagonal

$$\mathcal{D} = \{D_i | D_i = \{P_{0,i}, P_{1,(i+1) \bmod v}, \dots, P_{v-1,(i+v-1) \bmod v}\}, 0 \leq i < v\}.$$

The set system consisting of the set of all packets $P_{i,j}$ and the set of all blocks R_i and D_i is a complementary design of a $(v^2, v(v-1), v-1)$-covering. That is, if at most $v - 1$ packets are lost, then, we can still verify the authenticity of at least one block (i.e., v packets). The set system also has the following convenient properties:

- If one packet is lost, then we can still verify the authenticity of all $v^2 - 1$ packets that are not lost.
- If two packets are lost, then we can still verify the authenticity of at least $v^2 - 4$ packets.

We will use the following multiple message authentication code (see [1]) for our example. Let $\mathbf{a}_1, \mathbf{a}_2, \dots, \mathbf{a}_v$ be a sequence of v messages. The authentication tag for a message $\mathbf{a}_i$ is computed as

$$h_{x,y,z_i}(\mathbf{a}_i) = y(a_{i,0} + a_{i,1}x + \dots + a_{i,l-1}x^{l-1}) + z_i = yf_{\mathbf{a}_i}(x) + z_i,$$

where $x, y, z_i, a_{i,0}, \dots, a_{i,l-1} \in \mathbf{F}_q$ (q a prime power). The key parts x and y remain unchanged for all messages in the sequence. Only the part z_i is refreshed for each message.

In our case, the messages in the sequence are derived from the blocks by concatenating the packets in each block. We denote by $f_{P_{i,j}}(x)$ the sum $p_0 + p_1x + \dots + p_{l-1}x^{l-1}$, where $p_0, \dots, p_l$ are the letters of the packet $P_{i,j}$. Similarly, $f_{R_i}(x)$ is defined as

$$f_{R_i}(x) = f_{P_{i,0}}(x) + x^l f_{P_{i,1}}(x) + \dots + x^{l(v-1)} f_{P_{i,v-1}}(x).$$

The definition of $f_{D_i}(x)$ is straightforward. The authentication tag for the block R_i (resp., D_i) is computed as $yf_{R_i}(x) + z_{R_i}$ (resp., $yf_{D_i}(x) + z_{D_i}$). The tag for a given block is sent along with the last packet of the block.

Note that although each packet is contained in two blocks, we don't have to process the packets twice. We evaluate the polynomials $f_{P_{i,j}}(x)$ only for one value of x, and then, we combine the computed values to derive the authentication tags. Table 3 compares the example presented here to the case when each packet is authenticated separately. The complexities are expressed in number of multiplications over the finite field when the Horner's procedure is used for polynomial evaluation. We can see that in the example presented here, the number of keys (the part that changes) is significantly smaller when the number of packets increases. The price we pay is slightly increased time complexity and smaller erasure-tolerance.

Table 3. Comparison of the code from a complementary design of a $(v^2, v(v-1), v-1)$-covering with the case when an authentication tag is computed for each packet separately

code	multiplicative complexity	number of keys	erasures tolerated
from $(v^2, v(v-1), v-1)$-covering	$v^2l + 2v + \log l - 1$	$2v$	up to $v-1$
each packet separately	v^2l	v^2	up to $v^2 - 1$

5.2 η-Codes from Reed-Solomon Codes

Reed-Solomon codes have already been used to construct authentication codes [4]. Here, we present a construction of η-codes based on Reed-Solomon codes.

Let $a_1, a_2, \ldots, a_k$ be the sequence of plaintext letters, where each a_i is an element of the Galois field F_q. The message is constructed by appending $n-k$ authentication tags $\tau_1, \tau_2, \ldots, \tau_{n-k}$ ($\tau_i \in F_q, 1 \leq i \leq n-k$) to the plaintext. The authentication tags are computed as

$$\tau_i = y_i + \sum_{j=1}^{k} a_j x_i^j$$

where x_i, y_i are randomly selected and secret (i.e., part of the key). In addition, we require that $x_{i_1} \neq x_{i_2}$ for $i_1 \neq i_2$.

Now, assume that t_a letters of the plaintext and t_τ authentication tags are lost during the transmission. Note that $t_a + t_\tau$ cannot be greater than t. The receiver can use t_a of the received authentication tags and the $k - t_a$ received letters of the plaintext to construct a system of linear equations that can be always solved for the unknown letters of the plaintext. Once the missing letters of the plaintext are recovered, the receiver can check the authenticity using the remaining $n - k - t_a - t_\tau$ authentication tags. Assuming that each letter of the message can be erased with same probability p, the probability that in a sequence of v messages, there is no forged message accepted as valid, is lower bounded by the product

$$\prod_{i=0}^{t}(1 - \frac{k^{n-k-i}}{q^{n-k-i}})^{v \times \binom{n}{i}p^i(1-p)^{n-i}}.$$

The authentication tags in the example have dual role. They can be used to verify the authenticity of the plaintext or to recover some lost letters of the plaintext. Therefore, the η-codes described above, offer more security than the codes constructed by composing an authentication code and an erasure-resilient code on one hand, and they are more resilient to erasures than authentication codes on the other hand. This is illustrated in Table 4. The first row corresponds to a code that is derived by composing an authentication code and an erasure-resilient code that can recover one erasure. The second row corresponds to a code of length $n = k + 2$ constructed as above. Since only one erasure is tolerated, the condition $x_{i_1} \neq x_{i_2}$ for $i_1 \neq i_2$ is not necessary, and the complexity can be reduced from $2k$ to k multiplications by using a multiple message authentication code as in the previous example. The final code is an ordinary authentication code. The letters of the message in the last case are considered to be elements of a Galois field F_{q^2}.

Table 4. Comparison of an η-code from RS code with the current practices

code	multiplicative complexity	non-deception probability	erasure tolerance
composition	k multipl. in F_q	$(1 - \frac{k}{q})^v$	1 erasure
our η-code	k multipl. in F_q	$(1 - \frac{k}{q})^{pv}(1 - \frac{k^2}{q^2})^{(1-p)v}$	1 erasure
A-code	$k/2$ multipl. in F_{q^2}	$(1 - \frac{k}{2q^2})^v$	0 erasures

6   Conclusion and Future Work

We studied a more general authentication model where both unconditionally secure information authentication and loss-resilience are achieved at the same time by means of erasure-tolerant authentication codes (or η-codes). We derived results concerning some fundamental problems in the generalized model, and presented examples of η-codes. Our future research will concentrate on providing new constructions of η-codes and applying η-codes in multicast stream authentication over lossy channels.

Acknowledgments

We thank Yvo Desmedt for the helpful discussions on this work.

References

1. V. Afanassiev, C. Gehrmann and B. Smeets, "Fast Message Authentication Using Efficient Polynomial Evaluation," Proceedings of Fast Software Encryption Workshop 1997, pp. 190-204.
2. R. Anderson, F. Bergadano, B. Crispo, J. Lee, C. Manifavas, and R. Needham, "A New Family of Authentication Protocols," ACM Operating Systems Review 32(4), pp. 9-20, 1998.
3. F. Bergadano, D. Cavagnino, B. Crispo, "Chained Stream Authentication," Proceeding of Selected Areas in Cryptography 2000, pp. 142-155.
4. J. Bierbrauer, T. Johansson, G. Kabatianskii and B. Smeets, "On Families of Hash Functions Via Geometric Codes and Concatenation," Proceedings of Crypto '93, pp. 331-342.
5. J.L. Carter and M.N. Wegman, "Universal Classes of Hash Functions," Journal of Computer and System Sciences, Vol. 18, pp. 143-154, 1979.

6. S. Cheung, "An Efficient Message Authentication Scheme for Link State Routing," Proceedings of the 13th Annual Computer Security Application Conference, 1997.

7. Y. Desmedt, Y. Frankel and M. Yung, "Multi-Receiver/Multi-Sender Network Security: Efficient Authenticated Multicast/Feedback," Proceeding of INFOCOM 1992, pp.2045-2054.

8. R. Gennaro and P. Rohatgi, "How to Sign Digital Streams," Proceedings of Crypto '97, pp. 180-197.

9. T. Johansson, G. Kabatianskii and B. Smeets, "On the Relation Between A-codes and Codes Correcting Independent Errors," Proceedings of Eurocrypt 1993, pp. 1-11.

10. M. Luby, M. Mitzenmacher, M. A. Shokrollahi, D. A. Spielman, and V. Stemann, "Practical Loss-Resilient Codes," Proceedings of the 29th Symp. on Theory of Computing, 1997, pp. 150-159.

11. M.Luby, "LT codes," The 43rd IEEE Symposium on Foundations of Computer Science, 2002.

12. J.L. Massey, "Contemporary Cryptology: An Introduction," in Contemporary Cryptology, The Science of Information Integrity, ed. G.J. Simmons, IEEE Press, New York, 1992.

13. W.H.Mills, "Covering design I: coverings by a small number of subsets," Ars Combin. 8, pp. 199-315, 1979.

14. A. Perrig, R. Canneti, J. D. Tygar and D. Song, "Efficient Authentication and Signing of Multicast Streams Over Lossy Channels," Proceedings of the IEEE Security and Privacy Symposium, 2000.

15. M. Rabin, "Efficient Dispersal of Information for Security, Load Balancing, and Fault Tolerance," J. ACM 36(2), pp.335-348.

16. R.Rees, D.R.Stinson, R.Wei and G.H.J. van Rees, "An application of covering designs: Determining the maximum consistent set of shares in a threshold scheme," Ars Combin. 531, pp. 225-237, 1999.

17. P.Rogaway, "Bucket hashing and its application to fast message authentication," Proceedings of Crypto '95, pp. 29-42.

18. G.J. Simmons, "Authentication Theory / Coding Theory," Proceedings of Crypto '84, pp. 411-432.

19. G.J. Simmons, "A Survey of Information Authentication," in Contemporary Cryptology, The Science of Information Integrity, ed. G.J. Simmons, IEEE Press, New York, 1992.

20. D.R. Stinson, "Some Constructions and Bounds for Authentication Codes," Journal of Cryptology 1 (1988), pp. 37-51.

21. D.R. Stinson, "The Combinatorics of Authentication and Secrecy Codes," Journal of Cryptology 2 (1990), pp. 23-49.

22. D.R. Stinson, "Combinatorial Characterizations of Authentication Codes," Proceedings of Crypto '91, pp.62-73.

23. D.R. Stinson, "Universal Hashing and Authentication Codes," Proceedings of Crypto '91, pp. 74-85.

24. M.N. Wegman and J.L. Carter, "New Hash Functions and Their Use in Authentication and Set Equality," Journal of Computer and System Sciences, Vol. 22, pp. 265-279, 1981.

A   Proof of Theorem 1

We will prove only the lower bounds 3 and 4. The lower bounds 1 and 2 trivially follow from the discussion in Section 3 and the lower bounds for authentication codes given in [18].

3. The proof that we are going to present is an adaptation of the Massey's short proof [12] of the authentication channel capacity bound [18]. The probability that a particular message $y \in \mathcal{M}$ will be a valid message is given by the following expression:

$$P(y \text{ valid}) = \sum_z \chi(y, z) P_Z(z).$$

The probability of successful impersonation is

$$P_I = \max_y P(y \text{ valid}),$$

that is the best impersonation attack is when the adversary selects the fraudulent message to be the message that will be valid with maximum probability. Assume that y is a message from $\mathcal{M}$ that is valid with maximum probability ($P(y \text{ valid}) = P_I$) and assume that $y \notin \mathcal{M}_t$. Let $\hat{y} \in \mathcal{M}_t$ be a message derived from y by erasing some of its letters. Note that $\hat{y}$ is valid whenever y is valid, or equivalently, $\chi(y, z) = 1$ implies $\chi(\hat{y}, z) = 1$. In that case, we have

$$P_I \geq P(\hat{y} \text{ valid}) = \sum_z \chi(\hat{y}, z) P_Z(z) \geq \sum_z \chi(y, z) P_Z(z) = P(y \text{ valid}) = P_I.$$

Obviously, the probability that message $\hat{y}$ will be a valid message is P_I, and one best impersonation strategy is to choose always $\hat{y}$ as a possible fraudulent message.

Let Y_t be a random variable that takes values from $\mathcal{M}_t$ in a following manner: we randomly discard t letters from a message that is computed by the receiver from a randomly selected plaintext (according to the plaintext probability distribution) by applying a randomly selected encoding rule (according to the key probability distribution). It is clear that

$$P_I = P(\hat{y} \text{ valid}) = P(\hat{y} \text{ valid}) \times \sum_y P_{Y_t}(y) = \sum_y P_{Y_t}(y) P(\hat{y} \text{ valid})$$

$$\geq \sum_y P_{Y_t}(y) P(y \text{ valid}).$$

Note that equality holds only when the probabilities of validity are equal for all messages in $\mathcal{M}_t$. By substituting $P(y \text{ valid})$, we get

$$P_I \geq \sum_{y,z} P_{Y_t}(y) P_Z(z) \chi(y, z).$$

The joint probability $P_{Y_tZ}(y,z)$ is greater than zero if and only if $P_Z(z) > 0$ and $\chi(y,z) = 1$. Therefore, the relation above can be rewritten as

$$P_I \geq E\left[\frac{P_{Y_t}(y)P_Z(z)}{P_{Y_tZ}(y,z)}\right].$$

Using Jensen's inequality[4], we get

$$\log P_I \geq \log E\left[\frac{P_{Y_t}(y)P_Z(z)}{P_{Y_tZ}(y,z)}\right] \geq E\left[\log\frac{P_{Y_t}(y)P_Z(z)}{P_{Y_tZ}(y,z)}\right]$$
$$= H(Y_tZ) - H(Y_t) - H(Z) = -I(Y_t, Z).$$

The lower bound $P_d \geq 2^{-\inf_{Y_t} I(Y_t,Z)}$ trivially follows since the previous inequality holds for any probability distribution of Y_t. Now, we only need to show that $\inf_{Y_i} I(Y_i, Z) \leq \inf_{Y_{i-1}} I(Y_{i-1}, Z)$. Given a random variable Y_{i-1} that takes values from $\mathcal{M}_{i-1}$, let us construct a random variable Y_i that takes values from $\mathcal{M}_i$ as follows. If $y_{i-1} \in \mathcal{M}_{i-1}$ is the message that the receiver is supposed to get, we erase the first non-erased letter in y_{i-1} to get y_i. It is obvious that $I(Y_i, Z) \leq I(Y_{i-1})$ since anything that we can learn about the key given y_i we can also learn given y_{i-1} (e.g., we can erase one letter from y_{i-1} and guess the value of the key). Hence, for every probability distribution of Y_{i-1}, there is probability distribution of Y_i such that $I(Y_i, Z) \leq I(Y_{i-1}, Z)$, and therefore, the inequality $\inf_{Y_i} I(Y_i, Z) \leq \inf_{Y_{i-1}} I(Y_{i-1}, Z)$ will always hold.

4. For the probability of successful substitution, it holds that $\log P_S \geq -H(Z|Y_0)$ (see Theorem 5 [18]). Now, we have

$$P_d^2 \geq P_I P_S \geq 2^{-\inf_{Y_t} I(Y_t,Z)}2^{-H(Z|Y_0)}$$
$$= 2^{I(Y_0,Z)-\inf_{Y_t} I(Y_t,Z)-H(Z)}$$
$$P_d \geq \frac{2^{\frac{1}{2}(I(Y_0,Z)-\inf_{Y_t} I(Y_t,Z))}}{2^{\frac{1}{2}H(Z)}}$$
$$\geq \frac{2^{\frac{1}{2}(I(Y_0,Z)-\inf_{Y_t} I(Y_t,Z))}}{\sqrt{|\mathcal{E}|}}$$
$$\geq \frac{1}{\sqrt{|\mathcal{E}|}} \tag{1}$$

B Proof of Theorem 2

1. Assume that $d(y_1, y_2) \leq t$. Let y be the message derived from y_1 (resp., y_2) by erasing the letters where y_1 and y_2 differ and let $x = e_z^{-1}(y)$ be the

---

[4] If $f(x)$ is a convex function on an interval (a,b), $x_1, x_2, \ldots, x_n$ are real numbers $a < x_i < b$, and $w_1, w_2, \ldots, w_n$ are positive numbers with $\sum w_i = 1$, then

$$f\left[\sum_{i=1}^{n} w_i x_i\right] \leq \sum_{i=1}^{n} w_i f(x_i).$$

damaged plaintext corresponding to y. The plaintext x should be derivable from both x_1 and x_2 by erasing at most r letters. Therefore, the distance $d(x_1, x_2)$ is not greater than r, which is in contradiction with our assumption that $d(x_1, x_2) > r$.

2. Let us consider the code ς constructed as follows. For each codeword x in ζ, we put a codeword $y = e_z(x)$ in ς. Distance property of Theorem 2.1 implies that the distance of the code ς is greater than t. Clearly, the number of codewords in ς is equal to the number of codewords in ζ, and the theorem follows.

3. The third property follows from the previous property and the following observation. In the case of a full-recovery code, we have $r = 0$ and the set of all possible source states $\mathcal{S}$ forms a code with distance greater than r.

4. If the η-code is deterministic, then there is exactly one valid message given a source state and an encoding rule. If in addition the code is full-recovery, then the distance between any two intact valid messages should be greater than t.

5. The fact that p_s is greater than zero implies that for any two distinct messages y_1 and y_2 in $\mathcal{M}_0$, there is an encoding rule e_z such that y_1 and y_2 are both valid when the encoding rule e_z is used. According to the previous property, the distance between y_1 and y_2 must be greater than t.

C   Proofs for Section 4.2

C.1   Proof of Theorem 3

1. Let us consider the message $y = y_1 \| y_2$, where $y_1 \in \mathcal{M}'$ and $y_2 \in \mathcal{M}''$. Since the encoding rules are chosen independently, we have

$$P(y \text{ valid}) = P(y_1 \text{ valid}) P(y_2 \text{ valid}).$$

In that case, the probability $P(y \text{ valid})$ is less or equal than both P'_I and P''_I. Assume now that y_2 is not in $\mathcal{M}''$. The verifier will check only the validity of y_1. In this case, the probability that y is valid is $P(y \text{ valid}) = P(y_1 \text{ valid})$. Clearly, the probability $P(y \text{ valid})$ is again less or equal than P'_I, but we can select y_1 so that $P(y \text{ valid}) = P'_I$. Similarly, if we select y_1 so that there are more than t_1 erasures in it, the verifier will check the validity of y_2 only. The probability $P(y \text{ valid})$ will be less or equal than P''_I, and we can select y_2 so that $P(y \text{ valid}) = P''_I$. Therefore, the probability of successful impersonation of the new code is maximum of the probabilities of successful impersonation of the component codes (i.e., $P_I = \max(P'_I, P''_I)$).

An analogous argument holds for the probability of successful substitution. Let $y = y_1 \| y_2$ be the message intercepted by the adversary and let $\hat{y} = \hat{y}_1 \| \hat{y}_2$ be its substitution. Let us consider the case when $P'_S \geq P''_S$. It is not hard to show that the best substitution strategy is to substitute only the first part of the message (i.e., $\hat{y}_2 = y_2$ or $\hat{y}_2$ is derived by erasing more than t_2 letters). The probability that $\hat{y}$ is valid is equal to the probability that $\hat{y}_1$ is valid, and the probability of successful substitution P_S is equal to P'_S. Similarly, in the case when $P'_S \leq P''_S$, we get that $P_S = P''_S$. Therefore, $P_S = \max(P'_S, P''_S)$ and $P_d = \max(P'_d, P''_d)$.

2. First, we will show that $P_I \leq P'_I P''_I$. Let $y \in \mathcal{M}$ be an arbitrary message. For the probability that y is valid, we have

$$P(y \text{ valid}) = \sum_{z_1 z_2} \chi''(y, z_2) \chi'(e_{z_2}^{-1}(y), z_1) P(z_2) P(z_1)$$

$$= \sum_{z_2} \chi''(y, z_2) P(z_2) \sum_{z_1} \chi'(e_{z_2}^{-1}(y), z_1) P(z_1)$$

$$\leq P'_I \sum_{z_2} \chi''(y, z_2) P(z_2)$$

$$\leq P'_I P''_I.$$

From the last inequality, it follows that $P_I = \max_{y \in \mathcal{M}} P(y \text{ valid}) \leq P'_I P''_I$.

Now, let us consider the conditional probability $P(y'' \text{ valid} \mid y' \text{ valid})$, where y' and y'' are two distinct messages from $\mathcal{M}$. We have

$$P(y'' \text{ valid} \mid y' \text{ valid}) = \frac{P((y', y'') \text{ valid})}{P(y' \text{ valid})}$$

$$= \frac{\sum_{z_1 z_2} \phi''(y', y'', z_2) \phi'(e_{z_2}^{-1}(y'), e_{z_2}^{-1}(y''), z_1) P(z_2) P(z_1)}{\sum_{z_1 z_2} \chi''(y', z_2) \chi'(e_{z_2}^{-1}(y'), z_1) P(z_2) P(z_1)}$$

$$\leq \frac{\max_{y'_1, y''_1 \in \mathcal{M}'} P((y'_1, y''_1) \text{ valid}) \times \sum_{z_2} \phi''(y', y'', z_2) P(z_2)}{\sum_{z_2} \chi''(y', z_2) P(z_2) \sum_{z_1} \chi'(e_{z_2}^{-1}(y'), z_1) P(z_1)}$$

$$\leq \frac{\max_{y'_1, y''_1 \in \mathcal{M}'} P((y'_1, y''_1) \text{ valid}) \times \sum_{z_2} \phi''(y', y'', z_2) P(z_2)}{\min_{y'_1 \in \mathcal{M}'} P(y'_1 \text{ valid}) \times \sum_{z_2} \chi''(y', z_2) P(z_2)}$$

$$\leq \tilde{P}'_S P''_S.$$

From the previous inequality, it is obvious that $P_S \leq \tilde{P}'_S P''_S$.

C.2   Proof of Corollary 1

If the probabilities $P((y', y'') \text{ valid})$ and $P(y \text{ valid})$ are uniformly distributed, then

$$\tilde{P}'_S = \frac{\max_{y', y'' \in \mathcal{M}'} P((y', y'') \text{ valid})}{\min_{y \in \mathcal{M}'} P(y \text{ valid})} = \frac{P((y_1, y_2) \text{ valid})}{P(y_1 \text{ valid})} = P(y_2 \text{ valid} \mid y_1 \text{ valid})$$

where $y_1, y_2 \in \mathcal{M}'$ are two arbitrary messages. On the other hand, for the probability of successful substitution we have

$$P'_S = \max_{y_1, y_2 \in \mathcal{M}'} \frac{P((y_1, y_2) \text{ valid})}{P(y_1 \text{ valid})} = P(y_2 \text{ valid} \mid y_1 \text{ valid}).$$

Hence, $\tilde{P}'_S$ is equal to P'_S, and $P_S \leq P'_S P''_S$. From the last inequality and Theorem 3(2), it follows that $P_d \leq P'_d P''_d$.

D   Proof of Theorem 4

It is not difficult to show that if $d(\mathcal{M}_0) > t$, then $P_I = P_{uI}$ and $P_S = P_{uS}$. Clearly, if the underlying authentication code is without secrecy (resp., with secrecy), then the erasure-resilient code is without secrecy (resp., with secrecy) also.

Now, suppose that we have an erasure-resilient code without secrecy (resp., with secrecy) such that $P_I = \frac{|\mathcal{S}|}{|\mathcal{M}_0|} < 1$ and $P_S = \frac{|\mathcal{S}|}{|\mathcal{M}_0|}$ (resp., $P_S = \frac{|\mathcal{S}|-1}{|\mathcal{M}_0|-1}$). Since $P_I \geq P_{uI} \geq \frac{|\mathcal{S}|}{|\mathcal{M}_0|}$ and $P_S \geq P_{uS} \geq \frac{|\mathcal{S}|}{|\mathcal{M}_0|}$ (resp., $P_S \geq P_{uS} \geq \frac{|\mathcal{S}|-1}{|\mathcal{M}_0|-1}$), we have $P_{uI} = \frac{|\mathcal{S}|}{|\mathcal{M}_0|}$ and $P_{uS} = \frac{|\mathcal{S}|}{|\mathcal{M}_0|}$ (resp., $P_{uS} = \frac{|\mathcal{S}|-1}{|\mathcal{M}_0|-1}$). Therefore, the underlying authentication code has same impersonation and substitution probabilities as the erasure-tolerant authentication code. Obviously, if the erasure-tolerant authentication code is without secrecy (resp., with secrecy), then the underlying authentication code is without secrecy (resp., with secrecy) also.

We need to show now that $d(\mathcal{M}_0)$ is greater than t. For the underlying authentication code it holds that the probability $P(y \text{ valid})$ is $\frac{|\mathcal{S}|}{|\mathcal{M}_0|}$ for all $y \in \mathcal{M}_0$, and the probability $P(y_2 \text{ valid} \mid y_1 \text{ valid})$ is $\frac{|\mathcal{S}|}{|\mathcal{M}_0|}$ (resp., $\frac{|\mathcal{S}|-1}{|\mathcal{M}_0|-1}$) for any two distinct y_1 and y_2 in $\mathcal{M}_0$. Now, assume that there are two messages y_1 and y_2 in $\mathcal{M}_0$ such that $d(y_1, y_2) \leq t$. Let y be the message in $\mathcal{M}$ derived by erasing all the letters in y_2 that differ from the corresponding letters in y_1. Since $P(y_2 \text{ valid} \mid y_1 \text{ valid})$ is less than 1, there is an encoding rule e_z such that y_1 is a valid message under the encoding rule e_z, but y_2 is not a valid message under the encoding rule e_z. Therefore, we have $P(y \text{ valid}) > P(y_2 \text{ valid}) = \frac{|\mathcal{S}|}{|\mathcal{M}_0|}$, which is in contradiction with our assumption that $P_I = \frac{|\mathcal{S}|}{|\mathcal{M}_0|}$.

Generalized Strong Extractors
and Deterministic Privacy Amplification[*]

Robert König and Ueli Maurer

Department of Computer Science,
Swiss Federal Institute of Technology (ETH), Zurich,
CH-8092 Zurich, Switzerland
{rkoenig, maurer}@inf.ethz.ch

Abstract. Extracting essentially uniform randomness from a somewhat random source X is a crucial operation in various applications, in partic- ular in cryptography where an adversary usually possesses some partial information about X. In this paper we formalize and study the most general form of extracting randomness in such a cryptographic setting. Our notion of strong extractors captures in particular the case where the catalyst randomness is neither uniform nor independent of the actual extractor input. This is for example important for privacy amplification, where a uniform cryptographic key is generated by Alice and Bob sharing some partially secret information X by exchanging a catalyst R over an insecure channel accessible to an adversary Eve. Here the authentication information for R creates, from Eve's viewpoint, a dependence between X and R. We provide explicit constructions for this setting based on strong blenders. In addition, we give strong deterministic randomness extractors for lists of random variables, where only an unknown subset of the variables is required to have some amount of min-entropy.

1 Introduction

1.1 Extracting Uniform Randomness

Extracting essentially uniform randomness from somewhat random information is an important operation arising in different settings, ranging from the deran- domization of probabilistic algorithms, the design of pseudo-random generators, to the generation of information-theoretically secure cryptographic keys.

One can distinguish different variations of this problem, depending on whether the randomness extraction is deterministic or makes use of some catalyst ran- domness, and whether or not the generated random string must be protected from an adversary with side information, including the catalyst (cryptographic vs. non-cryptographic case).

Non-cryptographic randomness extraction has been studied extensively. A deterministic randomness extraction function $f : \Omega \rightarrow \Omega'$ is characterized by

---

[*] This work was partially supported by the Swiss National Science Foundation, project No. 200020-103847/1.

N.P. Smart (Ed.): Cryptography and Coding 2005, LNCS 3796, pp. 322–339, 2005.
© Springer-Verlag Berlin Heidelberg 2005

the set $\mathcal{S}$ of random variables (often called a source) X for which it generates an essentially uniform output (e.g., has distance at most ε from the uniform distribution). Such a function is called an $(\mathcal{S}, \varepsilon)$-extractor[1] [Dod00]. The question of the existence of such extractors and the problem of finding explicit constructions has been considered for a large number of sources and remains an important research topic [TV00]. Examples include various kinds of "streaming" sources (e.g., [vN51, Eli72, Blu84, SV86, Vaz87b, Vaz87c]), which produce a sequence of symbols (as for example Markov sources), families consisting of pairs or tuples of independent weak random variables (e.g., [CG88, DO03, DEOR04]), families generated by samplers (e.g., [TV00]), and various kinds of bit-fixing and symbol-fixing sources (e.g., [CGH+85, CW89, KZ03, GRS04]).

The term "extractor" is generally used for the probabilistic non-cryptographic case. In this case, the extractor takes as a second input an independent uniform random string R, which can be seen as a catalyst. The source X from which randomness is extracted is usually characterized by a lower bound on the min-entropy. A (k, ε)-extractor [NZ96] extracts ε-close to uniform randomness under the sole condition that the min-entropy of X is at least k.

Such a (k, ε)-extractor is called *strong* if the output is ε-close to uniform even when R is taken as part of the output. This is useful in a setting where R is communicated over an (authenticated) channel accessible to an adversary who should still be completely ignorant about the generated string. This setting, usually referred to as *privacy amplification*, is discussed in the following section.

Note that the concept of an $(\mathcal{S}, \varepsilon)$-extractor is a strict generalization of the concept of a (k, ε)-extractor if one views the catalyst as part of the input to the (then deterministic) extractor. In the same sense, the $(\mathcal{S}, \varepsilon)$-strong extractors defined in this paper are a strict generalization of (k, ε)-strong extractors. The output of an $(\mathcal{S}, \varepsilon)$-extractor is required to be close to uniform even given some additional piece of information, which does not necessarily have to be part of the input, but is characterized by the family $\mathcal{S}$.

1.2 Privacy Amplification

Classical privacy amplification, introduced by Bennett, Brassard, and Robert [BBR88] (and further analysed in [BBCM95]), refers to the following setting. Two parties Alice and Bob are connected by an authenticated but otherwise insecure communication channel, and they share a random variable X about which Eve has partial information, modeled by a random variable Y known to her.[2] The random variable X could for instance be the result of a quantum cryptography protocol or some other protocol.

Alice and Bob's goal is to generate an almost uniform random string S about which Eve has essentially no information, i.e., which is essentially uniform from

---

[1] Note that the term extractor usually refers to the probabilistic case, which is generally attributed to Nisan and Zuckerman [NZ96], see below.

[2] The setting is described by the joint probability distribution P_{XY} or, more precisely, by a set of such distributions. Actually, in the literature X is usually assumed to be a uniformly distributed bitstring, but this restriction is not necessary.

her point of view and can thus be used as a cryptographic key. This is achieved by Alice choosing a random string R, sending it to Bob (and hence also to Eve), and Alice and Bob applying a strong extractor (with catalyst randomness R) to obtain S. This works if the min-entropy of X, when given $Y = y$, is sufficiently high, with overwhelming probability over the values y that Y can take on. As mentioned above, the extractor must be strong since S must be uniform even when conditioned on R.

1.3   Contributions of This Paper

This privacy amplification setting can be unsatisfactory for two different reasons. First, in a practical setting, where the goal is to make as conservative and realistic assumptions as possible, one might worry that the catalyst randomness generated by one of the parties is neither uniform nor fully independent of X. Therefore, a natural question to ask is whether privacy amplification is possible without catalyst randomness, i.e., by a deterministic function. This problem is formalized in Section 3, where our new notion of strong extractors is introduced. We also provide a definition of strong condensers, which only guarantee some amount of min-entropy in the output and show how these concepts are related to each other.

A non-uniform and dependent catalyst can be seen as a special case of the above when viewed as part of the input to the (then deterministic) privacy amplification with two input random variables. In Section 4 we show that the amount of extractable uniform randomness is determined essentially by the difference of the amount of min-entropy and the degree of dependence. These results give rise to new sources allowing for conventional deterministic randomness extraction, in particular for *dependent* pairs of weak random variables, thus relaxing the independence condition needed in the constructions of [CG88, DO03, DEOR04]. As an important example, the type of dependence considered includes the case of outputs generated by a (hidden) Markov model, thus generalizing for example [Blu84].

In Section 5 we present strong extractors (or, equivalently, deterministic privacy amplification) for a setting where Alice and Bob share a list of random variables, some unknown subset of which contains sufficient min-entropy, and where the adversary also knows some unknown subset of them. Note that the problem of constructing such extractors was recently considered by Lee, Lu, Tsai and Tzeng [LLTT05] along with a different cryptographic application in the context of key agreement for group communication. One of our constructions is very similar to theirs, though our analysis is different. We stress that the problem considered here is different from the problem of constructing extractors for several independent sources (each having a specific amount of min-entropy). Concerning the latter problem, there has recently been a significant breakthrough by Barak et al. [BIW04].

A second generalization of standard privacy amplification is to get rid of the need for an authenticated communication channel between Alice and Bob.[3] In this case, the shared random variable X must also be used to authenticate the catalyst R, in addition to being the input to the extraction procedure. This creates a dependence, from the adversary's viewpoint, between X and R, thus requiring the use of our generalized strong extractors. This setting has been considered before [MW97, DO03, RW03]. Our results lead to a more general and modular treatment with simpler proofs, but this application is not discussed here.

1.4   Outline

Section 2 introduces some basic concepts used throughout the paper. We then present our general definition of strong extractors and strong condensers in Section 3.1, and show how these primitives and some of their basic properties are related to privacy amplification. In Section 3.2, we discuss how our definition of a strong extractor generalizes various known definitions of randomness extractors. In Section 3.3 we establish a relation between strong extractors and strong condensers. In Section 4, we show how to construct $(\mathcal{S}, \varepsilon)$-strong extractors for a non-trivial family $\mathcal{S}$ which consists of dependent pairs of random variables. Finally, in Section 5, we show how to construct strong extractors for tuples of independent weak random variables with certain properties.

2   Preliminaries

For $n \in \mathbb{N}$ we denote by $[n]$ the set $\{1, \dots, n\}$. If $x = (x_1 \cdots x_n) \in \{0,1\}^n$ is a bitstring and $S \subset [n]$ a set of indices, we write $x|_S$ for the concatenation of the bits x_i with $i \in S$.

We will denote by $\mathcal{P}(\Omega)$ the set of probability distributions[4] on an alphabet Ω. Moreover, $\mathcal{P}(\Omega_1) \times \mathcal{P}(\Omega_2) \subset \mathcal{P}(\Omega_1 \times \Omega_2)$ will be the set of independent distributions on $\Omega_1 \times \Omega_2$. If $X_1, X_2 \in \mathcal{P}(\Omega)$, we write $P_{X_1} \equiv P_{X_2}$ if $P_{X_1}(x) = P_{X_2}(x)$ for all $x \in \Omega$. For $(X_1, X_2) \in \mathcal{P}(\Omega_1 \times \Omega_2)$, the distribution $X_1 \times X_2 \in \mathcal{P}(\Omega_1) \times \mathcal{P}(\Omega_2)$ is defined by $P_{X_1 \times X_2}(x_1, x_2) := P_{X_1}(x_1) P_{X_2}(x_2)$ for all $(x_1, x_2) \in \Omega_1 \times \Omega_2$.

The statistical distance between two distributions P and P' over the same alphabet Ω is defined as $d(P, P') := \frac{1}{2} \sum_{z \in \Omega} |P(z) - P'(z)|$. Note that the statistical distance satisfies

$$d(P_1 \times Q, P_2 \times Q) = d(P_1, P_2) \tag{1}$$

and is strongly convex, i.e.,

$$d\left(\sum_i \lambda_i P_i, \sum_i \lambda_i Q_i\right) \le \sum_i \lambda_i d(P_i, Q_i) \qquad \text{if } \lambda_i \ge 0 \text{ and } \sum_i \lambda_i = 1. \tag{2}$$

---

[3] Actually, in most realistic settings the channel is completely insecure and authenticity must be guaranteed by use of a pre-distributed short secret key.

[4] The terms random variable and probability distribution will be used interchangeably.

Let $U_\Omega \in \mathcal{P}(\Omega)$ denote a random variable with uniform distribution on Ω. A random variable $Z \in \mathcal{P}(\Omega)$ is ε-close to uniform if $d(Z, U_\Omega) \leq \varepsilon$.

For a set $\mathcal{S} \subset \mathcal{P}(\Omega)$ of probability distributions, let

$$\mathcal{B}^\varepsilon(\mathcal{S}) := \{X \in \mathcal{P}(\Omega) \mid \exists Y \in \mathcal{S} : d(X, Y) \leq \varepsilon\}$$

denote the distributions which are ε-close to some distribution in $\mathcal{S}$. We write $\overline{\mathcal{S}}$ for the convex hull of $\mathcal{S}$, i.e., the set of distributions which can be written as a convex combination of distributions in $\mathcal{S}$.

For $(X, Y) \in \mathcal{P}(\Omega_1 \times \Omega_2)$, we define the min-entropy of X and the conditional min-entropy of X given Y as follows[5]:

$$H_\infty(X) := -\log_2(\max_x P_X(x)) \qquad H_\infty(X|Y) := \min_y H_\infty(X|Y = y).$$

We call a random variable X for which only a lower bound on its min-entropy is known (i.e., $H_\infty(X) \geq k$ for some k) a *weak* random variable. Finally, we use[6]

$$H_0(X) := \log_2(|\operatorname{supp}(X)|)$$

to measure the size of the support $\operatorname{supp}(X) := \{x \in \Omega_1 \mid P_X(x) > 0\}$ of X.

We will use the following property of the statistical distance, which we state as a lemma.

Lemma 1. *Let $(S, Y) \in \mathcal{P}(\Omega_1 \times \Omega_2)$ be an arbitrary pair of random variables. Then there exists[7] a random variable S' which is uniformly distributed on Ω_1, independent of Y, and satisfies $\mathbb{P}[S = S'] \geq 1 - d((S, Y), U_{\Omega_1} \times Y)$.*

Proof. Let S_y be a random variable with distribution $P_{S_y} \equiv P_{S|Y=y}$ and let $d_y := d(S_y, U_{\Omega_1})$. We use the following well-known fact.

For an arbitrary random variable $T \in \mathcal{P}(\Omega)$, there exists a random variable T' defined by a channel[8] $P_{T'|T}$ with the property that T' is uniformly distributed on Ω and $\mathbb{P}[T = T'] = 1 - d(T, U_\Omega)$. Applying this to S_y, we conclude that there exists a random variable S'_y defined by a channel $P_{S'_y|S_y}$ such that

$$\mathbb{P}[S'_y = S_y] = 1 - d_y \qquad \text{and} \qquad P_{S'_y} = P_{U_{\Omega_1}}.$$

Let us define S' by the conditional distributions $P_{S'|Y=y,S=s} := P_{S'_y|S_y=s}$ for all $(s, y) \in \Omega_1 \times \Omega_2$. Then we obtain for all $(s', y) \in \Omega_1 \times \Omega_2$

$$P_{S'|Y=y}(s') = \sum_{s \in \Omega_1} P_{S|Y=y}(s) P_{S'|Y=y,S=s}(s') = \sum_{s \in \Omega_1} P_{S_y}(s) P_{S'_y|S_y=s}(s') = P_{S'_y}(s')$$

---

[5] $H_\infty(X|Y = y)$ is to be understood as the min-entropy of the conditional distribution $P_{X|Y=y}$.

[6] For $0 < \alpha < \infty$, $\alpha \neq 1$, the *Rényi entropy of order α* is defined as $H_\alpha(X) := \frac{1}{1-\alpha} \log_2\left(\sum_x P_X(x)^\alpha\right)$. The quantities $H_0(X)$ and $H_\infty(X)$ are obtained by taking the limits $\alpha \to 0$ and $\alpha \to \infty$, respectively.

[7] "exists" is to be interpreted as follows: One can define a new random experiment with random variables S, S', and Y such that P_{SY} is equal in both experiments and such that S' satisfies the stated conditions.

[8] This means that T and T' are jointly distributed according to $P_{TT'}(t, t') := P_T(t) P_{T'|T=t}(t')$.

which implies that S' is indeed uniform and independent of Y. Moreover, we have
$\mathbb{P}[S = S'|Y = y] = \sum_{s \in \Omega_1} P_{S|Y=y}(s)P_{S'|Y=y,S=s}(s) = \sum_{s \in \Omega_1} P_{S_y}(s)P_{S'_y|S_y=s}(s)$,
hence $\mathbb{P}[S = S'|Y = y] = \mathbb{P}[S_y = S'_y]$ and $\mathbb{P}[S = S'|Y = y] = 1 - d_y$. But
$\underset{y \leftarrow Y}{\mathbb{E}}[d_y] = d((S,Y), U_{\Omega_1} \times Y)$. The statement now follows from

$$\underset{y \leftarrow Y}{\mathbb{E}}[\mathbb{P}[S = S'|Y = y]] = \mathbb{P}[S = S'].$$

3   Strong Extraction for General Families of Random Variables

3.1   Basic Definitions and the Relation to Privacy Amplification

In the general setting of privacy amplification described in the introduction, the two parties (possibly after communicating first) have a shared random string X, whereas the adversary holds some information about X which is summarized by a random variable Y. It is important to note that Y does not necessarily have to be a part of X, but may depend in some other more intricate way on X. As an example, if Alice and Bob used X to authenticate some message M using a MAC, then Eve might learn $Y = (M, MAC_X(M))$. As a consequence, we may usually only assume that the pair (X, Y) has some specific structure (depending on the particular setting), i.e., it is contained in some family $\mathcal{S}$ of distributions. The question is then what Alice and Bob have to do in order to *deterministically* extract a key S from X which is uniform from the point of view of the adversary.

According to Lemma 1, if for the extracted key S, the quantity $d((S,Y), U_{\Omega'} \times Y)$ is small, then S is with high probability identical to a perfectly uniformly distributed "ideal" key which is independent of the part Y known to the adversary. This motivates the following general definition.

Definition 1. *Let $\mathcal{S} \subset \mathcal{P}(\Omega_1 \times \Omega_2)$ be a set of probability distributions on $\Omega_1 \times \Omega_2$. A function $\mathrm{Ext} : \Omega_1 \to \Omega'$ is an $(\mathcal{S}, \varepsilon)$-strong extractor if for every pair $(X, Y) \in \mathcal{S}$,*

$$d((\mathrm{Ext}(X), Y), U_{\Omega'} \times Y) \leq \varepsilon.$$

Using this new terminology, Alice and Bob simply have to apply an appropriate $(\mathcal{S}, \varepsilon)$-strong extractor in order to obtain the desired result. The following lemma describes some intuitive properties of strong extractors which follow directly from the definition and properties of the statistical distance.

Lemma 2. *An $(\mathcal{S}, \varepsilon)$-strong extractor is*

(i). *an $(\overline{\mathcal{S}}, \varepsilon)$-strong extractor.*
(ii). *a $(\mathfrak{B}^\delta(\mathcal{S}), \varepsilon + \delta)$-strong extractor for every $\delta \geq 0$.*
(iii). *an $(\mathcal{S}', \varepsilon)$-strong extractor for the family of distributions*

$$\mathcal{S}' := \{(X, (Y, Z)) \mid P_{(X,Y)|Z=z} \in \mathcal{S} \text{ for all } z \in \mathrm{supp}(Z)\} .$$

(iv). *an (S', ε)-strong extractor for the family of distributions*

$$S' := \{(X, Z) \mid (X, Y) \in S, X \leftrightarrow Y \leftrightarrow Z\},$$

where the notation $X \leftrightarrow Y \leftrightarrow Z$ means that X, Y and Z form a Markov chain.

Property (i) expresses the obvious fact that an (S, ε)-strong extractor also works on any convex combination of distributions in S. Property (ii) implies that in the context of privacy amplification, Alice and Bob can obtain an almost perfect secret key even if the initial situation is only close to a situation for which the extractor is appropriate. Property (iii) states that any additional piece of information Z does not help the adversary if conditioned on every value that Z can take on, the extracted key is close to uniform from the adversary's view. Finally, Property (iv) asserts that the extracted key still looks uniform to the adversary even if he processes his piece of information Y to obtain some different random variable Z.

As a weakening of the concept of extractors, it is natural to consider also the concept of condensers (see e.g., [RSW00]). In the privacy amplification setting, this corresponds to a situation where Alice and Bob would like to obtain a key which has a large amount of min-entropy from the point of view of the adversary. This may be used for example in an authentication protocol. We are thus led to the following analogous definition.

Definition 2. *Let $S \subset P(\Omega_1 \times \Omega_2)$ be a set of probability distributions on $\Omega_1 \times \Omega_2$. A function $\mathrm{Cond} : \Omega_1 \to \Omega'$ is an (S, k, ε)-strong condenser if for every $(X, Y) \in S$, there exists a random variable S such that*

$$d\big((\mathrm{Cond}(X), Y), (S, Y)\big) \leq \varepsilon \qquad and \quad H_\infty(S|Y) \geq k.$$

3.2   Relation to Known Definitions

In this section, we show that known definitions of extractors are in fact special instances of (S, ε)-strong extractors. Let us begin with the following general notion of deterministic[9] extractors, first introduced by Dodis.

Definition 3 ([Dod00]). *Let $S \subset P(\Omega)$ be a set of probability distributions on Ω. A function $\mathrm{Ext} : \Omega \to \Omega'$ is an (S, ε)-extractor if for every $X \in S$, $d(\mathrm{Ext}(X), U_{\Omega'}) \leq \varepsilon$.*

Obviously, such an extractor corresponds to an (S', ε)-extractor for the family $S' := \{(X, \bot) \mid X \in S\}$ where $\bot$ denotes a constant random variable. Note that an (S, ε)-strong extractor is an (S, ε)-extractor according to Definition 3, a fact which follows from Property (iv) of Lemma 2.

Our definition also generalizes the concept of strong (k, ε)-extractors, which are defined as follows.

---

[9] In this paper, we generally use the term *deterministic* to refer to procedures which (contrary to probabilistic ones) do not require a seed consisting of truly random bits. Note, however, that a probabilistic extractor can be seen as a deterministic one in the sense of Definition 3.

Definition 4 ([NZ96]). *A strong* (k, ε)-*extractor is a function* $\mathrm{Ext} : \{0,1\}^n \times \{0,1\}^d \to \{0,1\}^m$ *such that for every* $X \in \mathcal{P}(\{0,1\}^n)$ *with* $H_\infty(X) \geq k$,

$$d\big((\mathrm{Ext}(X,R),R),U_{\{0,1\}^m} \times R\big) \leq \varepsilon.$$

A strong (k, ε)-extractor is an $(\mathcal{S}', \varepsilon)$-extractor for the family of distributions $\mathcal{S}' \subset \mathcal{P}\big((\{0,1\}^n \times \{0,1\}^d) \times \{0,1\}^d\big)$ given by

$$\mathcal{S}' := \Big\{\big((X,R),R\big) \mid H_\infty(X) \geq k,\ R \text{ independent of } X \text{ and } P_R \equiv P_{U_{\{0,1\}^d}}\Big\}.$$

Similarly, our concept generalizes the so-called strong blenders introduced by Dodis and Oliveira in [DO03]. To describe this type of strong extractors, it is convenient to introduce the following families of distributions, which we also use in Sections 4 and 5.

Definition 5 ([CG88]). *The set* $\mathbf{CG}(\Omega_1,)[k_1, \Omega_2]k_2$ *of so-called* Chor-Goldreich-sources *is the set of all pairs* $(X_1, X_2) \in \mathcal{P}(\Omega_1) \times \mathcal{P}(\Omega_2)$ *of independent random variables such that* $H_\infty(X_1) \geq k_1$ *and* $H_\infty(X_2) \geq k_2$.

*   The set* $\mathbf{CG}(\Omega_1, \Omega_2)[k]$ *is the set of all pairs of independent random variables* $(X_1, X_2) \in \mathcal{P}(\Omega_1) \times \mathcal{P}(\Omega_2)$ *satisfying* $H_\infty(X_1 X_2) \geq k$. *Furthermore, we define* $\mathbf{CG}(\Omega)[k] := \mathbf{CG}(\Omega, \Omega)[k]$.

To simplify the notation, we will sometimes refer to the set $\{0,1\}^n$ simply by n in these two definitions. For example, we will write $\mathbf{CG}(n_1, n_2)[k_1, k_2]$ instead of $\mathbf{CG}(\{0,1\}^{n_1}, \{0,1\}^{n_2})[k_1, k_2]$.

Definition 6 ([DO03]). *A* (k_1, k_2, ε)-*strong blender is a function* $\mathrm{Ble} : \{0,1\}^{n_1} \times \{0,1\}^{n_2} \to \{0,1\}^m$ *such that*

$$d\big((\mathrm{Ble}(X,Y),Y),U_{\{0,1\}^m} \times Y\big) \leq \varepsilon$$

for all pairs $(X,Y) \in \mathbf{CG}(n_1, n_2)[k_1, k_2]$.

With our new notion of strong extraction, a (k_1, k_2, ε)-strong blender is an $(\mathcal{S}, \varepsilon)$-strong extractor for the special family of distributions

$$\mathcal{S} := \big\{\big((X,Y),Y\big) \mid (X,Y) \in \mathbf{CG}(n_1, n_2)[k_1, k_2]\big\}. \tag{3}$$

We will reconsider strong blenders in Section 4.1. Note that in the definition of the family $\mathcal{S}$, the random variables X and Y are independent. In Section 4.2, we show how to construct strong extractors even in the case where X and Y depend on each other.

As already mentioned, our definition of strong extractors also applies to the more general situation where the "public" information Y given to the adversary is not simply a part of X. In particular, this is the case when it is unavoidable to leak certain additional information about X, for instance if we would like to provide some error tolerance with respect to X. A general solution to this problem is accurately modeled by the concept of fuzzy extractors introduced by Dodis, Reyzin and Smith (see [DRS04] for details). It is easy to see that our definition of strong extractors also generalizes these fuzzy extractors.

3.3   Strong Condensers from Strong Extractors

Intuitively, in the setting of privacy amplification, if Alice and Bob derive a secret key S by applying a strong extractor, this key will still have a high amount of min-entropy from the point of view of the adversary even if the adversary is given a (short) additional piece of information. This means that an $(\mathcal{S}, \varepsilon)$-strong extractor is in fact a strong condenser for a different family $\mathcal{S}'$, which models the situation where the adversary gets additional information. This is formally expressed by Lemma 4.

The proof relies on the following technical result, which appears in a more general form in [MW97] and is implicitly used in [NZ96]. For an arbitrary pair $(X, Z) \in \mathcal{P}(\Omega_1 \times \Omega_2)$ of random variables and $\delta > 0$,

$$\mathop{\mathbb{P}}_{z \leftarrow Z}\left[H_\infty(X|Z = z) \geq H_\infty(X) - H_0(Z) - \log_2 \tfrac{1}{\delta}\right] \geq 1 - \delta . \qquad (4)$$

We reformulate this statement in a way which is more useful for our purpose.

Lemma 3. *Let $(S, Y, Z) \in \mathcal{P}(\Omega_1 \times \Omega_2 \times \Omega_3)$ be arbitrary random variables and let $\delta > 0$. Then there exists a random variable S' defined by a channel $P_{S'|YZ}$ such that*

$$H_\infty(S'|(Y, Z)) \geq H_\infty(S|Y) - H_0(Z) - \log_2 \tfrac{1}{\delta} \qquad and \qquad (5)$$
$$d((S, Y, Z), (S', Y, Z)) \leq \delta. \qquad (6)$$

Proof. Let us define the set

$$\Gamma_\delta := \{(y, z) \in \Omega_2 \times \Omega_3 \mid H_\infty(S|Y = y, Z = z) \geq H_\infty(S|Y) - H_0(Z) - \log_2 \tfrac{1}{\delta}\}.$$

Then by identity (4),

$$\mathop{\mathbb{P}}_{z \leftarrow Z|Y=y}\left[(y, z) \in \Gamma_\delta\right] \geq 1 - \delta . \qquad (7)$$

We define S' by

$$P_{S'|Y=y, Z=z} := \begin{cases} P_{S|Y=y, Z=z} & \text{if } (y, z) \in \Gamma_\delta \\ P_{U_{\Omega_1}} & \text{otherwise.} \end{cases}$$

Statement (5) is now a consequence of the definition of Γ_δ.

As the quantity $d(P_{S|Y=y, Z=z}, P_{S'|Y=y, Z=z})$ is at most 1 if $(y, z) \notin \Gamma_\delta$ and 0 otherwise, we conclude, using (7), that

$$\sum_{z \in \Omega_3} P_{Z|Y=y}(z) d(P_{S|Y=y, Z=z}, P_{S'|Y=y, Z=z}) \leq \mathop{\mathbb{P}}_{z \leftarrow Z|Y=y}\left[(y, z) \notin \Gamma_\delta\right] \leq \delta .$$

Statement (6) then follows from

$$d((S, Y, Z), (S', Y, Z)) = \mathop{\mathbb{E}}_{y \leftarrow Y}\left[\sum_{z \in \Omega_3} P_{Z|Y=y}(z) d(P_{S|Y=y, Z=z}, P_{S'|Y=y, Z=z})\right] .$$

Lemma 3 allows us to derive the main result of this section.

Lemma 4. *Let* $\mathrm{Ext} : \Omega_1 \rightarrow \{0,1\}^{n_0}$ *be an* $(\mathcal{S}, \varepsilon)$-*strong extractor and let* $\delta > 0$. *Then* Ext *is an* $(\mathcal{S}', k, \varepsilon + \delta)$-*strong condenser for the family*

$$\mathcal{S}' := \left\{ (X, (Y, Z)) \mid (X, Y) \in \mathcal{S} \text{ and } H_0(Z) \leq n_0 - k - \log_2 \tfrac{1}{\delta} \right\}.$$

Proof. Let $S := \mathrm{Ext}(X)$. Then by Lemma 1 there is a random variable S' which is uniformly distributed and independent of Y such that $\mathbb{P}[S = S'] \geq 1 - \varepsilon$. In particular, we have $H_\infty(S'|Y) = n_0$. Therefore, by Lemma 3, there is a random variable S'' such that

$$H_\infty(S''|(Y, Z)) \geq n_0 - H_0(Z) - \log_2 \tfrac{1}{\delta} \geq k$$

and

$$d\big((S', (Y, Z)), (S'', (Y, Z))\big) \leq \delta .$$

The statement now follows from the triangle inequality of the statistical distance and the fact that

$$d\big((S, (Y, Z)), (S', (Y, Z))\big) \leq \varepsilon$$

which holds because $\mathbb{P}[S = S'] \geq 1 - \varepsilon$.

This result is implicitly used for example in a protocol by Renner and Wolf [RW03] for privacy amplification over a non-authenticated channel. Without elaborating this any further, we point out that our generalized concepts of condensers and extractors allow to simplify existing security proofs such as the one given in [RW03].

4 Strong Extraction with a Weak and Dependent Catalyst

An important special case of our generalized notion of (deterministic) strong extractors is when the input can be seen as consisting of two parts, an actual input X_1 and a non-uniform and dependent catalyst X_2 which is also part of the output. In this section we introduce a dependence measure for such pairs (X_1, X_2) of random variables and show that the amount of uniform randomness extractable from (X_1, X_2) is determined essentially by the difference of the min-entropies of X_1 and X_2 and the level of dependence. In Section 4.1 we reformulate the definition of strong blenders and in Section 4.2 we state our main result concerning strong extraction from dependent variables. The dependence measure we consider is defined as follows.

Definition 7. *The set of* m-*independent distributions* $\mathcal{I}_m(\Omega_1, \Omega_2) \subset \mathcal{P}(\Omega_1 \times \Omega_2)$ *is the set of all pairs* (X_1, X_2) *of random variables on* $\Omega_1 \times \Omega_2$ *which can be written as a convex combination of* m *independent distributions, i.e.,*

$$\mathcal{I}_m(\Omega_1, \Omega_2) := \left\{ \sum_{i \in [m]} \lambda_i P_i \times Q_i \,\middle|\, \forall i \in [m] : \lambda_i \geq 0, P_i \in \mathcal{P}(\Omega_1), Q_i \in \mathcal{P}(\Omega_2) \right\}$$

The dependence index *of a pair of random variables* $(X_1, X_2) \in \mathcal{P}(\Omega_1 \times \Omega_2)$ *is defined as the quantity*

$$dep(X_1, X_2) := \log_2 \left(\min\{m \in \mathbb{N} \mid (X_1, X_2) \in \mathcal{I}_m(\Omega_1, \Omega_2)\} \right).$$

Obviously, $\mathcal{I}_m(\Omega_1, \Omega_2) \subset \mathcal{I}_{m'}(\Omega_1, \Omega_2)$ for $m < m'$ and $dep(X_1, X_2) = 0$ if and only if X_1 and X_2 are independent. Note that an m-independent distribution is for example obtained by observing the output of a (hidden) Markov source with at most m states at subsequent time steps.

4.1   Strong Blenders

Strong blenders can be used to perform privacy amplification when Alice and Bob have a pair (X, Y) of independent weak random variables and the adversary is given Y (compare e.g., [DO03]). To model this situation using strong extractors, it is convenient to use a "copying" operator[10] **cc** which transforms a pair of random variables (X, Y) into a pair $((X, Y), Y)$. This models the fact that the adversary is given Y.

Note that this operator has the following simple property which can be verified by direct calculation. If $\sum_i \mu_i = 1$ with $\mu_i \geq 0$ and $P_i \in \mathcal{P}(\Omega_1 \times \Omega_2)$ for all i, then

$$\mathbf{cc}\left(\sum_i \mu_i P_i\right) = \sum_i \mu_i \, \mathbf{cc}(P_i) \tag{8}$$

With this definition, the family of distributions $\mathbf{cc}(\mathbf{CG}(n_1, n_2)[k_1, k_2])$ is exactly the family given in equation (3). In other words, a (k_1, k_2, ε)-strong blender according to Definition 6 is a $(\mathbf{cc}(\mathbf{CG}(n_1, n_2)[k_1, k_2]), \varepsilon)$-strong extractor. In the sequel, we will use the terms strong blender and $(\mathbf{cc}(\mathbf{CG}(n_1, n_2)[k_1, k_2]), \varepsilon)$-strong extractor interchangeably, depending on whether or not we would like to refer to the parameters explicitly.

Recently, new results concerning extraction from independent weak sources were obtained by Barak et al. ([BIW04, BKS+05]) and Raz[11] [Raz05]. We use these extractors in Section 4.2.

4.2 m-Independence and Strong Extraction

The following lemma states that every pair (X_1, X_2) of random variables is close to a convex combination of independent random variables having some specific amount of min-entropy which depends on $dep(X_1, X_2)$. An analogous statement holds for the pair $((X_1, X_2), X_2)$.

---

[10] Formally, the copying operator $\mathbf{cc} : \mathcal{P}(\Omega_1 \times \Omega_2) \rightarrow \mathcal{P}((\Omega_1 \times \Omega_2) \times \Omega_2)$ is defined as follows. If $P_{X_1 X_2} \in \mathcal{P}(\Omega_1 \times \Omega_2)$ and $P := \mathbf{cc}(P_{X_1 X_2})$, then $P((x_1, x_2), x_3) := P_{X_1 X_2}(x_1, x_2) \cdot \delta_{x_2, x_3}$ for all $x_i \in \Omega_i$, $i = 1, \ldots, 3$, where δ_{x_2, x_3} denotes the Kronecker-delta, which equals 1 if $x_2 = x_3$ and 0 otherwise.

[11] In particular, Raz [Raz05] presents $(\mathbf{CG}(n, n)[k_1, k_2], \varepsilon)$-extractors for parameters $k_1 = (\frac{1}{2} + \delta)n$ and $k_2 = \Theta(\log n)$ where $\delta > 0$ is an arbitrarily small constant.

Lemma 5. *Let $(X_1, X_2) \in \mathcal{P}(\Omega_1 \times \Omega_2)$ and $\delta_1, \delta_2 > 0$ be arbitrary and define*

$$k_i = H_\infty(X_i) - dep(X_1, X_2) - \log_2 \tfrac{1}{\delta_i} \qquad for\ i = 1, 2.$$

Then we have

$$(X_1, X_2) \in \mathfrak{B}^{\delta_1 + \delta_2}\left(\overline{\mathbf{CG}(\Omega_1, \Omega_2)[k_1, k_2]}\right) \qquad and$$
$$\mathbf{cc}(X_1, X_2) \in \mathfrak{B}^{\delta_1 + \delta_2}\left(\overline{\mathbf{cc}(\mathbf{CG}(\Omega_1, \Omega_2)[k_1, k_2])}\right).$$

Proof. Let $m := 2^{dep(X_1, X_2)}$. Then there is a distribution $(X_1', X_2', Z) \in \mathcal{P}(\Omega_1 \times \Omega_2 \times [m])$ such that $P_{X_1 X_2} \equiv P_{X_1' X_2'} \equiv \sum_{z \in [m]} P_Z(z) P_{X_1'|Z=z} P_{X_2'|Z=z}$. For $i = 1, 2$, applying identity (4) to the pair (X_i', Z) shows that there are two subsets $\mathcal{A}_1, \mathcal{A}_2 \subseteq [m]$ and distributions $\{P_j^i\}_{j \in [m]} \subset \mathcal{P}(\Omega_i)$ for $i = 1, 2$ such that $P_{X_1 X_2}$ has the form $P_{X_1 X_2} \equiv \sum_{j \in [m]} \lambda_j P_j^1 \times P_j^2$, where for every $i = 1, 2$, $\sum_{j \in \mathcal{A}_i} \lambda_j \geq 1 - \delta_i$ as well as $H_\infty(P_j^i) \geq k_i$ for all $j \in \mathcal{A}_i$. In particular, we may write

$$P_{X_1 X_2} \equiv \sum_{j \in \mathcal{A}_1 \cap \mathcal{A}_2} \lambda_j Q_j + \left(1 - \sum_{j \in \mathcal{A}_1 \cap \mathcal{A}_2} \lambda_j\right) Q \tag{9}$$

for some distributions $\{Q_j\}_{j \in \mathcal{A}_1 \cap \mathcal{A}_2} \in \mathbf{CG}(\Omega_1, \Omega_2)[k_1, k_2]$ and a distribution $Q \in \mathcal{P}(\Omega_1 \times \Omega_2)$, where $(1 - \sum_{j \in \mathcal{A}_1 \cap \mathcal{A}_2} \lambda_j) \leq \delta_1 + \delta_2$. By the strong convexity (2) of the statistical distance, the first statement follows. The second statement follows similarly by application of **cc** to both sides of (9), using (8) and (2).

Using Lemma 5, Lemma 2 and an explicit construction by Raz [Raz05], we immediately obtain[12] a strong extractor with a weak and dependent catalyst. Theorem 1 gives the exact parameters. Intuitively, it expresses the fact that the amount of required min-entropy rises with the amount of dependence there is. In the setting of privacy amplification, this result states that there is an (explicit) deterministic function which Alice and Bob can use to extract a secret key S in a situation where they initially share a pair of weak random variables (X_1, X_2), and where X_2 is known to the adversary. We emphasize that contrary to the setting analysed in [DO03], X_2 need not be completely independent of X_1.

Theorem 1. *For any parameters n, m, κ_1, κ_2 and $0 < \delta < \tfrac{1}{2}$ satisfying*

$$\kappa_1 \geq \left(\frac{1}{2} + \delta\right) \cdot n + 4 \log_2 n$$
$$\kappa_2 \geq 5 \log_2(n - \kappa_1)$$
$$m \leq \delta \cdot \min\left\{\frac{n}{8}, \frac{\kappa_1}{40}\right\} - 1,$$

where n is sufficiently large, there exists an (explicit) $(\mathcal{S}_{\kappa_1, \kappa_2}^m, \varepsilon)$-strong extractor Ext $: \{0,1\}^n \times \{0,1\}^n \to \{0,1\}^m$ *for the family of distributions*

$$\mathcal{S}_{\kappa_1, \kappa_2}^m := \left\{ ((X_1, X_2), X_2) \mid H_\infty(X_i) - dep(X_1, X_2) \geq \kappa_i + \frac{3}{2}m \text{ for } i = 1, 2 \right\}.$$

---

[12] Note that our techniques also apply, e.g., to the constructions provided by Barak et al. ([BIW04, BKS+05]).

with error $\varepsilon := 3 \cdot 2^{-\frac{3}{2}m}$. *Moreover, this extractor is also strong in the first input, i.e., it is an* $(\widehat{\mathcal{S}^m_{\kappa_1,\kappa_2}}, \varepsilon)$-*strong extractor for the family of distributions*

$$\widehat{\mathcal{S}^m_{\kappa_1,\kappa_2}} := \left\{ \left((X_1, X_2), X_1\right) \mid \left((X_1, X_2), X_2\right) \in \mathcal{S}^m_{\kappa_1,\kappa_2} \right\}.$$

5   Strong Extractors for the Family $\mathcal{T}^N_\Omega(k)$

In this section, we consider the following family of random variables, which is somewhat related to symbol-fixing sources (see [CGH+85, CW89, KZ03, GRS04]) since the "positions" having "good" randomness are unknown.

Definition 8. *For an N-tuple of random variables $(X_1, \ldots, X_N) \in \mathcal{P}(\Omega^N)$ and a subset $A \subset [N]$, let $X|_A$ denote the concatenation of those random variables X_i with $i \in A$. The family $\mathcal{T}^N_\Omega(k)$ is the set of all pairs of the form $\left((X_1, \ldots, X_N), X|_A\right)$ where $(X_1, \ldots, X_N) \in \mathcal{P}(\Omega)^n$ are independent random variables and $A \subset [N]$ is such that there exists two distinct indices $i, j \in [N]$ with $j \notin A$ and $H_\infty(X_i X_j) \geq k$.*

In the privacy amplification setting, this corresponds to a situation where Alice and Bob have a sequence of independent random variables and the adversary obtains a subsequence. The only thing guaranteed is that two of these random variables (say X_i and X_j) have joint min-entropy at least[13] k and at least one of the two (say X_j) is unknown to the adversary. Note that extractors for this family have been used for other applications than privacy amplification [LLTT05].

We give two new constructions for strong extractors for the family $\mathcal{T}^n_\Omega(k)$. The first construction is based on special group-theoretic strong blenders. It is presented in Section 5.1. The second construction is very similar to a recent construction due to Lee, Lu, Tsai and Tzeng [LLTT05] and related to the construction of strong blenders presented in [DEOR04]. Our proof proceeds along the lines of similar derivations in [DO03] and [DEOR04].

5.1   Group-Theoretic Extractors for the Family $\mathcal{T}^N_\Omega(k)$

Theorem 2 below shows how a $(\mathcal{T}^N_\Omega(k), \varepsilon)$-strong extractor can be constructed in a generic way, using the following simple observation. We omit the trivial proof.

Lemma 6. *Let $(\mathcal{G}, +)$ be a group and let $X_1, X_2 \in \mathcal{P}(\mathcal{G})$ be two independent random variables defined on $\mathcal{G}$. Then $H_\infty(X_1 + X_2) \geq \max\{H_\infty(X_1), H_\infty(X_2)\}$.*

Intuitively, this lemma says that taking a random step according to X_2 on the Cayley graph defined by the group $\mathcal{G}$, starting from a random position defined

---

[13] Note that usually, we have $k > \log_2 |\Omega|$, implying that both X_i and X_j must have some amount of min-entropy individually. As pointed out in the introduction, this problem is different than the extraction problem considered, e.g., by Barak et al. [BIW04, BKS+05].

by X_1, we end up with an element that is at least as random as the variable which contains more randomness. This observation allows us to give a generic construction of a $(T_\Omega^N(k), \varepsilon)$-strong extractor, solving this problem in a (at least conceptually) similar manner as the way Kamp and Zuckerman [KZ03] treat the problem of randomness extraction from symbol-fixing sources.

Theorem 2. *Let $(\mathcal{G}, +)$ be an abelian group and let* $\text{Ext} : \mathcal{G} \times \mathcal{G} \to \Omega$ *be a* $(cc(\mathbf{CG}(\mathcal{G})[k]), \varepsilon)$-*strong extractor of the form* $\text{Ext}(x_1, x_2) := f(x_1 + x_2)$. *Then the function* $F(x_1, \ldots, x_N) := f(\sum_{i \in [N]} x_i)$ *is a* $(T_\mathcal{G}^N(k), \varepsilon)$-*strong extractor.*

Proof. Let $((X_1, \ldots, X_N), X|_A) \in T_\mathcal{G}^N(k)$. Suppose $i, j \in [N]$ are such that $i \in A$, $j \notin A$ and $H_\infty(X_i X_j) \geq k$. Then $F(X_1, \ldots, X_N) = \text{Ext}(X', Y')$ where $X' := \sum_{\ell \in [N] \setminus A} X_\ell$ is independent of $Y' := \sum_{\ell \in A} X_\ell$ and $H_\infty(X') + H_\infty(Y') \geq k$ by assumption and Lemma 6. Because $(X_1, \ldots, X_N) \leftrightarrow X|_A \leftrightarrow Y'$ is a Markov chain, the statement follows from Lemma (iv). The case where $i \notin A$ is treated similarly. $\square$

Using the following function $\text{Ext} : \mathbb{Z}_p \times \mathbb{Z}_p \to \mathbb{Z}_{n_0}$ (originally proposed in [CG88] and later shown to be a strong blender in [DO03])

$$\text{Ext}(x_1, x_2) := \begin{cases} \log_g(x_1 + x_2 \mod p) \mod n_0 & \text{if } x_1 + x_2 \neq 0 \mod p \\ 0 & \text{otherwise,} \end{cases}$$

where $p > 2$ is a prime, g a generator of $\mathbb{Z}_p^*$ and n_0 a divisor of $p - 1$, Theorem 2 immediately gives a $(T_{\mathbb{Z}_p}^n(\log_2 p + 2 \log_2(\frac{1}{\varepsilon}) + 2 \log_2 n_0), \varepsilon)$-strong extractor for every $\varepsilon \geq \frac{2}{p}$. Note that for appropriately chosen parameters p, g and n_0, the resulting extractor is efficiently computable (see [CG88] for details).

5.2   More Extractors for the Family $T_\Omega^N(k)$

It is easy to see that any $(cc(\mathbf{CG}(n)[k]), \varepsilon)$-strong extractor which is symmetric in its arguments is a $(T_{\{0,1\}^n}^2(k), \varepsilon)$-strong extractor. This, combined with the result by [DO03] that the inner product is a strong blender gives a very simple $(T_{\{0,1\}^n}^2(k), \varepsilon)$-strong extractor.

Lemma 7 ([DO03]). *The inner product modulo 2, denoted* $\langle \cdot, \cdot \rangle : \{0,1\}^n \times \{0,1\}^n \to \{0,1\}$, *is a* $(T_{\{0,1\}^n}^2(k), \varepsilon)$-*strong extractor, where* $\log \frac{1}{\varepsilon} = \frac{k-n}{2} + 1$.

A slight extension of this statement allows us to construct a $(T_{\{0,1\}^n}^2(k), \varepsilon)$-strong extractor which extracts just a single bit. The construction given here is more general than necessary (the parameters a, b, c could be omitted), but allows to prove Lemma 9 more easily.

Lemma 8. *Let M be an invertible $n \times n$ matrix over $GF(2)$ and let $a, b \in \{0,1\}^n$ and $c \in \{0,1\}$ be arbitrary. Define the function* $\text{Ext}_M^{a,b,c}(x_1, x_2) := \langle x_1, M x_2 \rangle + \langle a, x_1 \rangle + \langle b, x_2 \rangle + c$ *where addition is modulo 2. Then the function* $\text{Ext}_M^{a,b,c}$ *is a* $(T_{\{0,1\}^n}^2(k), \varepsilon)$-*extractor, where* $\log(\frac{1}{\varepsilon}) = \frac{k-n}{2} + 1$.

In the following proofs, we use the *non-uniformity* $\delta(Z) := d(Z, U_\Omega)$ to denote the distance of a distribution $Z \in \mathcal{P}(\Omega)$ from the uniform distribution on Ω.

Proof. We have to prove that

$$\mathop{\mathbb{E}}_{x_2 \leftarrow X_2} [\delta(\mathrm{Ext}_M^{a,b,c}(X_1, x_2))] \le \varepsilon \quad \text{and} \quad \mathop{\mathbb{E}}_{x_1 \leftarrow X_1} [\delta(\mathrm{Ext}_M^{a,b,c}(x_1, X_2))] \le \varepsilon \tag{10}$$

for all pairs $(X_1, X_2) \in \mathbf{CG}(k)[n]$ with k as specified. Since the operation of adding a constant is a bijection, we have for every fixed $x_2 \in \{0,1\}^n$

$$\delta(\mathrm{Ext}_M^{a,b,c}(X_1, x_2)) = \delta(\mathrm{Ext}_M^{a,b,0}(X_1, x_2)) = \delta(\mathrm{Ext}_M^{a,0,0}(X_1, x_2)) .$$

Hence, as $\mathrm{Ext}_M^{a,0,0}(x_1, x_2) = \langle x_1, M x_2\rangle + \langle a, x_1\rangle = \langle x_1, M x_2 + a\rangle$, we obtain

$$\mathop{\mathbb{E}}_{x_2 \leftarrow X_2} [\delta(\mathrm{Ext}_M^{a,b,c}(X_1, x_2))] = \mathop{\mathbb{E}}_{x_2 \leftarrow X_2} [\delta(\langle X_1, M x_2 + a\rangle)] .$$

Define the random variable X_2' by $X_2' := M X_2 + a$. Since the mapping $x_2 \mapsto M x_2 + a$ is a bijection, we have

$$\mathop{\mathbb{E}}_{x_2 \leftarrow X_2} [\delta(\langle X_1, M x_2 + a\rangle)] = \mathop{\mathbb{E}}_{x_2' \leftarrow X_2'} [\delta(\langle X_1, x_2'\rangle)] .$$

This combined with the fact that $H_\infty(X_2') = H_\infty(X_2)$ and Lemma 7 proves the first inequality in (10). The second inequality then follows from the first with the identity $\mathrm{Ext}_M^{a,b,c}(x_1, x_2) = \mathrm{Ext}_{M^T}^{b,a,c}(x_2, x_1)$ and the fact that the transpose M^T is invertible if M is invertible. $\quad\square$

Lemma 8 allows us to obtain a $(\mathcal{T}_{\{0,1\}^n}^N(k), \varepsilon)$-strong extractor for $N > 2$ which again extracts only a single bit.

Lemma 9. *Let M be an invertible $n \times n$-matrix over $GF(2)$ and let $\mathrm{Ext}_M : (\{0,1\}^n)^N \to \{0,1\}$ be the function $\mathrm{Ext}_M(x_1, \ldots, x_N) := \sum_{s<t} \langle x_s, M x_t\rangle$, where $M x_t$ is the matrix-vector multiplication over $GF(2)$ and addition is modulo 2. Then Ext_M is a $(\mathcal{T}_{\{0,1\}^n}^N(k), \varepsilon)$-strong extractor, where $\log(\frac{1}{\varepsilon}) = \frac{k-n}{2} + 1$.*

Proof. Suppose that $((X_1, \ldots, X_N), X|_A) \in \mathcal{T}_{\{0,1\}^n}^N(k)$ and let $i, j \in [N]$ be such that $i \in A$, $j \notin A$ and $H_\infty(X_i X_j) \ge k$. Without loss of generality, we may assume that $i < j$, since $((X_{\pi(1)}, \ldots, X_{\pi(N)}), X|_A) \in \mathcal{T}_{\{0,1\}^n}^N(k)$ for every permutation $\pi \in S_N$. A straightforward calculation shows (compare [LLTT05]) that we can write $\mathrm{Ext}_M(x_1, \ldots, x_N) = \mathrm{Ext}_M^{a,b,c}(x_i, x_j)$ with a, b, c depending only on the variables x_ℓ with $\ell \notin \{i, j\}$. [14] Hence we have

$$d((\mathrm{Ext}_M(X_1, \ldots, X_N), X|_A), (U_{\{0,1\}}, X|_A)) =$$
$$\mathop{\mathbb{E}}_{\tilde{x} \leftarrow X_{A \setminus \{i\}}} \mathop{\mathbb{E}}_{x_i \leftarrow X_i} [\delta(\mathrm{Ext}_M^{a(\tilde{x}), b(\tilde{x}), c(\tilde{x})}(x_i, X_j))]$$

---

[14] More precisely, a, b and c are given by the expressions

$$a := M \sum_{t>i, t \neq j} x_t + M^T \sum_{s<i} x_s,$$
$$b := M \sum_{t>j} x_t + M^T \sum_{s<j, s \neq i} x_s$$
$$c := \sum_{s<t \in [m] \setminus \{i,j\}} \langle x_s, M x_t\rangle$$

for appropriately defined functions a, b, c from $(\{0,1\}^n)^{|A|-1}$ to $\{0,1\}^n$ and $\{0,1\}$, respectively. The claim then follows from Lemma 8.

To get an extractor which produces several bits, we use the following reformulation of Vaziranis parity lemma [Vaz87a].

Lemma 10. *Let $\mathcal{S} \subset \mathcal{P}(\Omega_1 \times \Omega_2)$ and let $\mathrm{Ext} : \Omega_1 \to \{0,1\}^m$ be such that the function $x \mapsto \langle v, \mathrm{Ext}(x) \rangle$ is an $(\mathcal{S}, \varepsilon)$-strong extractor for every $v \in \{0,1\}^m \backslash \{0^m\}$. Then Ext is an $(\mathcal{S}, 2^m \cdot \varepsilon)$-strong extractor.*

Proof. We use the following so-called parity lemma by Vazirani [Vaz87a]. For every $A \in \mathcal{P}(\{0,1\}^m)$, the non-uniformity $\delta(A)$ of A is bounded as follows: $\delta(A) \leq \sum_{v \in \{0,1\}^m \backslash \{0^m\}} \delta(\langle v, A \rangle)$. It implies that for any pair of random variables $(A, B) \in \mathcal{P}(\{0,1\}^m \times \Omega_2)$,

$$d((A,B), (U_{\{0,1\}^m}, B)) = \mathop{\mathbb{E}}_{b \leftarrow B}[\delta(A|_{B=b})]$$

$$\leq \sum_{v \in \{0,1\}^m \backslash \{0^m\}} \mathop{\mathbb{E}}_{b \leftarrow B}[\delta(\langle v, A|_{B=b} \rangle)]$$

$$= \sum_{v \in \{0,1\}^m \backslash \{0^m\}} d((\langle v, A \rangle, B), (U_{\{0,1\}}, B)),$$

where the linearity of the expectation was used. Applying this to $(A, B) = (\mathrm{Ext}(X), Y)$ with $(X, Y) \in \mathcal{S}$ immediately yields the claim.

Finally, this implies the main result of this section. Note that the efficient construction of suitable matrices $M_1, \ldots, M_m$ in the following theorem is discussed in [DEOR04].

Theorem 3. *Let $M_1, \ldots, M_m$ be $n \times n$-matrices over $GF(2)$ such that for every non-empty subset $I \subset [m]$ the matrix $\sum_{i \in I} M_i$ is invertible. Then the function $\mathrm{Ext} : (\{0,1\}^n)^N \to \{0,1\}^m$ defined by*

$$\mathrm{Ext}(x_1, \ldots, x_n) = \left(\sum_{s<t} \langle x_s, M_1 x_t \rangle, \ldots, \sum_{s<t} \langle x_s, M_m x_t \rangle \right)$$

is a $(T_{\{0,1\}^n}^N(k), \varepsilon)$-strong extractor, where $\log(\frac{1}{\varepsilon}) = \frac{k-n}{2} - m + 1$.

Proof. Let $v \in \{0,1\}^m \backslash \{0^m\}$ and $I(v) := \{i \in [m] \mid v_i = 1\}$. Then

$$\langle v, \mathrm{Ext}(x_1, \ldots, x_n) \rangle = \sum_{s<t} \langle x_s, (\sum_{i \in I(v)} M_i) x_t \rangle = \mathrm{Ext}_{M(v)}(x_1, \ldots, x_n) ,$$

where $M(v) := \sum_{i \in I(v)} M_i$ is invertible by assumption and $\mathrm{Ext}_{M(v)}$ is defined as in Lemma 9. Hence, by Lemma 9, the function $(x_1, \ldots, x_n) \mapsto \langle v, \mathrm{Ext}(x_1, \ldots, x_n) \rangle$ is a $(T_{\{0,1\}^n}^N(k), \varepsilon)$-strong extractor for every $v \in \{0,1\}^m \backslash \{0^m\}$, where $\log(\frac{1}{\varepsilon}) = \frac{k-n}{2} + 1$. Lemma 10 then implies the desired result.

References

[ABH+86] M. Ajtai, L Babai, P. Hajnal, J. Komlos, P. Pudlak, V. Rodl, E. Sze-meredi, and G. Turan. Two lower bounds for branching programs. In *ACM Symposium on Theory of Computing*, pages 30–38, 1986.

[BBCM95] C. Bennett, G. Brassard, C. Crépeau, and U. Maurer. Generalized privacy amplification. *IEEE Transaction on Information Theory*, 41(6):1915–1923, November 1995.

[BBR88] C. Bennett, G. Brassard, and J. Robert. Privacy amplification by public discussion. *SIAM Journal on Computing*, 17(2):210–229, 1988.

[BIW04] B. Barak, R. Impagliazzo, and A. Wigderson. Extracting randomness from few independent sources. In *IEEE Symposium on Foundations of Computer Science (FOCS)*, 2004.

[BKS+05] B. Barak, G. Kindler, R. Shaltiel, B. Sudakov, and A. Wigderson. Simulating independence: New constructions of condensers, ramsey graphs, dispersers, and extractors. In *STOC '05: Proceedings of the thirty-seventh annual ACM symposium on Theory of computing*, pages 1–10, 2005.

[Blu84] M. Blum. Independent unbiased coin flips from a correlated biased source: a finite state markov chain. *IEEE Symposium on the Foundations of Computer Science*, 1984.

[CG88] B. Chor and O. Goldreich. Unbiased bits from sources of weak randomness and probabilistic communication complexity. *SIAM Journal On Computing*, 17(2):230–261, April 1988.

[CGH+85] B. Chor, O. Goldreich, J. Håstad, J. Freidmann, S. Rudich, and R. Smolensky. The bit extraction problem or t-resilient functions. In *IEEE Symposium on Foundations of Computer Science (FOCS)*, 1985.

[CW89] A. Cohen and A. Wigderson. Dispersers, deterministic amplification, and weak random sources (extended abstract). In *IEEE Symposium on Foundations of Computer Science (FOCS)*, pages 14–19, 1989.

[DEOR04] Y. Dodis, A. Elbaz, R. Oliveira, and R. Raz. Improved randomness extraction from two independent sources. *International Workshop on Randomization and Approximation Techniques in Computer Science (RANDOM)*, August 2004.

[DO03] Y. Dodis and R. Oliveira. On extracting private randomness over a public channel. *International Workshop on Randomization and Approximation Techniques in Computer Science (RANDOM)*, pages 143–154, August 2003.

[Dod00] Y. Dodis. *Exposure-Resilient Cryptography*. PhD thesis, Massachussetts Institute of Technology, August 2000.

[DRS04] Y. Dodis, L. Reyzin, and A. Smith. Fuzzy extractors: How to generate strong keys from biometrics and other noisy data. In *Advances in Cryptology — EUROCRYPT 2004*, volume 3027 of *Lecture Notes in Computer Science*, pages 523–539, May 2004.

[DSS01] Y. Dodis, A. Sahai, and A. Smith. On perfect and adaptive security in exposure-resilient cryptography. *Lecture Notes in Computer Science, EUROCRYPT '01*, 2045:301–324, 2001.

[Eli72] P. Elias. The efficient construction of an unbiased random sequence. *Annals of Mathematics Statistics*, 43(3):865–870, 1972.

[GRS04] A. Gabizon, R. Raz, and R. Shaltiel. Deterministic extractors for bit-fixing sources by obtaining an independent seed. In *IEEE Symposium on Foundations of Computer Science (FOCS)*, 2004.

[KZ03]   J. Kamp and D. Zuckerman. Deterministic extractors for bit-fixing sources and exposure-resilient cryptography. In *IEEE Symposium on Foundations of Computer Science*, 2003.

[LLTT05]   C.J. Lee, C.J. Lu, S.C. Tsai, and W.G. Tzeng. Extracting randomness from multiple independent sources. *IEEE Transaction on Information Theory*, 51(6):2224–2227, June 2005.

[MW97]   U. Maurer and S. Wolf. Privacy amplification secure against active adversaries. In *Advances in Cryptology — CRYPTO '97*, volume 1294 of *Lecture Notes in Computer Science*, pages 307–321, August 1997.

[NZ96]   N. Nisan and D. Zuckerman. Randomness is linear in space. *Journal of Computer and System Sciences*, 52(1):43–52, 1996.

[Raz05]   R. Raz. Extractors with weak random seeds. In *STOC '05: Proceedings of the thirty-seventh annual ACM symposium on Theory of computing*, pages 11–20, 2005.

[RSW00]   O. Reingold, R. Shaltiel, and A. Wigderson. Extracting randomness via repeated condensing. In *IEEE Symposium on Foundations of Computer Science (FOCS)*, pages 22–31, 2000.

[RW03]   R. Renner and S. Wolf. Unconditional authenticity and privacy from an arbitrarily weak secret. In *Advances in Cryptology — CRYPTO 2003*, volume 2729 of *Lecture Notes in Computer Science*, pages 78–95, August 2003.

[Sak96]   M. Saks. Randomization and derandomization in space-bounded computation. In *SCT: Annual Conference on Structure in Complexity Theory*, 1996.

[Sha02]   R. Shaltiel. Recent developments in explicit constructions of extractors. *Bulletin of the European Association for Theoretical Computer Science*, 77:67–95, June 2002.

[SV86]   M. Santha and U.V. Vazirani. Generating quasi-random sequences from slightly random sources. *Journal of Computer and System Sciences*, 33:75–87, 1986.

[TV00]   L. Trevisan and S. P. Vadhan. Extracting randomness from samplable distributions. In *IEEE Symposium on Foundations of Computer Science (FOCS)*, pages 32–42, 2000.

[Vaz87a]   U. Vazirani. Strong communcation complexity or generating quasi-random sequences from two communicating semi-random sources. *Combinatorica*, 7(4):375–392, 1987.

[Vaz87b]   U. V. Vazirani. Efficiency considerations in using semi-random sources. In *Proceedings of the nineteenth annual ACM conference on Theory of computing*, pages 160–168, 1987.

[Vaz87c]   U. V. Vazirani. Strong communication complexity or generating quasirandom sequences from two communicating semirandom sources. *Combinatorica*, 7(4):375–392, 1987.

[vN51]   J. von Neumann. Various techniques used in connection with random digits. *Applied Math Series*, 12:36–38, 1951.

[Zuc90]   D. Zuckerman. General weak random sources. In *IEEE Symposium on Foundations of Computer Science (FOCS)*, pages 534–543, 1990.

[Zuc91]   D. Zuckerman. Simulating BPP using a general weak random source. In *IEEE Symposium on Foundations of Computer Science (FOCS)*, pages 79–89, 1991.

On Threshold Self-healing Key Distribution Schemes⋆

Germán Sáez

Dept. Matemàtica Aplicada IV, Universitat Politècnica de Catalunya,
C. Jordi Girona, 1-3, Mòdul C3, Campus Nord, 08034-Barcelona, Spain
german@ma4.upc.edu

Abstract. Self-healing key distribution schemes enables a large and dynamic group of users to establish a group key over an unreliable network. A group manager broadcasts in every session some packet of information in order to provide a common key to members in the session group. The goal of self-healing key distribution schemes is that even if in a certain session the broadcast is lost, the group member can recover the key from the broadcast packet received before and after the session. This approach to key distribution is quite suitable for wireless networks, mobile wireless ad-hoc networks and in several Internet-related settings, where high security requirements need to be satisfied.

In this work we provide a generalization of previous definitions in two aspects. The first one is to consider general monotone decreasing structures for the family of subsets of users that can be revoked instead of a threshold one. The objective of this generalization is to reach more flexible performances of the scheme. In the second one, the distance between the broadcasts used to supply the lost one is limited in order to shorten the length of the broadcast information by the group manager. After giving the formal definition of threshold self-healing key distribution schemes, we find some lower bounds on the amount of information used for the system. We also give a general construction that gives us a family of threshold self-healing key distribution schemes by means of a linear secret sharing scheme. We prove the security of the schemes constructed in this way and we analyze the efficiency.

Keywords: Group key, self-healing, dynamic groups, linear secret sharing schemes, broadcast.

1 Introduction

Self-healing key distribution schemes enable large and dynamic groups of users of an *unreliable* network to establish group keys for secure communications.

⋆ This work was done while the author was in the *Dipartimento di Informatica ed Applicazioni* at the *Università di Salerno*, Italy. The author would like to thank people in the Crypto Research Group for their kind hospitality and useful comments. Research supported in part by Spanish *Ministerio de Ciencia y Tecnología* under project CICYT TIC 2003–00866.

N.P. Smart (Ed.): Cryptography and Coding 2005, LNCS 3796, pp. 340–354, 2005.

In a self healing key distribution scheme, a group manager provide a key to each member of the group by using packets that he sends over a broadcast channel at the beginning of each session. Every user on the group computes the group key by means of this packet and some private information supplied by the group manager. Multiple groups can be started by the group manager for different sessions by joining or removing users from the initial group. The main goal of these schemes is the self-healing property: if during a certain session some broadcasted packet gets lost, then users are still capable of recovering the group key for that session simply by using the packets they have received during a previous session and the packets they will receive at the beginning of a subsequent one, without requesting additional transmission from the group manager.

This new approach to key distribution is very useful due to the self-healing property, supporting secure communications in wireless networks, mobile wireless ad-hoc networks, broadcast communications over low-cost channels (live-events transmissions, etc.) and in several Internet-related settings.

Self-healing key distribution schemes were introduced by Staddon et al. in [10] providing formal definitions, lower bounds on the resources required on the scheme as well as some constructions. In [6], Liu et al. generalised the above definition and gave some constructions. Blundo et al. in [1] modified the proposed definitions, gave some lower bounds on the resources required on the scheme, proposed some efficient constructions and showed some problems in previous constructions. Finally, Blundo et al. in [2] analysed previous definitions and showed that no protocol could exist for some of them, proposed a new definition, gave some lower bounds on the resources of such schemes and proposed some schemes. All of these papers mainly focused in unconditionally secure schemes.

The contributions of our paper are the following. First of all we present some background on Information Theory and secret sharing schemes in Section 2 in order to make easy the reading of the paper. We formally define threshold self-healing key distribution schemes in Section 3. This definition contains two main differences comparing with the one presented in [1]. The first one is to consider a monotone decreasing family of rejected subset of users instead of a monotone decreasing threshold structure and the second one is the modification of the self-healing condition allowing the recuperation of keys from broadcasts at t (the threshold) units of distance. The first modification allows us to consider more flexible self-healing key distribution schemes that can reach better properties. The reason of the second modification is to allow short broadcasts. As far as we know this is the first time to consider this kind of self-healing key distribution schemes. In Section 4 some lower bounds on the resources required on the scheme are presented. Comparing with previous proposals, these lower bounds give us the same result for the amount of information that users must hold and shorter information that the group manager must broadcast in every session. After that, in Section 5 a family of threshold self-healing key distribution schemes is presented. This construction follows in part the ideas of [1] but considering any possible linear secret sharing scheme instead of a threshold one and a shorter broadcast to perform the threshold self-healing capability. At the end of the paper we present a particular case of our general construction in which we have a short broadcast for small revocations of users. This particular construction of a threshold self-healing key distribution scheme together with an example of broadcast for $t = 3$ is presented in Section 6.

2   Background on Information Theory and Secret Sharing Schemes

In this section we briefly introduce some basic notions of Information Theory and secret sharing schemes. We give an introduction to the entropy function as a tool to formally define self-healing key distribution scheme (see [4] for more details on Information Theory). We also give some basics on secret sharing schemes in order to use them in the paper (see [11] for a complete introduction to secret sharing schemes).

Let us suppose that $\mathbf{X}$ is a random variable taking values in a set X, and characterized by its probability distribution $\{P_{\mathbf{X}}(x)\}_{x \in X}$ that assigns to every $x \in X$ the probability $P_{\mathbf{X}}(x) = Pr(\mathbf{X} = x)$ of the event that $\mathbf{X}$ takes on the value x. The *entropy* of $\mathbf{X}$, denoted by $H(\mathbf{X})$, is a real number that measures the uncertainty about the value of $\mathbf{X}$ when the underlying random experiment is carried out. It is defined by

$$H(\mathbf{X}) = -\Sigma_{x \in X} P_{\mathbf{X}}(x) \log P_{\mathbf{X}}(x),$$

assuming that the terms of the form $0 \log 0$ are excluded from the summation, and where the logarithm is relative to the base 2. The entropy satisfies $0 \leq H(\mathbf{X}) \leq \log |X|$, where $H(\mathbf{X}) = 0$ if and only if there exists $x_0 \in \mathbf{X}$ such that $Pr(\mathbf{X} = x_0) = 1$; meanwhile $H(\mathbf{X}) = \log |X|$ if and only if $Pr(\mathbf{X} = x) = 1/|\mathbf{X}|$, for all $x \in \mathbf{X}$.

Given two random variables $\mathbf{X}$ and $\mathbf{Y}$, taking values on sets $\mathbf{X}$ and $\mathbf{Y}$, respectively, according to a joint probability distribution $\{P_{\mathbf{XY}}(x, y)\}_{x \in \mathbf{X}, y \in \mathbf{Y}}$ on their Cartesian product, the conditional uncertainty of $\mathbf{X}$, given the random variable $\mathbf{Y}$, called *conditional entropy* and denoted by $H(\mathbf{X}|\mathbf{Y})$, is defined as

$$H(\mathbf{X}|\mathbf{Y}) = -\Sigma_{y \in Y} \Sigma_{x \in X} P_{\mathbf{Y}}(y) P_{\mathbf{X}|\mathbf{Y}}(x|y) \log P_{\mathbf{X}|\mathbf{Y}}(x|y)$$

It can be showed that

$$0 \leq H(\mathbf{X}|\mathbf{Y}) \leq H(\mathbf{X}) \tag{1}$$

where $H(\mathbf{X}|\mathbf{Y}) = 0$ if and only if $\mathbf{X}$ is a function of $\mathbf{Y}$ and $H(\mathbf{X}|\mathbf{Y}) = H(\mathbf{X})$ if and only if $\mathbf{X}$ and $\mathbf{Y}$ are independent.

The *conditional mutual information* between $\mathbf{X}$ and $\mathbf{Y}$ given $\mathbf{Z}$ (where $\mathbf{X}$, $\mathbf{Y}$ and $\mathbf{Z}$ are random variables), is a measure of the amount of information by which the uncertainty about $\mathbf{X}$ is reduced by learning $\mathbf{Y}$, given $\mathbf{Z}$. It is defined by

$$I(\mathbf{X}; \mathbf{Y}|\mathbf{Z}) = H(\mathbf{X}|\mathbf{Z}) - H(\mathbf{X}|\mathbf{Z}, \mathbf{Y})$$

verifying $I(\mathbf{X}; \mathbf{Y}|\mathbf{Z}) = I(\mathbf{Y}; \mathbf{X}|\mathbf{Z})$ that is

$$I(\mathbf{X}; \mathbf{Y}|\mathbf{Z}) = H(\mathbf{X}|\mathbf{Z}) - H(\mathbf{X}|\mathbf{Z}, \mathbf{Y}) = H(\mathbf{Y}|\mathbf{Z}) - H(\mathbf{Y}|\mathbf{Z}, \mathbf{X}). \qquad (2)$$

A useful equality, widely applied in information-theoretic proofs, is given by the so-called *chain rule*. It is stated as follows: given n random variables, $\mathbf{X}_1, \ldots, \mathbf{X}_n$, the entropy of $\mathbf{X}_1, \ldots, \mathbf{X}_n$, can be written as

$$H(\mathbf{X}_1, \ldots, \mathbf{X}_n) = H(\mathbf{X}_1) + H(\mathbf{X}_2|\mathbf{X}_1) + \cdots + H(\mathbf{X}_n|\mathbf{X}_1, \ldots, \mathbf{X}_{n-1}).$$

A variant of the chain rule is the following

$$H(\mathbf{X}_1, \ldots, \mathbf{X}_n|\mathbf{Y}) = H(\mathbf{X}_1|\mathbf{Y}) + H(\mathbf{X}_2|\mathbf{Y}, \mathbf{X}_1) + \cdots + H(\mathbf{X}_n|\mathbf{Y}, \mathbf{X}_1, \ldots, \mathbf{X}_{n-1}) \qquad (3)$$

for a random variable $\mathbf{Y}$.

In this work we will use Lemma 5.1 and Lemma 5.3 presented in [1]:

Lemma 1. *(Lemma 5.1 in [1]) Let $\mathbf{X}, \mathbf{Y}$ and $\mathbf{W}$ be three random variables. If $H(\mathbf{X}|\mathbf{Y}, \mathbf{W}) = 0$ and $H(\mathbf{X}|\mathbf{W}) = H(\mathbf{X})$, then*

$$H(\mathbf{Y}) \geq H(\mathbf{X}).$$

Lemma 2. *(Lemma 5.3 in [1]) Let $\mathbf{X}, \mathbf{Y}$ and $\mathbf{W}$ be three random variables. If $H(\mathbf{Y}|\mathbf{W}) = 0$ then*

$$H(\mathbf{X}|\mathbf{Y}, \mathbf{W}) = H(\mathbf{X}|\mathbf{W}).$$

Another result needed in our work is the following Lemma:

Lemma 3. *Let $\mathbf{X}, \mathbf{Y}$ and $\mathbf{W}$ be three random variables. If $H(\mathbf{X}|\mathbf{W}) = 0$ then*

$$H(\mathbf{X}, \mathbf{Y}|\mathbf{W}) = H(\mathbf{Y}|\mathbf{W}).$$

Proof. We use the chain rule (3) to compute $H(\mathbf{X}, \mathbf{Y}|\mathbf{W}) = H(\mathbf{X}|\mathbf{W}) + H(\mathbf{Y}|\mathbf{W}, \mathbf{X})$. Using the hypothesis and Lemma 2 we obtain: $H(\mathbf{X}, \mathbf{Y}|\mathbf{W}) = 0 + H(\mathbf{Y}|\mathbf{W}) = H(\mathbf{Y}|\mathbf{W})$. $\qquad\square$

Secret sharing schemes play an important role in distributed cryptography. In these schemes, a secret value is shared among a set $\mathcal{U} = \{1, \ldots, n\}$ of n players in such a way that only qualified subsets of players can reconstruct the secret from their shares. The family of qualified subsets is the *access structure*, denoted by Γ. This family $\Gamma \subset 2^{\mathcal{U}}$ must be *monotone increasing*, that is, if $A_1 \in \Gamma$ and $A_1 \subset A_2 \subset \mathcal{U}$, then $A_2 \in \Gamma$. The family of authorized subsets Γ is determined by the collection of minimal authorized subsets Γ_0 called the *basis* of the structure. The family of non-authorized subsets $\overline{\Gamma} = 2^{\mathcal{U}} - \Gamma$ is monotone decreasing. An structure $\mathcal{R}$ is *monotone decreasing* when $A_1 \in \mathcal{R}$ and $A_2 \subset A_1$ imply $A_2 \in \mathcal{R}$. A monotone decreasing structure $\mathcal{R}$ is determined by the collection of maximal subsets $\mathcal{R}_0$.

Probably, the most used monotone access structures are $(T, n)-$*threshold* access structures, defined by $\Gamma = \{A \subset \mathcal{U} : |A| \geq T\}$, for some threshold T in a set $\mathcal{U}$ of n participants. *Shamir's secret sharing scheme* [8] realizes $(T, n)-$threshold access structures by means of polynomial interpolation.

To share a secret k in a finite field $GF(q)$ with $q > n$ a prime power, a special participant D outside $\mathcal{U}$ called dealer chooses a random polynomial $f(x) = k + a_1 x + \cdots + a_{T-1} x^{T-1} \in GF(q)[x]$ of degree $T - 1$ and sends to the participant i his secret share $s_i = f(x_i)$, where $x_1, \ldots, x_n \in GF(q)$ are distinct non-zero elements. Let A be a qualified subset of T participants who want to recover the secret k. They have T different values of the polynomial $f(z)$, of degree $T-1$, so they can obtain the value $k = f(0)$. We have $k = f(0) = \sum_{i \in A} \lambda_{0,i} f(x_i)$, where $\lambda_{0,i}$ are the Lagrange interpolation coefficients. It can also be proved that any subset of less than T participants cannot obtain any information about the secret from the shares they hold.

The *vector space secret sharing scheme* was introduced by Brickell [3]. Let us suppose that the dealer is D and that there is a public map

$$\psi : \mathcal{U} \cup \{D\} \longrightarrow GF(q)^\ell$$

where q is a prime power and ℓ is a positive integer. This map induces the monotone increasing access structure Γ defined as follows: $A \in \Gamma$ if and only if the vector $\psi(D)$ can be expressed as a linear combination of the vectors in the set $\psi(A) = \{\psi(i) \,|\, i \in A\}$. An access structure Γ is said to be a *vector space access structure* if it can be defined in the above way. If Γ is a vector space access structure, we can construct a secret sharing scheme for Γ with set of secrets $GF(q)$ (see [3] for a proof). To distribute a secret value $k \in GF(q)$, the dealer takes at random an element $v \in GF(q)^\ell$, such that $k = v \cdot \psi(D)$. The share of a participant $i \in \mathcal{U}$ is $s_i = v \cdot \psi(i)$. Let A be an authorized subset, $A \in \Gamma$; then, $\psi(D) = \sum_{i \in A} \lambda_i \psi(i)$, for some $\lambda_i \in GF(q)$. In order to recover the secret, players in A compute

$$\sum_{i \in A} \lambda_i s_i = \sum_{i \in A} \lambda_i v \cdot \psi(i) = v \cdot \sum_{i \in A} \lambda_i \psi(i) = v \cdot \psi(D) = k.$$

A scheme constructed in this way is called a *vector space secret sharing scheme*. Shamir's (T, n)−threshold scheme [8] can be seen as a vector space secret sharing scheme by choosing $\psi(D) = (1, 0, \ldots, 0) \in GF(q)^T$ as the vector of the dealer and $\psi(i) = (1, x_i, \ldots, x_i^{T-1}) \in GF(q)^T$ for $i \in \mathcal{U}$ (with $q > n$).

Vector space secret sharing schemes are a particular case of *linear secret sharing schemes*, which are essentially equal to vector space ones we have just explained, but where every participant can be associated with more than one vector. Simmons, Jackson and Martin [9] proved that any access structure Γ can be realized by a linear secret sharing scheme.

3   Threshold Self-healing Key Distribution Schemes

The model presented in [1] and the one given in [10] implement a self-healing key distribution scheme with good properties. The main drawback of these proposals of self-healing key distribution schemes, when a high number of sessions is considered, is that the amount of broadcast information is to large. A possibility to solve this fault is to use a different broadcast in order to decrease the number of broadcast bits, but with less features.

Let $\mathcal{U} = \{1, \ldots, n\}$ be the finite universe of users of a network. A broadcast unreliable channel is available, and time is defined by a global clock. Suppose that there is a group manager who sets up and manages, by means of join and revoke operations, a communication group, which is a dynamic subset of users of $\mathcal{U}$. Let $G_j \subset \mathcal{U}$ be the communication group established by the group manager in session j. Each user $i \in G_j$ holds a personal key S_i, received from the group manager before or when joining G_j. The personal key S_i can be seen as a sequence of elements from a finite set.

We denote the number of sessions supported by the scheme, by m, the set of users revoked by the group manager in session j by R_j, and the set of users who join the group in session j by J_j. Hence, $G_j = (G_{j-1} \cup J_j) - R_j$ for $j \geq 2$ and by definition $R_1 = \emptyset$. Moreover, for $j = 1, \ldots, m$, let K_j be the session key chosen by the group manager for session j. For each $i \in G_j$, the key K_j is determined by B_j and the personal key S_i.

Let $\mathbf{S}_i, \mathbf{B}_j, \mathbf{K}_j$ be random variables representing the personal key for user i, the broadcast message B_j and the session key K_j for session j, respectively. The probability distributions according to whom the above random variables take values are determined by the key distribution scheme and the random bits used by the group manager. In particular, we assume that session keys K_j are chosen independently and according to the uniform distribution.

Given a subset of users $G = \{i_1, \ldots, i_g\} \subset \mathcal{U}$, with $i_1 < \cdots < i_g$, we denote as $\mathbf{X}_G$ the random variables $\mathbf{X}_{i_1}, \ldots, \mathbf{X}_{i_g}$. For instance $\mathbf{S}_R$ denotes the personal keys of all users in $R \subset \mathcal{U}$.

We define $(t, \mathcal{R})$-threshold self-healing scheme as follows:

Definition 1. *Let $\mathcal{U}$ be the universe of users of a network, let m be the maximum number of sessions, and let $\mathcal{R} \subset 2^{\mathcal{U}}$ be a monotone decreasing access structure of subsets of users that can be revoked by the group manager. Let t be a threshold $t < m$. A $(t, \mathcal{R})$-threshold self-healing key distribution scheme is a protocol satisfying the following conditions:*

1. *The scheme is a session key distribution scheme, meaning that:*
   (a) *For each member $i \in G_j$, the key K_j is determined by B_j and S_i. Formally, it holds that:*
   $$H(\mathbf{K}_j | \mathbf{B}_j, \mathbf{S}_i) = 0.$$

   (b) *What users learn from the broadcast B_j and their own personal key cannot be determined from the broadcast or personal keys alone. That is:*
   $$H(\mathbf{K}_1, \ldots, \mathbf{K}_m | \mathbf{B}_1, \ldots, \mathbf{B}_m) = H(\mathbf{K}_1, \ldots, \mathbf{K}_m | \mathbf{S}_{G_1 \cup \cdots \cup G_m}) =$$
   $$= H(\mathbf{K}_1, \ldots, \mathbf{K}_m).$$

2. *The scheme has $\mathcal{R}$-revocation capability. That is, for each session j, if $R = R_j \cup R_{j-1} \cup \cdots \cup R_2$ is such that $R \in \mathcal{R}$, then the group manager can generate*

a broadcast message B_j such that all revoked users in R cannot recover K_j (even knowing all the information broadcast in sessions $1, \ldots, j$). In other words:

$$H(\mathbf{K}_j | \mathbf{B}_j, \mathbf{B}_{j-1}, \ldots, \mathbf{B}_1, \mathbf{S}_R) = H(\mathbf{K}_j).$$

3. The scheme is $(t, \mathcal{R})$-self-healing. This means that the two following properties are satisfied:

(a) Every $i \in G_r$, who has not been revoked after session r and before session s can recover all keys K_ℓ for $\ell = r, \ldots, s$, from broadcasts B_r and B_s, where $1 \leq r < s \leq m$ with $s - r \leq t$. Formally, it holds that:

$$H(\mathbf{K}_r, \ldots, \mathbf{K}_s | \mathbf{S}_i, \mathbf{B}_r, \mathbf{B}_s) = 0 \text{ if } s - r \leq t.$$

(b) Let $B \subset R_r \cup R_{r-1} \cup \cdots \cup R_2$ be a coalition of users removed from the group before session r and let $C \subset J_s \cup J_{s+1} \cup \cdots \cup J_m$ be a coalition of users who join the group from session s with $r < s$. Suppose $B \cup C \in \mathcal{R}$. Then, such a coalition does not get any information about keys K_j, for any $r \leq j < s$. That is:

$$H(\mathbf{K}_r, \ldots, \mathbf{K}_{s-1} | \mathbf{B}_1, \ldots, \mathbf{B}_m, \mathbf{S}_B, \mathbf{S}_C) = H(\mathbf{K}_r, \ldots, \mathbf{K}_{s-1}).$$

This definition has two differences with respect to the one presented in [1]. First the family of rejected subsets in [1] is $\mathcal{R} = \{R \subset \mathcal{U} : |R| \leq T\}$ while in our definition we consider the general case of any possible monotone decreasing structure $\mathcal{R}$, not only threshold ones. This allows us to consider more general self-healing key distribution schemes, where, for instance, some users or subsets of users can be more revocable than others. And the second one is that self-healing condition is only guaranteed for broadcasts at t units of distance.

4   Lower Bounds

In this section we present some bounds for a $(t, \mathcal{R})$-threshold self-healing key distribution scheme. The first one is a lower bound on the size of personal keys and the second one is a lower bound on the size of the broadcast, both compared to q, the size of the key space.

Theorem 1. In any $(t, \mathcal{R})$-threshold self-healing key distribution scheme with key space of size q, for any user i belonging to the group since session j, it holds that

$$H(\mathbf{S}_i) \geq (m - j + 1) \log q.$$

Proof. Using condition 3.(a) and some basic properties of the entropy we have that $H(\mathbf{K}_j, \ldots, \mathbf{K}_m | \mathbf{B}_j, \mathbf{B}_{j+t}, \mathbf{B}_{j+2t}, \ldots, \mathbf{B}_m, \mathbf{S}_i) = 0$. From condition 1.(b) we can derive $H(\mathbf{K}_j, \ldots, \mathbf{K}_m | \mathbf{B}_j, \mathbf{B}_{j+t}, \mathbf{B}_{j+2t}, \ldots, \mathbf{B}_m) = H(\mathbf{K}_j, \ldots, \mathbf{K}_m)$. Then Lemma 1 gives us the inequality $H(\mathbf{S}_i) \geq H(\mathbf{K}_j, \ldots, \mathbf{K}_m)$. So:

$$H(\mathbf{S}_i) \geq H(\mathbf{K}_j, \ldots, \mathbf{K}_m) = H(\mathbf{K}_j) + \cdots + H(\mathbf{K}_m) = (m - j + 1) \log q,$$

taking into account that keys $K_j, \ldots, K_m$ are chosen independently and uniformly at random. □

Recalling that $H(\mathbf{S}_i) \leq \log |S_i|$, we derive $\log |S_i| \geq (m - j + 1) \log q$ for $j = 1, \ldots, m$. Therefore, every user added in session j must store a personal key of at least $(m - j + 1) \log q$ bits.

Secondly we present a lower bound on the size of the broadcast.

Theorem 2. *In any $(t, \mathcal{R})$-threshold self-healing key distribution scheme with key space of size q and $m \geq 2t$, the following inequality holds*

$$H(\mathbf{B}_j) \geq t \log q.$$

For any $(t, \mathcal{R})$-threshold self-healing key distribution scheme with $t < m < 2t$ the inequality $H(\mathbf{B}_j) \geq \min(t, \max(j - 1, m - j)) \log q$ holds.

Proof. In order to prove the result we are going to consider two preliminary cases. First let us suppose that there exists a session r such that $1 \leq r < j$ with $j - r \leq t$ and let $i \in G_r$ be a user. Using properties (1) and (2) of the entropy function (see Section 2) we have

$$H(\mathbf{B}_j) \geq H(\mathbf{B}_j | \mathbf{S}_i, \mathbf{B}_r) \geq H(\mathbf{B}_j | \mathbf{S}_i, \mathbf{B}_r) - H(\mathbf{B}_j | \mathbf{S}_i, \mathbf{B}_r, \mathbf{K}_r, \ldots, \mathbf{K}_j) =$$

$$= H(\mathbf{K}_r, \ldots, \mathbf{K}_j | \mathbf{S}_i, \mathbf{B}_r) - H(\mathbf{K}_r, \ldots, \mathbf{K}_j | \mathbf{S}_i, \mathbf{B}_r, \mathbf{B}_j) = H(\mathbf{K}_r, \ldots, \mathbf{K}_j | \mathbf{S}_i, \mathbf{B}_r).$$

We have used that $H(\mathbf{K}_r, \ldots, \mathbf{K}_j | \mathbf{S}_i, \mathbf{B}_r, \mathbf{B}_j) = 0$ which derives from 3.(a) condition. On the other hand, we can apply condition 1.(a), Lemma 3 and condition 1.(b) to obtain:

$$H(\mathbf{B}_j) \geq H(\mathbf{K}_r, \ldots, \mathbf{K}_j | \mathbf{S}_i, \mathbf{B}_r) = H(\mathbf{K}_{r+1}, \ldots, \mathbf{K}_j | \mathbf{S}_i, \mathbf{B}_r) =$$

$$= H(\mathbf{K}_{r+1}, \ldots, \mathbf{K}_j) = H(\mathbf{K}_{r+1}) + \cdots + H(\mathbf{K}_j) = (j - r) \log q,$$

taking into account the way in which keys $K_{r+1}, \ldots, K_j$ have been chosen.

Now, we consider the following second case: suppose that there is a session s such that $j < s \leq m$ with $s - j \leq t$ and let $i \in G_j$ be a user. Using properties (1) and (2) of the entropy function (see Section 2) we have

$$H(\mathbf{B}_j) \geq H(\mathbf{B}_j | \mathbf{S}_i, \mathbf{B}_s) \geq H(\mathbf{B}_j | \mathbf{S}_i, \mathbf{B}_s) - H(\mathbf{B}_j | \mathbf{S}_i, \mathbf{B}_s, \mathbf{K}_j, \ldots, \mathbf{K}_s) =$$

$$= H(\mathbf{K}_j, \ldots, \mathbf{K}_s | \mathbf{S}_i, \mathbf{B}_s) - H(\mathbf{K}_j, \ldots, \mathbf{K}_s | \mathbf{S}_i, \mathbf{B}_j, \mathbf{B}_s) = H(\mathbf{K}_j, \ldots, \mathbf{K}_s | \mathbf{S}_i, \mathbf{B}_s).$$

We have used that $H(\mathbf{K}_j, \ldots, \mathbf{K}_s | \mathbf{S}_i, \mathbf{B}_j, \mathbf{B}_s) = 0$ which derives from 3.(a) condition. Using condition 1.(a) we can apply Lemma 3 and condition 1.(b):

$$H(\mathbf{B}_j) \geq H(\mathbf{K}_j, \ldots, \mathbf{K}_s | \mathbf{S}_i, \mathbf{B}_s) = H(\mathbf{K}_j, \ldots, \mathbf{K}_{s-1} | \mathbf{S}_i, \mathbf{B}_s) =$$

$$= H(\mathbf{K}_j, \ldots, \mathbf{K}_{s-1}) = H(\mathbf{K}_j) + \cdots + H(\mathbf{K}_{s-1}) = (s - j) \log q$$

for the same reason as above.

In order to prove the result for $m \geq 2t$ we distinguish between $j \geq t+1$ and $j \leq t$. If $j \geq t+1$ we apply the first case with $r = j-t$ obtaining $H(\mathbf{S}_i) \geq t \log q$. If $j \leq t$ we apply the second case with $s = j+t$ obtaining $H(\mathbf{S}_i) \geq t \log q$. The result for the case $t < m < 2t$ is obtained by applying the first case with $r = 1$ when $j \leq t$ or with $r = j-t$ when $j \geq t+1$ and the second case with $s = m$ when $j \geq m-t+1$ or with $s = j+t$ when $j \leq m-t$. □

So, every broadcast message is at least $t \log q$ bits long (when $m \geq 2t$).

5   A Family of Threshold Self-healing Key Distribution Schemes

Following the idea of Scheme 2 in [1] we can find a family of proposals using linear secret sharing schemes instead of Shamir secret sharing scheme and using shorter broadcast than the one in [1]. In this Section we present this construction, prove the security and analyze the efficiency.

We suppose that the set of users is $\mathcal{U} = \{1, \ldots, n\}$. The life of the scheme is divided in sessions $j = 1, \ldots, m$. The communication group in session j is denoted by $G_j \subset \mathcal{U}$. The subset of users revoked in session $j \geq 2$ is $R_j \subset G_{j-1}$ and the set of users who join the group is $J_j \subset \mathcal{U} - G_{j-1}$ with $R_j \cap J_j = \emptyset$. In this way $G_j = (G_{j-1} \cup J_j) - R_j$ for $j \geq 2$ and by definition $R_1 = \emptyset$. Let q be a prime power and denote by $K_j \in GF(q)$ the session key for group G_j.

Let $\mathcal{R} \subset 2^{\mathcal{U}}$ be a monotone decreasing access structure of subsets of users that can be revoked by the group manager and let $\Gamma = 2^{\mathcal{U}} - \mathcal{R}$ be a monotone increasing access structure. Let us consider a linear secret sharing scheme realizing Γ over the set $\mathcal{U}$. For simplicity, we suppose that there exists a public map

$$\psi : \mathcal{U} \cup \{D\} \longrightarrow GF(q)^{\ell}$$

which defines Γ as a vector space access structure. But the construction that we present here can be extended in a natural way to work with a linear secret sharing scheme in which a participant is associated with more than one vector. The use of a specific ψ fixes the properties of the scheme.

Now we describe the different phases of the $(t, \mathcal{R})$-threshold self-healing key distribution scheme.

Set-up. Let $G_1 \subset \mathcal{U}$. The group manager chooses random vectors $u_1, \ldots, u_m \in GF(q)^{\ell}$ and session keys $K_1, \ldots, K_m \in GF(q)$. For each $j = 1, \ldots, m$ the group manager computes the scalar $z_j = K_j + \psi(D)^{\top} u_j \in GF(q)$. The group manager sends privately to user $i \in G_1$ the personal key $S_i = (\psi(i)^{\top} u_1, \ldots, \psi(i)^{\top} u_m) \in GF(q)^m$. Note that if we use a linear secret sharing scheme in which a user i is associated with $m_i \geq 1$ vectors, then his secret information S_i consists of $m m_i$ elements in $GF(q)$.

Full addition. In order to add users $J_j \subset \mathcal{U}$ in session j, the group manager sends privately $S_i = (\psi(i)^{\top} u_j, \psi(i)^{\top} u_{j+1}, \ldots, \psi(i)^{\top} u_m) \in GF(q)^{m-j+1}$ to every user $i \in J_j$ as his personal key.

Broadcast. Suppose $R_j \subset G_{j-1}$ with $R_1 \cup R_2 \cup \cdots \cup R_j \in \mathcal{R}$ if $j \geq 2$. By definition we have $R_1 = \emptyset$. The group manager chooses a maximal non-authorized subset of users $W_j \in \mathcal{R}_0 = \overline{\Gamma}_0$ such that $R_1 \cup R_2 \cup \cdots \cup R_j \subset W_j$ and $W_j \cap G_j = \emptyset$ with minimum cardinality. The broadcast B_j in session $j = 1, \ldots, m$ is given by $B_j = B_j^1 \cup B_j^2$. The first part of the broadcast is defined as follows: let us suppose that binary representation of z_j has an even number of bits, say 2ω, in such a way that $z_j = (x_j, y_j)$ where $x_j, y_j \in GF(2^\omega)$ are $\omega = \lceil \frac{1}{2} \log q \rceil$ bits long. Then $B_j^1 = (X_j, Y_j)$, where:

$$X_j = \begin{cases} x_j & \text{if } j = 1, 2 \\ x_1 + x_2, x_1 + x_3, \ldots, x_1 + x_{j-1}, x_j & \text{if } j = 3, \ldots, t+2 \\ x_1 + x_{j-t}, x_1 + x_{j-t+1}, \ldots, x_1 + x_{j-1}, x_j & \text{if } j = t+3, \ldots, m \end{cases},$$

$$Y_j = \begin{cases} y_j, y_m + y_{j+1}, y_m + y_{j+2}, \ldots, y_m + y_{j+t} & \text{if } j = 1, \ldots, m-t-2 \\ y_j, y_m + y_{j+1}, y_m + y_{j+2}, \ldots, y_m + y_{m-1} & \text{if } j = m-t-1, \ldots, m-2 \\ y_j & \text{if } j = m-1, m \end{cases}.$$

The addition considered to compute X_j, Y_j is the usual in $GF(2^\omega)$. If binary representation of z_j has an odd number of bits, say $2\omega + 1$, we can consider that $z_j = (x_j, y_j)$ where $x_j \in GF(2^\omega)$ and $y_j \in GF(2^{\omega+1})$.

The second part of the broadcast is defined as follows: for $j = 1, 2$

$$B_j^2 = \{(k, \psi(k)^\top u_j)\}_{k \in W_j},$$

and for $j \geq 3$

$$B_j^2 = B_{j-1}^2 \cup \{(k, \psi(k)^\top u_j)\}_{k \in W_j}.$$

Theorem 3. *The proposed scheme is a $(t, \mathcal{R})$-threshold self-healing key distribution scheme.*

Proof. Condition 1.(a) is satisfied because he can perform session key computation in the following way. Since user $i \in G_j$ has $\{(k, \psi(k)^\top u_j)\}_{k \in W_j}$ and his personal key, he computes $\psi(D)^\top u_j$ using $\{(k, \psi(k)^\top u_j)\}_{k \in W_j \cup \{i\}}$ because $W_j \cup \{i\} \in \Gamma$. In effect: as far as $W_j \cup \{i\} \in \Gamma$, then $\psi(D) = \sum_{k \in W_j \cup \{i\}} \lambda_k \psi(k)$, for some $\lambda_k \in GF(q)$. So $\psi(D)^\top u_j = \sum_{k \in W_j \cup \{i\}} \lambda_k \psi(k)^\top u_j$. From the broadcast information B_j, the user can compute $z_j = (x_j, y_j)$ and the session key as $K_j = z_j - \psi(D)^\top u_j$.

For condition 1.(b) we should note that session keys $K_1, \ldots, K_m$ and vectors $u_1, \ldots, u_m$ have been chosen independently at random, then session keys and personal keys are independent. Furthermore broadcasts $B_1, \ldots, B_m$ determine $z_1, \ldots, z_m$ but these scalars perfectly hide $K_1, \ldots, K_m$ by means of $\psi(D)^\top u_j$, because $z_j = K_j + \psi(D)^\top u_j$.

Condition 2 follows noticing that, a user $i \in R_j$ knows, from the broadcast B_j, vectors $\{(k, \psi(k)^\top u_j)\}_{k \in W_j}$ and from his personal keys $\psi(i)^\top u_j \in GF(q)$ where $i \in W_j \notin \Gamma$, and this does not get any information on $\psi(D)^\top u_j$. This is true because for any scalar $s \in GF(q)$ there exists at least one vector $u \in GF(q)^\ell$ such that:

$$\left. \begin{array}{l} \psi(D)^\top u = s \\ \psi(k)^\top u = \psi(k)^\top u_j \quad \text{for any } k \in W_j \end{array} \right\}$$

because $W_j \not\subseteq \Gamma$. Observe that the number of vectors u satisfying this system of equations is independent of the value s.

The self-healing condition 3.(a) holds because from any two broadcasts B_r and B_s with $r < s$ and $s - r \leq t$, user i can compute all keys K_j where $i \in G_j$ with $r \leq j \leq s$. The computations are performed in the same way as in session key computation, calculating $z_r, \ldots, z_s$ from the first part of the broadcasts B_r and B_s and using $\psi(i)^\top u_r, \ldots, \psi(i)^\top u_s$ from his personal key. The computation of $z_r, \ldots, z_s$ from broadcasts B_r and B_s with $r < s$ and $s - r \leq t$ is as follows. Consider, for instance, the case in which $t + 3 \leq r < s \leq m - t - 2$ (the other cases are similar). Broadcasts in this case are:

$$B_r^1 = (x_1 + x_{r-t}, x_1 + x_{r-t+1}, \ldots, x_1 + x_{r-1}, x_r, y_r, y_m + y_{r+1}, y_m + y_{r+2}, \ldots, y_m + y_{r+t}),$$

$$B_s^1 = (x_1 + x_{s-t}, x_1 + x_{s-t+1}, \ldots, x_1 + x_{s-1}, x_s, y_s, y_m + y_{s+1}, y_m + y_{s+2}, \ldots, y_m + y_{s+t}).$$

From B_r^1 we extract x_r, y_r. Taking into account that $s - t \leq r < s$, broadcast B_s^1 give us $x_1 + x_r$ and then we can compute x_1. From the two broadcasts we calculate $x_{r-t}, x_{r-t+1}, \ldots, x_s$. Now, from B_s^1 we extract x_s, y_s. Taking into account that $r < s \leq t + r$, broadcast B_r^1 give us $y_m + y_s$ and then we can compute y_m. From the two broadcasts we calculate $y_r, y_{r+1}, \ldots, y_{s+t}$. At the end we can determine completely $z_r, \ldots, z_s$ and we also know partial information of the other scalars, because we have $x_{r-t}, \ldots, x_{r-1}$ and $y_{s+1}, \ldots, y_{s+t}$. With the second part of the broadcast a non-revoked user belonging to the group is allowed to compute secret keys corresponding to sessions $r, r + 1, \ldots, s$, but no information on secret keys of the other sessions can compute. Observe that in the proposal presented in [1] from broadcast B_r and B_s it can be completely computed $z_1, \ldots, z_s$.

To check condition 3.(b) let us consider a coalition of rejected users $B \subset R_{r-1}$ and a coalition of new users $C \subset J_s \cup \cdots \cup J_m$ such that $B \cup C \not\subseteq \Gamma$. Secret information held by users in $B \cup C$ and information broadcast in all the sessions do not get any information about keys K_j for $j = r, \ldots, s - 1$. This is true because they know, in the worst case, $S_i = (\psi(i)^\top u_s, \psi(i)^\top u_{s+1}, \ldots, \psi(i)^\top u_m) \in GF(q)^{m-s+1}$ for $i \in C$ and $B_1, \ldots, B_m$. Since $B \subset W_j$ for any $j = r, \ldots, s - 1$ and personal keys of users in B are known for sessions $j \geq s$, it is easy to see that all the possible values for K_j with $j = r, \ldots, s - 1$ have the same probability. $\square$

Observe that our family of schemes verifies that every user i added in one of the sessions $j = r + 1, \ldots, s - 1$, cannot obtain from broadcasts B_r and B_s, where $1 \leq r < s \leq m$ with $s - r > t$, any information about the value of the keys K_ℓ for $\ell = r, \ldots, s$. That is, $H(\mathbf{K}_r, \ldots, \mathbf{K}_s | \mathbf{S}_i, \mathbf{B}_r, \mathbf{B}_s) = H(\mathbf{K}_r, \ldots, \mathbf{K}_s)$.

We analyze the efficiency of the family of the proposed threshold self-healing key distribution schemes in terms of memory storage and communication complexity. In our construction every user i has to store a personal key of size $|S_i| = (m - j + 1) \log q$, which is optimal with respect to Theorem 1 when the structure $\Gamma = 2^{\mathcal{U}} - \mathcal{R}$ is a vector space access structure. In our construction, the broadcast length depends on the particular function ψ used. The second part of the broadcast has the same form as the proposed in [1] and its purpose is to perform the rejection capability as well as the computation of the key. Its length depends on the history of rejected subsets $R_2, R_3, etc.$ The first part of the broadcast is shorter than the one proposed in [1]. The total number of bits broadcast on the X_j part is

$$\frac{1}{2}(1+1+2+3+\cdots+(t+1)+(m-t-2)(t+1))\log q = \frac{1}{2}(1+(t+1)(m-1-t/2))\log q$$

for a maximum length of $\frac{1}{2}(t+1)\log q$ bits. As usual in this kind of computations we consider that x_j and y_j are $\frac{1}{2}\log q$ bits long. The total number of bits and the maximum length of the Y_j part is the same. Then the total number of broadcast bits is $(1 + (t + 1)(m - 1 - t/2))\log q$, for a maximum length less or equal to $(t+1)\log q$. These lengths are shorter than the lengths of the first part of broadcasts in [1] with total number of broadcast bits $\frac{1}{2}(m^2 - m + 2)\log q$ and maximum length $(m - 1)\log q$. A particular example of this construction can be found in Section 6.

6   A Particular Example of the Construction

In the particular self-healing key distribution scheme obtained with our construction from Shamir secret sharing scheme, it can be revoked a subset of at most $T - 1$ users, that is $|R| = |R_2 \cup \cdots \cup R_j| \leq T - 1$. And the length of the second part of the broadcast is proportional to $T - 1$, the cardinality of subset W_j. This two characteristics are common with the scheme proposed in [1].

But some other schemes can be proposed using our construction. A particular construction in which we have a short broadcast for small revocations of users can be proposed. The example is based in a bipartite access structure [7] proposed in [5] for group key distribution schemes.

Suppose that we want to implement a self-healing key distribution scheme in a context in which most of the revocations consists of revoking a small number of users (less than J, for some positive integer $J \leq T - 1$), although we want to have the possibility to revoke up to $T - 1$ users in some special circumstances. This situation would correspond a self-healing key distribution schemes in which communication groups are very similar, but with the capability revoking a big number of users. If we implement the self-healing scheme with Shamir secret sharing scheme in such a situation, then the second part of the broadcast B_j^2 of every session must have a proportional amount of information to $T - 1$, despite only two or three users must be revoked.

However, if we consider a secret sharing scheme realizing a specific bipartite access structure defined in the set of users, this allows to improve the efficiency of these revocations of few users. Bipartite access structures were first presented in [7]. In such a structure Γ, there is a partition of the set of participants, $\mathcal{U} = X \cup Y$, such that all participants in the same class play an equivalent role in the structure. We associate any subset $A \subset \mathcal{U}$ with the point of non-negative integers $\pi(A) = (x(A), y(A)) \in \mathbb{Z}^+ \times \mathbb{Z}^+$, where $x(A) = |A \cap X|$, $y(A) = |A \cap Y|$ and the structure to a region:

$$\pi(\Gamma) = \{\pi(A) : A \in \Gamma\} \subset \mathbb{Z}^+ \times \mathbb{Z}^+.$$

Let us come back to our previous situation. Let n' be the total number of possible real users. We consider a set $\mathcal{U} = X \cup Y$, where $X = \{1, \ldots, n', n' + 1, \ldots, n' + T - J - 1\}$ contains the n' possible real users and $T - J - 1$ dummy users, and $Y = \{n' + T - J, \ldots, n' + T - 1\}$ contains J dummy users. So, the set $\mathcal{U}$ contains $n = n' + T - 1$ users. Let us consider the following bipartite access structure Γ defined in $\mathcal{U} = X \cup Y$:

$$\Gamma = \{A \subset X \cup Y : |A| \geq J+1 \text{ and } |A \cap Y| \geq 1\} \cup \{A \subset X \cup Y : |A \cap X| \geq T\},$$

which corresponds to the following region:

$$\pi(\Gamma) = \{(x,y) \in \mathbb{Z}^+ \times \mathbb{Z}^+ : (x \geq T) \text{ or } (x + y \geq J + 1 \text{ and } y \geq 1)\}.$$

The maximal non-authorized subsets in this structure are defined by the points $(T-1, 0), (J-1, 1), (J-2, 2), \ldots, (1, J-1), (0, J)$. Note that non-authorized subsets of a (T, n)-threshold structure are defined by $T - 1$ users. If a subset of $\omega \leq J - 1$ users has to be revoked, then the second part of the broadcast B_j^2 has only J values, ω values related to the revoked users, and $J - \omega$ values related to dummy users in Y. Otherwise, if a subset of $\omega \geq J$ users is revoked, with $\omega \leq T - 1$, then the second part of the broadcast B_j^2 has $T - 1$ values: ω values related to the revoked users, in addition to $T - 1 - \omega$ values related to dummy users in X. Note that if we put $J = T - 1$, we obtain the threshold case. With lower values of J, the revocation can be performed in a more efficient way.

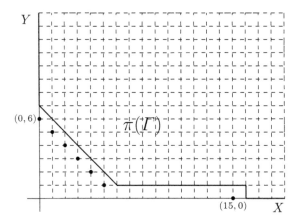

This bipartite access structure Γ can not be realized by a vector space secret sharing scheme (except in the threshold case $J = T - 1$, see [7] for the details), but by a linear one in which each participant is associated with two vectors instead of one. Therefore, each operation will have twice the cost of the same operation in the threshold case. In particular, the length of the personal keys are twice the length in the threshold case. If efficiency in the revocation of small subsets has priority, this scheme makes perfect sense.

The following picture shows an example of this situation, with $T = 16$ and $J = 6$. The region $\pi(\Gamma)$, delimited by thick lines, contains the points corresponding to the authorized subsets of the structure. The points corresponding to those subsets in $\mathcal{R}_0 = \overline{T}_0$, for example the one formed by 6 dummy users in Y, are marked with a circle.

In this example, if a subset of three users has to be revoked, it must be broadcast three values related to the three revoked users and three values related to three dummy users in Y. In this way, the broadcast information corresponds to a subset in $\mathcal{R}_0 = \overline{T}_0$ (symbolized by the point $(3,3)$ in the picture), and each non-revoked user can use the broadcast information and his personal key to compute the new group key.

The second novelty of our work is to consider that the distance between the broadcasts used to supply the lost one is limited in order to shorten the length of the broadcast information by the group manager. As an example, we can see the effect in the first part of the broadcast for self-healing key distribution for $z_j = (x_j, y_j)$ with $m = 10$ sessions and a threshold $t = 3$:

$$
\begin{aligned}
B_1^1 &= (x_1, & y_{10} + y_4, y_{10} + y_3, y_{10} + y_2, y_1) \\
B_2^1 &= (x_2, & y_{10} + y_5, y_{10} + y_4, y_{10} + y_3, y_2) \\
B_3^1 &= (x_1 + x_2, x_3, & y_{10} + y_6, y_{10} + y_5, y_{10} + y_4, y_3) \\
B_4^1 &= (x_1 + x_2, x_1 + x_3, x_4, & y_{10} + y_7, y_{10} + y_6, y_{10} + y_5, y_4) \\
B_5^1 &= (x_1 + x_2, x_1 + x_3, x_1 + x_4, x_5, & y_{10} + y_8, y_{10} + y_7, y_{10} + y_6, y_5) \\
B_6^1 &= (x_1 + x_3, x_1 + x_4, x_1 + x_5, x_6, & y_{10} + y_9, y_{10} + y_8, y_{10} + y_7, y_6) \\
B_7^1 &= (x_1 + x_4, x_1 + x_5, x_1 + x_6, x_7, & y_{10} + y_9, y_{10} + y_8, y_7) \\
B_8^1 &= (x_1 + x_5, x_1 + x_6, x_1 + x_7, x_8, & y_{10} + y_9, y_8) \\
B_9^1 &= (x_1 + x_6, x_1 + x_7, x_1 + x_8, x_9, & y_9) \\
B_{10}^1 &= (x_1 + x_7, x_1 + x_8, x_1 + x_9, x_{10}, y_{10})
\end{aligned}
$$

References

1. C. Blundo, P. D'Arco, A. De Santis and M. Listo. Design of Self-Healing Key Distribution Schemes. *Designs, Codes and Cryptography*, Vol. 32, pp. 15–44 (2004).
2. C. Blundo, P. D'Arco, A. De Santis and M. Listo. Definitions and Bounds for Self-Healing Key Distribution. In *31st International Colloquium on Automata, Languages and Programming (ICALP 04)*. Lecture Notes in Computer Science, **3142** (2004) 234–245.
3. E.F. Brickell. Some ideal secret sharing schemes. *Journal of Combinatorial Mathematics and Combinatorial Computing*, **9**, pp. 105–113 (1989).
4. T.M. Cover and J.A. Thomas. Elements of Information Theory. *John Wiley & Sons* (1991).
5. V. Daza, J. Herranz and G. Sáez. Constructing General Dynamic Group Key Distribution Schemes with Decentralized User Join. In *8th Australasian Conference on Information Security and Privacy (ACISP 03)*. Lecture Notes in Computer Science, **2727** (2003) 464–475.
6. D. Liu, P. Ning and K. Sun. Efficient Self-Healing Key Distribution with Revocation Capability. In *10th ACM Conference on Computer and Communications Security* (2003).

7. C. Padró and G. Sáez. Secret sharing schemes with bipartite access structure. *IEEE Transactions on Information Theory*, **46** (7), pp. 2596–2604 (2000).

8. A. Shamir. How to share a secret. *Communications of the ACM*, **22**, pp. 612–613 (1979).

9. G.J. Simmons, W. Jackson and K. Martin. The geometry of secret sharing schemes. *Bulletin of the ICA* **1**, pp. 71–88 (1991).

10. J. Staddon, S. Miner, M. Franklin, D. Balfanz, M. Malkin and D. Dean. Self-Healing Key Distribution with Revocation. *IEEE Symposium on Security and Privacy* (2002).

11. D.R. Stinson. An explication of secret sharing schemes. *Designs, Codes and Cryptography*, Vol. **2**, pp. 357–390 (1992).

Concrete Security of the Blum-Blum-Shub Pseudorandom Generator

Andrey Sidorenko and Berry Schoenmakers

Eindhoven University of Technology,
P.O. Box 513, 5600MB Eindhoven, The Netherlands
{a.sidorenko, l.a.m.schoenmakers}@tue.nl

Abstract. The asymptotic security of the Blum-Blum-Shub (BBS) pseudorandom generator has been studied by Alexi et al. and Vazirani and Vazirani, who proved independently that $O(\log \log N)$ bits can be extracted on each iteration, where N is the modulus (a Blum integer). The concrete security of this generator has been analyzed previously by Fischlin and Schnorr and by Knuth.

In this paper we continue to analyse the concrete security the BBS generator. We show how to select both the size of the modulus and the number of bits extracted on each iteration such that a desired level of security is reached, while minimizing the computational effort per output bit. We will assume à concrete lower bound on the hardness of integer factoring, which is obtained by extrapolating the best factorization results to date.

While for asymptotic security it suffices to give a polynomial time reduction a successful attack to factoring, we need for concrete security a reduction that is as efficient as possible. Our reduction algorithm relies on the techniques of Fischlin and Schnorr, as well as ideas of Vazirani and Vazirani, but combining these in a novel way for the case that more than one bit is output on each iteration.

1 Introduction

Generally speaking, a pseudorandom generator is a deterministic algorithm that, given a truly random binary sequence of length n, outputs a binary sequence of length $M > n$ that "looks random". The input to the generator is called the seed and the output is called the pseudorandom bit sequence. Security of a pseudorandom generator is a characteristic that shows how hard it is to tell the difference between the pseudorandom sequences and truly random sequences. For the Blum-Blum-Shub (BBS) pseudorandom generator [2] distinguishing these two distributions is as hard as factoring a large composite integer.

Although asymptotic security of the BBS generator is thoroughly analyzed [1,15] it has been uncertain how to select the size of the modulus and the number of bits extracted on each iteration such that a desired level of security is reached, while minimizing the computational effort per output bit. In this paper we answer this question. We construct an efficient reduction of a successful attack on the BBS generator to factoring. Then we assume a concrete lower bound

N.P. Smart (Ed.): Cryptography and Coding 2005, LNCS 3796, pp. 355–375, 2005.

on the hardness of integer factoring, which is obtained by extrapolating the best
factorization results to date. It gives a lower bound for the running time of the
successful attack on the BBS generator (Theorem 3). This lower bound is used
for selecting the optimal values for the size of the modulus and the number of
bits extracted on each iteration.

1.1 Notation

Throughout we use the following notation.

- $s \in_R S$ indicates that s is chosen uniformly at random from set S.
- N is a Blum integer, that is $N = pq$, where p, q are prime, $p \equiv q \equiv 3 \bmod 4$.
- n is the size (in bits) of N.
- $\mathbb{Z}_N(+1)$ is the set of integers of Jacobi symbol $+1$ modulo N.
- $\Lambda_N = \mathbb{Z}_N(+1) \cap (0, \frac{N}{2})$.
- $[y]_N = y \bmod N \in [0, N)$ for $y \in \mathbb{Z}$.
- $y \underline{\bmod} N \in \left(-\frac{N}{2}, \frac{N}{2}\right)$ denotes the smallest absolute residue of y modulo N.
- $\ell_i(y)$ denotes the i-th least significant bit of y, $i = 1, 2, \ldots$.
- $E_N(y) = |y^2 \underline{\bmod} N|$, which is referred to as the absolute Rabin function.

Note that the absolute Rabin function E_N permutes Λ_N [5].

1.2 Security of Pseudorandom Generators

Let G be a pseudorandom generator that produces binary sequences of length
M. Let $S \subset \{0, 1\}^M$ be the set of these sequences.

Consider a probabilistic algorithm A that, given a binary sequence $s = s_1 \ldots s_M$, outputs a bit $A(s) \in \{0, 1\}$. We may think of A as a statistical test of
randomness.

Definition 1. *A pseudorandom generator G passes statistical test A with toler-*
ance $\varepsilon > 0$ if

$$|\Pr(A(s) = 1 \mid s \in_R S) - \Pr(A(s) = 1 \mid s \in_R \{0, 1\}^M)| < \varepsilon.$$

Otherwise the pseudorandom generator G fails statistical test A with tolerance
ε. The probability is taken over all choices of s, and internal coin flips of A.

We refer to a sequence $s \in_R \{0, 1\}^M$ as a truly random binary sequence.
Throughout this paper we often call A an adversary that tries to distinguish
pseudorandom sequences from truly random sequences.

Note that the maximum possible value of ε is $\varepsilon = 1 - 2^{n-M}$. It corresponds
to the statistical test that with probability 1 outputs 1 if $s \in S$ and outputs 0 if
$s \in \{0, 1\}^M \backslash S$.

Definition 2. *A pseudorandom generator is asymptotically secure if it passes*
all polynomial time statistical tests with tolerance negligible in n.

The above definition originates from [16]. Asymptotic security guarantees that, as the seed length increases, no polynomial time statistical test can distinguish the pseudorandom sequences from truly random sequences with non-negligible probability. However, this definition says little about the security of the pseudorandom generator in practice for a particular choice of seed length and against adversaries investing a specific amount of computational effort. For practical considerations it is important to focus on *concrete security reductions* which give explicit bounds on running time and success probability of statistical tests. The following definition is due to [9,3,14].

Definition 3. *A pseudorandom generator is (T_A, ε)-secure if it passes all statistical tests with running time at most T_A with tolerance ε.*

We determine the values for T_A and ε such that the BBS pseudorandom generator defined below is (T_A, ε)-secure.

1.3   The BBS Generator

The following definition is due to [5].

Definition 4 (The BBS pseudorandom generator). *Let k, j be positive integers. Let $x_1 \in_R \Lambda_N$ be the seed. Consider a deterministic algorithm that transforms the seed into a binary sequence of length $M = jk$ by repeating the following steps for $i = 1, \ldots, k$.*

1. For $r = 1, \ldots, j$ output $b_{(i-1)j+r} = \ell_{j-r+1}(x_i)$.
2. $x_{i+1} = E_N(x_i)$.

We call this algorithm the BBS pseudorandom generator with j output bits per iteration.

In order to output a pseudorandom sequence (a BBS sequence) of length M the generator iterates the absolute Rabin function k times generating j bits on each iteration.

Remark 1. Strictly speaking, the above definition of the BBS differs from the original one presented in [2]. The original generator iterates the Rabin function $E_N^*(x) = x^2 \bmod N$ and outputs only one bit on each iteration ($j = 1$). We do not introduce a new name for the sake of simplicity.

1.4   Known Results and Our Contributions

Intuitively, the performance of the BBS generator can be improved in two ways. We can either use a modulus of a smaller size or extract more bits per iteration. However, in both cases the security of the algorithm is weakened. What are the optimal values for the parameters given that a certain level of security has to be reached? For instance, what is the optimal value for j?

The security of the BBS generator is proved by reduction. It is shown that if the generator is insecure then there exists an algorithm that factors the modulus. When analyzing asymptotic security the only requirement is that the reduction has to be polynomial time. In case of concrete security the reduction has to be as tight as possible. A tight reduction gives rise to a small modulus, which in turn ensures that the pseudorandom generator is efficient.

The following reductions are known for the BBS generator. The case $j = 1$ has been studied extensively. The tightest reduction is due to Fischlin and Schnorr [5], which gives rise to a rather efficient generator. For the case $j > 1$, the asymptotic security has been analyzed fully by Alexi et al. [1], and independently, by Vazirani and Vazirani [15], who proved that the BBS generator is secure if $j = O(\log \log N)$. However, using their reductions as a basis for the concrete security of the BBS generator would imply that in practical case it does not pay off to extract more than 1 bit on each iteration. Fischlin and Schnorr [5] already suggested ways to tighten the reductions for the case $j > 1$. However, as they point out, the first approach is completely impractical. The second one, though, is similar to our analysis, but they provide no sufficient detail.

Inspired by the ideas of [5] and [15] we construct a new security proof for the BBS generator with j output bits per iteration for $j > 1$. The new reduction is more efficient than all previously known reductions for $j > 1$.

We show how to select both the size of the modulus and the number of bits extracted on each iteration such that a desired level of security is reached, while minimizing the computational effort per output bit. Although the complexity of the reduction grows exponentially in j it does not mean that one should always choose $j = 1$. In Example 7.3 the optimal value is $j = 5$ rather than $j = 1$. We emphasize that the optimal parameter j depends on the length of the output sequence M and on the security parameters T_A, ε.

The rest of the paper is organized as follows. In Section 2 we describe a general idea of the security proof for the BBS generator. The result of [9,16] implies that if the generator is insecure then there exists an algorithm B that, given $E_N(x)$ for some $x \in \Lambda_N$, and $j-1$ least significant bits of x, guesses the j-th least significant bit $\ell_j(x)$. In Section 4 the algorithm B is used for inversion of the absolute Rabin function. Before that, in Section 3, we discuss a simplified inversion algorithm as a stepping stone to the general case. The simplified algorithm is of independent interest since it is almost optimal in terms of the running time. In Section 5 we analyze the success probability of the inversion algorithm of Section 4. We determine the complexity of this algorithm in Section 6. In Section 7 we state our main result about the concrete security of the BBS generator.

2   Security of the BBS Generator

In this section we describe a general idea of the security proof for the BBS generator.

Lemma 1. *Suppose the BBS generator is not (T_A, ε)-secure. Then there exists an algorithm B that, given $E_N(x)$ for some $x \in_R \Lambda_N$, $j-1$ least significant bits* of x, guesses the j-th least significant bit $\ell_j(x)$ with advantage $M^{-1}\varepsilon$. Here the probability is taken over all choices of $x \in_R \Lambda_N$, and internal coin flips of B. The running time $T_B \leq T_A + O(kn^2)$.

The proof of the above lemma can be found, for instance, in [9].

In Section 4 we show that the algorithm B can be used for the inversion of the absolute Rabin function. Before that, in Section 3, we show how to invert the absolute Rabin function using a "simpler" oracle. Section 3 serves as a stepping stone to the general case.

According to the following lemma inversion of the absolute Rabin function is as hard as factoring Blum integers.

Lemma 2 (Rabin). *Suppose there exists a probabilistic algorithm R that recovers $x \in \Lambda_N$ from $E_N(x)$ in expected time T_R. Then there exists an algorithm F that factors the modulus N in expected time $T_F = 2(T_R + 2\log_2 N)$.*

Since factoring Blum integers is assumed to be a hard problem (in Section 7.1 we will assume a concrete lower bound on the hardness of factoring) "attacking" BBS sequences is also a hard problem.

In practice the terms kn^2 and $\log_2 N$ are small in comparison with T_A and T_R respectively. We omit these terms in the further analysis.

3 The Simplified Inversion Algorithm

To complete our concrete security analysis of the BBS generator, we need to show how to invert the absolute Rabin function E_N, given an oracle of a particular type. However, in this section we will consider the related problem of inverting E_N given a more powerful oracle O_1, see below, and assuming that $2 \in \mathbb{Z}_N(+1)$ (which holds if $N \equiv 1 \bmod 8$, see e.g. [12]). The treatment of this case serves as a stepping stone to the general case, and, additionally, we will point out that in this case the reduction can be shown optimal up to a factor of $\ln n$.

The oracle O_1 is defined as a probabilistic algorithm that *for all $x \in \Lambda_N$,* given $E_N(x)$, guesses bit $\ell_1(x)$ with advantage $\delta > 0$, where the probability is taken over internal coin flips of O_1.

3.1 Binary Division

The main tool of the inversion algorithm is the *binary division* technique [5], which is a means to solve the following problem. The problem is to recover a value α, $0 \leq \alpha < N$, given $\ell_1(\alpha)$, $\ell_1([2^{-1}\alpha]_N)$, $\ldots$, $\ell_1([2^{-(n-1)}\alpha]_N)$, where n is the bit length of N.

The solution to this problem is given in terms of rational approximations. For a rational number β, $0 \leq \beta < 1$, we call βN a *rational approximation* of integer α, $0 \leq \alpha < N$, with *error* $|\alpha - \beta N|$. Given a rational approximation βN for α we can get a rational approximation $\beta_1 N$ for $\alpha_1 = [2^{-1}\alpha]_N$ for which the error is reduced by a factor of 2 as follows. If α is even, then $\alpha_1 = \alpha/2$

so put $\beta_1 = \beta/2$; otherwise, $\alpha_1 = (\alpha + N)/2$ so put $\beta_1 = (\beta + 1)/2$. Then we have $|\alpha_1 - \beta_1 N| = \frac{1}{2}|\alpha - \beta N|$. Note that to determine β_1, the only required information on α is its parity.

Given $\ell_1(\alpha), \ell_1([2^{-1}\alpha]_N), \ldots, \ell_1([2^{-(n-1)}\alpha]_N)$, the value of α can be recovered as follows. Put $\beta_0 = 1/2$, then $\beta_0 N$ is a rational approximation of α with error at most $N/2$. Next, we apply the above technique n times to obtain rational approximations $\beta_1 N, \ldots, \beta_n N$ for $[2^{-1}\alpha]_N, \ldots, [2^{-n}\alpha]_N$ respectively, at each step reducing the error by a factor of 2. We thus get a rational approximation $\beta_n N$ to $[2^{-n}\alpha]_N$, for which the error is less than $N/2^{n+1} < 1/2$. The closest integer to $\beta_n N$ is therefore equal to $[2^{-n}\alpha]_N$, and from this value we find α.

3.2   Majority Decision

The bits $\ell_1(\alpha), \ell_1([2^{-1}\alpha]_N), \ldots, \ell_1([2^{-(n-1)}\alpha]_N)$ used to recover α by means of the binary division technique will be obtained from the oracle O_1, which essentially outputs $\ell_1(\alpha)$ on input $E_N(\alpha)$ for $\alpha \in \Lambda_N$. However, since the output bit of O_1 is not always correct, we have to run O_1 several times and use some form of majority decision.

Suppose we know $E_N(\alpha)$ and our goal is to determine $\ell_1(\alpha)$ for some $\alpha \in \Lambda_N$. We run O_1 on input $E_N(\alpha)$ m times and assign the majority bit to $\ell_1(\alpha)$. We will show that for $m = \frac{1}{2}(\ln n + \ln p^{-1})\delta^{-2}$, where $0 < p < 1$, the majority decision errs with probability at most p/n.

Let $\tau_1, \ldots, \tau_m$ be the outputs of O_1. Without loss of generality, assume that $\ell_1(\alpha) = 0$. Then the majority decision errs if

$$\frac{1}{m}\sum_{i=1}^{m}\tau_i > \frac{1}{2}. \tag{1}$$

Since for each $\alpha \in \Lambda_N$ the probability that O_1 successfully guesses $\ell_1(\alpha)$ equals $\frac{1}{2} + \delta$ the expected value $\mathrm{E}[\tau_i] = \frac{1}{2} - \delta$, $i = 1, \ldots, m$. (1) implies that

$$\frac{1}{m}\sum_{i=1}^{m}\tau_i - \mathrm{E}[\tau_i] > \delta.$$

Since $\tau_1, \ldots, \tau_m$ are mutually independent Hoeffding's bound [8] gives

$$\Pr\left[\frac{1}{m}\sum_{i=1}^{m}\tau_i - \mathrm{E}[\tau_i] > \delta\right] \le \exp\left(-2m\delta^2\right).$$

It implies that for $m = \frac{1}{2}(\ln n + \ln p^{-1})\delta^{-2}$ the majority decision errs with probability p/n.

3.3   The Simplified Algorithm

Remark 2. Before describing the simplified inversion algorithm we point out an important fact about oracle O_1. For every $y \in \Lambda_N$ there always exist two different values x_1 and x_2 such that $E_N(x_i) = y$ and $x_i \in \mathbb{Z}_N(+1)$, $i = 1, 2$. Without loss of generality, let $x_1 < N/2$. Then $x_1 \in \Lambda_N$, $x_2 = N - x_1$. On input y oracle O_1 predicts $\ell_1(x_1)$ rather than $\ell_1(x_2)$. This property will be used on step 3 of the algorithm.

The inversion algorithm, given $E_N(x)$ for some $x \in \Lambda_N$ and parameter p, $0 < p < 1/2$, runs as follows.

1. Pick a random multiplier $a \in_R \mathbb{Z}_N(+1)$. Let $m = \frac{1}{2}(\ln n + \ln p^{-1})\delta^{-2}$.
2. Set $u_0 = 1/2$. $u_0 N$ is a rational approximation of $[ax]_N$ with error at most $N/2$. Set $l_{-1} = 0$.
3. For $t = 0, \ldots, n - 1$ do the following. Compute $E_N([a_t x]_N) = E_N(a_t)E_N(x)$ mod N. Run O_1 on input $E_N([a_t x]_N)$ m times. Let r_t be the majority output bit. Assign $l_t = r_t + l_{t-1} \bmod 2$. Let $a_{t+1} = [2^{-(t+1)}a]_N$. To determine a rational approximation $u_{t+1}N$ for $[a_{t+1}x]_N$, set $u_{t+1} = (u_t + l_t)/2$.
4. Compute $x' = a_n^{-1}\lfloor u_n N + 1/2 \rfloor \bmod N$. If $E_N(x') = E_N(x)$ output x', otherwise repeat the above procedure starting from step 1.

If no error occurs in the above algorithm we have $l_t = \ell_1([a_t x]_N)$ for $t = 0, \ldots, n - 1$. Setting $l_{-1} = 0$ at step 2 means that the algorithm works only if $\ell_1([2ax]_N) = 0$. Since a is chosen at random, $\ell_1([2ax]_N) = 0$ with probability $1/2$.

The goal of step 3 is to determine $\ell_1([a_t x]_N)$. The bit is obtained via the majority decision. Note that on input $E_N([a_t x]_N)$ O_1 predicts either $\ell_1([a_t x]_N)$ (if $\ell_1([a_{t-1} x]_N) = 0$) or $\ell_1(N - [a_t x]_N)$ (if $\ell_1([a_{t-1} x]_N) = 1$). Since $\ell_1(N) = 1$, we have $\ell_1([a_t x]_N) = \ell_1(N - [a_t x]_N) + 1 \bmod 2$. Therefore the majority bit r_t has to be added by l_{t-1} modulo 2 (see also Remark 2).

Since a single majority decision errs with probability at most p/n the probability that $E_N(x') = E_N(x)$ at step 4 is at least $1/2 - p$. Thus the expected running time of the inversion algorithm is at most $(1 - 2p)^{-1}n(\ln n + \ln p^{-1})\delta^{-2}T_{O_1}$, where T_{O_1} is the running time of O_1. For instance, for $p = 1/4$ the running time is essentially $2n(\ln n)\delta^{-2}T_{O_1}$.

Remark 3. The information-theoretic approach of Fischlin and Schnorr [5] implies that inversion of the absolute Rabin function needs to run O_1 at least $(\ln 2/4)n\delta^{-2}$ times. Therefore the running time of the above algorithm is optimal up to a factor of $\ln n$.

4   The Inversion Algorithm

In this section we show how to invert the absolute Rabin function E_N, having access to oracle B that, given $E_N(x)$ for some $x \in_R \Lambda_N$, $j - 1$ least significant bits of x, guesses j-th least significant bit $\ell_j(x)$ with advantage $M^{-1}\varepsilon$.

We build the inversion algorithm for $j \geq 1$ combining the inversion algorithm of [5] for $j = 1$ with the result of [15]. Basic idea of our inversion algorithm is the following. First B is converted into an algorithm O_{xor} that, given $E_N(x)$ for some $x \in \Lambda_N$, guesses the exclusive OR of some subset of first j least significant bits of x with advantage δ, where $\delta = (2^j - 1)^{-1}M^{-1}\varepsilon$. Then $E_N(x)$ is inverted using O_{xor} as an oracle.

4.1   Oracle for Exclusive OR

Let π be a subset of the set of positive integers. For an integer y, $y \geq 0$, let

$$\pi(y) = \sum_{i \in \pi} \ell_i(y) \bmod 2.$$

Note that the subset and the corresponding exclusive OR function are denoted by the same character π.

Let $y \in \Lambda_N$. On input $(\pi, E_N(y))$, where $\pi \subset \{1, \ldots, j\}$ is a nonempty subset, algorithm O_{xor} guesses $\pi(y)$ as follows

1. Select $r_1, \ldots, r_j \in_R \{0, 1\}$. Let r, $0 \leq r < 2^j$, be an integer such that $\ell_k(r) = r_k$, $k = 1, \ldots, j$.
2. Output $\pi(r)$ if $B(E_N(y), r_1, \ldots, r_{j-1}) = r_j$, otherwise output $\pi(r) + 1 \bmod 2$.

The below statement follows explicitly from the Computational XOR Proposition proposed by Goldreich [7].

Lemma 3. *For the above algorithm O_{xor} we have* $\Pr[O_{xor}(\pi, E_N(y)) = \pi(y)] = 1/2 + \delta$, $\delta = (2^j - 1)^{-1} M^{-1} \varepsilon$, *where the probability is taken over all choices of $y \in_R \Lambda_N$, nonempty subsets $\pi \subseteq \{1, \ldots, j\}$ with uniform probability distribution, and internal coin flips of O_{xor}.*

4.2   Inversion of the Absolute Rabin Function Using O_{xor}

The inversion algorithm described below is based on the same ideas as the simplified inversion algorithm of Section 3. The main difference between these two algorithms is due to the fact that, in comparison with O_1, the advantage of O_{xor} does not have to be the same for all input values. In order to use O_{xor} for the majority decision we have to randomize the input values. For this purpose we use two multipliers $a, b \in_R \mathbb{Z}_N$ and we call O_{xor} on inputs proportional to $E_N(c_{t,i}x)$, where $c_{t,i}$ is a function of a and b such that $c_{t,i}$'s for a fixed t are pairwise independent random variables.

Tightening the Rational Approximations. Suppose $E_N(x)$ is given for $x \in \Lambda_N$. The goal is to recover x.

Let $a, b \in_R \mathbb{Z}_N$. Let $u_0 N$ and vN be rational approximations of $[ax]_N$ and $[bx]_N$. In below algorithm we search through a certain number of quadruples $(u_0 N, vN, l_{a,0}, l_b)$, where $l_{a,0}, l_b \in \{0, 1\}$, so that for at least one of them

$$l_{a,0} = \ell_1([ax]_N), \quad l_b = \ell_1([bx]_N),$$
$$|[ax]_N - u_0 N| \leq \eta_a N, \quad |[bx]_N - vN| \leq \eta_b N, \tag{2}$$

where $\eta_a = 2^{-j-6}\delta^3, \eta_b = 2^{-j-4}\delta$ (these values result from the analysis of the inversion algorithm, which appears in the extended version of this paper). (2) implies that we have to try at most $\eta_a^{-1}\eta_b^{-1}$ quadruples.

Let $a_t = [2^{-t}a]_N, t = 1, \ldots, n$. By means of the binary division technique we construct rational approximations $u_t N$ for $[a_t x]_N$ so that if (2) holds then

$$|[a_t x]_N - u_t N| \leq \frac{\eta_a N}{2^t}, t = 1, \ldots, n.$$

For $t = n$ we have $|[a_n x]_N - u_n N| < 1/2$, i.e. the closest integer to $u_n N$ is $[a_n x]_N$. Therefore $x = [a_n^{-1} \lfloor u_n N + \frac{1}{2} \rfloor]_N$.

The binary division technique works only if for all $t = 0, \ldots, n-1$ the bits $\ell_1([a_t x]_N)$ are determined. Note that if (2) holds then $\ell_1([ax]_N) = l_{a,0}$. For $t = 1, \ldots, n-1$ we determine the bits $\ell_1([a_t x]_N)$ using oracle O_{xor}.

Finding $\ell_1([a_t x]_N)$ Via Majority Decision. Consider step t of the inversion algorithm for $1 \leq t < n$. At this step we know the rational approximation $u_t N$ for $[a_t x]_N$. The goal is to determine $\ell_1([a_t x]_N)$.

Let i be an integer from a multiset σ_t (we will define the multisets in the end of this section). Using O_{xor} we will determine $\ell_1([a_t x]_N)$ for a fraction of indices $i \in \sigma_t$ with probability slightly higher than $1/2$. Then the majority decision will provide us with a reliable value $\ell_1([a_t x]_N)$. The details follow.

Let $c_{t,i} = a_t(1 + 2i) + b$. Then

$$[c_{t,i} x]_N = [a_t x]_N (1 + 2i) + [bx]_N \bmod N.$$

Let $w_{t,i} = u_t(1 + 2i) + v$, $\tilde{w}_{t,i} = w_{t,i} \bmod 1$. Here $\tilde{w}_{t,i} N$ is an approximation of $[c_{t,i} x]_N$, whereas $w_{t,i} N$ is an approximation of $[a_t x]_N (1 + 2i) + [bx]_N$. Note that if the error of the rational approximation $w_{t,i} N$ is small enough we have

$$[2^j c_{t,i} x]_N = 2^j ([a_t x]_N (1 + 2i) + [bx]_N) - \lfloor 2^j w_{t,i} \rfloor N. \tag{3}$$

We will see that if (3) holds for a certain value of i then the i-th vote in the majority decision is correct with probability $1/2 + \delta$ (this probability cannot be higher since O_{xor} guesses correctly with probability $1/2 + \delta$). In Section 5 we analyze the probability that (3) holds. In the rest of this section we assume that (3) does hold.

It can be shown that if (3) holds then

$$[c_{t,i} x]_N = [a_t x]_N (1 + 2i) + [bx]_N - \lfloor w_{t,i} \rfloor N.$$

Since $\ell_1(N) = 1$, we get

$$\ell_1([c_{t,i} x]_N) = \ell_1([a_t x]_N) + \ell_1([bx]_N) + \lfloor w_{t,i} \rfloor \bmod 2. \tag{4}$$

If (2) holds then $\ell_1([bx]_N) = l_b$ and the only unknown components in (4) are $\ell_1([c_{t,i} x]_N)$ and $\ell_1([a_t x]_N)$. We will determine $\ell_1([c_{t,i} x]_N)$ through O_{xor} and then we will use (4) for the majority decision on $\ell_1([a_t x]_N)$.

Let $\pi_{t,i}$ be a random nonempty subset of $\{1, 2, \ldots, j\}$. Denote $r = \max\{k \mid k \in \pi_{t,i}\}$, $r \leq j$. Each time (for each values of t and i) a new random subset $\pi_{t,i}$ is selected. The value of r also depends on t and i. We write r instead of $r_{t,i}$ for the sake of simplicity. Denote

$$L_k(y) = y \bmod 2^k$$

for $y \in \mathbb{Z}$, $k = 1, 2, \dots$. If $y \geq 0$ $L_k(y)$ gives an integer that equals k least significant bits of y.

Lemma 4. *If (3) holds then*

$$\ell_1([c_{t,i}x]_N) = \pi_{t,i}([2^{r-1}c_{t,i}x]_N) + \pi_{t,i}(L_r(-\lfloor 2^{r-1}\tilde{w}_{t,i}\rfloor N)) \bmod 2.$$

We prove this lemma in Appendix A. Lemma 4 combined with (4) gives

$$\ell_1([a_t x]_N) = \pi_{t,i}([2^{r-1}c_{t,i}x]_N) + \pi_{t,i}(L_r(-\lfloor 2^{r-1}\tilde{w}_{t,i}\rfloor N)) + \\ \ell_1([bx]_N) + \lfloor w_{t,i}\rfloor \bmod 2.$$

If $[2^{r-1}c_{t,i}x]_N \in \Lambda_N$ in the above formula then we can replace $\pi_{t,i}([2^{r-1}c_{t,i}x]_N)$ by $O_{xor}(\pi_{t,i}, E_N([2^{r-1}c_{t,i}x]_N))$. However, since the output bit of O_{xor} is not always correct we have to use some form of majority decision to determine $\ell_1([a_t x]_N)$.

The majority decision on bit $\ell_1([a_t x]_N)$ works as follows. If for majority of indices $i \in \sigma_t$ such that $[2^{r-1}c_{t,i}x]_N \in \Lambda_N$

$$O_{xor}(\pi_{t,i}, E_N([2^{r-1}c_{t,i}x]_N)) + \pi_{t,i}(L_r(-\lfloor 2^{r-1}\tilde{w}_{t,i}\rfloor N)) + \\ \ell_1([bx]_N) + \lfloor w_{t,i}\rfloor = 0 \bmod 2, \tag{5}$$

the inversion algorithm decides that $\ell_1([a_t x]_N) = 0$, otherwise it decides that $\ell_1([a_t x]_N) = 1$.

Note that we can check if $[2^{r-1}c_{t,i}x]_N \in \Lambda_N$ as follows. By definition, $[2^{r-1}c_{t,i}x]_N \in \Lambda_N$ if $2^{r-1}c_{t,i} \in \mathbb{Z}_N(+1)$ and $[2^{r-1}c_{t,i}x]_N < \frac{N}{2}$. The first condition can be checked by computing Jacobi symbol of $2^{r-1}c_{t,i}$ modulo N. We check the second condition via the rational approximation of $[2^{r-1}c_{t,i}x]_N$. It can be shown that if (3) holds then for all r, $0 \leq r < j$, $[2^{r-1}c_{t,i}x]_N < \frac{N}{2}$ if and only if $\lfloor 2^r w_{t,i}\rfloor$ is even. If $[2^{r-1}c_{t,i}x]_N \notin \Lambda_N$ we discard the index i. Since $c_{t,i}$ is uniformly distributed in $\mathbb{Z}_N$, $[2^{r-1}c_{t,i}x]_N \in \Lambda_N$ with probability $1/4$ (each of the above conditions is satisfied with probability $1/2$).

Multisets σ_t. In this section we define the multisets σ_t, $t = 1, \dots, n-1$. For $t < \log_2 n + 4$ denote $m_t = 4 \cdot 2^t \delta^{-2}$. Let

$$\sigma_t = \{i \mid |1 + 2i| < m_t\}, \quad t = 1, \dots, \log_2 n + 3.$$

As t grows we choose a larger value for m_t. Therefore the majority decisions become more reliable as t grows. We cannot choose large m_t for small t for the following reason. For small t the error $|u_t N - [a_t x]_N|$ is large. If m_t is also large then $[a_t x]_N (1 + 2i) + [bx]_N$ can differ much from $w_{t,i} = u_t(1 + 2i) + v$ so that (3) does not hold and (5) cannot be used for the majority decision.

Define $\rho = \{i \mid |1 + 2i| < 2^6 n \delta^{-2}\}$. We randomly select $m = 8\delta^{-2} \log_2 n$ elements $\sigma = \{i_1, \ldots, i_m\}$ *with repetition* from ρ and let

$$\sigma_t = \sigma, \ m_t = m, \ t = \log_2 n + 4, \ldots, n - 1.$$

For $t = 1, \ldots, n - 1$ $|\sigma_t| = m_t$. For $t \geq \log_2 n + 4$ all the σ_t are the same, the number of elements (not necessarily different) in this multiset is m.

Note that there exist two basic bounds for error probabilities of majority decisions: Hoeffding's bound and Chebyshev's inequality. Hoeffding's bound (see also Section 3.2) is asymptotically stronger than Chebyshev's inequality. However, Hoeffding's bound requires mutual independence of the votes. For $t = \log_2 n + 4, \ldots, n - 1$ the multisets σ_t are chosen in such a way that Hoeffding's bound can be used. For $t < \log_2 n + 4$ the votes are just pairwise independent so only Chebyshev's inequality can be used. However, as mentioned above, the number of votes cannot be large for small t so in this case we cannot gain from using Hoeffding's bound rather than Chebyshev's inequality. This issue is addressed in more details in Section 5.

4.3   The Algorithm

In this section we formally describe the inversion algorithm. Suppose we know $E_N(x)$ for some $x \in \Lambda_N$. Let O_{xor} be an algorithm that, given $E_N(x)$ for some $x \in \Lambda_N$ and a subset $\pi \subset \{1, \ldots, j\}$, guesses $\pi(x)$ with advantage δ. The inversion algorithm that uses O_{xor} as an oracle and outputs x runs as follows.

```
Input E_N(x), N, j and oracle O_xor
---- First part: oracle calls ----
Select random integers a, b ∈ Z_N
For t = 1, ..., n do
    a_t = [2^{-t} a]_N
    For i ∈ σ_t do
        c_{t,i} = a_t(1 + 2i) + b
        If (2^{r-1} c_{t,i} / N) = +1 then
            Select a random nonempty subset π_{t,i} ⊂ {1, ...j}
            Set r = max{k | k ∈ π_{t,i}}
            g_{t,i} = O_xor(π_{t,i}, E_N([2^{r-1} c_{t,i} x]_N)), validity bit d_{t,i} = 1
        Else
            d_{t,i} = 0
        End if
    End do
End do
---- Second part: tightening the rational approximations ----
For ũ = 0, ..., ⌊η_a^{-1}/2⌋; ṽ = 0, ..., ⌊η_b^{-1}/2⌋; l_{a,0} = 0, 1; l_b = 0, 1 do
    Reset d_{t,i} with the values calculated in the first part
```

```
Rational u = 2η_a ũ, v = 2η_b ṽ, set u_0 = u
For t = 1, ..., n − 1 do
    Rational u_t = ½(l_{a,t−1} + u_{t−1})
    For i ∈ σ_t such that d_{t,i} = 1 do
        Rational w_{t,i} = u_t(1 + 2i) + v
        If ⌊2^r w_{t,i}⌋ = 0 mod 2 then
            Set r = max{k | k ∈ π_{t,i}}, assign w̃_{t,i} = w_{t,i} mod 1
            e_i = l_b + π_{t,i}(L_r(−⌊2^{r−1}w̃_{t,i}⌋N)) + ⌊w_{t,i}⌋ mod 2
        Else
            d_{t,i} = 0
        End if
    End do
    l_{a,t} = MajorityDecision(g_{t,∗} + e_∗ mod 2, d_{t,∗})
End do
x' = [a_n^{−1}⌊u_n N + ½⌋]_N
If (x'/N) = +1 and E_N(x') = E_N(x) then output x'
End do
```

On step t the goal of the algorithm is to determine $\ell_1([a_t x]_N)$. This bit is determined via majority decision. Note that $e_i = l_b + \pi_{t,i}(L_r(−⌊2^{r−1}w̃_{t,i}⌋N)) + ⌊w_{t,i}⌋ \bmod 2$ and $g_{t,i} = O_{xor}(\pi_{t,i}, E_N([2^{r−1}c_{t,i}x]_N))$ (see also (5)). If for a majority of indices $i ∈ σ_t$ such that $d_{t,i} = 1$ we have $g_{t,i} = e_i$ then the majority decision outputs 0, otherwise it outputs 1 (in terms of the above algorithm $d_{t,i} = 1$ if and only if $[2^{r−1}c_{t,i}x]_N ∈ Λ_N$). If the majority decision is correct then $l_{a,t} = \ell_1([a_t x]_N)$.

5   Analysis of the Inversion Algorithm

In this section we determine the success probability of the above inversion algorithm. More formally, we prove the following lemma.

Lemma 5. *The above algorithm, given $E_N(x)$ for $x ∈ Λ_N$, j, and N, outputs x with probability $2/9$, where the probability is taken over internal coin flips of the algorithm (which includes the coin flips of O_{xor}.*

Recall that the inversion algorithm works as follows. For $a, b ∈_R \mathbb{Z}_N$, we search through a certain number of quadruples $(u_0 N, vN, l_{a,0}, l_b)$ such that for at least one of them (2) holds, i.e. $l_{a,0} = \ell_1([ax]_N)$, $l_b = \ell_1([bx]_N)$; $|[ax]_N − u_0 N| ≤ η_a N$, $|[bx]_N − vN| ≤ η_b N$. Throughout this section we only consider a quadruple for which (2) holds (for the other quadruples we assume that the algorithm outputs x with probability 0).

At each step t, $1 ≤ t < n$, the goal of the inversion algorithm is to determine $\ell_1([a_t x]_N)$. Using O_{xor} this bit is determined via the majority decision, which depends on a certain number of votes. For $i ∈ σ_t$ such that $[2^{r−1}c_{t,i}x]_N ∈ Λ_N$, the i-th vote is set to 0 if (5) holds, otherwise it is set to 1. The decision on

$\ell_1([a_t x]_N)$ is set to the majority vote. The decision is correct if the majority of the votes is correct.

Consider step t, $1 \leq t < n$. Assume that for all $s < t$ we have determined correctly the bits $\ell_1([a_s x]_N)$. There exist two reasons why for some $i \in \sigma_t$ the i-th vote could be incorrect.

- The error of the rational approximation $w_{t,i} N$ is too large so that (3) does not hold.
- Oracle O_{xor} outputs a wrong bit (recall that it outputs the correct bit only with probability $1/2 + \delta$).

5.1   The Probability That (3) Does Not Hold

Lemma 6. *Assume that (2) holds and for all $s < t$ the bits $\ell_1([a_s x]_N)$ are determined correctly. Then the probability that (3) does not hold for some $i \in \sigma_t$ is at most $\delta/4$. Here the probability is taken over all choices of random multipliers $a, b \in_R \mathbb{Z}_N$.*

Proof. Let us rewrite (3) again:

$$[2^j c_{t,i} x]_N = 2^j \left([a_t x]_N (1 + 2i) + [bx]_N\right) - \lfloor 2^j w_{t,i} \rfloor N.$$

Intuitively, (3) does not hold if there exists a multiple of N between $2^j([a_t x]_N(1 + 2i) + [bx]_N)$ and $2^j w_{t,i} N$. Denote $\Delta_{t,i} = 2^j w_{t,i} N - 2^j([a_t x]_N(1 + 2i) + [bx]_N)$. Then (3) does not hold if and only if

$$\left|\Delta_{t,i}\right| \geq \left|2^j([a_t x]_N(1 + 2i) + [bx]_N)\right|_N = \left|2^j c_{t,i} x\right|_N,$$

where for $z \in \mathbb{Z}$ $|z|_N = \min([z]_N, N - [z]_N)$ denotes the distance from z to the closest multiple of N.

If (2) holds and for all $s < t$ we have determined correctly the bits $\ell_1([a_s x]_N)$ then

$$|[a_t x]_N - u_t N| = 2^{-t}([ax]_N - u_0 N) \leq 2^{-t-j-6} \delta^3 N,$$
$$|[bx]_N - v| = 2^{-j-4} \delta N.$$

Since $2^{-t} \delta^2 |1 + 2i| \leq 4$ for $i \in \sigma_t$ (see Section 4.2) the triangular inequality gives

$$|\Delta_{t,i}| = 2^j \left|u_t N(1 + 2i) - [a_t x]_N(1 + 2i) + vN - [bx]_N\right| \leq$$
$$\frac{\delta}{64}(2^{-t} \delta^2 |1 + 2i| + 4)N \leq \frac{\delta}{8} N.$$

Thus (3) does not hold only if $\left|2^j c_{t,i} x\right|_N \leq \delta N/8$. Since $c_{t,i}$ is uniformly distributed in $\mathbb{Z}_N$, the probability that (3) does not hold is at most $\delta/4$. It completes the proof of Lemma 6.

5.2   Error Probability of the Majority Decisions

Throughout this section we will refer to indices i such that $[2^{r-1}c_{t,i}x]_N \in \Lambda_N$ as *valid indices*. The i-th vote in the majority decision on $\ell_1([a_tx]_N)$ is correct if (3) holds and the reply of O_{xor} is correct. Following the notation of [5] we define boolean variables τ_i such that $\tau_i = 1$ only if the i-th vote is incorrect:

$$\tau_i = 1 \text{ if and only if (3) does not hold or } O_{xor}([c_{t,i}x]_N, \pi_{t,i}) \neq \pi_{t,i}([c_{t,i}x]_N).$$

It is shown [5] that for any fixed t, $1 \leq t < n$, the multipliers $c_{t,i}$ are pairwise independent. Thus boolean variables τ_i, $i \in \sigma_t$, are also pairwise independent.

Let μ_t be the number of valid indices $i \in \sigma_t$. The majority decision errs only if

$$\frac{1}{\mu_t} \sum_{\text{valid } i \in \sigma_t} \tau_i > \frac{1}{2}.$$

Due to the different choice of σ_t for $t < \log_2 n + 4$ and for $t \geq \log_2 n + 4$ (see Section 4.2) we divide our analysis into two parts.

Case $t < \log_2 n + 4$. Consider a step $t < \log_2 n + 4$. Since O_{xor} guesses correctly with probability $\frac{1}{2} + \delta$, Lemma 6 implies that the expected value $\mathrm{E}[\tau_i] \leq 1/2 - 3\delta/4$. The majority decision errs only if

$$\frac{1}{\mu_t} \sum_{\text{valid } i \in \sigma_t} \tau_i - \mathrm{E}[\tau_i] \geq \frac{3}{4}\delta.$$

Since the variance of any boolean variable is at most $1/4$, $\mathrm{Var}[\tau_i] \leq 1/4$. Chebyshev's inequality for μ_t pairwise independent random variables τ_i gives

$$\Pr\left[\frac{1}{\mu_t} \sum_{\text{valid } i \in \sigma_t} \tau_i - \mathrm{E}[\tau_i] \geq \frac{3}{4}\delta\right] \leq \left(\frac{3}{4}\delta\right)^{-2} \mathrm{Var}\left[\frac{1}{\mu_t} \sum_{\text{valid } i \in \sigma_t} \tau_i\right] \leq \frac{4}{9\mu_t\delta^2}.$$

Here the probability is taken over all choices of random multipliers $a, b \in_R \mathbb{Z}_N$, and internal coin flips of O_{xor}.

Since on average $\mu_t = m_t/4 = 2^t\delta^{-2}$, the majority decision for $\ell_1([a_tx]_N)$ errs with probability $\frac{4}{9}2^{-t}$. Thus the probability that at least one of the majority decisions for $t < \log_2 n + 4$ errs is at most $4/9$.

Case $t \geq \log_2 n + 4$. The technique we use for $t \geq \log_2 n + 4$ is called subsample majority decision. It is proposed by Fischlin and Schnorr [5].

Consider a step $t \geq \log_2 n + 4$. Instead of using indices from a large sample $\rho = \{i \mid |1 + 2i| < 2^6 n\delta^{-2}\}$ we use only indices from a small random subsample $\sigma = \{i_1, \ldots, i_m\} \subset \rho$, where $m = 8\delta^{-2}\log_2 n$ (see also Section 4.2). Although original τ_i, $i \in \rho$, are just pairwise independent $\tau_{i_1}, \ldots, \tau_{i_m}$ are mutually independent. Therefore for these random variables we can use a stronger bound instead of Chebyshev's inequality, namely Hoeffding's bound [8].

Let μ_t be the number of valid indices $i \in \sigma$ (the number of $i \in \sigma$ such that $[2^{r-1}c_{t,i}x]_N \in \Lambda_N$). The majority decision errs only if

$$\frac{1}{\mu_t} \sum_{\text{valid } i_s \in \sigma} \tau_{i_s} - \mathrm{E}[\tau_i] \geq \frac{3}{4}\delta,$$

Let ν_t denote the number of valid indices in ρ. Denote

$$\tau = \frac{1}{\nu_t} \sum_{\text{valid } i \in \rho} \tau_i,$$

where $|\rho| = 2^6 n \delta^{-2}$. The majority decision errs if either $\tau - \mathrm{E}[\tau] \geq \delta/4$ or

$$\frac{1}{\mu_t} \sum_{\text{valid } i_s \in \sigma} \tau_{i_s} - \tau \geq \frac{1}{2}\delta.$$

Chebyshev's inequality for pairwise independent τ_i, $i \in \rho$, gives $\Pr[\tau - \mathrm{E}(\tau) \geq \delta/4] \leq 4/(\nu_t \delta^2)$. Hoeffding's bound [8] implies that for fixed τ_i, $i \in \rho$, and a random subsample $\sigma \subset \rho$

$$\Pr\left[\frac{1}{\mu_t} \sum_{\text{valid } i_s \in \sigma} \tau_{i_s} - \tau \geq \frac{1}{2}\delta\right] \leq \exp\left(-2\mu_t \left(\frac{\delta}{2}\right)^2\right) = \exp\left(-\frac{1}{2}\mu_t \delta^2\right). \quad (6)$$

Since on average $\mu_t = m/4$ and $\nu_t = |\rho|/4$ (on average only $1/4$ of the indices are valid) the majority decision at each step $t \geq \log_2 n + 4$ errs with probability at most $16/(|\rho|\delta^2) + \exp(m\delta^2/8) = 1/(4n) + n^{-1/\ln 2} < 1/(3n)$ for $n > 2^9$. Thus the probability that at least one of the subsample majority decisions for $t \geq \log_2 n + 4$ errs is at most $1/3$.

Therefore the inversion algorithm of Section 4, given $E_N(x)$, j, and N, outputs x with probability at least $1 - (4/9 + 1/3) = 2/9$. It completes the proof of Lemma 5.

6   Complexity of the Inversion Algorithm

In this section we determine the running time of the inversion algorithm. The unit of time we use throughout this paper is a clock cycle.

The first part of the algorithm (oracle calls) consist of n steps $t = 1, \ldots, n$. On average we run the algorithm O_{xor} $m_t/2 = 2 \cdot 2^t \delta^{-2}$ times for $t < \log_2 n + 4$ and $m/2 = 4\delta^{-2} \log_2 n$ for $t \geq \log_2 n + 4$, therefore in total we run O_{xor} $32n\delta^{-2} + 4n(\log_2 n)\delta^{-2} \approx 4n(\log_2 n)\delta^{-2}$ times. Note that the number of oracle calls does not depend on the number of quadruples $(u, v, \ell_1([ax]_N), \ell_1([bx]_N))$.

In the second part (tightening the rational approximations) we do not use O_{xor} but we run a large exhaustive search cycle. The bottleneck of the second part is multiplication $\lfloor 2^{r-1}\tilde{w}_{t,i} \rfloor \cdot N$. The size (in bits) of $\tilde{w}_{t,i}$ is $\log_2(\eta_b^{-1}) = \log_2(\delta^{-1}) + j + 4$ and the size of N is n. For instance, for $\varepsilon = 1/2$, $M = 2^{20}$, $j = 5$ we have $\delta = 2^{-26}$ and hence $\log_2(\eta_b^{-1}) = 35$. Therefore we may assume that a single multiplication takes n clock cycles. Hence the complexity of the second part is at most the product of the following factors:

1. Number of quadruples $(u, v, \ell_1([ax]_N), \ell_1([bx]_N))$, that is $\eta_a^{-1}\eta_b^{-1}$;
2. Number of steps t, that is n;
3. Number of votes for the majority decision, that is $m/4 = 2\delta^{-2}\log_2 n$;
4. Complexity of the multiplication $\lfloor 2^{r-1}\tilde{w}_{t,i}\rfloor \cdot N$, that is n;

Since $\eta_a = 2^{-j-6}\delta^3$, $\eta_b = 2^{-j-4}\delta$, the complexity of the second part is $2^{11}\delta^{-6}n^2\log_2 n$ clock cycles. Recall that the running time of O_{xor} is essentially the same as the one of B. Thus the running time of the inversion algorithm is $4n(\log_2 n)\delta^{-2}(T_B + 2^{2j+9}n\delta^{-4})$. Lemma 5 implies that there exists algorithm R that inverts the absolute Rabin function in expected time

$$T_R \leq 18n(\log_2 n)\delta^{-2}(T_B + 2^{2j+9}n\delta^{-4}). \tag{7}$$

Remark 4. The argument of [5] implies that inversion of the absolute Rabin function needs at least $(\ln 2/4)n\delta^{-2}$ runs of B (see also Remark 3). Therefore the number of oracle runs in the above inversion algorithm is optimal up to a factor of $\log_2 n$. However, in (7) we also have a second component that cannot be neglected in practice.

7   Concrete Security of the BBS

In this section we state our main result about the concrete security of the BBS pseudorandom generator. We give a bound for running time T_A and advantage ε such that the BBS generator is (T_A, ε)-secure (Theorem 3).

Theorem 1. *Suppose the BBS pseudorandom generator is not (T_A, ε)-secure. Then there exist an algorithm F that factors the modulus N in expected time*

$$T_F \leq 36n(\log_2 n)\delta^{-2}(T_A + 2^{2j+9}n\delta^{-4}),$$

where $\delta = (2^j - 1)^{-1}M^{-1}\varepsilon$.

Proof. The statement follows from (7), Lemma 1, and Lemma 2.

Therefore a statistical test that distinguishes the BBS sequences from random sequences can be used to factor the modulus N. However, we observe that the reduction is not tight in the sense that for a practical choice of parameters $T_F \gg T_A$. Furthermore, Remark 4 implies that the reduction for the BBS generator cannot be significantly tighter. There is a large gap between security of this pseudorandom generator and the factoring problem.

In order to complete the concrete security analysis of the BBS generator we will assume a concrete lower bound on the hardness of integer factoring, which is obtained by extrapolating the best factorization results to date.

7.1   Hardness of Factoring

The fastest general-purpose factoring algorithm today is the general number field sieve. According to [4,11] on heuristic grounds the number field sieve is expected to require time proportional to $\gamma \exp((1.9229 + o(1))(\ln N)^{1/3}(\ln \ln N)^{2/3})$ for a constant γ. Following [4] we make an assumption that the $o(1)$-term can be treated as zero. From this we can calculate γ.

Let $L(n)$ be the number of clock cycles needed to factor an n-bit integer. We assume that $L(n) \approx \gamma \exp(1.9229(n \ln 2)^{1/3}(\ln(n \ln 2))^{2/3})$. Experience from available data points suggests that $L(512) \approx 3 \cdot 10^{17}$ clock cycles, therefore $\gamma \approx 2.8 \cdot 10^{-3}$ and

$$L(n) \approx 2.8 \cdot 10^{-3} \cdot \exp(1.9229(n \ln 2)^{1/3}(\ln(n \ln 2))^{2/3}). \tag{8}$$

Assumption 2. *No algorithm can factor a randomly chosen n-bit Blum-integer in expected time $T < L(n)$, where $L(n)$ is given by (8).*

All the results below hold under the above assumption.

Theorem 3 (Concrete security of the BBS). *Under Assumption 2, the BBS pseudorandom generator is (T_A, ε)-secure if*

$$T_A \leq \frac{L(n)}{36n(\log_2 n)\delta^{-2}} - 2^{2j+9}n\delta^{-4}, \tag{9}$$

where $\delta = (2^j - 1)^{-1}M^{-1}\varepsilon$.

7.2   Comparison with Known Results

Thus we have shown that there exist a reduction a successful attack on the BBS generator to factoring. While for asymptotic security it suffices to give a polynomial time reduction, we need for concrete security a reduction that is as efficient as possible. In this subsection we compare the complexity of our reduction, given by Theorem 3, with the results of [1,15,5].

A close look at the security proof of Alexi et al. [1] gives the following lemma.

Lemma 7 (Alexi et al.). *Under Assumption 2, the BBS pseudorandom generator is (T_A, ε)-secure if*

$$T_A \leq \frac{L(n)}{2^{27}2^{4j}n^3\varepsilon^{-8}M^8}. \tag{10}$$

Recall that M denotes the length of the output of the BBS generator (e.g., $M = 2^{20}$). Formula (10) has M^8 in the denominator whereas (9) has M^2. Thus our security proof is stronger than the one of [1].

A disadvantage of [15] is that this paper deals only with *deterministic* statistical tests, thus the result can not be expressed in terms of Definition 3. Furthermore [15] uses [1] as a building block so the complexity of the reduction proposed is of the same order.

The lemma below is due to Fischlin and Schnorr [5]. It establishes the concrete security of the BBS generator with 1 output bit per iteration.

Lemma 8 (Fischlin and Schnorr). *Under Assumption 2, the BBS pseudo-random generator with 1 output bit per iteration is* (T_A, ε)*-secure if*

$$T_A \leq \frac{L(n)}{6n(\log_2 n)\varepsilon^{-2}M^2} - 2^7 n\varepsilon^{-2}M^2 \log_2(8n\varepsilon^{-1}M). \tag{11}$$

Here the denominator of the first component in the righthand side is essentially the same as the one in (9) for $j = 1$. The second component in (11) is smaller in the absolute value than the second component in (9) by a factor of $\varepsilon^{-2}M^{-2}$. The reason is that there is a trick in the reduction [5] (namely, processing all approximate locations simultaneously) that allows to decrease the second component. We do not know if it is possible to apply this trick for $j > 1$ and we leave this question as an open problem. In the below example the factor of $\varepsilon^{-2}M^{-2}$ in the second component does not affect the final conclusion about the optimal value of j.

7.3   Example

An important application of Theorem 3 is that it can be used to determine the optimal values for the parameters of the BBS generator.

Suppose our goal is to generate a sequence of $M = 2^{20}$ bits such that no adversary can distinguish this sequence from truly random binary sequence in time $T_A = 2^{100}$ clock cycles with advantage $\varepsilon = 1/2$. The question is what length of the modulus n and parameter j should be used to minimize the computational effort per output bit.

Inequalities (9) and (11) connect the security parameters (T_A, ε) with parameters of the BBS (M, n, j) for $j \geq 1$ and $j = 1$ respectively. In order to find the optimal n and j we fix T_A, ε, M and consider n as a function of j.

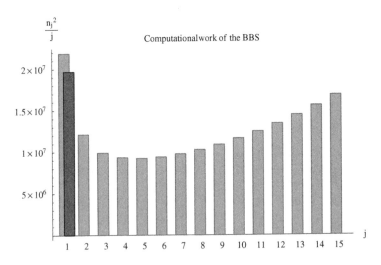

Fig. 1. The computational work of the BBS is minimal for $j = 5$

The computational work of the BBS (the running time needed to output a pseudorandom sequence) is proportional to n^2/j (each modular multiplication costs $O(n^2)$ binary operations so the generation of a BBS sequence takes $O(Mn^2/j)$ operations). Figure 1 displays the computational work of the BBS for $j \in [1, 15]$. There are two values of the computational work for $j = 1$ on the figure. The smaller value results from (11) and the larger one results from (9). For $j = 1$ the reduction [5] is more efficient. Nevertheless, the difference turns out to be not significant and we observe that extraction of 5 bits per iteration makes the BBS about 2 times faster in comparison with 1 bit case. However, even for $j = 5$ the BBS is quite slow since the corresponding length of the modulus $n = 6800$.

It is not true that it always pays off to extract more that 1 bit on each iteration. The optimal number of bits to be extracted on each iteration depends on the length of the output sequence M and on the security parameters T_A, ε. For instance, for $M = 2^{30}$ it turns out that the best choice is $j = 1$.

8   Concluding Remarks and Open Problems

Security of the BBS genarator has been thoroughly analyzed [1,2,5,6]. Nevertheless, it has been uncertain how to select the size of the modulus and the number of bits extracted on each iteration such that a desired level of security is reached, while minimizing the computational effort per output bit. In this paper we answer this question.

Generalizing the ideas of [5,15] we propose a new security proof for the BBS generator with several output bits per iteration. The new reduction is more efficient than all previously known reductions.

With minor changes our argument can be applied for the analysis of the RSA pseudorandom generator.

However, the new reduction is still not tight. There is a large gap between the security of BBS generator and the factoring problem. Moreover, using the information-theoretic approach, Fischlin and Schnorr [5] show that the security reduction for BBS cannot be significantly tighter. Searching for the constructions based on the factoring problem with tight security reduction is one of the challenging problems in the theory of provably secure pseudorandom generators.

Acknowledgements

We thank Natalia Chernova for very helpful discussions.

References

1. W. Alexi, B. Chor, O. Goldeich and C. P. Schnorr. RSA and Rabin functions: certain parts are as hard as the whole. SIAM Journal on Computing 17, 1988, pp. 194–209.
2. L. Blum, M. Blum and M. Shub. A Simple Unpredictable Pseudo-Random Number Generator. SIAM Journal on Computing 15, 1986, pp. 364–383.

3. M. Bellare and P. Rogaway. The exact security of digital signatures how to sign with RSA and Rabin. In Advances in Cryptology – Eurocrypt 1996, volume 1070 of Lecture Notes in Computer Science, pp. 399 – 416. Berlin: Springer-Verlag, 1996.
4. ECRYPT Yearly Report on Algorithms and Keysizes (2004). Available at `http://www.ecrypt.eu.org/documents/D.SPA.10-1.1.pdf`.
5. R. Fischlin and C. P. Schnorr. Stronger Security Proofs for RSA and Rabin Bits. Journal of Cryptology (2000) 13, pp. 221–244.
6. R. Gennaro. An improved pseudo-random generator based on discrete log. In Mihir Bellare, editor, Advances in Cryptology – Crypto 2000, volume 1880 of Lecture Notes in Computer Science, pp. 469 – 481. Springer-Verlag, 20–24 August 2000.
7. O. Goldreich. Three XOR-Lemmas – An Exposition. Technical report, ECCC TR95-056, 1995. Available at `ftp://ftp.eccc.uni-trier.de/pub/eccc/reports/1995/TR95-056/Paper.pdf`.
8. W. Hoeffding. Probability in Equations for Sums of Bounded Random Variables. Journal of the American Statistical Association 56 (1963), pp. 13–30.
9. D. E. Knuth. Seminumerical Algorithms, 3rd edition. Addison-Wesley, Reading, MA, 1997.
10. J. Katz, N. Wang. Efficiency Improvements for Signature Schemes with Tight Security Reductions. CCS'03, October 27-30, 2003, Washington, DC, USA.
11. A. K. Lenstra, E. R. Verheul. Selecting Cryptographic Key Sizes. Journal of Cryptology (2001) 14: 255–293.
12. A. J. Menezes, P. C. van Oorschot, S. A. Vanstone. Handbook of Appied Cryptography. CRC Press series on discrete mathematics and its applications, 2000.
13. M. O. Rabin. Digitalized signatures and public-key functions as intractible as factorization. Technical report, TR-212, MIT Laboratory for Computer Science, 1979.
14. V. Shoup. On the security of a practical identification scheme. In Advances in Cryptology – Eurocrypt 1996, volume 1070 of Lecture Notes in Computer Science, pp. 344 – 353. Berlin: Springer-Verlag, 1996.
15. U. V. Vazirani, V. V. Vazirani. Efficient and Secure Pseudo-Random Number Generation. Proceedings 25th Symposium on Foundations of Computing Science IEEE, pp. 458–463, 1984.
16. A. C. Yao. Theory and Application of Trapdoor Functions. Proceedings of IEEE Symposium on Foundations of Computer Science, pp. 80–91, 1982.

A   Proof of Lemma 4

Lemma 4 states that if (3) holds then

$$\pi_{t,i}([2^{r-1}c_{t,i}x]_N) = \ell_1([c_{t,i}x]_N) + +\pi_{t,i}(L_r(-\lfloor 2^{r-1}\tilde{w}_{t,i}\rfloor N)) \bmod 2.$$

To prove this lemma we will show that

$$\pi_{t,i}([2^{r-1}c_{t,i}x]_N) = \pi_{t,i}(2^{r-1}[c_{t,i}x]_N - \lfloor 2^{r-1}\tilde{w}_{t,i}\rfloor N) \tag{12}$$

and

$$\ell_1([c_{t,i}x]_N) + \pi_{t,i}(L_r(-\lfloor 2^{r-1}\tilde{w}_{t,i}\rfloor N)) = \\ \pi_{t,i}(2^{r-1}[c_{t,i}x]_N - \lfloor 2^{r-1}\tilde{w}_{t,i}\rfloor N) \bmod 2. \tag{13}$$

It can be shown that if (3) holds then for all r, $0 \le r \le j$,

$$[2^{r-1}c_{t,i}x]_N = 2^{r-1}[c_{t,i}x]_N - \lfloor 2^{r-1}\tilde{w}_{t,i} \rfloor N. \tag{14}$$

Applying function $\pi_{t,i}$ to both sides of (14) gives (12). To prove (13) we first note that

$$L_r(2^{r-1}[c_{t,i}x]_N - \lfloor 2^{r-1}\tilde{w}_{t,i} \rfloor N) = \\ (2^{r-1}\ell_1([c_{t,i}x]_N) + L_r(-\lfloor 2^{r-1}\tilde{w}_{t,i} \rfloor N)) \bmod 2^r. \tag{15}$$

From (14) follows that $2^{r-1}[c_{t,i}x]_N - \lfloor 2^{r-1}\tilde{w}_{t,i} \rfloor N \ge 0$. Hence in this case L_r corresponds to r least-significant bits. Thus applying function $\pi_{t,i}$ to the left-hand side of (15) gives

$$\pi_{t,i}(L_r(2^{r-1}[c_{t,i}x]_N - \lfloor 2^{r-1}\tilde{w}_{t,i} \rfloor N)) = \pi_{t,i}(2^{r-1}[c_{t,i}x]_N - \lfloor 2^{r-1}\tilde{w}_{t,i} \rfloor N). \tag{16}$$

Then we apply $\pi_{t,i}$ to the right-hand side of (15):

$$\pi_{t,i}((2^{r-1}\ell_1([c_{t,i}x]_N) + L_r(-\lfloor 2^{r-1}\tilde{w}_{t,i} \rfloor N)) \bmod 2^r) = \\ \pi_{t,i}(2^{r-1}\ell_1([c_{t,i}x]_N) + L_r(-\lfloor 2^{r-1}\tilde{w}_{t,i} \rfloor N)) = \\ \ell_1([c_{t,i}x]_N) + \pi_{t,i}(L_r(-\lfloor 2^{r-1}\tilde{w}_{t,i} \rfloor N)) \bmod 2, \tag{17}$$

since $\pi_{t,i} \subset \{1, \ldots, r\}$, $r \in \pi_{t,i}$. (15), (16), and (17) result in (13). It completes the proof of Lemma 4.

The Equivalence Between the DHP and DLP for Elliptic Curves Used in Practical Applications, Revisited

K. Bentahar

Dept. Computer Science, University of Bristol,
Merchant Venturers Building, Woodland Road,
Bristol, BS8 1UB, United Kingdom
bentahar@cs.bris.ac.uk

Abstract. The theoretical equivalence between the DLP and DHP problems was shown by Maurer in 1994. His work was then reexamined by Muzereau *et al.* [12] for the special case of elliptic curves used in practical cryptographic applications. This paper improves on the latter and tries to get the tightest possible reduction in terms of computational equivalence, using Maurer's method.

Keywords: DHP-DLP equivalence, Elliptic Curve Cryptosystems.

1 Introduction

Maurer and Wolf [7,9,8,11] proved that, for every cyclic group G with prime order p, the DLP and DHP over G are equivalent if there exists an elliptic curve, called *auxiliary elliptic curve*, over $\mathbb{F}_p$ with smooth order.

Muzereau *et al.* [12] showed that such auxiliary elliptic curves are highly likely to exist for almost all elliptic curve groups. It is however remarked that it gets extremely hard to construct them as the order of G increases. Auxiliary elliptic curves with smooth orders were built and explicitly presented for most of the curves in the SECG standard, hence making Maurer's proof applicable to most of the groups used in practical elliptic curve cryptography.

The idea behind the method introduced by Maurer [7] rests on the concept of *implicit representation*: The *implicit representation* of an integer a (modulo p) is defined to be $g^a \in G$. The algorithm proceeds by doing computations in the implicit representation instead of the usual explicit representation. For example, to compute $a + b$ in implicit form, $g^a \cdot g^b$ is computed instead which costs one multiplication. For $a - b$, we compute $g^a \cdot (g^b)^{-1}$ costing one inversion and one multiplication. To compute $a \cdot b$ in implicit form, one call to an *DH-oracle*, that computes g^{ab} given g^a and g^b, is needed. For the implicit form of a^{-1}, one uses the fact that $a^{p-1} = 1$, so $a^{p-2} = a^{-1}$, which would cost $\mathcal{O}(\lg p)$ calls to the DH-oracle. Hence, granted access to a DH-oracle for the group G, all algebraic algorithms can be converted to work in the implicit representation.

This paper builds on [12] by tightening the reduction and trying to extend the result to the remaining curves. Our goal is to show that, for the elliptic

N.P. Smart (Ed.): Cryptography and Coding 2005, LNCS 3796, pp. 376–391, 2005.

curve cryptosystems described in the various standards, the number of group operations and DH-oracle calls required to reduce the DLP to the DHP is reasonably "small." Say for example that this number is less than 2^r then, if we believe that the much more extensively studied DLP over the same group takes *at least* 2^ℓ operations to solve then an algorithm for solving the DHP, and thus breaking the DHP protocol, would require a minimum of $2^{\ell-r}$ group operations. Our target is therefore to minimise the value of r, in order to get the tightest possible security reduction.

Affine coordinates were used in [12] which requires division and hence a DH-inversion oracle was needed. This was implemented at the cost of $\mathcal{O}(\lg p)$ calls to a DH-oracle which is clearly an expensive choice as it leads to a large increase in the number of DH-oracle calls. We use *projective coordinates* instead to avoid this problem. As a further optimisation we use an optimised square root extraction algorithm.

One would also think that using *addition chains* may reduce the cost of exponentiation but it turns out that this saves very little and only adds complications. So it was decided to use traditional methods of exponentiation and concentrate on the more critical areas of the algorithm. Section 6 expands on this and justifies this decision.

The full version of this paper [2] providess a list of auxiliary elliptic curves giving almost the tightest possible reduction, using the Maurer method.

2   Notation and Definitions

Throughout the paper, we let G be a cyclic group with generator g and prime order $p > 3$. We begin by defining the problems DLP and DHP.

Definition 1 (DLP and DHP)

- *Given $h \in G$, the problem of computing an integer $\alpha \in [0, |G|)$ such that $g^\alpha = h$ is called the* Discrete Logarithm Problem *(DLP) with respect to g.*
- *Given two elements $g^a, g^b \in G$, we call the problem of computing g^{ab} the* Diffie-Hellman Problem *(DHP) with respect to g. (a, b are unknown)*

We also need to formalise the notion of a DH and DL oracles.

Definition 2 (DL and DH oracles)

- *A DH-oracle takes as input two elements $g^a, g^b \in G$ and returns g^{ab}. We write $\mathcal{DH}(g^a, g^b) = g^{ab}$.*
- *A DL-oracle takes as input an element $h = g^a \in G$ and returns $a \bmod |G|$. We write $\mathcal{DL}(h) = \mathcal{DL}(g^a) = a$.*

Both oracles return answers in unit time. (By definition of Oracles)

The equivalence between the two problems was theoretically established by Maurer and Wolf in the nineties [7,9,8,11], but it relies on the existence of some auxiliary elliptic curves whose orders must be smooth. These auxiliary elliptic curves are not necessarily easy to build and it seems they are exceptionally hard to find in general. Hence, a more concrete treatment for the elliptic curve groups used in practice proved necessary and this was done in [12]. The paper discussed the *computational equivalence* between the DLP and DHP, and it also presented an explicit list auxiliary elliptic curves needed for the reduction.

Note that, since solving any instance of the DHP given access to a DL-oracle is trivial[1], we only concentrate on the reverse implication for the equivalence to hold: If we suppose the DHP turns out to be easy, we wish to know if this implies that the DLP is easy as well.

The base 2 logarithm will be denoted by $\lg x$ (instead of $\log_2$). We will also use $\mathfrak{M}$ and $\mathfrak{I}$ to denote multiplications and inversions in G, respectively, and $\mathfrak{DH}$ for DH-oracle calls. Formulae of the form

$$x\mathfrak{DH} + y\mathfrak{I} + z\mathfrak{M}$$

mean: Cost is x DH-oracle calls, y inversions and z multiplications in G.

3   The Algorithm

Given $h \in G$, we want to find the unique α modulo p such that $h = g^\alpha$. We assume an elliptic curve E over $\mathbb{F}_p$ is given by the Weierstrass equation $y^2 = x^3 - 3x + b$, with smooth order given as a product of *coprime integers*

$$|E| = \prod_{j=1}^{s} q_j, \qquad (1)$$

with $q_j < B$ of roughly the same size, where B is some smoothness bound. This choice of the defining equation of E saves $1\mathfrak{DH}$ while adding points on it. The point at infinity on E is denoted by $\mathcal{O}$.

To solve a DLP in G, Maurer's approach is to use a DH-oracle and solve the problem in the implicit representation over E, which is supposed to have a smooth order. So, given $h = g^\alpha \in G$ and the elliptic curve E, as above, we check whether $g^{y^2} = g^{\alpha^3 - 3\alpha + b}$ can be solved for y. If so then we have found a point Q on E in its implicit form, otherwise we replace α by $\alpha + d$ for some random, small, integer d and do the checking again until we get a point Q on E.

Note that, at this stage, we know Q in its implicit representation only. The idea now is to solve $Q = kP$ over E, where P is a generator of E. Upon finding the value of k, we then compute kP in the explicit representation and hence recover the value of α, from the explicit first coordinate of Q. Given that E has a smooth order, we simply use the naive Pohlig-Hellman method of first solving the problem in the subgroups of E of prime power order, and then recovering k using the Chinese Remainder Theorem (CRT). The reader is referred to Algorithm 1 for the detailed description of the algorithm.

---

[1] Given $g^a, g^b \in G$, we compute $a = \mathcal{DL}(g^a)$ and then compute $g^{ab} = (g^b)^a$.

The crucial point to note is that we have a wide choice of curves over $\mathbb{F}_p$ that have sizes distributed in the Hasse interval $[p + 1 - \sqrt{p}, p + 1 + \sqrt{p}]$. So, with a bit of luck, one hopes that one of these sizes is smooth enough and hence the corresponding auxiliary elliptic curve would make solving our DLP easy. We draw the reader's attention to the fact that this is the same reason that makes the ECM factoring method so successful.

In the description of Algorithm 1, note that for the comparison step (12) to test whether a point $(X : Y : Z)$, in projective coordinates, is equal to a point (x, y), in affine coordinates, we simply check whether $xZ^2 = X$ and $yZ^3 = Y$. In implicit representation this becomes

$$(g^{Z^2})^x = g^X \quad \text{and} \quad (g^{Z^3})^y = g^Y.$$

This use of projective coordinates gives our greatest improvement over [12]. We also make some savings by storing precomputed values and using them throughtout the algorithm. The next two subsections will describe the improvements made.

Algorithm 1 Solve a DLP in a group G given access to a DH-oracle for G.
Input: A cyclic group $G = \langle g \rangle$ of prime order p, an elliptic curve $E/\mathbb{F}_p \colon y^2 = x^3 - 3x + b$, generated by P, $|E| = \prod_{j=1}^{s} q_j$ and $h = g^\alpha \in G$
Output: $\alpha = \mathcal{DL}(h)$

    Step 1. Compute a valid implicit x-coordinate related to the DL α
1: **repeat**
2:     Choose d randomly, and set $g^x \leftarrow hg^d$ $\langle g^x \leftarrow g^{\alpha+d} \rangle$
3:     $g^z \leftarrow g^{x^3 - 3x + b}$.
4: **until** $g^{z^{(p-1)/2}} = g$ $\langle$ *Test quadratic-residuosity of z (mod p)*$\rangle$
    Step 2. Compute g^y from $g^z = g^{y^2}$:
5: Extract the square root of z in implicit representation, to obtain g^y.
    Now, $Q = (x, y)$ is a point on E known in the implicit representation only (g^x, g^y).
    Step 3. Compute $k \colon Q = kP$ in $E(\mathbb{F}_p)$: $\langle$ *Use the Pohlig-Hellman simplification*$\rangle$
6: **for** $j = 1, \ldots, s$ **do**
7:     Compute $Q_j = (g^{u_j}, g^{v_j}, g^{w_j})$, where $(u_j, v_j, w_j) = \frac{|E|}{q_j} Q$ $\langle$ *Projective coordinates*$\rangle$
8:     Set $i \leftarrow 0$, $(u, v) \leftarrow \mathcal{O}$, $P_j \leftarrow \frac{|E|}{q_j} P$ $\langle$ *Affine coordinates*$\rangle$
9:     **repeat** $\langle$ *Solve $Q_j = k_j P_j$ in the subgroup of $E(\mathbb{F}_p)$ of order q_j*$\rangle$
10:       $i \leftarrow i + 1$.
11:       $(u, v) \leftarrow (u, v) + P_j$. $\langle (u, v) \leftarrow iP_j = i \frac{|E|}{q_j} P \rangle$
12:     **until** $(g^{w_j^2})^u = g^{u_j}$ and $(g^{w_j^3})^v = g^{v_j}$ $\langle$ *Test if (g^u, g^v) equals $(g^{u_j}, g^{v_j}, g^{w_j})$*$\rangle$
13:     $k_j \leftarrow i$.
14: **end for**
    Step 4. Construct α
15: Compute k (mod $|E|$) such that $\forall j \in \{1, \ldots, s\} \colon k \equiv k_j$ (mod q_j). $\langle$ *Use CRT*$\rangle$
16: Compute $kP = Q$ in affine coordinates.
17: Then x (mod p) is the abscissa of Q, and $\alpha = x - d$.

3.1   Square Root Extraction

We describe the special cases in the explicit notation. This algorithm is used by Algorithm 1, in the implicit representation, to compute g^y from $g^z = g^{y^2} = g^{x^3-3x+b}$, see Algorithm 2.

Suppose a is known to be a quadratic residue modulo p and we want to compute $x \in \mathbb{F}_p$ such that $x^2 \equiv a \pmod{p}$. Then, besides the general Tonelli and Shanks algorithm used in [12], we also treat two special cases:

1. If $p \equiv 3 \pmod 4$ then $x \equiv a^{(p+1)/4} \pmod{p}$,
2. If $p \equiv 5 \pmod 8$ then do the following: Compute $s = a^{(p-5)/8}$, $u = a \cdot s$, $t = s \cdot u$. If $t = 1$ then $x = u$ otherwise $x = 2^{(p-1)/4} \cdot u$.

Treating these special cases is worthwhile since half the primes are congruent to 3 modulo 4, and half of the remaining primes are congruent to 5 modulo 8. The only remaining primes are all congruent to 1 modulo 8. We gain no advantage by using similar methods for this case, so we simply use the Tonelli-Shanks algorithm for the remaining primes, see [4, p. 32].

Algorithm 2 Implicit square roots in a group G using a DH-oracle for G.
Input: A cyclic group $G = \langle g \rangle$ of odd prime order p, and $g^z = g^{y^2} \in G$.
Output: g^y.

1: **if** $p \equiv 3 \pmod 4$ **then**
2:    $g^y \leftarrow g^{z^{(p+1)/4}}$. $\langle$*First case:* $p \equiv 3 \pmod 4 \rangle$
3: **else if** $p \equiv 5 \pmod 8$ **then**
4:    $g^s \leftarrow g^{z^{(p-5)/8}}, g^u \leftarrow g^{zs}, g^t \leftarrow g^{su}$. $\langle$*Second case:* $p \equiv 5 \pmod 8 \rangle$
5:    **if** $g^t = g$ **then**
6:       $g^y \leftarrow g^u$.
7:    **else**
8:       $g^y \leftarrow g^{u \cdot 2^{(p-1)/4}}$.
9:    **end if**
10: **else**
11:    Write $p - 1 = 2^e \cdot w$, w odd. $\langle$*Tonelli and Shanks algorithm for* $p \equiv 1 \pmod 8 \rangle$
12:    Set $g^s \leftarrow g, r \leftarrow e, g^y \leftarrow g^{z^{(w-1)/2}}, g^b \leftarrow g^{zy^2}, g^y \leftarrow g^{zy}$. $\langle$*Initialise*$\rangle$
13:    **while** $g^b \not\equiv 1 \bmod p$. **do**
14:       Find the smallest $m \geq 1$ such that $g^{(b^{2^m})} \equiv 1 \bmod p$. $\langle$*Find exponent*$\rangle$
15:       Set $g^t \leftarrow g^{(s^{2^{r-m-1}})}, g^s \leftarrow g^{t^2}, r \leftarrow m, g^y \leftarrow g^{yt}, g^b \leftarrow g^{bs}$. $\langle$*Reduction*$\rangle$
16:    **end while**
17: **end if**

3.2   Explicit and Implicit Point Multiplication

As already remarked, we use the projective coordinate system in step 3 of Algorithm 1 instead of the affine coordinate system. The formulae for addition and doubling[2] in the implicit representation easily follow from their standard explicit counterparts. The cost of each operation is given in the following table.

	Doubling		Addition	
	Explicit	Implicit	Explicit	Implicit
$\mathfrak{D}\mathfrak{H}$		8		16
$\mathfrak{J}$		4		5
$\mathfrak{M}$	8	14	16	$\frac{3}{2}\lg p + \frac{13}{2}$

For the affine coordinates, note that we only need the explicit case (In the j-loop). The costs are:

$$1\mathfrak{J} + 4\mathfrak{M} \text{ for doubling and } 1\mathfrak{J} + 3\mathfrak{M} \text{ for addition.}$$

Since we will need to compute kP for different values of k but a fixed P, pre-computing the values $2^1 P, 2^2 P, \ldots, 2^{\lfloor \lg k \rfloor} P$ will save us some computation. Then, using the right-to-left binary method, we expect only $\frac{1}{2} \lg k$ elliptic curve additions. We now summarise the costs of exponentiation.

Implicit Exponentiation in Projective Coordinates: The cost of the pre-computation is about

$$(8\mathfrak{D}\mathfrak{H} + 4\mathfrak{J} + 14\mathfrak{M}) \lg k \tag{2}$$

and then each exponentiation would cost about

$$\left(8\mathfrak{D}\mathfrak{H} + \frac{5}{2}\mathfrak{J} + \frac{1}{4}(3\lg p + 13)\mathfrak{M}\right) \lg k. \tag{3}$$

Explicit Exponentiation in Affine Coordinates: The precomputation cost is

$$(1\mathfrak{J} + 4\mathfrak{M}) \lg k \tag{4}$$

and then each exponentiation would cost

$$\frac{1}{2}(1\mathfrak{J} + 3\mathfrak{M}) \lg k. \tag{5}$$

3.3   Complexity of the Algorithm

The complexity analysis of Algorithm 1, presented in Appendix A, yields the following theorem

Theorem 1. *Let G be a cyclic finite group of prime order p. Assume an elliptic curve E over $\mathbb{F}_p$ has been found, whose B-smooth order is*

$$|E| = \prod_{j=1}^{s} q_j,$$

---

[2] Doubling is the operation of adding a point to itself: $2P = P + P$.

*where q_j are not necessarily prime but are coprime of roughly the same size.
Then, solving a given instance of the DLP in G requires on average about*

$$\mathcal{O}\left(\frac{\log^2 p}{\log B}\right)\mathfrak{DH} + \mathcal{O}\left(\frac{B\log^2 p}{\log B}\right)\mathfrak{M}.$$

For comparison, we quote below the asymptotic costs obtained by [12]

$$\mathcal{O}\left(\frac{\log^3 p}{\log B}\right)\mathfrak{DH} + \mathcal{O}\left(\frac{B\log^2 p}{\log B}\right)\mathfrak{M}.$$

While the number of multiplications has remained the same, the number of DH-oracle calls has now become quadratic in the size of the group G instead of cubic.

Note that, in order to get a lower bound on the cost of solving a DHP instance, we no longer require the auxiliary elliptic curves' orders to be smooth. This is because as long as we assume that the DLP is an exponentially hard problem then we do not mind if the reduction from the DHP to the DLP is exponential too. This remark will allow us to choose $s = 3$ later, and then the task will be to find smooth elliptic curves whose orders are product of three coprime numbers. This is a significant relaxation of the smoothness condition.

4   Implications on the Security of the DHP

The implications of this reduction on the security of the DLP was treated in [12]. We only comment on its implications on the security of the DHP, as it is here where the work done in this paper matters most.

Let C_{DLP}, C_{DHP} denote the costs of solving the DLP and DHP on an elliptic curve of size p, respectively. By Maurer's reduction, we have $C_{DLP} = N_{\mathfrak{DH}} \cdot C_{DHP} + N_{\mathfrak{M}}$, where $N_{\mathfrak{DH}}, N_{\mathfrak{M}}$ are respectively the number of calls to the DH-oracle and number of multiplications in G. Hence, for $N_{\mathfrak{M}} \ll C_{DLP}$ we get

$$C_{DHP} = \frac{C_{DLP} - N_{\mathfrak{M}}}{N_{\mathfrak{DH}}} \sim \frac{C_{DLP}}{N_{\mathfrak{DH}}}.$$

Since solving the DLP on an elliptic curve E is believed to take at least $\sqrt{|E|}$ steps [3], in general, then setting

$$T_{DH} = \frac{\sqrt{|E|}}{N_{\mathfrak{DH}}},$$

we see that T_{DH} gives us a lower bound on the number of operations required to break the DHP, assuming $N_{\mathfrak{M}} \ll C_{DLP}$. Hence, it is the value of T_{DH} that gives the exact security result, given the best auxiliary elliptic curves we found.

The tightness of the security reduction is controlled by two values. The first being the number of field multiplications $N_{\mathfrak{M}}$, and second and most important is the value T_{DH} for the reason put forth earlier.

Tables 1 and 2 give the logarithms of these key values, $\lg N_{\mathfrak{M}}$ and $\lg N_{\mathfrak{DH}}$, for the curves in the SECG standard [17]. They also give $\lg \sqrt{|E|}$, the logarithm of the (believed) minimum cost of solving an instance of the DLP. The column headed adv gives the number of security bits gained on the previous results from [12]. The last rows of the tables are detached to indicate that the values are theoretical and that no auxiliary elliptic curves could be generated for them, mainly due to the sheer size of the numbers that needed to be factored.

Table 1. Summary of results for curves of large prime characteristic

secp curve	$\lg \sqrt{|E|}$	$\lg N_{\mathfrak{M}}$	$\lg N_{\mathfrak{DH}}$	$\lg T_{DH}$	adv
secp112r1	55.9	46.3	11.4	44.4	6.4
secp112r2	54.9	45.6	11.4	43.5	5.5
secp128r1	64.0	51.9	11.6	52.4	6.4
secp128r2	63.0	51.2	11.6	51.4	5.4
secp160k1	80.0	62.9	12.0	68.0	8.0
secp160r1	80.0	62.9	12.0	68.0	6.0
secp160r2	80.0	62.9	12.0	68.0	7.0
secp192k1	96.0	73.8	12.2	83.8	7.8
secp192r1	96.0	73.8	12.2	83.8	6.8
secp224k1	112.0	84.7	12.4	99.6	6.6
secp224r1	112.0	84.7	12.4	99.6	7.6
secp256k1	128.0	95.5	12.6	115.4	7.4
secp256r1	128.0	95.5	12.6	115.4	7.4
secp384r1	192.0	138.8	13.2	178.8	8.8
secp521r1	260.5	184.9	13.7	246.8	-

Now, given our estimates for the number of group operations and DH-oracle calls, we see that the smallest s for which $N_{\mathfrak{M}} \ll \sqrt{|E|}$ is $s = 3$. The reduction cost is then (see Appendix A for general s)

$$\left(\frac{149}{6}\lg p + \frac{55}{8}\right)\mathfrak{DH} + \left((\frac{3}{2}\lg p + \frac{13}{2})(3p^{1/3}) + (\frac{3}{2}\lg p + \frac{511}{4})\lg p\right)\mathfrak{M}.$$

As an illustration of the advantage gained over the previous results presented in [12], we consider the security of DHP for secp256r1: The DLP on this curve requires about 2^{128} computational steps, employing the currently known methods. Using our auxiliary elliptic curve, we deduce that the DHP cannot be solved in less than $2^{115.4}$ computational steps, as opposed to 2^{108} from the previous paper. That is a gain factor of about $2^{7.4}$ over the previously reported value in [12], see Table 1.

Since an amount of computation of about $2^{115.3} \approx 5 \cdot 10^{34}$ group operations is infeasible with today's computational power, one can draw the conclusion that a

Table 2. Summary of results for curves of even characteristic

sect curve	$\lg \sqrt{|E|}$	$\lg N_{\mathfrak{M}}$	$\lg N_{\mathfrak{DH}}$	$\lg T_{DH}$	adv
sect113r1	56.0	46.4	11.4	44.6	6.6
sect113r2	56.0	46.4	11.4	44.6	6.6
sect131r1	65.0	52.6	11.7	53.3	6.3
sect131r2	65.0	52.6	11.7	53.3	6.3
sect163k1	81.0	63.5	12.0	69.0	7.0
sect163r1	81.0	63.5	12.0	69.0	7.0
sect163r2	81.0	63.5	12.0	69.0	7.0
sect193r1	96.0	73.8	12.2	83.8	6.8
sect193r2	96.0	73.8	12.2	83.8	6.8
sect233k1	115.5	87.0	12.5	103.0	7.0
sect233r1	116.0	87.4	12.5	103.5	7.5
sect239k1	118.5	89.1	12.5	106.0	8.0
sect283k1	140.5	104.0	12.8	127.7	8.7
sect283r1	141.0	104.3	12.8	128.2	7.2
sect409k1	203.5	146.5	13.3	190.2	8.2
sect409r1	204.0	146.9	13.3	190.7	-
sect571k1	284.5	201.0	13.8	270.7	-
sect571r1	285.0	201.3	13.8	271.2	-

secure implementation of a protocol whose security depends on the intractability of the DHP on the curve secp256r1 can safely be used, provided the DLP is really of the conjectured complexity.

Note that the SECG standard [17] includes all the curves in the NIST [13] and the most used ones in the ANSI [1] standards, covering the most commonly used elliptic curves in practice.

5   Building the Auxiliary Elliptic Curves

By the argument presented in the previous section, we need to construct elliptic curves whose order is a product of three coprime numbers of roughly the same size. That is $q_i \approx p^{1/3}$. Muzereau *et el.* [12] used the Complex Multiplication (CM) technique to build auxiliary elliptic curves with smooth orders but this does not perform very well as p gets larger, due to the prohibitive precision then needed for the calculations. In our case, it proved to be computationally more efficient to generate random elliptic curves and then test if their sizes are of the required form.

Let us estimate the probability that a number in a large interval centred around p is a product of three co-primes of roughly the same size.

Given three randomly chosen (positive) integers, we first want to compute the probability that they are *pairwise* coprime. Let p be prime. The probability that p divides two of these integers but not the third is $3/p^2 \cdot (1 - 1/p)$ and the probability that p divides all of them at once is $1/p^3$. So, the probability that p is not a common divisor of any two of these integers is

$$1 - \frac{3}{p^2}\left(1 - \frac{1}{p}\right) - \frac{1}{p^3} = 1 - \frac{3}{p^2} + \frac{2}{p^3}.$$

Hence, the probability that three randomly chosen integers are pairwise coprime is

$$\prod_{p \text{ prime}} \left(1 - \frac{3}{p^2} + \frac{2}{p^3}\right) = \prod_{p \text{ prime}} \left(1 - \frac{1}{p}\right)^2 \left(1 + \frac{2}{p}\right) \approx 0.2867474.$$

The infinite product is clearly convergent but a closed form of its value could not be obtained by the author. The numerical approximation 0.2867474 was obtained using PARI, [14].

For a large interval (m, n), the product should be taken only for $p \leq m - n$. Now, since $1 - 3/p^2 + 2/p^3$ is positive, strictly increasing approaching 1 from below, we deduce that the above estimate is a *lower bound* to the actual probability we want.

In practice, for large p and corresponding Hasse intervals, the above value proved to be a good estimate and it matched nicely with a Monte Carlo simulation to estimate this probability over large Hasse intervals.

For most cryptographic groups G from the SECG standard, auxiliary elliptic curve E of the form $y^2 = x^3 - 3x + b$ were successfully generated by finding a suitable value of b. The full version of this paper [2] specifies these values together with the (prime) size of the group G, which is the characteristic of the prime finite-field over which E is defined. The size of the elliptic curve group $|E|$ is given as a product of three coprime numbers of roughly equal size.

When trying to generate the auxiliary elliptic curves, the main difficulty was to actually factor $|E|$. For large $|G|$, factorisation fails most of the time and another random value of b is tried without any success. This is the main reason for failing to produce the necessary data for the three curves secp521r1, sect571r1 and sect571k1. However, two missing auxiliary elliptic curves from [12], viz. secp224k1 and sect409r1 were successfully found. While first appears to have been just forgotten, the second was due to the difficulty of generating the auxiliary elliptic curves using the CM method.

6   Can We Do Better Using Maurer's Approach

Here, it is argued that not much improvement can be made using Maurer's reduction, as described in Algorithm 1.

Just computing g^{x^3} and (twice) checking the quadratic residuosity of $g^{x^3 - 3x + b}$ will cost at least

$$(2 + 2 \times \lg(p/2))\mathfrak{DH}.$$

For $s = 3$ we find that the ratio of the estimated cost of this paper to this bound is

$$\frac{\frac{149}{6}\lg p + \frac{55}{8}}{2\lg p} \sim \frac{149}{12} \approx 2^{3.6}.$$

Step 2 is not independent from the first so its cost can be reduced even further, but the third step does not seem to have any corelation with the previous steps. If we say that step 3 costs at least one exponentiation, to compute one of the $(|E|/q_j)Q$, then the ratio drops to

$$\frac{\frac{149}{6}\lg p + \frac{55}{8}}{(2 + 2/3)\lg p} \sim \frac{149}{16} \approx 2^{3.2}.$$

Hence, it turns out that about 3 bits of security is all that can be hoped for above the current work.

7   Conclusion

Assuming the DLP is an exponentially hard problem, we have shown that the Maurer-Wolf reduction with naive search yields a concrete security assurance for the elliptic curves recommended by the current standards, for which we could generate the auxiliary elliptic curves.

We have found two new auxiliary elliptic curves, missing from [12], viz. secp224k1 and sect409r1. It remains open to find auxiliary elliptic curves for the curves secp521r1, sect571r1 and sect571k1. These will have sizes larger than 500 bits, which presents the current factoring algorithms with a big challenge.

Acknowledgement

I wish to thank Nigel Smart and all the members of the Security Group at Bristol University for their great help and insightful discussions. Special thanks go to Dan Page for helping me run my code on a cluster.

References

1. ANSI. X9.62 – Public Key Cryptography for the Financial Services Industry: The Elliptic Curve Digital Signature Algorithm (ECDSA). 1999.
2. K. Bentahar. The Equivalence Between the DHP and DLP for Elliptic Curves Used in Practical Applications, Revisited. Cryptology ePrint Archive, Report 2005/307, 2005. http://eprint.iacr.org/.
3. I.F. Blake, G. Seroussi and N.P. Smart. Elliptic curves in cryptography. Cambridge University Press, 1999.
4. H. Cohen. A Course In Computational Algebraic Number Theory. Springer-Verlag, GTM 138, 1993.

5. D. Hankerson, A. Menezes, S. Vanstone. *Guide to elliptic curve cryptography.* Springer-Verlag, 2003.
6. D. E. Knuth. *The art of computer programming, vol. 2.* Addison Wesley Longman, 1998.
7. U. M. Maurer. *Towards the equivalence of breaking the Diffie-Hellman protocol and computing discrete logarithms.* In *Advances in Cryptology — CRYPTO 1994,* Springer LNCS 839, 271-281, 1994.
8. U. M. Maurer and S. Wolf. *On the difficulty of breaking the DH protocol.* Technical Report #24, Department of Computer Science, ETH Zurich, 1996.
9. U. M. Maurer and S. Wolf. *Diffie-Hellman Oracles.* In *Advances in Cryptology — CRYPTO 1996,* Springer LNCS 1109, 268-282, 1996.
10. U. M. Maurer and S. Wolf. *The relationship between breaking the Diffie-Hellman protocol and computing discrete logarithms.* *SIAM Journal on Computing,* **28,** 1689–1721, 1999.
11. U. M. Maurer and S. Wolf. *The Diffie-Hellman protocol.* Designs, Codes, and Cryptography, **19,** 147–171, 2000.
12. A. Muzereau, N.P. Smart and F. Vrecauteren *The equivalence between the DHP and DLP for elliptic curves used in practical applications.* LMS J. Comput. Math. 7(2004) 50-72, 2004.
13. NIST. *FIPS 186.2 Digital Signature Standard (DSS).* 2000.
14. PARI/GP, version 2.1.3, Bordeaux, 2000. http://pari.math.u-bordeaux.fr/.
15. H.-G. Rück. *A note on elliptic curves over finite fields.* Math. Comp., **49,** 301–304, 1987.
16. R. Schoof. *Elliptic curves over finite fields and the computation of square roots mod p.* Math. Comp., **44,** 483–494, 1985.
17. SECG. *SEC2 : Recommended Elliptic Curve Domain Parameters.* See http://www.secg.org/, 2000.
18. W.C. Waterhouse. *Abelian varieties over finite fields.* Ann. Sci. École Norm. Sup., **2,** 521–560, 1969.

A   Complexity Analysis of Algorithm 1

To simplify this task, each step of Algorithm 1 will be studied separately and then the results will be added up to obtain the total average cost.

Step 1: We first precompute g^{2^i} for $i = 1, \ldots, \lfloor \lg p \rfloor$. This will allow us to compute any power g^k with an average cost of $\frac{1}{2} \lg k \mathfrak{M}$, using the double-and-add algorithm of exponentiation. The precomputation requires $\lfloor \lg p \rfloor$ squarings, which costs

$$\lg p \mathfrak{M}.$$

Without loss of generality, we set $d = 0$ at the start of this step. Then, evaluating $g^z \leftarrow g^{x^3 - 3x + b} = g^{x^3} \cdot ((g^x)^3)^{-1} \cdot g^b$ requires

$$2 \mathfrak{DH} + {} + 1 \mathfrak{I} + (4 + \frac{1}{2} \lg b) \mathfrak{M}.$$

Note that

$$g^{(x+d)^3 - 3(x+d) + b} = g^{x^3 - 3x + b} \cdot (g^{x^2})^{3d} \cdot (g^x)^{3d^2} \cdot g^{d^3 - 3d}.$$

So for a second evaluation, we only need an extra

$$\left(3 + \frac{3}{2}\lg(3d) + \frac{3}{2}\lg(3d^2) + \frac{1}{2}\lg(d^3 - 3d)\right)\mathfrak{M} \sim (3 + 3\lg 3 + 6\lg d)\mathfrak{M}.$$

For the quadratic residuosity check we need to compute $g^{z^{(p-1)/2}}$. First pre-compute $g^{z^{2^i}}$ for $i = 1, \ldots, \lfloor \lg \frac{p}{2} \rfloor$, then the total cost is

$$\left(\lg\frac{p}{2} + \frac{1}{2}\lg\frac{p-1}{2}\right)\mathfrak{D}\mathfrak{H} \sim \left(\frac{3}{2}\lg p - \frac{3}{2}\right)\mathfrak{D}\mathfrak{H}.$$

Now, let ν be the number of iterations for step 1. Since $\mathbb{F}_p$ has $(p-1)/2$ quadratic non-residues, the probability for having $\nu = k$ iterations is

$$\Pr[\nu = k] = \left(\frac{p-1}{2p}\right)^{k-1} \cdot \frac{p+1}{2p}.$$

Hence, the expected number $\bar{\nu}$ of iterations for step 1 is

$$\bar{\nu} = \sum_{k=1}^{\infty} k \cdot \Pr[\nu = k] = \frac{p+1}{2p}\sum_{k=1}^{\infty} k\left(\frac{p-1}{2p}\right)^{k-1} = \frac{2p}{p+1} \approx 2.$$

Thus the total average cost of this first step is $\lg p\mathfrak{M} + [2\mathfrak{D}\mathfrak{H} + 1\mathfrak{I} + (4 + \frac{1}{2}\lg b)\mathfrak{M}] + [(3 + 3\lg 3 + 6\lg d)\mathfrak{M}] + 2 \times (\frac{3}{2}\lg p - \frac{3}{2})\mathfrak{D}\mathfrak{H}$. That is

$$(3\lg p - 1)\mathfrak{D}\mathfrak{H} + 1\mathfrak{I} + (\lg p + \frac{1}{2}\lg b + 6\lg d + 7 + 3\lg 3)\mathfrak{M}. \tag{6}$$

Step 2: Following Algorithm 2, we treat three cases:

1. If $p \equiv 3 \pmod 4$ then, using the precomputations from the previous step, we can compute $g^{z^{(p+1)/4}}$ in an average of

$$\frac{1}{2}\lg\frac{p+1}{4}\mathfrak{D}\mathfrak{H} \sim \left(\frac{1}{2}\lg p - 1\right)\mathfrak{D}\mathfrak{H}.$$

2. If $p \equiv 5 \pmod 8$ then the computation of $g^{z^{(p-5)/8}}, g^{zs}$ and g^{su} costs $(2 + \frac{1}{2}\lg\frac{p-5}{8})\mathfrak{D}\mathfrak{H} \sim (\frac{1}{2}\lg p + \frac{1}{2})\mathfrak{D}\mathfrak{H}$ on average.
   If $t = 1$ then no further computation is needed and the total cost is $(\frac{1}{2}\lg p + \frac{1}{2})\mathfrak{D}\mathfrak{H}$. Otherwise, $t \neq 1$ and then computing

$$g^{u \cdot 2^{(p-1)/4}} = \mathcal{D}\mathcal{H}(g^u, g^{(2^{(p-1)/4})} \mod p)$$

   will cost an extra $1\mathfrak{D}\mathfrak{H} + (\frac{3}{2}\lg\frac{p-1}{4} + \frac{1}{2}\lg p)\mathfrak{M}$.
   Since t behaves like a random variable, the average cost for this case is then

$$\left(\frac{1}{2}\lg p + \frac{1}{2}\right)\mathfrak{D}\mathfrak{H} + \frac{1}{2}(1\mathfrak{D}\mathfrak{H} + (2\lg p - 3)\mathfrak{M}).$$

3. Otherwise, we use the general (implicit) Tonelli and Shanks algorithm. We first write $p - 1 = 2^e \cdot w$, where w is odd.

The initialisation step requires roughly $(\frac{1}{2} \lg \frac{w-1}{2} + 2)\mathfrak{D}\mathfrak{H}$. Finding the exponent and reducing it requires $(r+2)\mathfrak{D}\mathfrak{H}$ per iteration, and at most e iterations are expected. Since $r \leq e$, we will need $e \cdot (r + 2) \leq e \cdot (e + 2)$ calls to the DH-oracle. Hence, the total number of the DH-oracle calls is about

$$\left(\frac{1}{2} \lg \frac{w - 1}{2} + 2 + (e + 2)e\right) \mathfrak{D}\mathfrak{H}.$$

Since p is odd, we can easily see that the expected value of e is

$$\sum_{k=1}^{\infty} k \cdot \Pr[e = k] = \sum_{k=1}^{\infty} k \cdot (1/2)^k = 2.$$

Bearing this in mind, we get $w = p/2^e = p/4$ and the total cost is then estimated to be

$$\left(\frac{1}{2} \lg p + \frac{17}{2}\right)\mathfrak{D}\mathfrak{H}.$$

Note: When concluding, we will use the weighted average of the costs above, which is

$$\left(\frac{1}{2} \lg p + \frac{15}{8}\right)\mathfrak{D}\mathfrak{H} + \frac{1}{8}(2 \lg p - 3)\mathfrak{M}. \tag{7}$$

Step 3: Before entering the j-loop, we first pre-compute $2^i Q$ for $i = 1, \ldots,$ $\lfloor \lg |E|^{1-1/s} \rfloor$. This is enough since q_j are of roughly the same size, so $q_j \approx |E|^{1/s}$ and then $\frac{|E|}{q_j} \approx |E|^{1-1/s}$.

Using equation (2), the cost of precomputation is found to be about

$$(8\mathfrak{D}\mathfrak{H} + 4\mathfrak{J} + 14\mathfrak{M})\left(1 - \frac{1}{s}\right) \lg |E|.$$

We also pre-compute $2^i P$ for $i = 1, \ldots, \lfloor \lg |E| \rfloor$ in affine coordinates[3]. According to equation (4), this costs about

$$(1\mathfrak{J} + 4\mathfrak{M}) \lg |E|.$$

Now, let j be fixed (We want to analyse the cost of one j-loop). The cost for computing $Q_j = (g^{u_j}, g^{v_j}, g^{w_j})$ such that $(u_j, v_j, w_j) = \frac{|E|}{q_j} Q$, given by equation (3), is about

$$\left(8\mathfrak{D}\mathfrak{H} + \frac{5}{2}\mathfrak{J} + \frac{1}{4}(3 \lg p + 13)\mathfrak{M}\right) \gamma_j,$$

where we have set $\gamma_j = \lg(|E|/q_j)$. For the evaluation of $P_j = \frac{|E|}{q_j} P$, in affine coordinates, equation (5) gives

$$\left(\frac{1}{2}\mathfrak{J} + \frac{3}{2}\mathfrak{M}\right) \gamma_j.$$

---

[3] We need i up to $\lg |E|$ as we will use these precomputed values in step 4 too.

For the i-loop, we note that $g^{w_j^2}$ and $g^{w_j^3}$ need to be computed only once for each j-loop, which costs $2\mathfrak{D}\mathfrak{H}$.

Now fix i. Computing $iR = (i-1)R + R$, in affine coordinates, can be achieved with one elliptic curve addition costing $1\mathfrak{I} + 3\mathfrak{M}$, since $(i-1)R$ has been computed and $1R = R$ is trivial.

The cost of comparison is about $2 \times \frac{3}{2}\lg p\mathfrak{M} = 3\lg p\mathfrak{M}$.

On average there will be $q_j/2$ i-loops for each j-loop, and therefore the average cost of the i-loop is

$$\frac{q_j}{2}\left(1\mathfrak{I} + 3(\lg p + 1)\mathfrak{M}\right).$$

Hence, the cost per one j-loop is

$$(8\gamma_j + 2)\mathfrak{D}\mathfrak{H} + \left(\frac{1}{2}q_j + 3\gamma_j\right)\mathfrak{I} + \left(\frac{3}{2}(\lg p + 1)q_j + \frac{1}{4}(3\lg p + 19)\gamma_j\right)\mathfrak{M}.$$

Noting that

$$\sum_{j=1}^{s}\gamma_j = \sum_{j=1}^{s}\lg\frac{|E|}{q_j} = (s-1)\lg|E|,$$

we find that the total cost for step 3, without the precomputation costs, is on average

$$(8(s-1)\lg|E| + 2s)\mathfrak{D}\mathfrak{H} + \left(\frac{1}{2}\sum_{i=1}^{s}q_j + 3(s-1)\lg|E|\right)\mathfrak{I}+$$
$$+\left(\frac{3}{2}(\lg p + 1)\sum_{i=1}^{s}q_j + \frac{1}{4}(3\lg p + 19)(s-1)\lg|E|\right)\mathfrak{M}.$$

Adding the precomputation costs, we finally get the total cost of step 3

$$(8(s-1/s)\lg|E| + 2s)\mathfrak{D}\mathfrak{H} + \left(\frac{1}{2}\sum_{i=1}^{s}q_j + (3s+2-\frac{4}{s})\lg|E|\right)\mathfrak{I}+$$
$$\left(\frac{3}{2}(\lg p + 1)\sum_{i=1}^{s}q_j + (\frac{1}{4}(3\lg p + 19)(s-1) + 18 - \frac{14}{s})\lg|E|\right)\mathfrak{M}. \quad (8)$$

Step 4: We use the Chinese Remainder Theorem to reconstruct $k \bmod |E|$ from $k \equiv k_j \pmod{q_j}$, $j = 1, \ldots, s$. We compute

$$k = \sum_{j=1}^{s} k_j \cdot \frac{|E|}{q_j} \cdot \hat{q}_j \pmod{|E|},$$

where $\hat{q}_j = \left(\frac{|E|}{q_j}\right)^{-1} \bmod q_j$. This requires $s\mathfrak{I} + 2s\mathfrak{M}$ operations. Note that inversions are computed in $\mathbb{F}_{q_1}, \ldots, \mathbb{F}_{q_s}$.

For computing kP, in affine coordinates, we use the previously precomputed values of $2^i P$. So this exponentiation would cost only $(1\mathfrak{I} + 3\mathfrak{M})\frac{1}{2}\lg k$. Taking $k \bmod |E|$ to be $\frac{|E|}{2}$ on average, we find the average cost of step 4 to be

$$\frac{1}{2}(\lg|E| - 1)\mathfrak{I} + \frac{3}{2}(\lg|E| - 1)\mathfrak{M}. \tag{9}$$

Conclusion: We conclude that the total cost for Algorithm 1 is

$$\left(8\left(s - \frac{1}{s}\right)\lg|E| + \frac{7}{2}\lg p + 2s + \frac{7}{8}\right)\mathfrak{D}\mathfrak{H}$$

$$+ \left(\frac{1}{2}\sum_{i=1}^{s} q_j + \left(3s + \frac{5}{2} - \frac{4}{s}\right)\lg|E| + \frac{1}{2}\right)\mathfrak{I}$$

$$+ \left(\begin{array}{c} \frac{3}{2}(\lg p + 1)\sum_{i=1}^{s} q_j + \left(\frac{1}{4}(3\lg p + 19)(s-1) + \frac{39}{2} - \frac{14}{s}\right)\lg|E| + \\ + \frac{5}{4}\lg p + \frac{1}{2}\lg b + 6\lg d + 3\lg 3 + \frac{41}{8} \end{array}\right)\mathfrak{M}.$$

Neglecting small terms and making the approximation[4] $|E| \approx p$ and $b \approx p/2$, the average cost of Algorithm 1 is then found to be

$$\left\{\left(8s - \frac{8}{s} + \frac{7}{2}\right)\lg p + 2s + \frac{7}{8}\right\}\mathfrak{D}\mathfrak{H} + \left(\frac{1}{2}\sum_{i=1}^{s} q_j + \left(3s + \frac{5}{2} - \frac{4}{s}\right)\lg p\right)\mathfrak{I} +$$

$$+ \left\{\frac{3}{2}(\lg p + 1)\sum_{i=1}^{s} q_j + \left(\frac{1}{4}(3\lg p + 19)(s-1) + \frac{85}{4} - \frac{14}{s}\right)\lg p\right\}\mathfrak{M}.$$

Note that if we take q_j to be of roughly the same size and fix B to be of this size then

$$s \approx \frac{\log|E|}{\log B} \approx \frac{\log p}{\log B}$$

and then

$$\sum_{j=1}^{s} q_j \approx \sum_{j=1}^{s} B = sB \approx \frac{B\log p}{\log B} = \frac{B\lg p}{\lg B}.$$

In practice, the cost of an inversion is at most $10\mathfrak{M}$, see [3, p. 37]. Using this fact we have now established Theorem 1, stated on page 381.

---

[4] $|E| = p+1-t$ where $t \in [-\sqrt{p}, \sqrt{p}]$ is the Frobenius trace, so $E = p(1+(1-t)/p) \approx p$, $b \approx p/2$ is the average value of b, and d is small.

Pairings on Elliptic Curves over Finite Commutative Rings

Steven D. Galbraith and James F. McKee[*]

Department of Mathematics,
Royal Holloway, University of London,
Egham, Surrey TW20 0EX, UK
{Steven.Galbraith, James.McKee}@rhul.ac.uk

Abstract. The Weil and Tate pairings are defined for elliptic curves over fields, including finite fields. These definitions extend naturally to elliptic curves over $\mathbb{Z}/N\mathbb{Z}$, for any positive integer N, or more generally to elliptic curves over any finite commutative ring, and even the reduced Tate pairing makes sense in this more general setting.

This paper discusses a number of issues which arise if one tries to develop pairing-based cryptosystems on elliptic curves over such rings. We argue that, although it may be possible to develop some cryptosystems in this setting, there are obstacles in adapting many of the main ideas in pairing-based cryptography to elliptic curves over rings.

Our main results are: (i) an oracle that computes reduced Tate pairings over such rings (or even just over $\mathbb{Z}/N\mathbb{Z}$) can be used to factorise integers; and (ii) an oracle that determines whether or not the reduced Tate pairing of two points is trivial can be used to solve the quadratic residuosity problem.

Keywords: Elliptic curves modulo N, pairings, integer factorisation, quadratic residuosity.

1 Introduction

Pairings are a major topic in elliptic curve public key cryptography, following the success of cryptosystems such as Joux's three-party key exchange protocol [10] and the Boneh-Franklin identity-based encryption scheme [1]. Recall that if E is an elliptic curve over a field K and if r is coprime to the characteristic of K then the Weil pairing maps $E[r] \times E[r]$ to μ_r, where μ_r is the group of r-th roots of unity in the field $\overline{K}$. The Tate pairing [5, 6] is usually used in practical implementations for efficiency reasons, though it is necessary to consider the so-called reduced Tate pairing which takes values in μ_r.

Elliptic curves modulo composite integers N have also been proposed for cryptography. The motivation is that security can also rely on the integer factorisation problem and that new functionalities might be possible due to the extra trapdoor. It is therefore a natural problem to try to develop pairing-based cryptosystems on

---

[*] This research was funded by EPSRC grant GR/R84375.

N.P. Smart (Ed.): Cryptography and Coding 2005, LNCS 3796, pp. 392–409, 2005.

elliptic curves modulo composite integers N. The fundamental question is whether pairings can be computed on elliptic curves over rings, and whether other aspects of pairing-based cryptography can be generalised to this situation. Indeed, the first author has been asked by several researchers whether this is possible.

We might imagine a system in which the factorisation of N is secret, but N is public, and a user is required to compute a pairing on some elliptic curve over $\mathbb{Z}/N\mathbb{Z}$ (or some extension ring). The security of the cryptosystem is presumed to rely on the hardness of factoring and possibly some other computational problems.

We will explain that the Weil pairing can be computed successfully in this setting (as long as certain data is provided). On the other hand, we show in Theorem 3 that the reduced Tate pairing cannot be computed without knowing the factorisation of N. As a companion to this result we show in Theorem 4 that even just being able to detect whether or not the reduced Tate pairing of two points is trivial would allow one to solve the quadratic residuosity problem. We will also argue that certain operations which are essential to many pairing-based cryptosystems (such as hashing to a point) cannot be performed with elliptic curves over rings if the factorisation is unknown.

Our opinion is that the use of pairings on elliptic curves over rings will not be as successful as the case of elliptic curves over finite fields. We believe that the potential secure and practical applications are, at best, limited to a few special situations. One might imagine, for example, a scheme where only the holder of secret information is supposed to be able to compute pairings.

The structure of the paper is as follows. We recall some results for elliptic curves over $\mathbb{Z}/N\mathbb{Z}$, making some observations concerning quadratic residuosity and the natural generalisations to finite extensions of $\mathbb{Z}/N\mathbb{Z}$. There follows an example which introduces some of the issues which arise when considering pairings on elliptic curves over rings. In particular, there are issues concerning the splitting of $N = pq$ where $p - 1$ and $q - 1$ (or $p + 1$ and $q + 1$) share some common factor: we recall some techniques for integer factorisation of numbers of this form. Then we gather some well-known results concerning the equivalence of factoring and extracting roots, put in the setting of surjective homomorphisms from $(\mathbb{Z}/N\mathbb{Z})^*$ to roots of unity. We then define what we mean by pairings for elliptic curves over $\mathbb{Z}/N\mathbb{Z}$ and over more general finite commutative rings, and prove the results stated above (Theorems 3 and 4).

When giving complexity estimates for algorithms over $\mathbb{Z}/N\mathbb{Z}$, the computation of the greatest common divisor of two numbers between 1 and N is counted as a single 'ring operation'.

2   Elliptic Curves Modulo N

Let N be an integer greater than 1 with $\gcd(N, 6) = 1$ (this restriction on N is not essential, but it simplifies the exposition in places). An elliptic curve E over the ring $\mathbb{Z}/N\mathbb{Z}$ is the set of solutions $(x : y : z)$ in projective space over $\mathbb{Z}/N\mathbb{Z}$ (insisting that $\gcd(x, y, z, N) = 1$) to a Weierstrass equation

$$y^2 z = x^3 + a_4 x z^2 + a_6 z^3, \tag{1}$$

where the discriminant of the cubic on the right, namely $4a_4^3 + 27a_6^2$, has no prime factor in common with N. There is a group law on $E(\mathbb{Z}/N\mathbb{Z})$ given by explicit formulae which can be computed without knowledge of the factorisation of N. The identity element is $0 = (0 : 1 : 0)$. We refer to Lenstra [13, 14] for details about elliptic curves over rings.

If the prime factorisation of N is $N = \prod_{i=1}^m p_i^{a_i}$, then $E(\mathbb{Z}/N\mathbb{Z})$ is isomorphic as a group to the direct product of elliptic curve groups $\prod_{i=1}^m E(\mathbb{Z}/p_i^{a_i}\mathbb{Z})$. If we let E_i be the reduction of E modulo p_i, then E_i is an elliptic curve over the field $\mathbb{F}_{p_i}$. One finds that

$$\#E(\mathbb{Z}/p_i^{a_i}\mathbb{Z}) = p_i^{a_i-1}\#E_i(\mathbb{F}_{p_i})$$

(see, for example, [7]).

The first proposal to base cryptosystems on elliptic curves over the ring $\mathbb{Z}/N\mathbb{Z}$ was by Koyama, Maurer, Okamoto and Vanstone [12]. Other proposals have been given by Demytko [4], Meyer and Mueller [21], and Vanstone and Zuccherato [31]. The security of such cryptosystems is related to the difficulty of factorising N.

The following theorem was established in [11].

Theorem 1. *Let N be a composite integer satisfying $\gcd(N, 6) = 1$. Given an oracle that takes as its input an elliptic curve over $\mathbb{Z}/N\mathbb{Z}$, and outputs the number of points on the curve, one can factorise N in random polynomial time.*

We remark that an oracle that merely tells us whether or not two elliptic curves over $\mathbb{Z}/N\mathbb{Z}$ have the same number of points can be used to solve the quadratic residuosity problem.

Theorem 2. *Suppose that $\mathcal{O}$ is an oracle that determines whether or not two elliptic curves over $\mathbb{Z}/N\mathbb{Z}$ have the same number of points. Let $a \in (\mathbb{Z}/N\mathbb{Z})^*$ be such that $\left(\frac{a}{N}\right) = 1$. Then there is a randomised polynomial time algorithm that makes one call to the oracle and returns a guess as to whether or not a is actually a square in $(\mathbb{Z}/N\mathbb{Z})^*$, with the following probabilities of success:*

- *if a is a square in $(\mathbb{Z}/N\mathbb{Z})^*$, then the algorithm will return the guess 'square';*
- *if a is not a square in $(\mathbb{Z}/N\mathbb{Z})^*$, then the algorithm will return the guess 'not a square' with probability $1 - \epsilon$, where*

$$\epsilon = O(\log p \log \log p / \sqrt{p}),$$

with p being any prime dividing N such that $\left(\frac{a}{p}\right) = -1$.

Proof. Suppose that we are given $a \in (\mathbb{Z}/N\mathbb{Z})^*$ with $\left(\frac{a}{N}\right) = 1$. We choose random a_4, a_6 in $\mathbb{Z}/N\mathbb{Z}$ such that equation (1) defines an elliptic curve E over $\mathbb{Z}/N\mathbb{Z}$. If a is a square in $(\mathbb{Z}/N\mathbb{Z})^*$, then the twisted curve $E^{(a)}$ with equation

$$y^2 z = x^3 + a_4 a^2 x z^2 + a_6 a^3 z^3$$

has the same number of points as E. Otherwise, let p be any prime dividing N such that $\left(\frac{a}{p}\right) = -1$. Suppose that $N = p^r n$, with $\gcd(p, n) = 1$. We can write the number of points on E over $\mathbb{Z}/N\mathbb{Z}$ as

$$\#E(\mathbb{Z}/N\mathbb{Z}) = (p + 1 - t)m_1 ,$$

where $p + 1 - t = \#E(\mathbb{Z}/p\mathbb{Z})$ and $m_1 = p^{r-1}\#E(\mathbb{Z}/n\mathbb{Z})$. Then the number of points on $E^{(a)}$ is

$$\#E^{(a)}(\mathbb{Z}/N\mathbb{Z}) = (p + 1 + t)m_2 ,$$

where $m_2 = p^{r-1}\#E^{(a)}(\mathbb{Z}/n\mathbb{Z})$. The numbers of points on E and $E^{(a)}$ are different unless

$$t = (p + 1)(m_1 - m_2)/(m_1 + m_2) .$$

Conditioning on the value of the pair (m_1, m_2) this value of t occurs with probability

$$O(\log p \log \log p / \sqrt{p})$$

(Theorem 2 in [18]). The implied constant here is absolute, not depending on (m_1, m_2), or on a. The result follows immediately.   □

More generally, one might not have access to an oracle as in Theorems 1 and 2, but might simply be given a single elliptic curve over $\mathbb{Z}/N\mathbb{Z}$ with a known number of points, M. Two complementary approaches for attempting to factorise N from this limited information have appeared in the literature:

1. One method is to multiply a point on a random quadratic twist by M. In general, it is hard to find points in $E(\mathbb{Z}/N\mathbb{Z})$ (for example, choosing an x-coordinate, putting $z = 1$, and solving for y requires taking a square root). However, there are formulae for performing point multiplications which use x-coordinates only (on the affine piece $z = 1$: see the *Formulary* of Cassels, and the exercises at the end of chapter 26, in [3]).
   A random $x \in \mathbb{Z}/N\mathbb{Z}$ is the x-coordinate of a point $(x : y : 1)$ on E_i with probability roughly $1/2$. If this is the case then $M(x : y : 1) = 0$ on E_i. On the other hand, if x is not a valid x-coordinate then there is a corresponding point $(x : y : 1)$ on a quadratic twist $E^{(d)}$ over $\mathbb{F}_{p_i}$ and for most curves we would not expect $M(x : y : 1) = 0$ on $E^{(d)}(\mathbb{F}_{p_i})$.
   The algorithm to factorise N is to take random x-coordinates (with $z = 1$) and to multiply by M, using x-coordinates only. With high probability the resulting point will be the identity modulo some, but not all, primes p dividing N. So taking the gcd of the resulting z-coordinate with N will split N. Details of this method are given in Okamoto and Uchiyama [24].
2. Another way to obtain this result is to mimic the standard randomised reduction from knowing $\#(\mathbb{Z}/N\mathbb{Z})^*$ to factoring (similar to the Miller-Rabin primality test). We write

$$M = \#E(\mathbb{Z}/N\mathbb{Z}) = 2^m M'$$

where M' is odd. We then choose random x-coordinates (with $z = 1$) and multiply by M' and then compute the sequence of doublings of this point. The details are given in [17]. In [11] it is noted that the prime 2 can be replaced by a larger prime.

3   Extension Rings

In practice, especially when considering pairings, we may be interested in extending our ring $\mathbb{Z}/N\mathbb{Z}$. For example, we may wish to force the full r-torsion onto our curve for some r. To this end we consider elliptic curves over rings of the form

$$R_{N,f} = (\mathbb{Z}/N\mathbb{Z})[x]/(f(x)),$$

where $f(x)$ is some polynomial in $(\mathbb{Z}/N\mathbb{Z})[x]$. The splitting of $f(x)$ may be different modulo different prime divisors of N. The above results readily extend to this setting, twisting by random elements of R. If N is squarefree, and $f(x)$ is squarefree modulo every prime dividing N, then $R = R_{N,f}$ is a product of finite fields.

4   An Example

To clarify the discussion above, and to introduce some of the issues which arise when considering pairings, we give an example.

Let p_1, p_2 be primes congruent to 2 modulo 3 and let r be a prime such that $r \mid (p_i + 1)$, $r^2 \nmid (p_i + 1)$ for $i = 1, 2$. Let E be the elliptic curve $y^2 = x^3 + 1$. For each $i = 1, 2$ let E_i be the reduction of E modulo p_i. It is well known that E_i is supersingular, that $\#E(\mathbb{F}_{p_i}) = p_i + 1$, and that $E[r] \subset E(\mathbb{F}_{p_i^2})$. Let $P, Q \in E_i[r]$. Then the Weil (or reduced Tate) pairing gives an r-th root of unity

$$e_r(P, Q) \in \mathbb{F}_{p_i^2}^* .$$

The field $\mathbb{F}_{p_i^2}$ can be defined as $\mathbb{F}_{p_i}(\theta)$ where $\theta^2 + \theta + 1 = 0$. If P and Q are points of order r such that $P \neq 0$ lies in $E_i(\mathbb{F}_{p_i})$ and Q does not lie in $E(\mathbb{F}_{p_i})$ then it is easy to show that $e_r(P, Q) \neq 1$.

Define $N = p_1 p_2$ and define the ring

$$R = (\mathbb{Z}/N\mathbb{Z})[\theta]/(\theta^2 + \theta + 1).$$

Then

$$R \cong \mathbb{F}_{p_1^2} \times \mathbb{F}_{p_2^2} .$$

Let E be the elliptic curve above. Then

$$\#E(R) = (p_1 + 1)^2 (p_2 + 1)^2 .$$

Note that there seems to be no reason to use large embedding degrees for elliptic curves over rings since the ground ring is already large due to the factoring problem. Indeed, one might prefer embedding degree 1, but then $r \mid (p_i - 1)$, and in the case of the Weil pairing also $r^2 \mid \#E(\mathbb{F}_{p_i})$, for each prime p_i dividing N, and so the attacks of Section 5 must be borne in mind.

Let P and Q be points of order r in $E(R)$ such that $P \neq 0$ lies in $E(\mathbb{Z}/N\mathbb{Z})$ and $Q \notin E(\mathbb{Z}/N\mathbb{Z})$. Then, as before, $e_r(P, Q)$ is a non-trivial r-th root of unity in R^*. The distortion map

$$\psi(x, y) = (\theta x, y)$$

can be used to map points $P \in E(\mathbb{Z}/N\mathbb{Z})$ into points in $E(R)$ and so we can obtain a non-trivial pairing between points of order r in $E(\mathbb{Z}/N\mathbb{Z})$.

We will argue in Section 8 that the reduced Tate pairing cannot be computed without knowing the factorisation of N. But, if E, R, P, Q and r are given, then one can compute the Weil pairing using Miller's algorithm [22, 23] as

$$e_r(P, Q) = (-1)^r F_{r,P}(Q)/F_{r,Q}(P)$$

without knowing the factorisation of N, where $F_{r,P}$ is a function on E with divisor

$$(F_{r,P}) = r(P) - r(0)$$

(see [23]).

Hence, one can solve decision Diffie-Hellman problems in $E[r]$ and one can implement Joux's three-party key exchange protocol [10] in this setting. It is therefore plausible that cryptosystems can be developed based on elliptic curves over rings which exploit both the hardness of the integer factorisation problem as well as aspects of pairing-based cryptography. Such systems might have potential functionalities which cannot be realised using elliptic curves over finite fields.

However, there are a number of issues which differ from the case of elliptic curves over finite fields which should be considered before one can develop such cryptosystems:

- The value r must play a symmetric role for all primes $p_i \mid N$.
  For example, if $\zeta \in R^*$ is such that $\zeta \pmod{p_1}$ has order r but $\zeta \equiv 1 \pmod{p_2}$ then one can split N by computing $\gcd(\zeta - 1, N)$. Similarly, if P is a point which has order r on E_1 but order coprime to r on E_2 then one can split N by multiplying E by r and computing gcds.
- The point order r must be provided to compute pairings.
  This is because Miller's algorithm essentially involves the operation of point multiplication by r. We know of no way to compute pairings if r is not provided. Since we are obliged to use the Weil pairing, both points must have order r. The information on r can be used to improve certain factoring algorithms (see Section 5 below). Hence, r should be chosen to be much smaller than the primes p_i (for the above example we would recommend choosing r to have 160 bits and the p_i to have at least 512 bits).
- Generating points on elliptic curves over rings is a hard problem.
  This is not an obstacle to a cryptosystem such as Joux's three-party key exchange if a base-point P is included in the system parameters. However, for

some protocols it may be necessary for users to find random points $Q \in E(R)$ without simply multiplying an existing point by some integer.

There are two traditional solutions to this problem. The first, used in the KMOV cryptosystem [12], is to choose $P = (x_P, y_P)$ and to modify the curve equation so that P lies on E. This solution could be used in pairing applications, though note that it does not preserve existing points. The second solution, due to Demytko [4], is to work with x-coordinates only. It is not possible to compute pairings exactly using x-coordinates only, but it may be possible to compute traces of pairings, as done by Scott and Barreto [28]. A different solution is to choose x_P and extend the ring as $R' = R(\sqrt{x_P^3 + a_4 x_P + a_6})$. It is unclear whether any of these solutions would be practical for pairing-based cryptosystems.

- Hashing to a point of order r is hard.
  In many pairing-based cryptosystems it is necessary to hash to a point of order r. The hashing process first involves finding a random point in $E(R)$, which as mentioned above is already a potential difficulty. Further, to obtain a point of exact order r it is necessary to multiply the point by a cofactor m. Since r is public then, once m is also given, the exponent of the group $E(R)$ or $E(\mathbb{Z}/N\mathbb{Z})$ is known, and we can therefore hope to factorise N using the methods discussed at the end of Section 2.

  Due to this issue, it seems unlikely that cryptosystems such as the Boneh-Franklin identity-based encryption scheme [1] or the Boneh-Lynn-Shacham signature scheme [2] can be developed for elliptic curves over extensions of $\mathbb{Z}/N\mathbb{Z}$ without revealing the factorisation of N.

5   Factoring If r Is Known

From time to time (e.g., [9], [15]) authors propose variants of RSA, with or without elliptic curves, in which $N = pq$ and there is some r greater than 1 such that both $r \mid (p-1)$ and $r \mid (q-1)$. If r is small, then it cannot be kept secret, as observed in [20]: it will be a factor of $N - 1$, and Lenstra's Elliptic Curve Method [13] or Pollard's ρ method [25] can be used to recover r. Even if r is secret, it cannot be too large: applying Pollard's ρ method with the 'random' map

$$x \mapsto x^{N-1} + 1 \pmod{N}$$

will produce a sequence that repeats modulo p after $O(\sqrt{p/r})$ terms, on average (this observation also appeared in [20]), so that if r is too large then the factorisation of N will be found.

If r is known, then more powerful factorisation attacks are possible. Let us assume that $N = pq$ with p and q of similar size. Following [20], we can now employ a variant of Lehmer's method (described in [19]). Write

$$p = xr + 1, \quad q = yr + 1.$$

Then

$$(N - 1)/r = xyr + (x + y) = ur + v$$

where u and v $(0 \le v < r)$ are known and x, y are unknown. We have

$$x + y = v + cr, \qquad xy = u - c,$$

where c is the (unknown) carry in expressing $(N-1)/r$ in base r as above. Since $cr \le x + y$ and both x and y are of size about $\sqrt{N}/r$, there are of order $\sqrt{N}/r^2$ values of c to test. A candidate for c can be tested quickly, since

$$r^2 c^2 + (2rv + 4)c + v^2 - 4u = (x - y)^2$$

must be a square.

Again following [20] (see also [16] for the same idea in a different setting), we can improve this $O(\sqrt{N}/r^2)$ attack to one that takes only $O(N^{1/4}/r)$ ring operations (still assuming that $N = pq$ with p and q roughly equal). Note that the price for this improvement in speed is either to have increased storage or a heuristic algorithm. We observe that the exponent of $(\mathbb{Z}/N\mathbb{Z})^*$ is given by

$$\mathrm{lcm}(p - 1, q - 1) = \mathrm{lcm}(xr, yr),$$

and so divides xyr. Take random $a \in (\mathbb{Z}/N\mathbb{Z})^*$. Then

$$a^{ur} = a^{xyr+cr} = a^{cr}.$$

Putting $b = a^r$, we have

$$b^u = b^c$$

in $(\mathbb{Z}/N\mathbb{Z})^*$. Since c has magnitude $\sqrt{N}/r^2$, we can recover c (modulo the order of b, which with high probability will have order nearly as large as $xy \approx N/r^2$) in $O(N^{1/4}/r)$ ring operations, either using the baby-step giant-step method of Shanks [29] or Pollard's λ method [26].

Similar remarks hold if both $r \mid (p+1)$ and $r \mid (q+1)$, as in Section 4. Small r can be spotted as a factor of $N - 1$. Large r make N vulnerable to Pollard's ρ method with the map

$$x \mapsto x^{N-1} + 1 \pmod{N}.$$

If r is known, we can determine $q + p$ modulo r^2, and hence perform a similar attack to the above that will split N in $O(N^{1/4}/r)$ ring operations if p and q are of similar size.

Finally in this context we should consider the implications of knowing a divisor d of the group order of $E(\mathbb{Z}/N\mathbb{Z})$. From Section 2, we must imagine that the group order M is secret. We might hope to split N using Lenstra's Elliptic Curve Method [13] or Pollard's λ method [26], with the curve E and the base point $d(x : y : 1)$ for random x. The worst case complexity of these attacks is

$$O\left(\min_{p \mid n} \sqrt{\#E(\mathbb{F}_p)/ \gcd(d, \#E(\mathbb{F}_p))}\right)$$

ring operations.

6   Computing a Surjective Homomorphism from $(\mathbb{Z}/N\mathbb{Z})^*$ to Certain Roots of Unity Is as Hard as Factoring

It is well-known that computing square-roots modulo a composite is as hard as factoring [27], and indeed the same applies to rth roots if r is not too large and

$$\gcd(r, p-1) > 1$$

for some prime p dividing N. The ability to extract rth roots modulo N implies the ability to generate random rth roots of unity, and it is of course the latter that allows us to split N. We record these remarks here in a few Lemmas.

Let $N > 1$ be an odd integer that is not a prime power, and let $r > 1$ be any integer. Let $G_{N,r}$ be the unique maximal subgroup of $(\mathbb{Z}/N\mathbb{Z})^*$ having exponent dividing r, i.e.,

$$G_{N,r} = \{a \in (\mathbb{Z}/N\mathbb{Z})^* \mid a^r = 1\}.$$

Another way of saying this is that $G_{N,r}$ contains all the rth roots of unity in $(\mathbb{Z}/N\mathbb{Z})^*$.

Any oracle which computes a surjective group homomorphism from $(\mathbb{Z}/N\mathbb{Z})^*$ to $G_{N,r}$ can be used to factor N if r is not too large and $G_{N,r}$ is non-trivial: this statement is made precise in the Lemmas below. The homomorphic property simply ensures that the preimage of each element of $G_{N,r}$ is the same size: any map with this property, or something close to it, would suffice.

Let $\mathcal{O}(N, r, a)$ be an oracle which takes as input integers N and r and an element $a \in (\mathbb{Z}/N\mathbb{Z})^*$ and returns the image of a under a surjective group homomorphism from $(\mathbb{Z}/N\mathbb{Z})^*$ to $G_{N,r}$, where the homomorphism depends on N and r, but not a.

For example, if N is squarefree and $r \mid (p_i - 1)$ for all i then $\mathcal{O}(N, r, a)$ might return the Chinese remainder of the values

$$a^{(p_i-1)/r} \pmod{p_i}$$

for $1 \le i \le m$ (but any surjective homomorphism would do). In the case $r = 2$ and N odd and squarefree, this particular choice of oracle returns the Chinese remainder of the Legendre symbols for all p dividing N. We stress that this is not the same thing as the Jacobi symbol (which is the product of the Legendre symbols, and does not give a surjective homomorphism to $G_{N,2}$).

In the following Lemmas, we fix the notation

$$N = \prod_{i=1}^{m} p_i^{a_i}, \tag{2}$$

where $p_1, \ldots p_m$ are distinct odd primes, and $m \ge 2$. Then

$$G_{N,r} \cong \prod_{i=1}^{m} G_i, \tag{3}$$

where each G_i is cyclic of order dividing r.

Lemma 1. *Let N be as in (2) and let r be a positive integer greater than 1 such that*

$$r \mid p^{a_i - 1}(p_i - 1)$$

for $1 \leq i \leq m$. Let $\mathcal{O}$ be an oracle computing a surjective homomorphism from $(\mathbb{Z}/N\mathbb{Z})^*$ to $G_{N,r}$.

There is a randomised algorithm with negligible storage that will find a non-trivial factorisation of N in expected time $O(r)$ ring operations, using $O(r)$ oracle calls, on average.

Proof. The algorithm is simply to choose random $a \in (\mathbb{Z}/N\mathbb{Z})^*$ and to obtain

$$b = \mathcal{O}(N, r, a).$$

One can then compute $\gcd(b - 1, N)$. This will split N as long as there are two primes p and q dividing N such that

$$b \equiv 1 \pmod{p} \quad \text{but } b \not\equiv 1 \pmod{q}.$$

In (3), each G_i now has order r. Of the r^2 possibilities for the image of a in $G_1 \times G_2$, $2(r - 1)$ of them will split N, regardless of the image of a in the other G_i $(3 \leq i \leq m)$. For random a, the probability that we split N is therefore at least $2(r - 1)/r^2$, so that the expected number of oracle calls is $O(r)$. □

The running time of the above algorithm can be reduced at the expense of some storage.

Lemma 2. *Let N be as in (2) and let r be a positive integer greater than 1 such that*

$$r \mid p_i^{a_i - 1}(p_i - 1)$$

for $1 \leq i \leq m$. Let $\mathcal{O}$ be an oracle computing a surjective homomorphism from $(\mathbb{Z}/N\mathbb{Z})^*$ to $G_{N,r}$.

There is a randomised algorithm requiring $O(\sqrt{r} \log N)$ storage that will find a non-trivial factorisation of N in expected time $O(\sqrt{r} \log r \log \log r)$ ring operations, using $O(\sqrt{r})$ oracle calls, on average.

Proof. The algorithm chooses random $a \in (\mathbb{Z}/N\mathbb{Z})^*$ and forms a list of values $\mathcal{O}(N, r, a)$. When the list has length $O(\sqrt{r})$, one checks for repeats in the list modulo some but not all prime factors of N by standard fast polynomial evaluation techniques [30]. It is likely that there is a repeat modulo some prime dividing N whilst being very unlikely that there is a repeat modulo N. If r is not known, or if no repeat has been found, one can repeatedly double the length of the list until success. □

With the extra hypothesis that $\mathcal{O}(N, r, a) \pmod{p}$ depends only on $a \pmod{p}$, there is a low-storage variant using Pollard's ρ method [25]. Not all homomorphic oracles have this property. For example, with $N = pq$ and $r = 2$, we could perversely map a to the Chinese remainder of

$$\left(\frac{a}{p}\right) \pmod{q} \quad \text{and} \quad \left(\frac{a}{q}\right) \pmod{p}.$$

Lemma 3. *Let N be as in (2) and let r be a positive integer greater than 1 such that*

$$r \mid p_i^{a_i-1}(p_i - 1)$$

for $1 \leq i \leq m$. Let $\mathcal{O}$ be an oracle computing a surjective homomorphism from $(\mathbb{Z}/N\mathbb{Z})^*$ to $G_{N,r}$ satisfying the additional property that $\mathcal{O}(N, r, a) \pmod{p}$ depends only on $a \pmod{p}$ (for each prime p dividing N).

There is a heuristic algorithm requiring negligible storage that will find a non-trivial factorisation of N in heuristic expected time $O(\sqrt{r})$ ring operations, using $O(\sqrt{r})$ oracle calls, on average.

Proof. Apply Pollard's ρ method [25] with the 'random' map

$$x \mapsto \mathcal{O}(N, r, x) + 1.$$

(We add 1 to improve the pseudorandom behaviour of the map, by mixing addition with the multiplicative nature of our homomorphic oracle.) Described simply (but not optimally), we compute (but do not store) sequences x_n and $y_n = x_{2n}$, starting with (say) $x_0 = y_0 = 1$, and using the rule

$$x_{n+1} = \mathcal{O}(N, r, x_n) + 1.$$

At each step we compute

$$\gcd(x_n - y_n, N)$$

until we find a non-trivial factor of N.

The complexity is as standard for Pollard's ρ method.   □

The situation is of course much more trivial if r is a prime such that

$$r \mid p_i^{a_i-1}(p_i - 1)$$

for some but not all primes p_i dividing N.

Lemma 4. *Let N be as in (2) and let r be a prime such that*

$$r \mid p_i^{a_i-1}(p_i - 1)$$

for at least one but not all of the p_i. Let $\mathcal{O}$ be an oracle computing a surjective homomorphism from $(\mathbb{Z}/N\mathbb{Z})^*$ to $G_{N,r}$.

Then there is a randomised algorithm to find a non-trivial factorisation of N which runs in expected time $O(1)$ ring operations, using $O(1)$ oracle calls, on average.

Proof. The algorithm is simply to choose random $a \in (\mathbb{Z}/N\mathbb{Z})^*$ and to obtain

$$b = \mathcal{O}(N, r, a).$$

One can then compute

$$\gcd(b - 1, N).$$

Since there is at least one prime p_i dividing N such that

$$r \nmid p_i^{a_i - 1}(p_i - 1)$$

we know that

$$b \equiv 1 \pmod{p_i^{a_i}}.$$

Hence, all that is required is that

$$b \not\equiv 1 \pmod{p_j}$$

for some other prime p_j dividing N. The probability of this event is at least $(r-1)/r$, so that the expected number of oracle calls is at most 2. $\qquad\square$

Similar results hold for the rings $R_{N,f}$, but the performance of the analogous algorithms has worse dependence on r. If the r-torsion of the p-component of $R_{N,f}$ has order r^{n_p} (for p a prime dividing N), then the expected running time for the analogue of Lemmas 1, 2 and 3 has r replaced by $r^{\min(n_p)}$. The analogue of Lemma 4 remains just as trivial.

The case $r = 2$ in Lemma 1 is particularly attractive, since if N is odd then we know that $r \mid (p-1)$ for all p dividing N.

7 The Tate Pairing on Curves over Finite Commutative Rings

One of the main results of this paper is to argue that there is no way to compute reduced Tate pairings on general $E(\mathbb{Z}/N\mathbb{Z})$ without knowing the factorisation of N. We achieve this by showing that if $\mathcal{O}$ is an oracle for computing reduced Tate pairings on such curves then one can use $\mathcal{O}$ to factorise N.

We first recall the Tate pairing for elliptic curves over finite fields (see Frey and Rück [5, 6] for details). Let E be an elliptic curve over $\mathbb{F}_q$ (q being a power of a prime) and let r be coprime to q. Let $k \in \mathbb{N}$ be minimal such that $r \mid (q^k - 1)$. The integer k depends on both q and r (indeed, it is the order of q modulo r) and is often called the 'embedding degree'. Define by $E[r]$ the set of points $P \in E(\mathbb{F}_{q^k})$ such that $rP = 0$ (we assume that $\#E[r] > 1$). The Tate pairing is a non-degenerate pairing

$$\langle \cdot, \cdot \rangle_r : E[r] \times E(\mathbb{F}_{q^k})/rE(\mathbb{F}_{q^k}) \longrightarrow \mathbb{F}_{q^k}^*/(\mathbb{F}_{q^k}^*)^r .$$

In practice the Tate pairing is computed using an algorithm due to Miller [22, 23]. If $P \in E[r]$ and $Q \in E(\mathbb{F}_{q^k})$ then there is a function $F_{r,P}$ having divisor

$$(F_{r,P}) = r(P) - r(0) .$$

Miller's algorithm builds up this function $F_{r,P}$ in stages in a way analogous to the double and add algorithm for point exponentiation. Then

$$\langle P, Q \rangle_r = F_{r,P}(Q + S)/F_{r,P}(S)$$

for a suitable auxiliary point $S \in E(\mathbb{F}_{q^k})$. Different choices of auxiliary point S will give different values in $\mathbb{F}_{q^k}^*$, but they are all equivalent in the quotient group $\mathbb{F}_{q^k}^*/(\mathbb{F}_{q^k}^*)^r$.

Henceforth we shall not insist that the values of r and k satisfy $r \mid (q^k - 1)$. Provided that $P \in E[r]$, and that we can find a suitable auxiliary point S, we can still perform Miller's algorithm. The pairing may no longer be non-degenerate, and of course

$$(\mathbb{F}_{q^k}^*)^r = (\mathbb{F}_{q^k}^*)^{\gcd(r, q^k - 1)}.$$

Let $N = \prod_{i=1}^m p_i$ be a squarefree positive integer. As with other applications of elliptic curves, the natural way to generalise the Tate or Weil pairings to points on elliptic curves over $\mathbb{Z}/N\mathbb{Z}$ is to use the Chinese remainder theorem to piece together the values of the pairing over $\mathbb{F}_p$ for the various p dividing N. If the factorisation of N is known, then of course this computation can be done. One can readily generalise this to curves over the rings $R_{N,f}$, considered in section 3, provided that $R_{N,f}$ is a product of fields. Further generalisations are considered below.

If the factorisation of N is not known, then one can still compute the Weil pairing (of points P and Q of known order), or the 'raw' Tate pairing (where only the order of P is needed) over $\mathbb{Z}/N\mathbb{Z}$ simply by following Miller's algorithm, working modulo N throughout.

For cryptographic applications the fact that the Tate pairing assumes values defined modulo rth powers is intolerable. Hence, returning first to the case of finite fields, one uses the 'reduced' Tate pairing

$$e(P, Q) = \langle P, Q \rangle_r^{(q^k - 1)/\gcd(r, q^k - 1)}.$$

There is some choice in how to generalise this reduced Tate pairing to elliptic curves over rings. We give two definitions below. We will show in Theorem 3 that (with either definition) computing the reduced Tate pairing is as hard as factoring.

For clarity, we give both definitions for the simplest case where N is squarefree and we do not extend the ring $\mathbb{Z}/N\mathbb{Z}$. Generalisations will be discussed immediately afterwards.

Definition 1. *Suppose that $N = \prod_{i=1}^m p_i$ is a squarefree positive integer, and that r is a positive integer. Let E be an elliptic curve over $\mathbb{Z}/N\mathbb{Z}$ defined by (1). Suppose that P and Q are points on E, with the order of P dividing r. Let $\langle P, Q \rangle_{r,i}$ be the raw Tate pairing of P and $Q + rE$ over $\mathbb{Z}/p_i\mathbb{Z}$. Define*

$$e_i(P, Q) = \langle P, Q \rangle_{r,i}^{(p_i - 1)/\gcd(r, p_i - 1)}.$$

Then the reduced Tate pairing $e(P, Q)$ is defined to be the unique element of $(\mathbb{Z}/N\mathbb{Z})^*$ satisfying

$$e(P, Q) \equiv e_i(P, Q) \pmod{p_i}$$

for each i $(1 \le i \le m)$.

Definition 2. *Suppose that* $N = \prod_{i=1}^{m} p_i$ *is a squarefree positive integer, and that* r *is a positive integer. Let* E *be an elliptic curve over* $\mathbb{Z}/N\mathbb{Z}$ *defined by (1). Suppose that* P *and* Q *are points on* E, *with the order of* P *dividing* r. *Let* $\langle P, Q \rangle_r$ *be the raw Tate pairing of* P *and* $Q + rE$ *over* $\mathbb{Z}/N\mathbb{Z}$. *Then the reduced Tate pairing* $e(P, Q)$ *is defined by*

$$e(P, Q) = \langle P, Q \rangle_r^{g/\gcd(r,g)},$$

where g *is the exponent of* $(\mathbb{Z}/N\mathbb{Z})^*$.

The first definition seems more natural because it behaves well under reduction, but we recognise that there is a choice here.

If N is not squarefree, then $\mathbb{Z}/N\mathbb{Z}$ is not a product of fields. How can we interpret any of our pairings in this setting? To clarify this issue consider the case of an elliptic curve E over $\mathbb{Z}/p^2\mathbb{Z}$, not anomalous over $\mathbb{Z}/p\mathbb{Z}$. Then $\#E(\mathbb{Z}/p^2\mathbb{Z}) = p\#E(\mathbb{F}_p)$ and $E[p]$ is cyclic of order p. Define $\mu_p = \{1 + xp : 0 \le x < p\} \subset (\mathbb{Z}/p^2\mathbb{Z})^*$. One can certainly define a bilinear pairing on $E[p]$ taking values in μ_p, but both the geometric theory and Miller's algorithm break down for curves over such rings: the entire p-torsion lies on a straight line.

As a result, we propose to map each $\mathbb{Z}/p^a\mathbb{Z}$ to its residue field $\mathbb{Z}/p\mathbb{Z}$, and hence to map $E(\mathbb{Z}/N\mathbb{Z})$ to a product of curves over finite prime fields (the E_i in Section 3). Our pairings are defined for such curves, and we can define the pairing over $\mathbb{Z}/N\mathbb{Z}$ to be any preimage of the gluing together of these pairings. (In the context of oracles, we insist that the choice of preimage is made in a deterministic way.) As in the squarefree case, this can be further generalised to curves over the rings $R_{N,f}$ considered in Section 3. This generalisation is essential if we wish to consider embedding degrees greater than 1. For the reduced Tate pairing, the second definition extends in the obvious way: g is replaced by the exponent of $(R_{N,f})^*$. For the first definition, a little care is required: in each local factor we power up by the exponent of the local multiplicative group divided by its gcd with r, then we glue together the local values by the Chinese Remainder Theorem. Again we comment that this generalisation of the first definition behaves well under reduction.

We can even abstract this further, and consider an elliptic curve over any finite commutative ring R. (This includes rings such as $R_{N,f}$.) Such a ring is a product of local rings, each having prime power order. Each local factor has a residue field (of prime power order), and we can map our curve over R to a curve over the product of these residue fields. Then, as above, we can define a pairing value in each residue field, glue these together, and finally take a preimage in R. Again both definitions of the reduced Tate pairing extend equally naturally.

8   Reduced Tate Pairing Oracles

What is the minimum amount of information that we must feed to an oracle for computing reduced Tate pairings over rings of the form $R = R_{N,f}$? Certainly we must supply the oracle with N and f, and an elliptic curve E defined over the ring R, and two points P and Q on $E(R)$ whose pairing is to be returned. We might also supply the value of r (with P, but not necessarily Q, supposed to have order dividing r: something that the oracle can easily check), or we might leave it to the oracle to compute suitable r. (One pleasing feature of the reduced Tate pairing is that the pairing value coming from the minimal possible r is the same as that computed using a multiple of it: see Section 6 of [8].) If the curve has no points of order r over R, or if R does not contain an element of order r, then the pairing value is still defined, but may well be trivial.

We have therefore two flavours of oracle, depending on whether or not we know the value of r.

Oracle 1. *This oracle takes as its input $N \in \mathbb{N}$, $f \in \mathbb{Z}[X]$, E defined by equation (1) for some a_4, a_6 in $R = R_{N,f}$, points P and Q in $E(R)$, and $r \in \mathbb{N}$.*

The oracle performs the following checks, and returns 'fail' if any of them fail:

- $\gcd(N, 6) = 1$;
- $\gcd(N, 4a_4^3 + 27a_6^2) = 1$;
- P and Q are in $E(R)$;
- $rP = 0$.

If all of these consistency checks are passed, then the oracle returns the reduced Tate pairing of P and Q, with value in R.

Oracle 2. *This oracle takes as its input $N \in \mathbb{N}$, $f \in \mathbb{Z}[X]$, E defined by equation (1) for some a_4, a_6 in $R = R_{N,f}$, and points P and Q in $E(R)$.*

The oracle performs the following checks, and returns 'fail' if any of them fail:

- $\gcd(N, 6) = 1$;
- $\gcd(N, 4a_4^3 + 27a_6^2) = 1$;
- P and Q are in $E(R)$.

If all of these consistency checks are passed, then the oracle chooses (but does not reveal) r such that $rP = 0$, and returns the reduced Tate pairing of P and Q, with value in R.

If one can factor N then, by Miller's algorithm, one can implement either of these oracles in polynomial time (to find suitable r for the second, we can use fast point-counting techniques).

We now claim that such oracles can be used to build an integer factorisation algorithm.

Theorem 3. *Let $\mathcal{O}$ be a reduced-Tate-pairing oracle, either of the form Oracle 1 or of the form Oracle 2.*

Given a composite integer N, not a prime power, with $\gcd(N, 6) = 1$, we can use the oracle $\mathcal{O}$ to find a non-trivial factorisation of N in expected time $O(1)$ ring operations, using $O(1)$ oracle calls, on average.

Proof. We work with Definition 1 of the reduced Tate pairing in the proof, and remark afterwards how the proof adapts trivially if one prefers Definition 2.

Choose a random integer a in the range $1 < a < N$, and define

$$E_a : y^2 z = x(x - z)(x - az) = x^3 - (a + 1)x^2 z + axz^2$$

which has discriminant $16a^2(a - 1)^2$. (We could transform this equation into Weierstrass form, as in (1), if desired.) This is an elliptic curve over $\mathbb{Z}/N\mathbb{Z}$ as long as $\gcd(N, 2a(a - 1)) = 1$. We take $r = 2$ and note that the points $P = (0 : 0 : 1), Q = (1 : 0 : 1)$ and $R = (a : 0 : 1)$ all have order 2.

One can compute the Tate pairing over $\mathbb{Q}$ of P with itself. The function $F = x$ satisfies $(F) = 2(P) - 2(0)$. If Q is taken to be the auxiliary point then the Tate pairing of P and $P + 2E$ is

$$\langle P, P \rangle_2 = F(P + Q)/F(Q) = x(R)/x(Q) = a/1 = a.$$

If another auxiliary point $(u : v : 1)$ is used then

$$P + (u : v : 1) = (a/u : -av/u^2 : 1)$$

and so the pairing value is a/u^2.

Calling the oracle $\mathcal{O}$ (with arguments N, $f = 1$, $P = Q = (0 : 0 : 1)$, and $r = 2$ (if needed)) performs exactly the operation of the oracle $\mathcal{O}(N, 2, a)$ in Section 6. Hence we can use the reduced Tate pairing oracle $\mathcal{O}$ to find a non-trivial factorisation of N in $O(1)$ ring operations, as in Lemma 1 (with $r = 2$).   □

We remark that if instead the reduced Tate pairing were defined by Definition 2, then the reduction to integer factorisation is just as simple. If the power of 2 dividing $p_i - 1$ is the same for each prime p_i dividing N, then the identical argument works. If not, then the reduced pairing is guaranteed to be trivial modulo at least one but not all of the p_i, and a similar argument goes through, analogous to Lemma 4.

The idea of the proof can be generalised to other small values of r, starting from a curve over $\mathbb{Q}$ (or a low-degree number field) with an r-torsion point.

Inspired by the quadratic residuosity observations in section 2, we note that the situation is still more favourable here, at least if we work with our preferred definition of the reduced Tate pairing.

Theorem 4. *Let N be an odd, composite integer, and let $\mathcal{O}$ be an oracle that tells us whether or not the reduced Tate pairing (as in Definition 1) of two points on an elliptic curve over $\mathbb{Z}/N\mathbb{Z}$ is trivial.*

Given $a \in (\mathbb{Z}/N\mathbb{Z})^*$ satisfying $\left(\frac{a}{N}\right) = 1$, we can use the oracle $\mathcal{O}$ to determine whether or not a is a square in $(\mathbb{Z}/N\mathbb{Z})^*$ with a single call to the oracle $\mathcal{O}$.

Proof. As above, we take the curve

$$E_a : y^2 z = x(x - z)(x - az) = x^3 - (a + 1)x^2 z + axz^2,$$

and ask the oracle $\mathcal{O}$ whether or not the reduced Tate pairing of $P = (0 : 0 : 1)$ with itself is trivial. The answer is 'yes' precisely when a is a square in $(\mathbb{Z}/N\mathbb{Z})^*$.   □

Acknowledgments

The authors gratefully acknowledge the insight of Jorge Villar, who pointed out some inadequacies in an early version of this paper, and the comments of the referees.

References

1. D. Boneh and M. Franklin, Identity-based encryption from the Weil pairing, in J. Kilian (ed.), CRYPTO 2001, Springer LNCS 2139 (2001) 213–229.
2. D. Boneh, B. Lynn and H. Shacham, Short signatures from the Weil pairing, *J. Crypt.*, **17**, No. 4 (2004) 297–319.
3. J.W.S. Cassels, *Lectures on Elliptic Curves*, LMS Student Texts **24**, Cambridge (1991).
4. N. Demytko, A new elliptic curve based analogue of RSA, in T. Helleseth (ed.), EUROCRYPT 1993, Springer LNCS 765 (1994) 40–49.
5. G. Frey and H.-G. Rück, A remark concerning m-divisibility and the discrete logarithm problem in the divisor class group of curves, *Math. Comp.*, **52** (1994) 865–874.
6. G. Frey, M. Müller and H.-G. Rück, The Tate pairing and the discrete logarithm applied to elliptic curve cryptosystems, *IEEE Trans. Inf. Th.*, **45** (1999) 1717–1719.
7. S.D. Galbraith, Elliptic curve Paillier schemes, *J. Crypt.*, **15**, No. 2 (2002) 129–138.
8. S.D. Galbraith, K. Harrison and D. Soldera, Implementing the Tate pairing, in C. Fieker and D.R. Kohel (eds.), ANTS V, Springer LNCS 2369 (2002) 324–337.
9. M. Girault, An Identity-Based Identification Scheme Based on Discrete Logarithms Modulo a Composite Number, in I.B. Damgard (ed.), EUROCRYPT 1990, Springer LNCS 473 (1991) 481–486.
10. A. Joux, A One Round Protocol for Tripartite Diffie-Hellman, in W. Bosma (ed.), ANTS IV, Springer LNCS 1838 (2000) 385–394.
11. N. Kunihiro and K. Koyama, Equivalence of counting the number of points on elliptic curve over the ring $\mathbb{Z}_n$ and factoring n, in K. Nyberg (ed.), EUROCRYPT 1998, Springer LNCS 1403 (1998) 47–58.
12. K. Koyama, U.M. Maurer, T. Okamoto and S.A. Vanstone, New public-key schemes based on elliptic curves over the ring $\mathbb{Z}_n$, in J. Feigenbaum (ed.), CRYPTO 1991, Springer LNCS 576 (1992) 252–266.
13. H.W. Lenstra Jr., Factoring integers with elliptic curves, *Annals of Mathematics*, **126** (1987) 649–673.
14. H.W. Lenstra Jr., Elliptic curves and number theoretic algorithms, *Proc. International Congr. Math.*, Berkeley 1986, AMS (1988) 99–120.
15. C.H. Lim and P.J. Lee, Security and performance of server-aided RSA computation protocols, in D. Coppersmith (ed.), CRYPTO 1995, Springer LNCS 963 (1995) 70–83.
16. W. Mao, Verifiable partial sharing of integer factors, in S. Tavares and H. Meijer (eds.), SAC 1998, Springer LNCS 1556 (1998) 94–105.
17. S. Martin, P. Morillo and J.L. Villar, Computing the order of points on an elliptic curve modulo N is as difficult as factoring N, *Applied Math. Letters*, **14** (2001) 341–346.
18. J.F. McKee, Subtleties in the distribution of the numbers of points on elliptic curves over a finite prime field, *J. London Math. Soc.* (2), **59** (1999) 448–460.

19. J.F. McKee and R.G.E. Pinch, Old and new deterministic factoring algorithms, in H. Cohen (ed.), ANTS II, Springer LNCS 1122 (1996) 217–224.
20. J.F. McKee and R.G.E. Pinch, Further attacks on server-aided RSA cryptosystems, unpublished manuscript (1998).
21. B. Meyer and V. Mueller, A public key cryptosystem based on elliptic curves over $\mathbb{Z}/n\mathbb{Z}$ equivalent to factoring, in U.M. Maurer (ed.), EUROCRYPT 1996, Springer LNCS 1070 (1996) 49–59.
22. V.S. Miller, Short programs for functions on curves, unpublished manuscript (1986).
23. V.S. Miller, The Weil pairing, and its efficient calculation, *J. Crypt.*, **17**, No. 4 (2004) 235–261.
24. T. Okamoto and S. Uchiyama, Security of an identity-based cryptosystem and the related reductions, in K. Nyberg (ed.), EUROCRYPT 1998, Springer LNCS 1403 (1998) 546–560.
25. J.M. Pollard, A Monte Carlo method for factorisation, *BIT*, **15** (1975) 331–334.
26. J.M. Pollard, Monte Carlo methods for index computations (mod p), *Math. Comp.*, **32** (1978) 918–924.
27. M.O. Rabin, Digitalized signatures and public-key functions as intractable as factorization, Technical report TR-212, MIT Laboratory for Computer Science (1979).
28. M. Scott and P.S.L.M. Barreto, Compressed pairings, in M. K. Franklin (ed.), CRYPTO 2004, Springer LNCS 3152 (2004) 140–156.
29. D. Shanks, Class number, a theory of factorisation and genera, in D.J. Lewis (ed.), Number theory institute 1969, Proceedings of symposia in pure mathematics, vol. 20, Providence RI, AMS (1971) 415–440.
30. J.W.M. Turk, Fast arithmetic operations on numbers and polynomials, in H.W. Lenstra Jr. and R. Tijdeman (eds.), Computational methods in number theory, Part 1, Mathematical Centre Tracts 154, Amsterdam (1984).
31. S.A. Vanstone and R.J. Zuccherato, Elliptic curve cryptosystems using curves of smooth order over the ring $\mathbb{Z}_n$, *IEEE Trans. Inform. Theory*, **43**, No.4 (1997) 1231–1237.

A Key Encapsulation Mechanism for NTRU

Martijn Stam

Dept. Computer Science, University of Bristol, Merchant Venturers Building,
Woodland Road, Bristol, BS8 1UB, United Kingdom
stam@cs.bris.ac.uk

Abstract. In this article we present a key encapsulation mechanism (KEM) for NTRU. The KEM is more efficient than a naive approach based on NAEP and resistant against the decryption failures that may occur when using NTRU. We also introduce plaintext awareness for KEMs and use it to tighten a security result by Dent.

Keywords: NTRU, Key Encapsulation, Random Oracle Model, Plaintext Awareness.

1   Introduction

Although the invention of public key encryption systems such as RSA and ElGamal was a major breakthrough, it was quickly realized that the straightforward use of these primitives was neither secure nor very efficient. The security issue was tackled with the introduction of a proper security model, such as IND-CCA2, and modes of operations that would reduce the by now well-defined security of the scheme to some mathematically hard problem. The efficiency issue is dealt with by using hybrid encryption. The data itself is encrypted under a symmetric cipher with a one-time key. This one-time key is freshly generated by the sender and encrypted under the public key scheme. Curiously it took rather long for the security model to catch up with this practice.

In 1998 Cramer and Shoup [5] published a breakthrough article containing the first efficient IND-CCA2 public key encryption scheme under only standard assumptions (DDH). En passant they formalized hybrid encryption schemes. Such a scheme consists of two parts, a data encryption mechanism (DEM) and a key encapsulation mechanism (KEM). The DEM is a symmetric primitive that encrypts (and decrypts) the data under a fresh session key. The KEM is a public key mechanism to generate the one-time session key for the DEM and encapsulate (and decapsulate) it. Cramer and Shoup demonstrate that the security properties of KEMs and DEMs can be considered independently of each other (as long as the session keys output by the KEM are suitable for the DEM). In this article we will only discuss KEMs.

The NTRU cryptosystem [9] was introduced as a compact and efficient alternative to systems based on RSA or discrete logarithms. Apart from its speed, it has the important advantage that as yet no efficient quantum algorithm is known to break it, contrary to factoring and computing discrete logarithms, both easy tasks for a quantum computer.

N.P. Smart (Ed.): Cryptography and Coding 2005, LNCS 3796, pp. 410–427, 2005.

NTRU has the rather unusual property that its underlying one-way trapdoor function does not always have a correct inverse: in particular the function is not guaranteed to be injective, inevitably resulting in decryption errors. This necessitates the use of specialized protocols to prevent weaknesses based on decryption errors that can occur when plugging NTRU in a standard construction based on a one-way trapdoor permutation.

Indeed, in the standard definition of secure encryption, an adversary cannot gain anything from asking for the decryption of a message it has properly encrypted (this notion is intuitively so strong that it led to the formalization of plaintext-awareness— valid encryptions imply knowledge of the plaintext). However, in the case of NTRU valid encryptions do not always decrypt to the encrypted message. Thus an adversary can learn something new even by querying the decryption oracle on messages it has just encrypted itself.

For functionality purposes one could allow a negligible number of decryption failures. However, in the case of an attack even a small number of decryption errors can be harmful if these are somehow related to the secret key and if the corresponding ciphertexts can be found efficiently (given the public key). Indeed, for NTRU this is the case, as was demonstrated by Proos, who presented an attack on an early version of NTRU encryption [13].

In the security proofs, the presence of decryption failures changes the behaviour of the decryption oracle. One needs to simulate this oracle for the adversary without knowledge of the secret key. Typically this simulation is based on calling the random oracle for valid encryptions, so the list of all inputs to the random oracle will also reveal the correct decryption of a valid ciphertext. If the adversary has not called the random oracle for a certain ciphertext, this ciphertext will be incorrect with overwhelming probability (over the 'future' choices of the random oracle), so declaring it so in the simulation only introduces a negligible error. As a result, the adversary cannot detect that it is interacting with a simulated world and hence its behaviour in the real world and the simulated world will be identical. In specific, its success probability will be identical.

In case of decryption failures, this gives a simulation that is too good—the simulation manages to correctly decrypt ciphertexts that would cause failures in the real world. This difference might be easily noticeable, thus the adversary might behave completely differently in the real world and in this simulated world.

As a reaction to the exposed vulnerabilities due to decryption failures [13,10], the people behind NTRU designed NAEP [11]. A disadvantage of NAEP is that it is not completely clean cut: instead of determining the full input to the one-way function and then applying it, a two tier method is used, where part of the one-way function is applied to part of the input and this intermediate result is hashed to derive the remainder of the input.

However, little is known about KEMs for NTRU, although they have been considered in the past. In this work we examine the restrictions decryption failures put on KEM-designs and show that under reasonable assumptions one of Dent's constructions [6] can serve as a KEM. This yields a more efficient and natural KEM than the one based on NAEP as presented by Whyte [17,18].

2   Preliminaries

2.1   Trapdoor One-Way Functions

Before we formally introduce NTRU, we revisit some other related notions and notations. We let λ denote the security parameter and λ_1 the number of random bits used by the key generation. A trapdoor one-way function consists of a triple of algorithms:

Key Generation. A probabilistic algorithm $\mathsf{Gen} : \{0,1\}^{\lambda_1} \to \mathcal{K}_p \times \mathcal{K}_s$ that produces the public and private key of the one-way function. We will also write $(pk, sk) \leftarrow \mathsf{Gen}(1^\lambda)$ or just (pk, sk), which is still understood to be probabilistic based on a uniform sampling of $\{0,1\}^{\lambda_1}$.

The one-way function. A deterministic algorithm $E : \mathcal{K}_p \times \mathcal{M} \times \mathcal{R} \to \mathcal{C}$, also $e \leftarrow E_{pk}(m, r)$, that on input a public key pk, maps (m, r) to an image e.

The trapdoor. A deterministic algorithm $D : \mathcal{K}_s \times \mathcal{C} \to (\mathcal{M} \times \mathcal{R}) \cup \{\bot\}$ that on input a secret key and some image e produces a preimage. The $\bot$-symbol is used to denote failure, for instance if e is not in the range of E. We will write $(m, r) \leftarrow D_{sk}(e)$, so possibly $(m, r) = \bot$ abusing notation.

A proper trapdoor satisfies $D_{sk}(E_{pk}(m, r)) = (m, r)$ for all $(m, r) \in \mathcal{M} \times \mathcal{R}$ and $(pk, sk) \in [\mathsf{Gen}(1^\lambda)]$, where $[\cdot]$ denotes the support of a probability distribution. Some functions do not have a full trapdoor but only a partial one. That is a function $D^{\mathcal{M}} : \mathcal{K}_s \times \mathcal{C} \to \mathcal{M} \cup \{\bot\}$ satisfying $D_{sk}^{\mathcal{M}}(E(m, r)) = m$.

An inversion failure occurs when the relevant equality above does not hold. For a given key we let $\mathcal{S}_{(pk,sk)}$ denote the set of all originals that cause a decryption failure, i.e., $\mathcal{S}_{(pk,sk)} = \{(m, r) | D_{sk}(E_{pk}(m, r)) \neq (m, r)\}$. We also define $\mathcal{S}_{(pk,sk)}(m) = \{r | D_{sk}(E_{pk}(m, r)) \neq (m, r)\}$ and for use with partial trapdoors $\mathcal{S}_{(pk,sk)}^{\mathcal{M}} = \{(m, r) | D_{sk}^{\mathcal{M}}(E_{pk}(m, r)) \neq m\}$. The latter definition implies that for a function with only a partial trapdoor, one can have pairs (m, r) and (m, r') yielding the same ciphertext but with neither being categorized as an inversion failure (whereas for a full trapdoor it is clear that the decryption cannot possibly return both originals so at least one will be qualified as causing a failure).

Given these sets we can also determine the probability that decryption failures occur. The relevant quantities are defined as $\epsilon_{(pk,sk)}^{\bot} = |\mathcal{S}_{(pk,sk)}|/|\mathcal{M} \times \mathcal{R}|$, $\epsilon_{(pk,sk)}^{\bot}(m) = |\mathcal{S}_{(pk,sk)}(m)|/|\mathcal{R}|$, and $\epsilon_{(pk,sk)}^{\bot,\mathcal{M}} = |\mathcal{S}_{(pk,sk)}^{\mathcal{M}}|/|\mathcal{M} \times \mathcal{R}|$. All these probabilities are per key; we let $\bar{\epsilon}^{\bot}$ etc. denote the probabilities averaged over key generation as well. For a proper functioning of the trapdoor one-way primitive the very least we require is that $\bar{\epsilon}^{\bot}$ is sufficiently small.

The trapdoor one-way assumption relates to the problem of inverting E without the secret key (trapdoor). An adversary A is given a randomly selected key and a randomly selected image and has to find a preimage. More precisely, define A's advantage as

$$\mathsf{Adv}_A^{OW}(\lambda) = \Pr[E_{pk}(A(pk, e)) = e | e \leftarrow E_{pk}(m, r), (m, r) \leftarrow \mathcal{M} \times \mathcal{R}, (pk, sk)] .$$

We call A a (t, ϵ) adversary if it has runtime t and advantage ϵ. The one-way assumption states that for all PPT adversaries A the advantage is negligible.

Of relevance, especially when only using the partial inverse, is the k-partial one-wayness assumption. Here the adversary is given the public key pk and an image e, and is allowed to output a list of k elements in $\mathcal{M}$. If for one of the elements m_i in its list there exists an $r \in \mathcal{R}$ such that $E_{pk}(m_i, r) = e$ the adversary is deemed successful. The k-partial one-way assumption states that all efficient adversaries have negligible success probability.

2.2   The NTRU Primitive

Central to NTRU [9] is the almost one-way trapdoor function

$$E(m, r) = m + rh$$

which is defined over the ring $R_q = \mathbb{Z}_q[X]/(X^N - 1)$ where the integers q and N are part of the system parameters and $h \in R_q$ is the public key. We will assume that all elements in R_q are represented by a polynomial of degree at most $N - 1$ and with coefficients in $\{0, \ldots, q-1\}$. This representation gives us a canonical map of R_q into $R = \mathbb{Z}[X]/(X^N - 1)$ and allows us to identify these polynomials with N-tuples of integers in $\{0, \ldots, q-1\}$.

NTRU makes ample use of sparse polynomials, especially those with exactly d coefficients set to 1 and the rest to 0. These sets will be denoted $R_q(d)$. Let $\tilde{R}_q$ be the union of $R_q(d)$ for fairly balanced d's, i.e., $|N/2 - d| \leq q/2$. The obvious map from the bitstrings $\{0,1\}^N$ to the binary polynomials of degree $< N$ is also used to map into $R_q(d)$, allowing for embedding failures. This map is injective, easy to invert and gives a uniform distribution over its range.

Given q and N, let p be a small element in R_q coprime to q (as a polynomial over the integers). Typically p will equal 2 or $2 + X$ and in both cases $\mathbb{Z}[X]$-polynomials reduced modulo both $X^N - 1$ and p will be represented with polynomials of degree smaller than N and whose coefficients lie in $\{0, 1\}$.

In the literature several NTRU key generation algorithms have appeared, sometimes tailored for a specific set of parameters. Only recently an algorithm has been presented that actually generates both parameters and keys [12]. We will assume that the relevant parameters (including d_f and d_g used to denote sparsity of some secret polynomials used during key generation) have been properly generated and concentrate on key generation.

Algorithm 1: NTRU Key Generation.

Let the parameters p, q, N and $0 < d_f, d_g < N$ be given. This algorithm generates a public key $h \in R_q$ and a private key $f \in R_q$, a small factor of h.

1.   [Picking a candidate trapdoor] Pick at random $f' \in R_q(d_f)$.
2.   [Checking the candidate] Set $f \leftarrow 1 + pf'$ and compute f^{-1} over R_q. If this inverse does not exist, return to the previous step.
3.   [Pick the blinding factor] Pick an invertible $g \in R_q(d_g)$ at random.
4.   [Compute the public key] Set $h \leftarrow pf^{-1}g$ (again over R_q). The public key is h and the corresponding private key is f.

For the trapdoor function let a further parameter $0 < d_r < N$ be given. Define the function

$$E : \tilde{R}_q \times R_q(d_r) \rightarrow R_q$$

by $E(m, r) = m + rh$. The one-way assumption on this function is that for properly chosen parameters, keys generated according to the above and $e = E(m, r)$ for uniformly chosen m and r, finding a preimage (m', r') such that $e = E(m', r')$ is hard.

Recovery of m given $e = E(m, r)$ and the trapdoor f works as follows. Compute $ef = mf + rhf = mf + prg$. Note that this is done in R_q. However, because all polynomials in this expression are small, with high probability we will be able to derive $a = mf + prg$ as polynomials in R, so without the reduction modulo q (this is where the decryption failures are introduced) But then $a \equiv m \bmod p$ since $prg = 0 \bmod p$ and $f_p^{-1} = 1$ by choice of parameters. Since $m \bmod p$ is the same as m in $\tilde{R}_q$, this results in recovery of the message. Obviously one can, given m, also reconstruct rh if so desired. Since h has an inverse in R_q, this will also lead to r itself. Thus we can recover both m and r, where recovering just m takes essentially one product computation plus some modular arithmetic, but recovery of r as well takes an additional product computation.

Only recently an algorithm was proposed to generate N, p, q, d_f, d_g, and d_r as a function of the security parameter λ. Before that one usually worked with a fixed parameter set. For security comparable to 1024-bit RSA, the set $N = 251, q = 239, d_f = 145$, and $d_g = d_r = 72$ was recommended. For this parameter set the probability of decryption failures has been estimated [16]. It can be shown that with the right centering algorithm, decryption failures can only be due to something called a gap failure, where one computes $a = fm + prg$ over $\mathbb{Z}[X]/(X^N - 1)$ and finds that there are at least two coefficients of a that differ by at least q. It has been estimated that the probability (over the choice of the keys, m and r) is approximately 2^{-156}, hence $\bar{\epsilon}^{\perp} \approx 2^{-156}$. If we assume that the failure probability is more or less independent of the choice of the key, so $\bar{\epsilon}^{\perp} \approx \epsilon_{(pk,sk)}^{\perp}$ for all keys, we can estimate the size of $\mathcal{S}_{(pk,sk)} \approx 2^{-156} 2^{212} 2^{250} = 2^{206}$. For other parameter sets (for instance when q is much smaller, say $q = 128$) much higher failure probabilities have been reported, as high as 2^{-25}. Note that these probabilities are taken over uniformly random choices of m and r. If an adversary can somehow deviate from these uniform distributions, he might be able to induce higher failure probabilities.

3   Padding Schemes for Encryption

In this section we introduce RAEP, robust asymmetric encryption padding, which turns a trapdoor one-way primitive suffering from decryption failures into a fully fledged IND-CCA2 secure encryption scheme. Basically it is a slight modification of NAEP.

The overall framework we will use is a keyed trapdoor one-way function E with a full inverse D, as described in Section 2.1. We allow for a small failure probability, that is $\bar{\epsilon}^{\perp}$ is small. The encryption scheme also uses a padding scheme pad : $\{0,1\}^{\lambda_3} \times \{0,1\}^{\lambda_2} \to \mathcal{M} \times \mathcal{R}$ such that pad is easily invertible (without errors or trapdoors) on its first element and the range of pad is a subset of $\mathcal{M} \times \mathcal{R}$ of negligible size that is easily verifiable. All these ingredients are mixed into an encryption scheme in a straightforward manner: for encryption, given a message $M \in \{0,1\}^{\lambda_3}$, pick uniform randomness $w \in \{0,1\}^{\lambda_2}$, compute the padding $(m,r) \leftarrow$ pad(M,w) and apply the one-way function E_{pk} to (m,r) to obtain the ciphertext c.

Decryption more or less follows the same pattern in reverse and with inverses. Given a ciphertext c, compute a tentative $(m,r) \leftarrow D_{sk}(c)$. If $(m,r) =\perp$ or if (m,r) does not lie in the range of pad output failure $(\perp)$, otherwise retrieve M from (m,r) and return M.

The IND-CCA2 security of such a scheme not only depends on the one-way function but also on the padding and their interaction. Typically the padding scheme is based on a one or two-round Feistel-network using hash functions that are modelled as random oracles in the proof.

The scheme is IND-CPA by arguing that for all M and over the randomness of w (and implicitly that of the random oracle), pad(M,w) is computationally indistinguishable from the uniform distribution over $\mathcal{M} \times \mathcal{R}$, unless the oracle is called on w. But in that case distinguishing between $E \circ$ pad applied to two different messages implies knowledge of a (partial) preimage of E, contradicting the (partial) one-way assumption.

In the absence of decryption failures, the padding scheme is plaintext aware by arguing that the only way for an adversary to create some (m,r) in the range of pad is by calling the random oracle on the embedded message M. All other pairs (m,r) not created thusly by the adversary can be deemed invalid.

In the presence of decryption failures, additional work is needed to obtain a plaintext aware scheme. Ideally the probability of a decryption failure is given over a uniform input distribution and indeed the padding is such that for honest participants this is the relevant probability. However, an adversary might deviate from the protocol resulting in a different probability distribution.

It should be difficult for the adversary to find an element that is both in the range of the padding scheme and in this set $\mathcal{S}_{(pk,sk)}$. Assuming a non-empty intersection, this can be achieved by forcing the adversary to use a call to the random oracle to obtain any information about an element in pad's range. If the fraction of pad's range that intersects $\mathcal{S}$ is sufficiently small, the adversary needs to make too many calls to the random oracle to be effective (and efficient).

The first padding scheme presented to be secure in the ROM was OAEP [2], although it was later famously shown by Shoup [14,15] to have a flawed proof of security. For the RSA and Rabin primitives an alternative proof was constructed [8] and for arbitrary functions E a patch called OAEP+ was presented [14]. Both OAEP and OEAP+ can be regarded as two round Feistel networks. Boneh [4] presented SAEP+, a more efficient one-round padding scheme.

Boneh proved that SAEP+ is secure in the ROM under a partial one-way assumption. The introduction of decryption failures—in the real world— has its

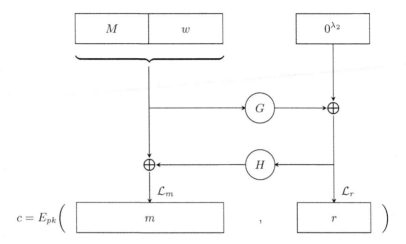

Fig. 1. RAEP

ramifications for the security of SAEP+. Recall that, given the public key, the set of ciphertexts causing failures is fixed. Note that, without further assumptions, we cannot exclude an adversary capable of testing membership of this set and constructing elements in this set, solely based on a public key. For instance, the padding is such that it gives an adversary complete control over the second input to E. Now suppose that $(m, r) \in \mathcal{S}_{(pk,sk)}$. Then the simulator used in the proof of plaintext awareness fails if the adversary asks for decryption of $E_{pk}(m, r)$: in the real world a failure will occur, whereas in the simulated world decryption to M will miraculously succeed.

This problem is countered in NAEP by changing the order of randomness and padding. We use a variation, called RAEP (for robust asymmetric encryption padding) as shown in Figure 1. Both RAEP and NAEP can be considered as a two-round Feistel network with all-zero initial vector.

Let $\lambda_2, \lambda_3, \lambda_4$ be security parameters and assume we are given a (keyed) function E acting on $\mathcal{M} \times \mathcal{R}$ (the range of E is less relevant). We assume that there are efficient invertible mappings $L_{\mathcal{M}} : \{0,1\}^{\lambda_3+\lambda_4} \to \mathcal{M} \cup \{\bot\}$ and $L_{\mathcal{R}} : \{0,1\}^{\lambda_2} \to \mathcal{R} \cup \{\bot\}$ such that the uniform distribution on the input yields a uniform distribution on the output, conditioned on $L_{\mathcal{M}}$ not outputting $\bot$. If $\lambda_3 + \lambda_4 = \lceil \log_2 |\mathcal{M}| \rceil$ and $\lambda_2 = \lceil \log_2 |\mathcal{R}| \rceil$ these embeddings will typically exist with small embedding error ($\leq 1/2$, so on average two tries suffice to find an element that embeds properly). Let $G : \{0,1\}^{\lambda_3} \times \{0,1\}^{\lambda_4} \to \{0,1\}^{\lambda_2}$ and $H : \{0,1\}^{\lambda_2} \to \{0,1\}^{\lambda_3+\lambda_4}$ be hash-functions. Key generation for the encryption scheme is identical to that of the one-way function.

Algorithm 2: RAEP Encryption.

This is the RAEP encryption which takes as input a message $M \in \{0,1\}^{\lambda_3}$ and a public key pk and outputs a ciphertext c.

1. [Pick randomness] Pick $w \in \{0,1\}^{\lambda_4}$ at random.
2. [Determine r] Set $\hat{r} \leftarrow G(M, w)$ and $r \leftarrow L_{\mathcal{R}}(\hat{r})$.
3. [Determine m] Set $a \leftarrow (M\|w) \oplus H(\hat{r})$ and $m = L_{\mathcal{M}}(a)$.
4. [Check embeddings] If $m = \bot$ or $r = \bot$ return to Step 1.
5. [Invoke one-way function] Set $c \leftarrow E_{pk}(m, r)$.

Algorithm 3: RAEP Decryption.

This is the RAEP decryption that on input a ciphertext $c \in \mathcal{C}$ and a private key sk outputs either a message $M \in \{0,1\}^{\lambda_3}$ or a failure symbol $\bot$.

1. [Invert one-way function] Set $(m, r) \leftarrow D_{sk}(c)$. If this results in failure, output $\bot$.
2. [Undo embeddings] Set $\hat{r} \leftarrow L_{\mathcal{R}}^{-1}(r)$ and $a \leftarrow L_{\mathcal{M}}^{-1}(m)$.
3. [Undo one-time pad] Set $(M\|w) = a \oplus H(\hat{r})$.
4. [Verify $\hat{r}$] Compute $\hat{r}' = G(M, w)$. If $\hat{r} = \hat{r}'$ output M, else return $\bot$.

The security proof of RAEP follows the same lines as that of NAEP and SAEP. However we will briefly discuss its resilience against decryption failures. From a high level perspective, pad maps the set $\{0,1\}^{\lambda_3 + \lambda_4}$ into a subset of $\mathcal{M} \times \mathcal{R} \cup \{\bot\}$, where we use $\bot$ to allow for embedding failures that would require rerandomization. If pad were a completely random map, the probability $\epsilon_{(pk,sk)}^{\bot}$ of a decryption failure would carry over from $\mathcal{M} \times \mathcal{R}$ to pad's range. For the parameter set introduced at the end of Section 2.2 this means there are approximately 2^{94} elements in pad's range that cause a failure. Now this also means there are at most 2^{94} dangerous elements in G's range. Since G's range has cardinality 2^{212}, it takes an adversary an expected 2^{116} oracle queries to find an input that will cause a decryption failure (indeed, the adversary has to be lucky that if the r-part corresponds to a failure, the m-part fits with it). Note that RAEP's padding is not completely random, but we feel it is sufficiently close to random. Howgrave-Graham et al. [11] use a slightly different argument that also shows that finding a point in the intersection would take the adversary too many oracle queries.

4   Key Encapsulation Mechanism

4.1   Introduction

Often a public key encryption scheme is used only to encrypt a session key for a symmetric cipher. In this case one actually requires a key encapsulation mechanism, a primitive that is weaker than encryption.

Definition 4. *A Key Encapsulation Mechanism (KEM) is a triple of algorithms*

Key generation. *A probabilistic algorithm* Gen $: \{0,1\}^{\lambda_1} \to \mathcal{K}_p \times \mathcal{K}_s$ *that produces the public and private key of the cryptosystem. We will also write* $(pk, sk) \leftarrow$ Gen(1^{λ}) *or just* (pk, sk)*, which is still understood to be probabilistic based on a uniform sampling of* $\{0,1\}^{\lambda_1}$*.*

Encapsulation. *A deterministic algorithm* $\mathsf{Enc} : \mathcal{K}_p \times \{0,1\}^{\lambda_2} \rightarrow \mathcal{C} \times \{0,1\}^{\lambda_3}$ *that on input a public key and a seed generates a key for the symmetric cipher and an encapsulation of that ephemeral key under pk. We will also write* $(c,k) \leftarrow \mathsf{Enc}_{pk}(U_{\lambda_2})$.

Decapsulation. *A deterministic algorithm* $\mathsf{Dec} : \mathcal{K}_s \times \mathcal{C} \rightarrow \{0,1\}^{\lambda_3} \cup \{\bot\}$ *that on input a secret key and an encapsulation of an ephemeral key, produces that ephemeral key. The* $\bot$*-symbol is used to denote failure, for instance if* $c \notin [\mathsf{Enc}_{pk}]$. *We will write* $k \leftarrow \mathsf{Dec}_{sk}(c)$.

Note the use of the word encapsulation rather than encryption. Indeed, c need not be an encryption of k in the traditional sense: it might not be possible to exert any control over the key k whatsoever.

The use of bitstrings for the randomness and the session key is perhaps slightly restrictive, but not really so: typically one assumes a source of uniform randomness of bitstrings is available and then transforms this into whatever distribution one needs. As long as the distribution native to the one-way problem and that arising from transforming a uniform distribution over bitstrings are statistically close, this only introduces small deviations in the security reductions.

We require that the KEM works, i.e., that decapsulation and encapsulation yield identical session keys. For all $(pk, sk) \in [\mathsf{Gen}(1^\lambda)]$ we define

$$\epsilon_{(pk,sk)}^{\bot,KEM} = \Pr[k \neq k' | k' \leftarrow \mathsf{Dec}_{sk}(c), (c,k) \leftarrow \mathsf{Enc}_{pk}(U_{\lambda_2})]$$

as the probability of a decapsulation failure for the key (pk, sk). We require that for all occurring keys $\epsilon_{(pk,sk)}^{\bot,KEM}$ is negligible in λ. For most primitives there are no failures at all but for the case we are interested in, NTRU, failures can occur.

There are two known security notions for a KEM. The first, IND-CPA, covers security against a passive adversary and the second, IND-CCA2, covers security against an active adversary.

A KEM is called IND-CPA if one cannot efficiently distinguish between valid tuples (c, k_0) as output by Enc and dummy tuples (c, k_1) where k_0 is a replaced by an unrelated k_1 uniformly chosen at random. On input the public key and a tuple (c, k_b) the adversary outputs a guess b' for b. Let

$$\mathsf{Adv}_A^{IND-CPA}(\lambda) = |\Pr[b' = b | b' \leftarrow A(pk, c, k_b), b \leftarrow \{0,1\}] - 1/2|$$

be the advantage of adversary A. Then we require that for all PPT adversaries A this advantage is negligible in λ.

For a KEM to be IND-CCA2 the indistinguishability should hold even against an adversary that has a decapsulation oracle at its command, as described in the following game.

Game 1: IND-CCA2 Security.

This game describes the type of attack an adversary may mount.

1. [Provide key] Set $(pk, sk) \leftarrow \mathsf{Gen}(1^\lambda)$ and give pk to the adversary.
2. [First oracle phase] The adversary is alllowed to ask queries to a decapsulation oracle. When the adversary has had enough queries it signifies it is ready to receive a challenge.
3. [Receive challenge] A challenge is created by setting $(c, k_0) \leftarrow \mathsf{Enc}_{pk}(U_{\lambda_2})$ and picking $k_1 \leftarrow \{0, 1\}^{\lambda_3}$ at random. A bit b is chosen at random and the challenge (c, k_b) is presented to the adversary.
4. [Second oracle phase] The adversary is alllowed to ask queries to a decapsulation oracle. However, the oracle refuses to answer query c.
5. [Output guess] The adversary outputs b', its best guess for b.

Let $\mathsf{Adv}_A^{IND-CCA2}(\lambda) = |\Pr[b = b'|b, b' \text{ from the game above}] - 1/2|$ be the advantage of the adversary in the game above. For IND-CCA2 security we require that this advantage is negligible for all PPT adversaries A. Note that the adversary has no influence over the challenge it gets.

4.2 Adaptive Plaintext Awareness

In this section we introduce plaintext awareness[1] for KEMs by adapting the framework for public key encryption as laid out by Bellare and Palacio [1]. PA2 captures the property that all ways (for an adversary) to make valid encapsulations require knowledge of what is encapsulated in the first place. This property should also hold if the adversary has an external encapsulator at hand (where output of the oracle obviously does not count as a valid encapsulation). The underlying idea is that whatever use an adversary makes of the decapsulation oracle, it could have computed herself. This is modelled by an efficient simulator of the oracle's output. The simulator is (obviously) not given the private key sk, however it is given the random coins of the adversary, corresponding to its subconscious.

Game R: Ideal Adaptive Plaintext Awareness.

This game describes the real game, where the adversary talks with a true decapsulation oracle.

1. [Generate key] Set $(pk, sk) \leftarrow \mathsf{Gen}(1^\lambda)$.
2. [Initialize] Set $CList \leftarrow []$, pick random coins $R(C)$ and run the adversary on input pk and $R(C)$.
3. [Query Phase] In this phase the adversary can consult its oracles: on a decapsulation query for $c \notin CList$, return $\mathsf{Dec}_{sk}(c)$ to the adversary; on an encapsulation query set $c \leftarrow \mathsf{Enc}_{pk}(U_{\lambda_2})$, append c to $CList$ and return c to the adversary.
4. [Adversary's Output] The adversary terminates with output x.

---

[1] We opt for the term plaintext awareness in keeping with the terminology for encryption schemes, although one could argue that key awareness would be more to the point.

5. [Distinguishing] Run the distinguisher on input x and let $D(x)$ be the outcome of the game.

In the ideal world we replace the decapsulation oracle with a simulation based on the random coins of the adversary. This gives rise to the following game:

Game S: Simulated Adaptive Plaintext Awareness.

This game describes the ideal game, where the adversary talks with a simulator instead of with a true decapsulation oracle.

1. [Generate key] Set $(pk, sk) \leftarrow \mathsf{Gen}(1^\lambda)$.
2. [Initialize] Set $CList \leftarrow []$, pick random coins $R(C)$ and $R(S)$. Set $St \leftarrow (pk, R(C), R(S))$ and run the adversary on input pk and $R(C)$.
3. [Query Phase] In this phase the adversary can consult its oracles: on a decapsulation query for $c \notin CList$, run $(St, k) \leftarrow S(St, c, Clist)$ and return k to the adversary; on an encapsulation query set $c \leftarrow \mathsf{Enc}_{pk}(U_{\lambda_2})$, append c to $CList$ and return c to the adversary.
4. [Adversary's Output] The adversary terminates with output x.
5. [Distinguishing] Run the distinguisher on input x and let $D(x)$ be the outcome of the game.

If we denote the outcome of the real game as D_{real} and of the ideal game as D_{ideal}, we can define A's advantage relative to S and D as

$$\mathsf{Adv}_{S,D}^{PA2}(A) = \Pr[D_{ideal} = 1 | \text{using } S, D] - \Pr[D_{real} = 1 | \text{using } \mathsf{Dec}, D] \ .$$

We call a KEM adaptive plaintext aware (PA2) if for all PPT A there exists an efficient simulator S such that for all PPT distinguishers D the advantage $\mathsf{Adv}_{S,D}^{PA2}(A)$ is negligible. We call an adversary A a (t, ϵ)-PA2 adversary if it runs in time t and there exists a PPT simulator S such that for all distinguishers running in time t the advantage of A relative to S and D is at most ϵ.

Note that, due to the nature of KEMs, we no longer need a plaintext creator as featured in Bellare and Palacio's definition. At first glance it might look overly complicated to have a stateful simulator, to provide the simulator with $Clist$ and to use a distinguisher. However, if $Clist$ were not given to the simulator, the adversary A could use the encapsulation oracle as a source of randomness unavailable to S. We have kept the distinguisher in this definition. Bellare and Palacio remark that this makes the definition stronger, as the simulator does not get to see the random coins of the distinguisher. However, a stronger notion of PA2 makes the theorem that IND-CPA + PA2 security implies IND-CCA2 security weaker. Luckily, Bellare and Palacio only use a deterministic distinguisher and weakening the definition (to strengthen the theorem) in this respect would not cause any problems.

We stress that correct simulation of the decryptor includes correct simulation of decryption failures. It turns out that correct simulation of a deterministic decapsulator is greatly aided by allowing a state to be passed on. Imagine a KEM with the property that sk and sk' are both secret keys to the same public key and that both (c, m) and (c, m') can be output by the encapsulator. If $\mathsf{Dec}(c, sk) = m$ and $\mathsf{Dec}(c, sk') = m'$, the simulator has to guess which secret key was used (requiring randomness) and he has to stick with this guess (requiring a state). Also note that the guess has to be probabilistic from C's point of view, or else the distinguisher at the end would notice. Finally, correct simulation of decryption failures is crucial in the proof that IND-CPA + PA2 implies IND-CCA2. If plaintext awareness was defined along the lines that, for all challenges put forward by the adversary, one can simulate a preimage, decryption failures would invalidate this implication.

In analogy to Bellare and Palacio, we would like to show that IND-CPA + PA2 security implies IND-CCA2 security *in the standard model*, as formalized in the conjecture below.

Conjecture 5. *Let A be a (t, ϵ)-IND-CCA2 adversary, then there exist a (t_1, ϵ_1)-IND-CPA adversary A_1 and a (t_2, ϵ_2)-PA2 adversary A_2 with $\epsilon_1 + 2\epsilon_2 \geq \epsilon, t_1 = poly(t)$ and $t_2 \approx t$.*

Unfortunately, adapting Bellare and Palacio's proof to the setting of KEMs fails due to a technical detail pointed out below. What we can show is that the conjecture holds in the non-programmable random oracle model for a suitably strong simulator. This result suffices for our needs later on, since the KEM constructions we use have security proofs in the non-programmable random oracle model to begin with. Dent [7] recently gave a stronger notion of plaintext awareness for KEMs that does allow a proof in the standard model.

We call a scheme PA2' secure (in the non-programmable random oracle model) if there is an efficient universal simulator S such that for all PPT adversaries the advantage

$$\mathsf{Adv}_A^{PA2'}(\lambda) = |\Pr[A^{\mathsf{Dec}_{sk}(\cdot), \mathsf{Enc}_{pk}, \mathcal{O}} = 1|(pk, sk)] - \Pr[A^{S, \mathsf{Enc}_{pk}, \mathcal{O}} = 1|(pk, sk)]|$$

is negligible, where $\mathcal{O}$ is the random oracle, $\mathsf{Dec}, \mathsf{Enc}$, and S have also access to the oracle $\mathcal{O}$ and S gets to see all queries made by the adversary to $\mathcal{O}$ (at the time they are made). However, S does not get the randomness of the adversary A (this would be of no use anyway, since the simulator is chosen before the adversary). Given some universal simulator, we say that an adversary is a (t, ϵ)-PA2' adversary if it runs in time t with advantage ϵ.

Theorem 6. *Let A be a (t, ϵ)-IND-CCA2 adversary (in the random oracle model), then there exist a (t_1, ϵ_1)-IND-CPA adversary A_1 and (t_2, ϵ_2)-PA2' and (t_3, ϵ_3)-PA2' adversaries A_2 and A_3 with $\epsilon_1 + \epsilon_2 + \epsilon_3 \geq \epsilon, t_1 = poly(t)$ and $t_3 \approx t_2 = t$.*

Proof: For the proof we use a slightly different, but equivalent notion of IND-CCA2 security. Namely, we let E_0 be a proper encapsulation oracle and E_1 be a fake encapsulation oracle returning a pair (C, K) where C is a properly generated encapsulation, but K is a randomly and independently generated bitstring (of the correct length). Then the advantage of A can be reformulated as

$$\mathsf{Adv}_A^{IND-CCA2}(\lambda)$$
$$= |\Pr[A^{\mathsf{Dec}_{sk}(\cdot), E_0}(pk) = 1|(pk, sk)] - \Pr[A^{\mathsf{Dec}_{sk}(\cdot), E_1}(pk) = 1|(pk, sk)]| \, ,$$

where the oracle E_b is called only once and with the usual restriction that Dec_{sk} is not called on E_b's output. We can now relate these two probabilities with those that arise from replacing the decapsulation oracle with a simulator, as depicted in the following commutative diagram:

$$
\begin{array}{ccc}
A^{\mathsf{Dec}_{sk}(\cdot), E_0} & \xrightarrow{\mathsf{Adv}_{A_2}^{PA2'}} & A^{S(\cdot), E_0} \\
\Big\downarrow {\scriptstyle \mathsf{Adv}_A^{IND-CCA2}} & & \Big\downarrow {\scriptstyle \mathsf{Adv}_{A_1}^{IND-CPA}} \\
A^{\mathsf{Dec}_{sk}(\cdot), E_1} & \xleftarrow{\mathsf{Adv}_{A_3}^{PA2'}} & A^{S(\cdot), E_1}
\end{array}
$$

The diagram shows that the IND-CCA advantage can be upper bounded by the sum of the three other advantages. A more detailed proof can be found in the full version of this paper. $\qquad$ *Q.E.D.*

In the standard model the proof fails, since we are no longer guaranteed that there is a universal simulator that works for both A_2 and for A_3, although it seems highly unlikely that a simulator that works for one would fail for the other for a scheme that is IND-CPA secure. Note that for encryption schemes the problem is resolved by the quantification over all plaintext creators, i.e., a simulator is universal for all plaintext creators, in particular to a plaintext creator that given two messages always picks the first, respectively the second. Dent [7] has shown a similar approach works for KEMs, yet proving our conjecture based on our definition of plaintext awareness remains a tantalizing open problem.

5   Padding Schemes for Key Encapsulation

Dent [6] has provided several constructions to build an IND-CCA2 secure KEM out of an arbitrary trapdoor one-way function. Some of his constructions rely on additional assumptions, such as efficient plaintext-ciphertext checkers for the underlying one-way function. However, his fourth construction is fairly generally applicable[2] and the one we describe below. Key generation for the KEM is identical to that of the one-way function.

Let $G : \{0,1\}^{\lambda_2} \to \{0,1\}^{\lambda_4}$ and $\mathsf{KDF} : \{0,1\}^{\lambda_2} \to \{0,1\}^{\lambda_3}$ be a hash-function, respectively a key derivation function. Let $L_{\mathcal{M}} : \{0,1\}^{\lambda_2} \to \mathcal{M}$ and $L_{\mathcal{R}} : \{0,1\}^{\lambda_4} \to \mathcal{R}$ be embeddings as before, so $\lambda_2 = \lceil \log_2 |\mathcal{M}| \rceil$ and $\lambda_4 = \lceil \log_2 |\mathcal{R}| \rceil$.

---

[2] Dent claims his construction only works for probabilistic functions, but for most deterministic functions the domain can be split in two, a message part and a random part, thus turning it into a probabilistic function.

Algorithm 7: Dent's Encapsulation.

This encapsulation is based on Dent's KEM construction. It basically leaves out the randomization of the message.

1. [Pick seed] Choose $w \in \{0, 1\}^{\lambda_2}$ at random and set $m \leftarrow L_\mathcal{M}(w)$. Repeat until $m \neq \bot$.
2. [Determine randomness] Set $r \leftarrow L_\mathcal{R}(G(w))$. If $r = \bot$ return to the previous step.
3. [Invoke one-way primitive] Compute $e \leftarrow E(m, r)$.
4. [Derive key] Compute $k \leftarrow \mathsf{KDF}(w)$.
5. [Output] Return k as session key and e as encapsulation thereof.

This is decapsulation given by Dent. It is based on a partial trapdoor only.

Algorithm 8: Dent's Original Decapsulation.

Inputs are a secret key sk and an encapsulation e. It outputs a session key k.

1. [Invert] Set $m \leftarrow D_{sk}^m(e)$ and $w \leftarrow L_\mathcal{M}^{-1}(m)$.
2. [Reconstruct randomness] Set $r \leftarrow L_\mathcal{R}(G(w))$.
3. [Recompute the encapsulation] Set $e' \leftarrow E(m, r)$.
4. [Verify encapsulation] If $e \neq e'$ terminate with output $\bot$.
5. [Output session key] Compute $k \leftarrow \mathsf{KDF}(w)$. Return k as session key.

Below is an alternative decapsulation that works if E permits a full inverse. It is easy to see that these two decapsulators are functionally equivalent when both apply (and the inverses D and D^m behave as expected). One might be more efficient than the other depending on the application.

Algorithm 9: Alternative Decapsulation.

Inputs are a secret key sk and an encapsulation e. It outputs a session key k.

1. [Invert] Set $(m, r) \leftarrow D_{sk}(e)$ and $w \leftarrow L_\mathcal{M}^{-1}(m)$.
2. [Recompute randomness] Set $r' \leftarrow L_\mathcal{R}(G(w))$.
3. [Verify encapsulation] If $r \neq r'$ terminate with output $\bot$.
4. [Output session key] Compute $k \leftarrow \mathsf{KDF}(w)$. Return k as session key.

The question is whether for NTRU existing techniques to construct a KEM out of a trapdoor one-way function remain secure or not. For this we use a mathematically more elegant description of the scheme (closer to what Dent described, but further removed from the gritty bit-oriented reality), where $G : \mathcal{M} \to \mathcal{R}$ and $\mathsf{KDF} : \mathcal{M} \to \{0, 1\}^{\lambda_2}$ and encryption starts with picking $m \in \mathcal{M}$ uniformly at random.

We will now prove that Dent's scheme is IND-CPA and PA2 and thus IND-CCA2 secure. The two indistinguishability claims are nothing new, but we believe our proof separating IND-CPA and PA2 sheds some new light on the protocol, in particular on its reliance on the random oracle model in the proof and on possible avenues to reduce this reliance [?]. For IND-CPA decryption failures are hardly an issue and for IND-CCA2 they are taken into account separately.

Theorem 10. *Let A be a (t, ϵ)-IND-CPA adversary in the non-programmable random oracle model making q_G and q_{KDF} oracle calls to G respectively to KDF. Then there exists a (t', ϵ')-k-partial inverter for the underlying one-way function, where $t \approx t', \epsilon' \geq \epsilon$ and $k \leq q_G + q_{KDF}$.*

Proof: Let A be said IND-CPA adversary. Now construct an inverter as follows.

Algorithm 11: Inverter.

This algorithm gets as input a public key pk and an image y and computes a list containing a partial preimage of y.

1.   [Initialize A] Generate $k \leftarrow U$ and feed $(pk, (y, k))$ to the adversary.
2.   [Run A] Run A; upon oracle queries from A to either G or KDF, forward the query to an independent non-programmable random oracle and relay the answer back to the adversary. Keep a single list of all queries made (G and KDF have the same domain).
3.   [Output inverse] After A terminates (disregarding its output), output the list of all oracle queries.

It is clear that the algorithm runs in time $\approx t$ and that its output contains at most $q_G + q_{KDF}$ elements. We now argue that with probability at least ϵ, the algorithm above outputs a partial inversion. Recall that the advantage of A relates to distinguishing true encapsulations ($b = 0$) from false ones ($b = 1$). In the latter case, the adversary is however presented with a valid encapsulation of something, just not of the key accompanying it. Let $E^{-1}(y) = \{(m, r)|E(m, r) = y\}$ and suppose the random oracle for G satisfies, for all relevant m that $G(m) = r$. For $b = 0$ we define $KDF(m) = k$ whereas for $b = 1$ we let $KDF(m)$ be random. With these oracles the input to the adversary corresponds to the distribution it expects, however the oracles are no longer fully random. The only way for an adversary to find out that either the oracles are improper or with which set it is dealing, is by querying it on some $m \in E^{-1}(y)$.   *Q.E.D.*

For the plaintext awareness, we begin by describing a simulator S. Let A be a PA2' adversary. If A makes a random oracle query to G, forward, relay and store. If the adversary makes a decapsulation query on e, check whether there is a pair $(m, G(m))$ such that $E(m, G(m)) = e$. If so, return $KDF(m)$ (just by computing the function), otherwise return $\perp$.

Theorem 12. *Let A be a PA2' adversary in the ROM making q_D decapsulation queries. Then its advantage against simulator S is at most $q_D/|R| + \tilde{\epsilon}^{\perp}$, where $\tilde{\epsilon}^{\perp}$ is defined below and only G is modelled as a (non-programmable) random oracle.*

Proof: Clearly the probability of an adversary and a distinguisher's advantage is upper bounded by the probability that $S(e) \neq D_{sk}(e)$ for some queried e. Note that if $S(e) = w$, and in the absence of decryption failures, we also have $D_{sk}(e) = w$, so of relevance is the probability that there exists a e with $S(e) = \perp$ but with $D_{sk}(e) \neq \perp$. Assuming uniqueness of inverse $(w, r) = D_{sk}(e)$, this means

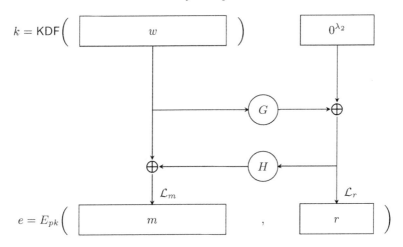

$$k = \mathsf{KDF}\left(\boxed{\quad\quad\quad w \quad\quad\quad} \right) \quad\quad \boxed{0^{\lambda_2}}$$

$$e = E_{pk}\left(\boxed{\quad\quad\quad m \quad\quad\quad} \quad , \quad \boxed{\quad r \quad} \right)$$

Fig. 2. RAEP-based KEM

that $G(w) = r$. This probability (over the choices made by the oracle) equals $1/|R|$. Since there are q_D queries that can go wrong, the total error is upper bounded by $q_D/|R|$.

We now deal with the issue of decryption failures. For Dent's construction to work, the probability that an adversary can find $(w, G(w)) \in S$ should be small, where the adversary has full control over w. This is implied by a small value of $\tilde{\epsilon}^{\perp} = \mathbb{E}_{(pk,sk)} \max_w S_{(pk,sk)}(w)/|R|$.   *Q.E.D.*

We can combine all of the above in the following theorem.

Theorem 13. *Let A be a (t, ϵ)-IND-CCA2 adversary in the non-programmable random oracle model. Then there exists a (t', ϵ')-inverter for the first part of E with $t' \approx t$ and with $\epsilon' \geq (\epsilon - q_D/|\mathcal{R}| - \tilde{\epsilon}_{\perp})/(q_{\mathsf{KDF}} + q_G)$.*

If we compare the version without decryption failures, i.e., with $\tilde{\epsilon}_{\perp} = 0$ with the result of Dent, we see that we have tightened the reduction.

Unfortunately for NTRU the value $\tilde{\epsilon}^{\perp}$ can be quite large (indeed, one cannot possibly maintain that the padding scheme results in a random subset of $\mathcal{M} \times \mathcal{R}$). The situation is not quite satisfactory, similar to the case of using SAEP+ directly for NTRU. There are several possible solutions to this problem. A trivial solution is to use RAEP (or NAEP) without adaptations, where the message that is encrypted serves as the session key (so no need for a key derivation function). The downside of this approach is that potentially some bandwidth is lost. An alternative solution based on NAEP is given by Whyte [17,18].

In Figure 2 the corresponding RAEP version is given. Here what used to be message M and randomness w in the encryption scheme are joint together to form the seed (again w) for a key derivation function (it is easy to see that a KDF is necessary here, because if $k = w$ then given k it is easy to compute the unique encapsulation e, breaching IND-CPA security). For completeness we state the encapsulation and decapsulation algorithms. Note that this differs in several places with the NTRU-KEM appearing on slide 19 [18].

Algorithm 14: Encapsulation.

This is encapsulation based on RAEP.

1.   [Pick seed] Choose w at random.
2.   [Determine randomness] Compute $\hat{r} \leftarrow G(w)$ and $r \leftarrow L_{\mathcal{R}}(\hat{r})$.
3.   [Randomize message] Compute $m \leftarrow L_{\mathcal{M}}(w \oplus H(\hat{r}))$.
4.   [Invoke NTRU primitive] Compute $e \leftarrow E(m, r)$.
5.   [Derive key] Set $k \leftarrow \mathsf{KDF}(w)$.
6.   [Output] Return k as session key and e as encapsulation thereof.

Algorithm 15: Decapsulation.

This is decapsulation based on RAEP.

1.   [Invert one-way function] Compute $(m, r) \leftarrow D_{sk}(e)$. It this results in failure, output $\bot$.
2.   [Undo embeddings] Compute $\hat{m} \leftarrow L_{\mathcal{M}}^{-1}(m)$ and $\hat{r} \leftarrow L_{\mathcal{R}}(r)$.
3.   [Retrieve seed] Compute $w \leftarrow \hat{m} \oplus H(\hat{r})$.
4.   [Recompute randomness] Set $\hat{r}' \leftarrow G(w)$.
5.   [Verify encapsulation] If $\hat{r} \neq \hat{r}'$ output $\bot$.
6.   [Derive key] Set $k \leftarrow \mathsf{KDF}(w)$.
7.   [Output] Return k as session key.

We present a third, different solution that is more efficient: there is no need for an extra hash-function. The downside is that our solution expands the key size and sofar is only heuristic (but well motivated). Recall the way RAEP dealt with decryption failures. Because pad was sufficiently random, $\tilde{\epsilon}_{\bot}$ applied to its range as well, giving only a small number of failures with respect to the range of G. To be more precise, the probability of G mapping an element into a potential decryption failure is at most $\epsilon_{(pk,sk)}^{\bot}|\mathcal{M}|/|\mathcal{R}|$, so the relevant term $\tilde{\epsilon}_{\bot} \leq q_G \bar{\epsilon}^{\bot}|\mathcal{M}|/|\mathcal{R}|$.

Our observation is that the same argument could be applied if the set $\mathcal{S}_{(pk,sk)}$ would be sufficiently random. But randomization of $\mathcal{S}_{(pk,sk)}$ can effectively be achieved by applying a keyed smoothing function straight after the padding (and before the embedding into $\mathcal{M} \times \mathcal{R}$). We conjecture that for NTRU a combination of standard diffusion and confusion does the trick, i.e., XOR'ing with a key followed by a random key-driven permutation of the bits (coordinates). This would add $(\lambda_2 + \lambda_4)(1 + \log_2(\lambda_2 + \lambda_4))$ bits to the key, but the transformation is considerably faster to perform than the hash function H used in the NAEP/RAEP-based solutions.

6   Conclusion

In this work we have expanded the theory of KEMs to include plaintext awareness and resilience against decryption errors. The use of plaintext awareness allowed us to tighten the security reduction for an existing KEM given by Dent. Resilience against decryption errors is relevant when used in conjunction with (certain variants of) NTRU. We have described a novel approach to deal with decryption failures that outperforms existing art.

The author would like to thank Nigel Smart, Joe Silverman, Alex Dent and the anonymous referees for valuable comments and discussions.

References

1. M. Bellare and A. Palacio. Towards plaintext-aware public-key encryption without random oracles. *Asiacrypt'04*, LNCS **3329**, pages 48–62.
2. M. Bellare and P. Rogaway. Optimal asymmetric encryption. *Crypto'94*, LNCS **839**, pages 92–111.
3. A. Boldyreva and M. Fischlin. Analysis of Random Oracle Instantiation Scenarios for OAEP and Other Practical Schemes *Crypto'05*, LNCS **3621**, pages 412–429.
4. D. Boneh. Simplified OAEP for the RSA and Rabin functions. *Crypto'01*, LNCS **2139**, pages 275–291.
5. R. Cramer and V. Shoup. A practical public key cryptosystem provably secure against adaptive chosen ciphertext attack. *Crypto'98*, LNCS **1462**, pages 13–25.
6. A. W. Dent. A designer's guide to KEMs. *CCC'03*, LNCS **2898**, pages 133–151.
7. A. W. Dent. Cramer-Shoup is plaintext-aware in the standard model. Technical Report 261, IACR's ePrint Archive, 2005.
8. E. Fujisaki, T. Okamoto, D. Pointcheval, and J. Stern. RSA-OAEP is secure under the RSA assumption. *Crypto'01*, LNCS **2139**, pages 260–274.
9. J. Hoffstein, J. Pipher, and J. Silverman. NTRU: A new high speed public key cryptosystem. *ANTS-III*, LNCS **1423**, pages 267–288.
10. N. Howgrave-Graham, P. Nguyen, D. Pointcheval, J. Proos, J. Silverman, A. Singer, and W. Whyte. The impact of decryption failures on the security of NTRU encryption. *Crypto'03*, LNCS **2729**, pages 226–246.
11. N. Howgrave-Graham, J. Silverman, A. Singer, and W. Whyte. NAEP: Provable security in the presence of decryption failures. Technical Report 172, IACR's ePrint Archive, 2003.
12. N. Howgrave-Graham, J. H. Silverman, and W. Whyte. Choosing parameter sets for NTRUEncrypt with NAEP and SVES-3. *CT-RSA'05*, LNCS **3376**, pages 118–135.
13. J. Proos. Imperfect decryption and an attack on the NTRU encryption scheme. Technical Report 2, IACR's ePrint Archive, 2003.
14. V. Shoup. OAEP reconsidered. *Crypto'01*, LNCS **2139**, pages 239–259.
15. V. Shoup. OAEP reconsidered. *Journal of Cryptology*, 15:223–249, 2002.
16. J. Silverman and W. Whyte. Estimating decryption failure probabilities for NTRUEncrypt. Technical Report 18, NTRU Cryptosystems, 2003.
17. W. Whyte. X9.98 top N issues, 2003. Slides available online.
18. W. Whyte. Choosing NTRUEncrypt parameters, 2004. Slides available online.

Efficient Identity-Based Key Encapsulation to Multiple Parties

M. Barbosa[1,*] and P. Farshim[2]

[1] Departamento de Informática, Universidade do Minho,
Campus de Gualtar, 4710-057 Braga, Portugal
`mbb@di.uminho.pt`

[2] Department of Computer Science, University of Bristol,
Merchant Venturers Building, Woodland Road,
Bristol BS8 1UB, United Kingdom
`farshim@cs.bris.ac.uk`

Abstract. We introduce the concept of identity based key encapsulation to multiple parties (mID-KEM), and define a security model for it. This concept is the identity based analogue of public key KEM to multiple parties. We also analyse possible mID-KEM constructions, and propose an efficient scheme based on bilinear pairings. We prove our scheme secure in the random oracle model under the Gap Bilinear Diffie-Hellman assumption.

Keywords: Key Encapsulation Mechanism (KEM), Hybrid Encryption, Identity Based Cryptography, Provable Security.

1 Introduction

A traditional public key encryption scenario involves communication between two users. One party, Alice, wishes to communicate securely with another party, Bob. Alice uses Bob's public key, which she (somehow) knows to be authentic, to create a message that only Bob can decrypt using his private key.

Due to the large computational cost associated with public key encryption algorithms, most practical applications use *hybrid encryption* to handle large plaintext messages. Rather than using the public key encryption algorithm directly over the plaintext, one generates a random *session key* and uses it to encrypt the message under a more efficient symmetric algorithm. The (small) session key is then encrypted using the public key encryption algorithm, and the message is transferred as the combination of both ciphertexts.

The hybrid public key encryption paradigm has gained strength in recent years with the definition and formal security analysis of the generic KEM-DEM construction [7,8,17]. In this approach to hybrid encryption one defines a symmetric *data encapsulation mechanism* (DEM) which takes a key k and a message

---

* Funded by scholarship SFRH/BPD/20528/2004, awarded by the Fundação para a Ciência e Tecnologia, Ministério da Ciência e do Ensino Superior, Portugal.

N.P. Smart (Ed.): Cryptography and Coding 2005, LNCS 3796, pp. 428–441, 2005.

M and computes $C \leftarrow \text{DEM}_k(M)$. Given knowledge of k one can also recover M via $M \leftarrow \text{DEM}_k^{-1}(C)$.

The key k is transferred to the recipient of the ciphertext in the form of an encapsulation. To create this encapsulation, the sender uses a *key encapsulation mechanism* (KEM). This is an algorithm which takes as input a public key pk and outputs a session key k plus an encapsulation E of this session key under the public key. We write this as

$$(k, E) \leftarrow \text{KEM}(\text{pk}).$$

Notice that the session key is not used as input to the KEM. The recipient recovers the key k using his private key sk via the decapsulation mechanism. We denote this by

$$k \leftarrow \text{KEM}^{-1}(E, \text{sk}).$$

The full ciphertext of the message M is then given by $E||C$.

The use of the KEM-DEM philosophy allows the different components of a hybrid encryption scheme to be designed in isolation, leading to a simpler analysis and hopefully more efficient schemes. In fact, Cramer and Shoup [7] established that an hybrid encryption scheme is IND-CCA2 secure, if both the KEM and DEM modules are *semantically secure* [9] against *adaptive chosen ciphertext attacks* [14] (see Theorems 4 and 5 in [7]).

Smart [16] extends this notion to a setting in which Alice wants to send the same message to multiple parties. He considers the question: is it possible, in this situation, to avoid carrying out n instances of the KEM-DEM construction? Smart [16] proposes a multiple KEM (or mKEM) primitive whereby the sender can create an encapsulation of the session key k under n public keys. The purpose of this type of construction is to save $n-1$ data encapsulations overall: a single DEM invocation using k suffices to obtain valid ciphertexts for all recipients. The entire message encryption process becomes

$$(k, E) \leftarrow \text{mKEM}(\text{pk}_1, \ldots, \text{pk}_n)$$
$$C \leftarrow \text{DEM}_k(M)$$

and each recipient i will get the pair (E, C). For decryption, user i simply performs the following two operations:

$$k \leftarrow \text{mKEM}^{-1}(E, \text{sk}_i)$$
$$M \leftarrow \text{DEM}_K^{-1}(C).$$

Identity-based encryption is a form of public key encryption in which users' public keys can take arbitrary values such as e-mail addresses. A central trusted authority manages the public keys in the system, providing legitimate users with private keys that allow recovering messages encrypted under their identities.

Boneh and Franklin [5] introduced a secure and efficient identity-based encryption scheme based on pairings on elliptic curves. Lynn [11] mentions that the encryption algorithm proposed by Boneh and Franklin is unlikely to be used, since in practice one will use a form of identity based key encapsulation mechanism. Bentahar et al. [4] call such an encapsulation mechanism ID-KEM. Lynn [11] describes a possible ID-KEM construction, but he gives no security model or proof. Bentahar et al. [4] formalise the notion of key encapsulation for the identity-based setting, define a security model and propose several provably secure ID-KEM constructions.

In this work we bring together the concepts of ID-KEM and mKEM and propose mID-KEM: identity-based key encapsulation to multiple users.

This paper is organised as follows. In Sections 2 and 3 we briefly discuss background on pairings, ID-KEMs and mKEMs. The concept and security of an mID-KEM primitive is defined in Section 4. In Section 5 we present an efficient mID-KEM construction, and prove it is secure in Section 6. We conclude the paper by analysing the efficiency of our scheme in Section 7.

2   Background on Pairings

Our constructions will use bilinear maps and bilinear groups. We briefly review the necessary facts about these here. Further details may be found in [5,6].

Let G_1, G_2 and G_T be groups with the following properties:

- G_1 and G_2 are additive groups of prime order q.
- G_1 has generator P_1 and G_2 has generator P_2.
- There is an isomorphism ρ from G_2 to G_1, with $\rho(P_2) = P_1$.
- There is a bilinear map $\hat{t} : G_1 \times G_2 \to G_T$.

In many cases one can set $G_1 = G_2$ as is done in [5]. When this is so, we can take ρ to be the identity map; however, to take advantage of certain families of groups [12], we do not restrict ourselves to this case.

The map $\hat{t}$, which is usually derived from the Weil or Tate pairings on an elliptic curve, is required to satisfy the following conditions:

1. Bilinear: $\hat{t}(aQ, bR) = \hat{t}(Q, R)^{ab}$, for all $Q \in G_1$, $R \in G_2$ and $a, b \in \mathbb{Z}_q$.
2. Non-degenerate: $\hat{t}(P_1, P_2) \neq 1$.
3. Efficiently computable.

Definition 1 (Bilinear groups). *We say that G_1 and G_2 are bilinear groups if there exists a group G_T with $|G_T| = |G_1| = |G_2| = q$, an isomorphism ρ and a bilinear map $\hat{t}$ satisfying the conditions above; moreover, the group operations in G_1, G_2 and G_T and the isomorphism ρ must be efficiently computable.*

A *group description* $\Gamma = [G_1, G_2, G_T, \hat{t}, \rho, q, P_1, P_2]$ describes a given set of bilinear groups as above. There are several hard problems associated with a group description Γ which can be used for building cryptosystems. These have their origins in the work of Boneh and Franklin [5].

Notation. If S is a set then we write $v \leftarrow S$ to denote the action of sampling from the uniform distribution on S and assigning the result to the variable v. If A is a probabilistic polynomial time (PPT) algorithm we denote the action of running A on input I and assigning the resulting output to the variable v by $v \leftarrow A(I)$. Note that since A is probabilistic, $A(I)$ is a probability space and not a value.

Bilinear Diffie-Hellman Problem (BDH). Consider the following game for a group description Γ and an adversary A.

1. $a, b, c \leftarrow \mathbb{Z}_q^*$
2. $t \leftarrow A(\Gamma, aP_1, bP_1, cP_2)$

The advantage of the adversary is defined to be

$$\text{Adv}_{\text{BDH}}(A) = \Pr[t = \hat{t}(P_1, P_2)^{abc}]. \tag{1}$$

Notice there are various equivalent formulations of this: given xP_2 one can compute xP_1 via the isomorphism ρ. This particular formulation allows for a clearer exposition of the security proof in Section 6.

Decisional Bilinear Diffie-Hellman Problem (DBDH). Consider Γ and the following two sets

$$\mathcal{D}_\Gamma = \{(aP_1, bP_1, cP_2, \hat{t}(P_1, P_2)^{abc}) : a, b, c \in [1, \ldots, q]\},$$
$$\mathcal{R}_\Gamma = G_1 \times G_1 \times G_2 \times G_T.$$

The goal of an adversary is to be able to distinguish between the two sets. This idea is captured by the following game.

1. $a, b, c \leftarrow \mathbb{Z}_q^*$
2. $d \leftarrow \{0, 1\}$
3. If $d = 0$ then $\alpha \leftarrow G_T$
4. Else $t \leftarrow \hat{t}(P_1, P_2)^{abc}$
5. $d' \leftarrow A(\Gamma, aP_1, bP_1, cP_2, t)$

We define the advantage of such an adversary by

$$\text{Adv}_{\text{DBDH}}(A) = |2\Pr[d = d'] - 1|.$$

The Gap Bilinear Diffie-Hellman Problem (GBDH). Informally, the *gap bilinear Diffie-Hellman problem* is the problem of solving BDH with the help of an oracle which solves DBDH. The use of such relative or "gap" problems was first proposed by Okamoto and Pointcheval [13].

Let $\mathcal{O}$ be an oracle that, on input $\beta \in \mathcal{R}_\Gamma$, returns 1 if $\beta \in \mathcal{D}_\Gamma$ and 0 otherwise. For an algorithm A, the advantage in solving GBDH, which we denote by $\text{Adv}_{\text{GBDH}}(A, q_G)$, is defined as in (1) except that A is granted oracle access to $\mathcal{O}$ and can make at most q_G queries.

3   Review of ID-KEMs and mKEMs

An identity based KEM (ID-KEM) scheme is specified by four polynomial time algorithms:

- $\mathsf{G}_{\text{ID-KEM}}(1^t)$. A PPT algorithm which takes as input 1^t and returns the master public key M_{pk} and the master secret key M_{sk}. We assume that M_{pk} contains all system parameters, including the security parameter t.

- $\mathbb{X}_{\text{ID}-\text{KEM}}(\text{ID}, M_{\text{sk}})$. A deterministic algorithm which takes as input M_{sk} and an identifier string $\text{ID} \in \{0,1\}^*$, and returns the associated private key S_{ID}.
- $\mathbb{E}_{\text{ID}-\text{KEM}}(\text{ID}, M_{\text{pk}})$. This is the PPT encapsulation algorithm. On input of ID and M_{pk} this outputs a pair (k, C) where $k \in \mathbb{K}_{\text{ID}-\text{KEM}}(M_{\text{pk}})$ is a key in the space of possible session keys at a given security level, and $C \in \mathbb{C}_{\text{ID}-\text{KEM}}(M_{\text{pk}})$ is the encapsulation of that key.
- $\mathbb{D}_{\text{ID}-\text{KEM}}(S_{\text{ID}}, C)$. This is the deterministic decapsulation algorithm. On input of C and S_{ID} this outputs k or a failure symbol $\perp$.

The soundness of an ID-KEM scheme is defined as:

$$\Pr\left(k = \mathbb{D}_{\text{ID}-\text{KEM}}(S_{\text{ID}}, C) \middle| \begin{array}{l} (M_{\text{pk}}, M_{\text{sk}}) \leftarrow \mathbb{G}_{\text{mID}-\text{KEM}}(1^t) \\ \text{ID} \leftarrow \{0,1\}^* \\ (k, C) \leftarrow \mathbb{E}_{\text{ID}-\text{KEM}}(\text{ID}, M_{\text{pk}}) \\ S_{\text{ID}} \leftarrow \mathbb{X}_{\text{ID}-\text{KEM}}(\text{ID}, M_{\text{sk}}) \end{array} \right) = 1$$

Consider the following two-stage game between an adversary A of the ID-KEM and a challenger.

IND-CCA2 Adversarial Game
1. $(M_{\text{pk}}, M_{\text{sk}}) \leftarrow \mathbb{G}_{\text{ID}-\text{KEM}}(1^t)$
2. $(s, \text{ID}^*) \leftarrow A_1^{\mathcal{O}}(M_{\text{pk}})$
3. $(k_0, C^*) \leftarrow \mathbb{E}_{\text{ID}-\text{KEM}}(\text{ID}^*, M_{\text{pk}})$
4. $k_1 \leftarrow \mathbb{K}_{\text{ID}-\text{KEM}}(M_{\text{pk}})$
5. $b \leftarrow \{0,1\}$
6. $b' \leftarrow A_2^{\mathcal{O}}(M_{\text{pk}}, C^*, s, \text{ID}^*, k_b)$

In the above s is some state information and $\mathcal{O}$ denotes oracles to which the adversary has access. In this model the adversary has access to two oracles: a private key extraction oracle which, on input of $\text{ID} \neq \text{ID}^*$, will output the corresponding value of S_{ID}; and a decapsulation oracle with respect to any identity ID of the adversary's choosing. The adversary has access to this decapsulation oracle, subject to the restriction that in the second phase A is not allowed to call it with the pair (C^*, ID^*).

The adversary's advantage in the game is defined to be

$$\text{Adv}_{\text{ID}-\text{KEM}}^{\text{IND}-\text{CCA2}}(A) = |2\Pr[b' = b] - 1|.$$

An ID-KEM is considered to be IND-CCA2 secure if for all PPT adversaries A, the advantage in this game is a negligible function of the security parameter t.

The notion of KEM is extended in [16] to multiple parties. An mKEM is defined by the following three polynomial time algorithms:

- $\mathbb{G}_{\text{mKEM}}(1^t)$ which is a probabilistic key generation algorithm. On input of 1^t and the domain parameters, this outputs a public/private key pair (pk, sk).

- $\mathbb{E}_{\text{mKEM}}(\mathcal{P})$ which is a probabilistic encapsulation algorithm. On input of a set of public keys $\mathcal{P} = \{\text{pk}_1, \ldots, \text{pk}_n\}$ this algorithm outputs an encapsulated key-pair (k, C), where $k \in \mathbb{K}_{\text{mKEM}}(t)$ is the session key and C is an encapsulation of the key k under the public keys $\{\text{pk}_1, \ldots, \text{pk}_n\}$.
- $\mathbb{D}_{\text{mKEM}}(\text{sk}, C)$ which is a decapsulation algorithm. This takes as input an encapsulation C and a private key sk, and outputs a key k or a special symbol $\perp$ representing the case where C is an invalid encapsulation with respect to the private key sk.

The soundness of an mKEM scheme is defined as:

$$\Pr\left(k = \mathbb{D}_{\text{mKEM}}(\text{sk}_j, C) \left| \begin{array}{l} (\text{pk}_i, \text{sk}_i) \leftarrow \mathbb{G}_{\text{mKEM}}(1^t), 1 \leq i \leq n \\ (k, C) \leftarrow \mathbb{E}_{\text{mKEM}}(\text{pk}_1, \ldots, \text{pk}_n) \\ j \leftarrow \{1, \ldots, n\} \end{array} \right. \right) = 1$$

Security of an mKEM is defined in a similar manner to a KEM via the following game.

> (m, n)-IND-CCA2 Adversarial Game
> 1. $(\text{pk}_i, \text{sk}_i) \leftarrow \mathbb{G}_{\text{mKEM}}(1^t)$ for $1 \leq i \leq n$
> 2. $\mathcal{P} \leftarrow \{\text{pk}_1, \ldots, \text{pk}_n\}$
> 3. $(s, \mathcal{P}^*) \leftarrow A_1^{\mathcal{O}}(\mathcal{P})$ where $\mathcal{P}^* \subseteq \mathcal{P}$ and $m = |\mathcal{P}^*| \leq n$
> 4. $(k_0, C^*) \leftarrow \mathbb{E}_{\text{mKEM}}(\mathcal{P}^*)$
> 5. $k_1 \leftarrow \mathbb{K}_{\text{mKEM}}(t)$
> 6. $b \leftarrow \{0, 1\}$
> 7. $b' \leftarrow A_2^{\mathcal{O}}(\text{pk}, C^*, s, \mathcal{P}^*, k_b)$

In the above s is some state information and $\mathcal{O}$ denotes the decapsulation oracle to which the adversary has access in rounds 1 and 2. As it is noted in [16], restricting access to this oracle by simply disallowing the query (C^*, pk^*) for some $\text{pk}^* \in \mathcal{P}^*$ would be too lenient. Consider an mKEM in which an encapsulation is of the form $C = c_1 || \ldots || c_n$ where each c_i is intended for the recipient with public key pk_i. On reception of a challenge encapsulation of this form, the adversary could query the decapsulation oracle with a pair (c_i^*, pk_i^*), which is *not* disallowed, and win the game.

To avoid this type of problem, Smart [16] defines more restrictive, but still reasonable, oracle access. The restriction imposed in the model is that in the second stage the adversary is only allowed to query the decapsulation oracle with those C which decapsulate to a different key from that encapsulated by C^*. We will come back to this in the next section.

The advantage of the adversary A is defined as

$$\text{Adv}_{\text{mKEM}}^{(m,n)-\text{IND-CCA2}}(A) = |2\Pr[d = d'] - 1|.$$

An mKEM is considered to be (m, n)-IND-CCA2 secure if for all PPT adversaries A, the above advantage is a negligible function of the security parameter t.

4   ID-KEM to Multiple Parties

In this section we propose a new cryptographic primitive: identity based key encapsulation to multiple parties (mID-KEM). This primitive is a direct adaptation of the mKEM primitive in [16] to the identity-based setting.

An mID-KEM scheme is a four-tuple of polynomial time algorithms:

- $\mathbb{G}_{\mathtt{mID-KEM}}(1^t)$. A PPT algorithm which takes as input a security parameter 1^t and returns the master public key M_{pk} and the master secret key M_{sk}. We assume that M_{pk} contains all system parameters, including the security parameter t.
- $\mathbb{X}_{\mathtt{mID-KEM}}(\mathtt{ID}, M_{\mathrm{sk}})$. A deterministic algorithm which takes as input M_{sk} and an identifier string $\mathtt{ID} \in \{0,1\}^*$, and returns the associated private key $S_{\mathtt{ID}}$.
- $\mathbb{E}_{\mathtt{mID-KEM}}(\mathcal{I}, M_{\mathrm{pk}})$. This is the PPT encapsulation algorithm. On input of a tuple $\mathcal{I} = (\mathtt{ID}_1, \ldots, \mathtt{ID}_n)$ of n identities and M_{pk}, this outputs a pair (k, C), where $k \in \mathbb{K}_{\mathtt{mID-KEM}}(M_{\mathrm{pk}})$ is a session key and $C = (c_1, \ldots, c_n)$, with each $c_i \in \mathbb{C}_{\mathtt{mID-KEM}}(M_{\mathrm{pk}})$ an encapsulation of k for $\mathtt{ID}_i \in \mathcal{I}$.
- $\mathbb{D}_{\mathtt{mID-KEM}}(S_{\mathtt{ID}_i}, c_i)$. This is the deterministic decapsulation algorithm. On input of the private key $S_{\mathtt{ID}_i}$ for identity $\mathtt{ID}_i$ and c_i, this outputs k or a failure symbol $\perp$.

The soundness of an mID-KEM scheme is defined as:

$$\Pr\left[k = \mathbb{D}_{\mathtt{mID-KEM}}(S_{\mathtt{ID}_j}, c_j, M_{\mathrm{pk}}) \left| \begin{array}{l} (M_{\mathrm{pk}}, M_{\mathrm{sk}}) \leftarrow \mathbb{G}_{\mathtt{mID-KEM}}(1^t) \\ \mathtt{ID}_i \leftarrow \{0,1\}^*, 1 \le i \le n \\ (k, c_1, \ldots, c_n) \leftarrow \mathbb{E}_{\mathtt{mID-KEM}}(\mathtt{ID}_1, \ldots, \mathtt{ID}_n, M_{\mathrm{pk}}) \\ S_{\mathtt{ID}_i} \leftarrow \mathbb{X}_{\mathtt{mID-KEM}}(\mathtt{ID}_i, M_{\mathrm{sk}}), 1 \le i \le n \\ j \leftarrow \{1, \ldots, n\} \end{array} \right. \right] = 1$$

There is a subtle difference between this and the mKEM definition presented in the previous section. Recall that Smart [16] views encapsulation for multiple users as an algorithm taking a set $\mathcal{P}$ of public keys and producing a *single* token C. Decapsulation must then recover the encapsulated key whenever C is provided together with the secret key for one of the users in the set. This is the most natural extension of KEM to the multi-user setting.

In our approach, however, C is seen as an n-tuple, where the i-th component is an encapsulation specific to $\mathtt{ID}_i$, and the decapsulation algorithm does not accept the whole of C as input, but only c_i. We take this approach because it simplifies the security model (note the definition and access restrictions of the decapsulation oracle below) and allows for a clearer explanation of the security proof in Section 6.

Conceptually, these two approaches are equivalent. Any C in the mKEM model can be recast as an n-tuple by setting $c_i = C$ for all i. Conversely an n-tuple in the mID-KEM model is just a special case of the first approach that provides more information, as it implies a semantic interpretation of the encapsulation.

We define the semantic security of an mID-KEM scheme according to the following two-stage game between an adversary A and a challenger. [1]

n-IND-CCA2 Adversarial Game
1. $(M_{pk}, M_{sk}) \leftarrow \mathbb{G}_{mID-KEM}(1^t)$
2. $(s, \mathcal{I}^*) \leftarrow A_1^{\mathcal{O}}(M_{pk})$
3. $(k_0, C^*) \leftarrow \mathbb{E}_{mID-KEM}(\mathcal{I}^*, M_{pk})$
4. $k_1 \leftarrow \mathbb{K}_{mID-KEM}(M_{pk})$
5. $b \leftarrow \{0, 1\}$
6. $b' \leftarrow A_2^{\mathcal{O}}(M_{pk}, C^*, s, \mathcal{I}^*, k_b)$

In the above, $\mathcal{I}^* = (\mathtt{ID}_1^*, \ldots, \mathtt{ID}_n^*)$ denotes the identities chosen by the adversary to be challenged on, $C^* = (c_1^*, \ldots, c_n^*)$ denotes the challenge encapsulations, s is some state information and $\mathcal{O}$ denotes oracles to which the adversary has access.

In this model the adversary has access to a private key extraction oracle which, on input of $\mathtt{ID} \notin \mathcal{I}^*$, will output the corresponding value of $S_{\mathtt{ID}}$. The adversary can also query a decapsulation oracle with respect to any identity $\mathtt{ID}$ of its choice. It has access to this decapsulation oracle, subject to the restriction that in the second phase A is not allowed to call $\mathcal{O}$ with any pair $(c_i^*, \mathtt{ID}_i^*)$. [2]

The adversary's advantage in the game is defined to be

$$\mathrm{Adv}_{mID-KEM}^{n-IND-CCA2}(A) = |2 \Pr[b' = b] - 1|.$$

An mID-KEM is considered to be n-IND-CCA2 secure, if for all PPT adversaries A, the advantage in the game above is a negligible function of the security parameter t.

5   mID-KEM Schemes

In this section we present two mID-KEM schemes, one of which is a simple construction that uses identity based encryption. The second scheme is non-trivial in the sense that it provides a significant improvement in efficiency.

5.1   A Simple mID-KEM Scheme

One might think that an obvious way to construct an mID-KEM scheme would be to use n instances of an ID-KEM scheme. However this is not valid since it would not guarantee that the same key k is encapsulated to all users. In fact, this would imply using n parallel instances of the associated DEM, thereby subverting the principle of efficient hybrid encryption to multiple users.

---

[1] Similarly to what is done in [16], one could define an (m, n)-IND-CCA2 notion of security for mID-KEMs. Our definition is equivalent to (n, n)-IND-CCA2, which is the worst case scenario.

[2] Note that this accounts for cases where the adversary includes repetitions in $\mathcal{I}^*$.

An mID-KEM scheme can be easily built using a generic construction similar to that used in [16], whereby an IND-CCA2 secure public key encryption algorithm is used to build a secure mKEM.

In the identity based setting, one would take a secure identity based encryption algorithm, for instance Boneh and Franklin's Full-Ident scheme [5], as the starting point. The resulting mID-KEM construction would operate as follows.

$\mathbb{E}_{\texttt{mID-KEM}}(\texttt{ID}_1, \ldots, \texttt{ID}_n, M_{\text{pk}})$
- $m \leftarrow \mathbb{M}_{\text{IBE}}(M_{\text{pk}})$
- $k \leftarrow \text{KDF}(m)$
- $c_i \leftarrow \mathbb{E}_{\text{IBE}}(m, \texttt{ID}_i, M_{\text{pk}})$, for $1 \leq i \leq n$
- Return $(k, c_1, \ldots, c_n)$

$\mathbb{D}_{\texttt{mID-KEM}}(S_{\texttt{ID}_i}, c_i, M_{\text{pk}})$
- $m \leftarrow \mathbb{D}_{\text{IBE}}(S_{\texttt{ID}_i}, c_i, M_{\text{pk}})$
- $k \leftarrow \text{KDF}(m)$
- Return k

where KDF is a key derivation function that maps the message space of the identity based encryption algorithm to the key space of the DEM.

The security proof in [16] for the generic mKEM construction builds on the work by Bellare et al. [3], who established that a public key encryption algorithm which is secure in the single user setting, is also secure when multiple users are involved. Given that the authors have no knowledge of such a result for identity based encryption algorithms, proving the security of the previous construction will require extending Bellare et al.'s results. Since the main result of this paper is a more efficient scheme proposed in the next section, we do not further debate the security of this simpler construction.

5.2 An Efficient mID-KEM from Bilinear Pairings

The scheme we propose in this section is inspired by Smart's OR constructions in [15]. Our construction uses a parameter generator $\mathbb{G}_{\texttt{mID-KEM}}(1^t)$ similar to that of most identity based schemes. On input of a security parameter 1^t, it outputs the master public key M_{pk} and the master secret key M_{sk}, where $M_{\text{sk}} \leftarrow \mathbb{Z}_q^*$ and $M_{\text{pk}} \leftarrow M_{\text{sk}} \cdot P_1$.

Private key extraction is also performed in the usual way. On input of an identity ID and the master secret key M_{sk}, the extraction algorithm $\mathbb{X}_{\texttt{mID-KEM}}$ returns $S_{\texttt{ID}} \leftarrow M_{\text{sk}} \cdot H_1(\texttt{ID})$, where $H_1 : \{0,1\}^* \to G_2$ is a cryptographic hash function.

The key encapsulation and decapsulation algorithms work as follows.

$\mathbb{E}_{\texttt{mID-KEM}}(\texttt{ID}_1, \ldots, \texttt{ID}_n, M_{\text{pk}})$
- $r \leftarrow \mathbb{Z}_q^*$
- $s \leftarrow \mathbb{Z}_q^*$
- $U_r \leftarrow rP_1$
- $U_s \leftarrow sP_1$
- $Q_{\texttt{ID}_i} \leftarrow H_1(\texttt{ID}_i)$
- For $i = 1$ to n
  $\quad U_i \leftarrow r(Q_{\texttt{ID}_i} - sP_2)$
  $\quad c_i \leftarrow (U_r, U_s, U_i)$
- $T \leftarrow \hat{t}(M_{\text{pk}}, rsP_2)$
- $k \leftarrow H_2(T\|U_r\|U_s)$
- Return $(k, c_1, \ldots, c_n)$

$\mathbb{D}_{\texttt{mID-KEM}}(S_{\texttt{ID}_i}, c_i, M_{\text{pk}})$
- $t \leftarrow \hat{t}(U_r, S_{\texttt{ID}_i})$
- $t' \leftarrow \hat{t}(M_{\text{pk}}, -U_i)$
- $T \leftarrow t \cdot t'$
- $k \leftarrow H_2(T\|U_r\|U_s)$
- Return k

In the above, H_2 is a cryptographic hash function that takes elements in $G_T || G_1 || G_1$ to the DEM key space.

6   Proof of Security

Theorem 1 describes our result on the security of the mID-KEM scheme in Section 5.2 with respect to the security model defined in Section 4. The technique we use in the proof might allow one to establish the security of a suitable modification of the scheme in [15], for which no proof is provided.

Theorem 1. *The mID-KEM construction described in Section 5.2 is n-IND-CCA2 secure under the hardness of GBDH in the random oracle model. Specifically, for any PPT algorithm A that breaks this mID-KEM scheme, there exists a PPT algorithm B with*

$$\mathrm{Adv}_{\mathtt{mID-KEM}}^{n-\mathtt{IND-CCA2}}(A) \leq 2q_T^n \cdot \mathrm{Adv}_{\mathtt{GBDH}}(B, q_D^2 + 2q_D q_2 + q_2)$$

where q_1, q_2, q_X and q_D denote the maximum number of queries the adversary places to the H_1, H_2, private key extraction and decapsulation oracles respectively, and $q_T = q_1 + q_X + q_D$ is the total number of queries placed to H_1.

Proof. Let S be the event that A wins the n-IND-CCA2 adversarial game in Section 4. Let also $T^* || U_r^* || U_s^*$ denote the value that must be passed to H_2 to obtain the correct key encapsulated in the challenge ciphertext. Then, we have

$$\Pr[\mathsf{S}] = \Pr[\mathsf{S} \wedge \mathsf{Ask}] + \Pr[\mathsf{S} \wedge \neg\mathsf{Ask}] \leq \Pr[\mathsf{S} \wedge \mathsf{Ask}] + \frac{1}{2} \tag{2}$$

where Ask denotes the event that A queries H_2 with $T^* || U_r^* || U_s^*$ during the simulation. This follows from the fact that if A does not place this query, then it can have no advantage.

We argue that in the event $\mathsf{S} \wedge \mathsf{Ask}$, there is an adversary B that uses A to solve the GBDH problem with probability q_T^{-n} while making at most $q_D^2 + 2q_D q_2 + q_2$ DBDH oracle queries. We implement algorithm B as follows.

For a given GBDH problem instance $(\Gamma, aP_2, bP_2, cP_2)$ the strategy is to embed the problem into the mID-KEM challenge presented to the adversary. Namely we construct the simulation so that the value of the pairing associated with the challenge encapsulation is $T^* = \hat{t}(P_1, P_2)^{abc}$. Hence, when A places a query with the correct value of T^* to H_2, it provides us with the solution to the GBDH problem instance.

We pass Γ and $M_{\mathtt{pk}} = cP_1 = \rho(cP_2)$ on to A as the global public parameters. We also maintain three lists that allow us to provide consistent answers to A's oracle calls: $L_1 \subset \{0,1\}^* \times G_2 \times \mathbb{F}_q$; $L_2 \subset G_T \times G_1 \times G_1 \times \mathbb{K}_{\mathtt{mID-KEM}}(M_{\mathtt{pk}})$; and $L_D \subset \{0,1\}^* \times G_1 \times G_1 \times \mathbb{K}_{\mathtt{mID-KEM}}(M_{\mathtt{pk}})$.

Initially, we generate a list of n random indices $i_1^*, \ldots, i_n^*$, with $1 \leq i_k^* \leq q_T$. These indices will determine the set of n identities which B is guessing the adversary will include in its challenge query $\mathcal{I}^*$. Namely, these indices determine the way in which the hash function H_1 is simulated.

H_1 **queries:** $H_1(\mathrm{ID}_i)$. On the i-th query to H_1, we check whether i matches one of the indices i_k^*. If this is the case, we generate random $r_{i_k^*} \in \mathbb{Z}_q^*$, return $Q_{\mathrm{ID}_{i_k^*}} = r_{i_k^*}P_2 + bP_2$ and add $(\mathrm{ID}_{i_k^*}, Q_{\mathrm{ID}_{i_k^*}}, r_{i_k^*})$ to L_1. Otherwise, we generate random $r_i \in \mathbb{Z}_q^*$, return $Q_{\mathrm{ID}_i} = r_i P_2$ and add $(\mathrm{ID}_i, Q_{\mathrm{ID}_i}, r_i)$ to L_1.

Algorithm B requires that at the end of phase one, the adversary outputs a list of identities $\mathcal{I}^* = (\mathrm{ID}_1^*, \ldots, \mathrm{ID}_n^*)$ such that $\mathrm{ID}_k^* = \mathrm{ID}_{i_k^*}$. This happens with probability at least q_T^{-n}. If this is not the case, the simulation fails. Otherwise, the mID-KEM challenge encapsulation is created as

$$
\begin{aligned}
U_r &\leftarrow \rho(aP_2) \\
U_s &\leftarrow \rho(bP_2) \\
U_k &\leftarrow r_{i_k^*} aP_2 \\
c_k^* &\leftarrow (U_r, U_s, U_k)
\end{aligned}
$$

for each $\mathrm{ID}_k^* \in \mathcal{I}^*$. Notice that this is a well-formed encapsulation:

$$
U_k = a(Q_{\mathrm{ID}_k^*} - bP_2) = a(r_{i_k^*}P_2 + bP_2 - bP_2) = r_{i_k^*}aP_2.
$$

When A outputs $\mathcal{I}^*$, algorithm B responds with $C^* = (c_1^*, \ldots, c_n^*)$ constructed as above plus a random challenge key k^*.

Given that A is not allowed to obtain the private keys associated with the identities in the challenge, the extraction oracle is easily simulated as follows.

Extraction queries: $\mathcal{O}_X(\mathrm{ID}_j)$. On input ID_j, we obtain Q_{ID_j} by calling H_1 first. If Q_{ID_j} is of the form $r_i P_2$, we look in L_1 for the corresponding r_i and return $S_{\mathrm{ID}_i} = r_i(cP_2)$. If during the first stage of the game, A should query for the extraction of a private key corresponding to a Q_{ID} of the form $r_{i_k^*}P_2 + bP_2$, the simulation would fail.

Finally, the decapsulation and H_2 oracles are simulated as follows.

Decapsulation queries: $\mathcal{O}_D(\mathrm{ID}_m, U_{r_m}, U_{s_m}, U_m)$. On queries in which we can calculate a value T_m, we do so, and call H_2 with parameters (T_m, U_{r_m}, U_{s_m}) to obtain the return value. This will happen when $\mathrm{ID}_m \notin \mathcal{I}^*$, in which case we query the private key extraction oracle, and compute T_m as in a standard decapsulation:

$$
T_m = \hat{t}(U_{r_m}, S_{\mathrm{ID}_m})\hat{t}(\rho(cP_2), -U_m).
$$

When $\mathrm{ID}_m \in \mathcal{I}^*$ we first search L_2 for an entry such that $U_{r_m} = U_{r_n}$, $U_{s_m} = U_{s_n}$ and the DBDH oracle returns 1 when queried with $(U_{r_m}, U_{s_m}, cP_2, T_n)$. If this entry exists, we return the corresponding k_n. If not, we search L_D for a tuple matching the parameters $(\mathrm{ID}_m, U_{r_m}, U_{s_m})$. If it exists, we return the associated key k_m. Otherwise, we generate a random k_m, return it and update L_D by adding the entry $(\mathrm{ID}_m, U_{r_m}, U_{s_m}, k_m)$.

H_2 **queries:** $H_2(T_n, U_{r_n}, U_{s_n})$. We first check if T_n is the solution we are looking for by calling the DBDH oracle with (aP_1, bP_1, cP_2, T_n). If it returns 1, the solution is found, we output T_n and terminate. Otherwise, we proceed to search L_2 for a tuple matching the parameters (T_n, U_{r_n}, U_{s_n}). If it exists, we return the corresponding key k_n. If not, we search L_D for an entry such that $U_{r_m} = U_{r_n}$, $U_{s_m} = U_{s_n}$ and the DBDH oracle returns 1 when queried with $(U_{r_m}, U_{s_m}, cP_2, T_n)$. If this entry exists, we return the corresponding k_m. Otherwise, we generate a random k_n, return it and update L_2 by appending the entry $(T_n, U_{r_n}, U_{s_n}, k_n)$.

To complete the proof, it suffices to notice that B will only fail when the identities $\mathcal{I}^*$ chosen by A at the end of the first round do not match the random indices chosen at the beginning. Since this happens with probability at least q_T^{-n}, we conclude

$$\Pr[\mathsf{S} \wedge \mathsf{Ask}] \leq q_T^n \cdot \mathrm{Adv}_{\mathrm{GBDH}}(B, q_D^2 + 2q_D q_2 + q_2). \tag{3}$$

The total number of DBDH oracle calls $(q_D^2 + 2q_D q_2 + q_2)$ follows from the way we simulate the decapsulation and H_2 oracles. Theorem 1 now follows from (2) and (3).

We note that the tightness of the security reduction in Theorem 1 decreases exponentially with the number of recipients. For practical purposes, and in line with the discussions on practice-oriented provable security in [2,10], it is important to emphasise that our result provides theoretical security guarantees only in scenarios where messages are encrypted to a small number of recipients. A slightly better result, although still exponential in n, can be obtained by optimising the simulation in the proof above. We leave it as an open problem to find a sub-exponential security reduction for this type of scheme.

7   Efficiency Considerations

Execution time and bandwidth usage are two important factors affecting the efficiency of a KEM. In this section we present a comparison, with respect to these factors, between the simple scheme described in Section 5.1 and the scheme proposed in Section 5.2.

The former scheme requires n identity based encryption computations for encapsulation, and one identity based decryption computation for decapsulation. Assuming that we use the Full-Ident scheme of Boneh and Franklin [5], this means n pairing computations for encapsulation and another for decapsulation. On the other hand, as it can be seen from the description of the latter scheme, it takes only one pairing computation to encapsulate and two more to decapsulate. Note that one could essentially eliminate the pairing computation needed in encapsulation by pre-computing the value $\hat{t}(M_{\mathrm{pk}}, P_2)$.

The bandwidth usage of the scheme in Section 5.2 is $2\ell_1 + \ell_2$ bits for each recipient, where ℓ_1 and ℓ_2 are the bit lengths of the representations of elements of G_1 and G_2. Using the Full-Ident encryption scheme in [5], the simpler scheme uses $\ell_1 + 2\ell_k$ bits per user where ℓ_k is the length of the session key. For example, using supersingular elliptic curves over characteristic three we could take $\ell_k = 128$ and $\ell_1 = \ell_2 \approx 190$ bits. This means a bandwidth usage of 570 bits for the second scheme versus 446 bits for the first one.

To conclude this discussion we note the n-fold gain in computational weight obtained in Section 5.2 is achieved at the cost of a small increase in bandwidth usage, and also by reducing the security of the scheme to the hardness of GBDH, which is a stronger assumption than the hardness of BDH used in [5].

Acknowledgements

The authors would like to thank Nigel Smart and John Malone-Lee for their input during this work, and the anonymous reviewers whose comments helped to improve it.

References

1. M. Abdalla, M. Bellare and P. Rogaway. DHAES : An encryption scheme based on the Diffie-Hellman problem. *Cryptology ePrint Archive*, Report 1999/007. 1999.
2. M. Bellare. Practice-priented provable-security. *Proceedings of First International Workshop on Information Security*, LNCS 1396. Springer-Verlag, 1998.
3. M. Bellare, A. Boldyreva and S. Micali. Public-key encryption in the multi-user setting: security proofs and improvements. *Advances in Cryptology – EUROCRYPT 2000*, LNCS 1807:259-274. Springer-Verlag, 2000.
4. K. Bentahar, P. Farshim, J. Malone-Lee and N. P. Smart. Generic constructions of identity-based and certificateless KEMs. *Cryptology ePrint Archive*, Report 2005/058. 2005.
5. D. Boneh and M. Franklin. Identity based encryption from the Weil pairing. *SIAM Journal on Computing*, 32:586–615. 2003.
6. D. Boneh, B. Lynn and H. Shacham. Short signatures from the Weil pairing. *Advances in Cryptology – ASIACRYPT 2001*, LNCS 2248:514–532. Springer-Verlag, 2001.
7. R. Cramer and V. Shoup. Design and analysis of practical public-key encryption schemes secure against adaptive chosen-ciphertext attack. *SIAM Journal on Computing*, 33:167–226. 2003.
8. A. W. Dent. A designer's guide to KEMs. *Coding and Cryptography*, LNCS 2898:133-151. Springer-Verlag, 2003.
9. S. Goldwasser and S. Micali. Probabilistic encryption. *Journal of Computer and Systems Sciences*, 28:270–299. 1984.
10. N. Koblitz and A. Menezes. Another look at provable security. *Cryptology ePrint Archive*, Report 2004/152. 2004.
11. B. Lynn. Authenticated identity-based encryption. *Cryptology ePrint Archive*, Report 2002/072. 2002.
12. A. Miyaji, M. Nakabayashi and S. Takano. New explicit conditions of elliptic curve traces for FR-reduction. *IEICE Transactions on Fundamentals of Electronics, Communications and Computer Sciences*, E84-A:1234–1243. 2001.
13. T. Okamoto and D. Pointcheval. The gap-problems: A new class of problems for the security of cryptographic schemes. *Public Key Cryptography – PKC 2001*, LNCS 1992:104–118. Springer-Verlag, 2001.

14. C. Rackoff and D. R. Simon Non-interactive zero-knowledge proof of knowledge and chosen ciphertext attack. *Advances in Cryptology – CRYPTO 1991*, LNCS 576:433-444. Springer-Verlag, 1991
15. N. P. Smart. Access control using pairing based cryptography. *Topics in Cryptology – CT-RSA 2003*, LNCS 2612:111–121. Springer-Verlag, 2003.
16. N. P. Smart. Efficient key encapsulation to multiple parties. *Security in Communication Networks*, LNCS 3352:208–219. Springer-Verlag, 2005.
17. V. Shoup. A proposal for an ISO standard for public key encryption (version 2.1). *Preprint*. 2001.

Security Proof of Sakai-Kasahara's Identity-Based Encryption Scheme

Liqun Chen[1] and Zhaohui Cheng[2]

[1] Hewlett-Packard Laboratories, Bristol, UK
liqun.chen@hp.com
[2] School of Computing Science, Middlesex University,
The Burroughs Hendon, London NW4 4BT, UK
m.z.cheng@mdx.ac.uk

Abstract. Identity-based encryption (IBE) is a special asymmetric encryption method where a public encryption key can be an arbitrary identifier and the corresponding private decryption key is created by binding the identifier with a system's master secret. In 2003 Sakai and Kasahara proposed a new IBE scheme, which has the potential to improve performance. However, to our best knowledge, the security of their scheme has not been properly investigated. This work is intended to build confidence in the security of the Sakai-Kasahara IBE scheme. In this paper, we first present an efficient IBE scheme that employs a simple version of the Sakai-Kasahara scheme and the Fujisaki-Okamoto transformation, which we refer to as SK-IBE. We then prove that SK-IBE has chosen ciphertext security in the random oracle model based on a reasonably well-explored hardness assumption.

1 Introduction

Shamir in 1984 [34] first formulated the concept of Identity-Based Cryptography (IBC) in which a public and private key pair is set up in a special way, i.e., the public key is the identifier (an arbitrary string) of an entity, and the corresponding private key is created by using an identity-based key extraction algorithm, which binds the identifier with a master secret of a trusted authority. In the same paper, Shamir provided the first key extraction algorithm that was based on the RSA problem, and presented an identity-based signature scheme. By using varieties of the Shamir key extraction algorithm, more identity-based signature schemes and key agreement schemes were proposed (e.g., [23,24]). However, constructing a practical Identity-Based Encryption (IBE) scheme remained an open problem for many years.

After nearly twenty years, Boneh and Franklin [4], Cocks [16] and Sakai *et al.* [31] presented three IBE solutions in 2001. The Cocks solution is based on the quadratic residuosity. Both the Boneh and Franklin solution and the Sakai *et al.* solution are based on bilinear pairings on elliptic curves [35], and the security of their schemes is based on the Bilinear Diffie-Hellman (BDH) problem [4]. Their schemes are efficient in practice. Boneh and Franklin defined a well-formulated security model for IBE in [4]. The Boneh-Franklin scheme (BF-IBE for short)

N.P. Smart (Ed.): Cryptography and Coding 2005, LNCS 3796, pp. 442–459, 2005.
© Springer-Verlag Berlin Heidelberg 2005

has received much attention owing to the fact that it was the first IBE scheme to have a proof of security in an appropriate model.

Both BF-IBE and the Sakai *et al.* IBE solution have a very similar private key extraction algorithm, in which an identity string is mapped to a point on an elliptic curve and then the corresponding private key is computed by multiplying the mapped point with the master private key. This key extraction algorithm was first shown in Sakai *et al.*'s work [30] in 2000 as the preparation step of an identity-based key establishment protocol. Apart from BF-IBE and the Sakai *et al.* IBE scheme [31], many other identity-based cryptographic primitives have made use of this key extraction idea, such as the signature schemes [10,22], the authenticated key agreement schemes [12,36], and the signcryption schemes [7,13]. The security of these schemes were scrutinized (although some errors in a few reductions were pointed out recently but fixed as well, e.g., [15,19]).

Based on the same tool, the bilinear pairing, Sakai and Kasahara in 2003 [29] presented a new IBE scheme using another identity-based key extraction algorithm. The idea of this algorithm can be tracked back to the work in 2002 [28]. This algorithm requires much simpler hashing and therefore improves performance. More specifically, it maps an identity to an element $h \in \mathbb{Z}_q^*$ instead of a point on an elliptic curve. The corresponding private key is generated as follow: first, compute the inverse of the sum of the master key (a random integer from $\mathbb{Z}_q^*$) and the mapped h; secondly, multiply a point of the elliptic curve (which is the generator of an order q subgroup of the group of points on the curve) with the inverse (obtained in the first step). After the initial paper was published, a number of other identity-based schemes based on this key extraction idea have been published, for examples [26,27].

However, these schemes are either unproven or their security proof is problematic (e.g., [14]). In modern cryptography, a carefully scrutinized security reduction in a formal security model to a hardness assumption is desirable for any cryptographic scheme. Towards this end, this work is intended to build confidence in the security of the Sakai and Kasahara IBE scheme.

The remaining part of the paper is organized as follows. In next section, we recall the existing primitive, some related assumptions and the IBE security model. In Section 3, we first employ a simple version of the Sakai and Kasahara IBE scheme from [29] and the Fujisaki-Okamoto transformation [17] to present an efficient IBE scheme (we refer to it as SK-IBE). We then prove that SK-IBE has chosen ciphertext security in the random oracle model. Our proof is based on a reasonably well-explored hardness assumption. In Section 4, we show some possible improvements of SK-IBE, both on security and performance. In Section 5, we compare between SK-IBE and BF-IBE. We conclude the paper in Section 6.

2   Preliminaries

In this section, we recall the existing primitives, including bilinear pairings, some related assumptions and the security model of IBE.

2.1   Bilinear Groups and Some Assumptions

Here we review the necessary facts about bilinear maps and the associated groups using a similar notation of [5].

- $\mathbb{G}_1$, $\mathbb{G}_2$ and $\mathbb{G}_T$ are cyclic groups of prime order q.
- P_1 is a generator of $\mathbb{G}_1$ and P_2 is a generator of $\mathbb{G}_2$.
- ψ is an isomorphism from $\mathbb{G}_2$ to $\mathbb{G}_1$ with $\psi(P_2) = P_1$.
- $\hat{e}$ is a map $\hat{e} : \mathbb{G}_1 \times \mathbb{G}_2 \to \mathbb{G}_T$.

The map $\hat{e}$ must have the following properties.

Bilinear: For all $P \in \mathbb{G}_1$, all $Q \in \mathbb{G}_2$ and all $a, b \in \mathbb{Z}$ we have $\hat{e}(aP, bQ) = \hat{e}(P, Q)^{ab}$.

Non-degenerate: $\hat{e}(P_1, P_2) \neq 1$.

Computable: There is an efficient algorithm to compute $\hat{e}(P, Q)$ for all $P \in \mathbb{G}_1$ and $Q \in \mathbb{G}_2$.

Note that following [37], we can either assume that ψ is efficiently computable or make our security proof relative to some oracle which computes ψ.

There are a batch of assumptions related to the bilinear groups. Some of them have already been used in the literature and some are new variants. We list them below and show how they are related to each other. We also correct a minor inaccuracy in stating an assumption in the literature. Recently it has come to our attention that some other related assumptions were discussed in [40].

We use a unified naming method; in particular, provided that X stands for an assumption, sX stands for a stronger assumption of X, which implies that the problem corresponding to sX would be easier than the problem corresponding to X. In the following description, $\alpha \in_R \beta$ denotes that α is an element chosen at random from a set β.

Assumption 1 (Diffie-Hellman (DH)). *For $x, y \in_R \mathbb{Z}_q^*$, $P \in \mathbb{G}_1^*$, given (P, xP, yP), computing xyP is hard.*

Assumption 2 (Bilinear DH (BDH)) [4]. *For x, y, $z \in_R \mathbb{Z}_q^*$, $P_2 \in \mathbb{G}_2^*$, $P_1 = \psi(P_2)$, $\hat{e} : \mathbb{G}_1 \times \mathbb{G}_2 \to \mathbb{G}_T$, given $(P_1, P_2, xP_2, yP_2, zP_2)$, computing $\hat{e}(P_1, P_2)^{xyz}$ is hard.*

Assumption 3 (Decisional Bilinear DH (DBDH)). *For x, y, z, $r \in_R \mathbb{Z}_q^*$, $P_2 \in \mathbb{G}_2^*$, $P_1 = \psi(P_2)$, $\hat{e} : \mathbb{G}_1 \times \mathbb{G}_2 \to \mathbb{G}_T$, distinguishing between the distributions $(P_1, P_2, xP_2, yP_2, zP_2, \hat{e}(P_1, P_2)^{xyz})$ and $(P_1, P_2, xP_2, yP_2, zP_2, \hat{e}(P_1, P_2)^r)$ is hard.*

Assumption 4 (DH Inversion (k-DHI)) [28]. *For an integer k, and $x \in_R \mathbb{Z}_q^*$, $P \in \mathbb{G}_1^*$, given $(P, xP, x^2P, \ldots, x^kP)$, computing $\frac{1}{x}P$ is hard.*

Theorem 1. (Mitsunari et al. [28]). *DH and 1-DHI are polynomial time equivalent, i.e., if there exists a polynomial time algorithm to solve DH, then there exists a polynomial time algorithm for 1-DHI, and if there exists a polynomial time algorithm to solve 1-DHI, then there exists a polynomial time algorithm for DH.*

Assumption 5 (Collision Attack Assumption 1 (k-CAA1)). *For an integer k, and $x \in_R \mathbb{Z}_q^*$, $P \in \mathbb{G}_1^*$, given $(P, xP, h_0, (h_1, \frac{1}{h_1+x}P), \ldots, (h_k, \frac{1}{h_k+x}P))$ where $h_i \in_R \mathbb{Z}_q^*$ and distinct for $0 \le i \le k$, computing $\frac{1}{h_0+x}P$ is hard.*

Theorem 2. *If there exists a polynomial time algorithm to solve (k-1)-DHI, then there exists a polynomial time algorithm for k-CAA1. If there exists a polynomial time algorithm to solve (k-1)-CAA1, then there exists a polynomial time algorithm for k-DHI.*

The proof is presented in a full version of this paper, which is available at [11].

Assumption 6 (Collision Attack Assumption 2 (k-CAA2) [28]). *For an integer k, and $x \in_R \mathbb{Z}_q^*$, $P \in \mathbb{G}_1^*$, given $(P, h_0, (h_1, \frac{1}{h_1+x}P), \ldots, (h_k, \frac{1}{h_k+x}P))$ where $h_i \in_R \mathbb{Z}_q^*$ and distinct for $0 \le i \le k$, computing $\frac{1}{h_0+x}P$ is hard.*

Mitsunari *et al.* established the relation between k-CAA2 and k-DHI (also called k-wDHA) in [28], while in the definition of k-CAA2 the value h_0 was not given as input. However, when consulting their proof of Theorem 3.5 [28], we note that h_0 has to be given as part of the problem.

Theorem 3. (Mitsunari et al. [28]). *There exists a polynomial time algorithm to solve (k-1)-DHI if and only there exists a polynomial time algorithm for k-CAA2.*

Assumption 7 (Strong CAA (k-sCAA1) [41]). *For an integer k, and $x \in_R \mathbb{Z}_q^*$, $P \in \mathbb{G}_1^*$, given $(P, xP, (h_1, \frac{1}{h_1+x}P), \ldots, (h_k, \frac{1}{h_k+x}P))$ where $h_i \in_R \mathbb{Z}_q^*$ and distinct for $1 \le i \le k$, computing $(h, \frac{1}{h+x}P)$ for some $h \in \mathbb{Z}_q^*$ but $h \notin \{h_1, \ldots, h_k\}$ is hard.*

Zhang *et al.*'s short signature proof [41] and Mitsunari *et al.*'s traitor tracing scheme [28] used this assumption. However, the traitor tracing scheme was broken by Tô et al. in [38] because it was found to be in fact based on a "slightly" different assumption, which does not require to output the value of h. Obviously, if one does not have to demonstrate that he knows the value of h, the problem is not hard. He can simply choose a random element from $\mathbb{G}_1$ that is not shown in the problem as the answer, because $\mathbb{G}_1$ is of prime order q and any $r \in \mathbb{Z}_q^*$ satisfies $r = \frac{1}{h+x} \bmod q$ for some h.

Assumption 8 (Strong DH (k-sDH) [2]). *For an integer k, and $x \in_R \mathbb{Z}_q^*$, $P \in \mathbb{G}_1^*$, given $(P, xP, x^2P, \ldots, x^kP)$, computing $(h, \frac{1}{h+x}P)$ where $h \in \mathbb{Z}_q^*$ is hard.*

Theorem 4. *If there exists a polynomial time algorithm to solve (k-1)-sCAA1, then there exists a polynomial time algorithm for k-sDH. If there exists a polynomial time algorithm to solve (k-1)-sDH, then there exists a polynomial time algorithm for k-sCAA1.*

The proof is presented in [11].

Assumption 9 (Exponent Problem ((k+1)-EP) [41]). *For an integer k, and $x \in_R \mathbb{Z}_q^*$, $P \in \mathbb{G}_1^*$, given $(P, xP, x^2P, \ldots, x^kP)$, computing $x^{k+1}P$ is hard.*

Theorem 5. (Zhang et al. [41]). *There exists a polynomial time algorithm to solve k-DHI if and only if there exists a polynomial time algorithm for (k+1)-EP.*

Assumption 10 (Bilinear DH Inversion (k-BDHI) [1]). *For an integer k, and $x \in_R \mathbb{Z}_q^*$, $P_2 \in \mathbb{G}_2^*$, $P_1 = \psi(P_2)$, $\hat{e} : \mathbb{G}_1 \times \mathbb{G}_2 \to \mathbb{G}_T$, given $(P_1, P_2, xP_2, x^2P_2, \ldots, x^kP_2)$, computing $\hat{e}(P_1, P_2)^{1/x}$ is hard.*

Assumption 11 (Decisional Bilinear DH Inversion (k-DBDHI)). *For an integer k, and $x, r \in_R \mathbb{Z}_q^*$, $P_2 \in \mathbb{G}_2^*$, $P_1 = \psi(P_2)$, $\hat{e} : \mathbb{G}_1 \times \mathbb{G}_2 \to \mathbb{G}_T$, distinguishing between the distributions $(P_1, P_2, xP_2, x^2P_2, \ldots, x^kP_2, \hat{e}(P_1, P_2)^{1/x})$ and $(P_1, P_2, xP_2, x^2P_2, \ldots, x^kP_2, \hat{e}(P_1, P_2)^r)$ is hard.*

Theorem 6. *BDH and 1-BDHI are polynomial time equivalent, i.e., if there exists a polynomial time algorithm to solve BDH, then there exists a polynomial time algorithm for 1-BDHI, and if there exists a polynomial time algorithm to solve 1-BDHI, then there exists a polynomial time algorithm for BDH.*

Proof: If there is a polynomial time algorithm $\mathcal{A}$ to solve the BDH problem, we construct a polynomial time algorithm $\mathcal{B}$ to solve the 1-BDHI problem. Given an instance of 1-BDHI problem (Q_1, Q_2, yQ_2), $\mathcal{B}$ works as follow to compute $\hat{e}(Q_1, Q_2)^{1/y}$.

1. Set $x = 1/y$, which $\mathcal{B}$ does not know.
2. Set $P_1 = Q_1$, $P_2 = yQ_2$ and $xP_2 = Q_2$.
3. Pass $\mathcal{A}$ the BDH challenge, $(P_1, P_2, xP_2, xP_2, P_2)$, and get $T = \hat{e}(P_1, P_2)^{x^2}$ $= \hat{e}(Q_1, yQ_2)^{(1/y)^2} = \hat{e}(Q_1, Q_2)^{1/y}$.

If there is a polynomial time algorithm $\mathcal{A}$ to solve the 1-BDHI problem, we construct a polynomial time algorithm $\mathcal{B}$ to solve the BDH problem. Given an instance of BDH problem $(P_1, P_2, aP_2, bP_2, cP_2)$, $\mathcal{B}$ works as follow to compute $\hat{e}(P_1, P_2)^{abc}$.

1. (a) Set $d = 1/(a + b + c)$, which $\mathcal{B}$ does not know.
   (b) Set $Q_1 = (a + b + c)P_1 = \psi((a + b + c)P_2)$, $Q_2 = (a + b + c)P_2$ and $dQ_2 = P_2$.
   (c) Pass $\mathcal{A}$ the 1-BDHI challenge, (Q_1, Q_2, dQ_2), and get $T_1 = \hat{e}(Q_1, Q_2)^{1/d}$ $= \hat{e}(P_1, P_2)^{(a+b+c)^3}$.
2. Follow Item 1 (a) - (c) to get $T_2 = \hat{e}(P_1, P_2)^{a^3}$, $T_3 = \hat{e}(P_1, P_2)^{b^3}$, $T_4 = \hat{e}(P_1, P_2)^{c^3}$, $T_5 = \hat{e}(P_1, P_2)^{(a+b)^3}$, $T_6 = \hat{e}(P_1, P_2)^{(a+c)^3}$, $T_7 = \hat{e}(P_1, P_2)^{(b+c)^3}$.
3. Compute $\hat{e}(P_1, P_2)^{abc} = (\frac{T_1 \cdot T_2 \cdot T_3 \cdot T_4}{T_5 \cdot T_6 \cdot T_7})^{1/6}$.   □

Assumption 12 (Bilinear CAA 1 (k-BCAA1)). *For an integer k, and $x \in_R \mathbb{Z}_q^*$, $P_2 \in \mathbb{G}_2^*$, $P_1 = \psi(P_2)$, $\hat{e} : \mathbb{G}_1 \times \mathbb{G}_2 \to \mathbb{G}_T$, given $(P_1, P_2, xP_2, h_0, (h_1, \frac{1}{h_1+x}P_2), \ldots, (h_k, \frac{1}{h_k+x}P_2))$ where $h_i \in_R \mathbb{Z}_q^*$ and distinct for $0 \leq i \leq k$, computing $\hat{e}(P_1, P_2)^{1/(x+h_0)}$ is hard.*

Theorem 7. *If there exists a polynomial time algorithm to solve (k-1)-BDHI, then there exists a polynomial time algorithm for k-BCAA1. If there exists a polynomial time algorithm to solve (k-1)-BCAA1, then there exists a polynomial time algorithm for k-BDHI.*

The proof is presented in [11].

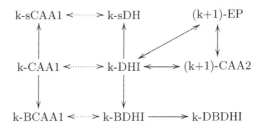

k-A $\longrightarrow$ k-B: if k-A is polynomial-time solvable, so is k-B;
k-A $\dashrightarrow$ k-B: if (k-1)-A is polynomial-time solvable, so is k-B.

Fig. 1. Relation among the assumptions

The relation among these assumptions can be described by Fig. 1. In the literature, the k-DBDHI assumption was used in [1] to construct a selective-identity secure IBE scheme (see next section for definition) without random oracles [6] and k-sDH is used to construct a short signature [2] without random oracles, while k-sCAA1 is used by [41] to construct a short signature with random oracles and to build a traitor tracing scheme [28].

2.2   IBE Schemes and Their Security Model

Let k be a security parameter, and $\mathbb{M}$ and $\mathbb{C}$ denote the message and cipher-text spaces respectively. An IBE scheme is specified by four polynomial time algorithms:

- **Setup** takes as input 1^k, and returns a master public key M_{pk} and a master secret key M_{sk};
- **Extract** takes as input M_{pk}, M_{sk} and $ID_A \in \{0,1\}^*$, an identifier string for entity A, and returns the associated private key d_A;
- **Encrypt** takes as input M_{pk}, ID_A and a message $m \in \mathbb{M}$, and returns a ciphertext $C \in \mathbb{C}$; and

- **Decrypt** takes as input M_{pk}, ID_A, d_A and C, and returns the corresponding value of the plaintext m or a failure symbol $\perp$.

The security of an IBE scheme is defined by the following game between a challenger $\mathcal{C}$ and an adversary $\mathcal{A}$ formalized in [4].

- Setup. $\mathcal{C}$ takes a security parameter k and runs the Setup algorithm. It gives $\mathcal{A}$ M_{pk} and keeps M_{sk} to itself.
- Phase 1. $\mathcal{A}$ issues queries as one of follows:
  - Extraction query on ID_i. $\mathcal{C}$ runs the Extract algorithm to generate d_{ID_i} and passes it to $\mathcal{A}$.
  - Decryption query on (ID_i, C_i). $\mathcal{C}$ decrypts the ciphertext by finding d_{ID_i} first (through running Extract if necessary), and then running the Decrypt algorithm. It responds with the resulting plaintext.
- Challenge. Once $\mathcal{A}$ decides that Phase 1 is over, it outputs two equal length plaintexts $m_0, m_1 \in \mathbb{M}$, and an identity ID_{ch} on which it wishes to be challenged. The only constraint is that $\mathcal{A}$ must not have queried the extraction oracle on ID_{ch} in Phase 1. $\mathcal{C}$ picks a random bit $b \in \{0,1\}$ and sets C_{ch}=Encrypt(M_{pk}, ID_{ch}, m_b) $\in \mathbb{C}$. It sends C_{ch} as the challenge to $\mathcal{A}$.
- Phase 2. $\mathcal{A}$ issues more queries as in Phase 1 but with two restrictions: (1) Extraction queries cannot be issued on ID_{ch}; (2) Decryption queries cannot be issued on (ID_{ch}, C_{ch}).
- Guess. Finally, $\mathcal{A}$ outputs a guess $b' \in \{0,1\}$ and wins the game if $b' = b$.

We refer to this type of adversary as an IND-ID-CCA adversary. If $\mathcal{A}$ cannot ask decryption queries, we call it an IND-ID-CPA adversary. The advantage of an IND-ID-CCA adversary $\mathcal{A}$ against an IBE scheme $\mathcal{E}_{ID}$ is the function of security parameter k defined as: $Adv_{\mathcal{E}_{ID},\mathcal{A}}(k) = |Pr[b' = b] - 1/2|$.

Definition 1. *An identity-based encryption scheme $\mathcal{E}_{ID}$ is IND-ID-CCA secure if for any IND-ID-CCA adversary, $Adv_{\mathcal{E}_{ID},\mathcal{A}}(k)$ is negligible.*

Canetti *et al.* formulated a weaker IBE notion, selective-identity adaptive chosen ciphertext attacks secure scheme (IND-sID-CCA for short), in which, an adversary has to commit the identity on which it wants to be challenged before it sees the public system parameters (the master public key) [8]. The latest work [20] provides some formal security analysis of this formulation.

3   SK-IBE

In this section, we investigate the security strength of SK-IBE. We choose the simplest variant of the Sakai and Kasahara IBE scheme [29] as the basic version of SK-IBE. This basic version was also described by Scott in [32]. To achieve security against adaptive chosen ciphertext attacks, we make use of the Fujisaki-Okamoto transformation [17] as it was used in BF-IBE [4].

3.1   Scheme

SK-IBE is specified by four polynomial time algorithms:

Setup. Given a security parameter k, the parameter generator follows the steps.

1. Generate three cyclic groups $\mathbb{G}_1$, $\mathbb{G}_2$ and $\mathbb{G}_T$ of prime order q, an isomorphism ψ from $\mathbb{G}_2$ to $\mathbb{G}_1$, and a bilinear pairing map $\hat{e} : \mathbb{G}_1 \times \mathbb{G}_2 \to \mathbb{G}_T$. Pick a random generator $P_2 \in \mathbb{G}_2^*$ and set $P_1 = \psi(P_2)$.
2. Pick a random $s \in \mathbb{Z}_q^*$ and compute $P_{pub} = sP_1$.
3. Pick four cryptographic hash functions $H_1 : \{0,1\}^* \to \mathbb{Z}_q^*$, $H_2 : \mathbb{G}_T \to \{0,1\}^n$, $H_3 : \{0,1\}^n \times \{0,1\}^n \to \mathbb{Z}_q^*$ and $H_4 : \{0,1\}^n \to \{0,1\}^n$ for some integer $n > 0$.

The message space is $\mathbb{M} = \{0,1\}^n$. The ciphertext space is $\mathbb{C} = \mathbb{G}_1^* \times \{0,1\}^n \times \{0,1\}^n$. The master public key is $M_{pk} = (q, \mathbb{G}_1, \mathbb{G}_2, \mathbb{G}_T, \psi, \hat{e}, n, P_1, P_2, P_{pub}, H_1, H_2, H_3, H_4)$, and the master secret key is $M_{sk} = s$.

Extract. Given an identifer string $ID_A \in \{0,1\}^*$ of entity A, M_{pk} and M_{sk}, the algorithm returns $d_A = \frac{1}{s+H_1(ID_A)} P_2$.

Remark 1. The result of the Extract algorithm is a short signature d_A on the message ID_A signed under the private signing key s. As proved in Theorem 3 of [41], this signature scheme is existentially unforgeable under chosen-message attack [21] in the random oracle model [6], provided that the k-sCAA1 assumption is sound in $\mathbb{G}_2$.

Encrypt. Given a plaintext $m \in \mathbb{M}$, ID_A and M_{pk}, the following steps are performed.

1. Pick a random $\sigma \in \{0,1\}^n$ and compute $r = H_3(\sigma, m)$.
2. Compute $Q_A = H_1(ID_A)P_1 + P_{pub}$, $g^r = \hat{e}(P_1, P_2)^r$.
3. Set the ciphertext to $C = (rQ_A, \sigma \oplus H_2(g^r), m \oplus H_4(\sigma))$.

Remark 2. In the Encrypt algorithm, the pairing $g = \hat{e}(P_1, P_2)$ is fixed and can be pre-computed. It can further be treated as a system public parameter. Therefore, no pairing computation is required in Encrypt.

Decrypt. Given a ciphertext $C = (U, V, W) \in \mathbb{C}$, ID_A, d_A and M_{pk}, follow the steps:

1. Compute $g' = \hat{e}(U, d_A)$ and $\sigma' = V \oplus H_2(g')$
2. Compute $m' = W \oplus H_4(\sigma')$ and $r' = H_3(\sigma', m')$.
3. If $U \neq r'(H_1(ID_A)P_1 + P_{pub})$, output $\bot$, else return m' as the plaintext.

3.2   Security of SK-IBE

Now we evaluate the security of SK-IBE. We prove that the security of SK-IBE can reduce to the hardness of the k-BDHI problem. The reduction is similar to the proof of BF-IBE [4]. However, we will take into account the error in Lemma 4.6 of [4] corrected by Galindo [19].

Theorem 8. *SK-IBE is secure against IND-ID-CCA adversaries provided that $H_i(1 \leq i \leq 4)$ are random oracles and the k-BDHI assumption is sound. Specifically, suppose there exists an IND-ID-CCA adversary $\mathcal{A}$ against SK-IBE that has advantage $\epsilon(k)$ and running time $t(k)$. Suppose also that during the attack $\mathcal{A}$ makes at most q_D decryption queries and at most q_i queries on H_i for $1 \leq i \leq 4$ respectively (note that H_i can be queried directly by $\mathcal{A}$ or indirectly by an extraction query, a decryption query or the challenge operation). Then there exists an algorithm $\mathcal{B}$ to solve the q_1-BDHI problem with advantage $Adv_\mathcal{B}(k)$ and running time $t_\mathcal{B}(k)$ where*

$$Adv_\mathcal{B}(k) \geq \frac{1}{q_2(q_3+q_4)}[(\frac{\epsilon(k)}{q_1}+1)(1-\frac{2}{q})^{q_D}-1]$$
$$t_\mathcal{B}(k) \leq t(k) + O((q_3+q_4)\cdot(n+\log q) + q_D \cdot \mathcal{T}_1 + q_1^2 \cdot \mathcal{T}_2 + q_D \cdot \chi)$$

where χ is the time of computing pairing, $\mathcal{T}_i$ is the time of a multiplication operation in $\mathbb{G}_i$, and q is the order of $\mathbb{G}_1$ and n is the length of σ. We assume the computation complexity of ψ is trivial.

Proof: The theorem follows immediately by combining Lemma 1, 2 and 3. The reduction with three steps can be sketched as follow. First we prove that if there exists an IND-ID-CCA adversary, who is able to break SK-IBE by launching the adaptive chosen ciphertext attacks as defined in the security model of Section 2.2, then there exists an IND-CCA adversary to break the **BasicPub**$^\text{hy}$ scheme defined in Lemma 1 with the adaptive chosen ciphertext attacks. Second, if such IND-CCA adversary exists, then we show (in Lemma 2) that there must be an IND-CPA adversary that breaks the corresponding **BasicPub** scheme by merely launching the chosen plaintext attacks. Finally, in Lemma 3 we prove that if the **BasicPub** scheme is not secure against an IND-CPA adversary, then the corresponding k-BDHI assumption is flawed.   □

Lemma 1. *Suppose that H_1 is a random oracle and that there exists an IND-ID-CCA adversary $\mathcal{A}$ against SK-IBE with advantage $\epsilon(k)$ which makes at most q_1 distinct queries to H_1 (note that H_1 can be queried directly by $\mathcal{A}$ or indirectly by an extraction query, a decryption query or the challenge operation). Then there exists an IND-CCA adversary $\mathcal{B}$ which runs in time $O(time(\mathcal{A}) + q_D \cdot (\chi + \mathcal{T}_1))$ against the following* **BasicPub**$^\text{hy}$ *scheme with advantage at least $\epsilon(k)/q_1$ where χ is the time of computing pairing and $\mathcal{T}_1$ is the time of a multiplication operation in $\mathbb{G}_1$.*

BasicPub$^\text{hy}$ *is specified by three algorithms:* **keygen**, **encrypt** *and* **decrypt**.
keygen: *Given a security parameter k, the parameter generator follows the steps.*

1. *Identical with step 1 in Setup algorithm of SK-IBE.*
2. *Pick a random $s \in \mathbb{Z}_q^*$ and compute $P_{pub} = sP_1$. Randomly choose different elements $h_i \in \mathbb{Z}_q^*$ and compute $\frac{1}{h_i+s}P_2$ for $0 \leq i < q_1$.*
3. *Pick three cryptographic hash functions: $H_2 : \mathbb{G}_T \to \{0,1\}^n$, $H_3 : \{0,1\}^n \times \{0,1\}^n \to \mathbb{Z}_q^*$ and $H_4 : \{0,1\}^n \to \{0,1\}^n$ for some integer $n > 0$.*

The message space is $\mathbb{M} = \{0,1\}^n$. The ciphertext space is $\mathbb{C} = \mathbb{G}_1^* \times \{0,1\}^n \times \{0,1\}^n$. The public key is $K_{pub} = (q, \mathbb{G}_1, \mathbb{G}_2, \mathbb{G}_T, \psi, \hat{e}, n, P_1, P_2, P_{pub}, h_0, (h_1, \frac{1}{h_1+s}P_2), \ldots, (h_i, \frac{1}{h_i+s}P_2), \ldots, (h_{q_1-1}, \frac{1}{h_{q_1-1}+s}P_2), H_2, H_3, H_4)$ and the private key is $d_A = \frac{1}{h_0+s}P_2$. Note that $\hat{e}(h_0P_1 + P_{pub}, d_A) = \hat{e}(P_1, P_2)$.
encrypt: *Given a plaintext $m \in \mathbb{M}$ and the public key K_{pub},*

1. *Pick a random $\sigma \in \{0,1\}^n$ and compute $r = H_3(\sigma, m)$, and $g^r = \hat{e}(P_1, P_2)^r$.*
2. *Set the ciphertext to $C = (r(h_0P_1 + P_{pub}), \sigma \oplus H_2(g^r), m \oplus H_4(\sigma))$.*

decrypt: *Given a ciphertext $C = (U, V, W)$, K_{pub}, and the private key d_A, follow the steps.*

1. *Compute $g' = \hat{e}(U, d_A)$ and $\sigma' = V \oplus H_2(g')$,*
2. *Compute $m' = W \oplus H_4(\sigma')$ and $r' = H_3(\sigma', m')$,*
3. *If $U \neq r'(h_0P_1 + P_{pub})$, reject the ciphertext, else return m' as the plaintext.*

Proof: We construct an IND-CCA adversary $\mathcal{B}$ that uses $\mathcal{A}$ to gain advantage against **BasicPub^{hy}**. The game between a challenger $\mathcal{C}$ and the adversary $\mathcal{B}$ starts with the challenger first generating a random public key K_{pub} by running algorithm keygen of **BasicPub^{hy}** ($\log q_1$ is part of the security parameter of **BasicPub^{hy}**). The result is $K_{pub} = (q, \mathbb{G}_1, \mathbb{G}_2, \mathbb{G}_T, \psi, \hat{e}, n, P_1, P_2, P_{pub}, h_0, (h_1, \frac{1}{h_1+s}P_2), \ldots, (h_i, \frac{1}{h_i+s}P_2), \ldots, (h_{q_1-1}, \frac{1}{h_{q_1-1}+s}P_2), H_2, H_3, H_4)$, where $P_{pub} = sP_1$ with $s \in \mathbb{Z}_q^*$, and the private key $d_A = \frac{1}{h_0+s}P_2$. The challenger passes K_{pub} to adversary $\mathcal{B}$. Adversary $\mathcal{B}$ mounts an IND-CCA attack on the **BasicPub^{hy}** scheme with the public key K_{pub} using the help of $\mathcal{A}$ as follows.

$\mathcal{B}$ chooses an index I with $1 \leq I \leq q_1$ and simulates the algorithm Setup of SK-IBE for $\mathcal{A}$ by supplying $\mathcal{A}$ with the SK-IBE master public key $M_{pk} = (q, \mathbb{G}_1, \mathbb{G}_2, \mathbb{G}_T, \psi, \hat{e}, n, P_1, P_2, P_{pub}, H_1, H_2, H_3, H_4)$ where H_1 is a random oracle controlled by $\mathcal{B}$. The master secret key M_{sk} for this cryptosystem is s, although $\mathcal{B}$ does not know this value. Adversary $\mathcal{A}$ can make queries on H_1 at any time. These queries are handled by the following algorithm H_1-**query**.
H_1-**query** (ID_i): $\mathcal{B}$ maintains a list of tuples (ID_i, h_i, d_i) indexed by ID_i as explained below. We refer to this list as H_1^{list}. The list is initially empty. When $\mathcal{A}$ queries the oracle H_1 at a point ID_i, $\mathcal{B}$ responds as follows:

1. If ID_i already appears on the H_1^{list} in a tuple (ID_i, h_i, d_i), then $\mathcal{B}$ responds with $H_1(ID_i) = h_i$.
2. Otherwise, if the query is on the I-th distinct ID and $\bot$ is not used as d_i (this could be inserted by the challenge operation specified later) by any existing tuple, then $\mathcal{B}$ stores $(ID_I, h_0, \bot)$ into the tuple list and responds with $H_1(ID_I) = h_0$.

3. Otherwise, $\mathcal{B}$ selects a random integer $h_i (i > 0)$ from K_{pub} which has not been chosen by $\mathcal{B}$ and stores $(ID_i, h_i, \frac{1}{h_i+s}P_2)$ into the tuple list. $\mathcal{B}$ responds with $H_1(ID_i) = h_i$.

Phase 1: $\mathcal{A}$ launches Phase 1 of its attack, by making a series of requests, each of which is either an extraction or a decryption query. $\mathcal{B}$ replies to these requests as follows.

Extraction query (ID_i): $\mathcal{B}$ first looks through list H_1^{list}. If ID_i is not on the list, then $\mathcal{B}$ queries $H_1(ID_i)$. $\mathcal{B}$ then checks the value d_i: if $d_i \neq \perp$, $\mathcal{B}$ responds with d_i; otherwise, $\mathcal{B}$ aborts the game (**Event 1**).

Decryption query (ID_i, C_i): $\mathcal{B}$ first looks through list H_1^{list}. If ID_i is not on the list, then $\mathcal{B}$ queries $H_1(ID_i)$. If $d_i = \perp$, then $\mathcal{B}$ sends the decryption query $C_i = (U, V, W)$ to $\mathcal{C}$ and simply relays the plaintext got from $\mathcal{C}$ to $\mathcal{A}$ directly. Otherwise, $\mathcal{B}$ decrypts the ciphertext by first computing $g' = \hat{e}(U, d_i)$, then querying $\zeta = H_2(g')$ (H_2 is controlled by $\mathcal{C}$), and computing $\sigma' = V \oplus \zeta, m' = W \oplus H_4(\sigma')$ and $r' = H_3(\sigma', m')$. Finally $\mathcal{B}$ checks the validity of C_i as step 3 of algorithm decrypt and returns m', if C_i is valid, otherwise the failure symbol $\perp$.

Challenge: At some point, $\mathcal{A}$ decides to end Phase 1 and picks ID_{ch} and two messages (m_0, m_1) of equal length on which it wants to be challenged. Based on the queries on H_1 so far, $\mathcal{B}$ responds differently.

1. If the I-th query on H_1 has been issued,
   - if $ID_I = ID_{ch}$ (and so $d_{ch} = \perp$), $\mathcal{B}$ continues;
   - otherwise, $\mathcal{B}$ aborts the game (**Event 2**).
2. Otherwise,
   - if the tuple corresponding to ID_{ch} is on list H_1^{list} (and so $d_{ch} \neq \perp$), then $\mathcal{B}$ aborts the game (**Event 3**);
   - otherwise, $\mathcal{B}$ inserts the tuple $(ID_{ch}, h_0, \perp)$ into the list and continues (this operation is treated as an H_1 query in the simulation).

Note that after this point, it must have $H_1(ID_{ch}) = h_0$ and $d_{ch} = \perp$. $\mathcal{B}$ passes $\mathcal{C}$ the pair (m_0, m_1) as the messages on which it wishes to be challenged. $\mathcal{C}$ randomly chooses $b \in \{0, 1\}$, encrypts m_b and responds with the ciphertext $C_{ch} = (U', V', W')$. Then $\mathcal{B}$ forwards C_{ch} to $\mathcal{A}$.

Phase 2: $\mathcal{B}$ continues to respond to requests in the same way as it did in Phase 1. Note that following the rules, the adversary will not issue the extraction query on ID_{ch} (for which $d_{ch} = \perp$) and the decryption query on (ID_{ch}, C_{ch}). And so, $\mathcal{B}$ always can answer other queries without aborting the game.

Guess: $\mathcal{A}$ makes a guess b' for b. $\mathcal{B}$ outputs b' as its own guess.

Claim: If the algorithm $\mathcal{B}$ does not abort during the simulation then algorithm $\mathcal{A}$'s view is identical to its view in the real attack.

Proof: $\mathcal{B}$'s responses to H_1 queries are uniformly and independently distributed in $\mathbb{Z}_q^*$ as in the real attack because of the behavior of algorithm keygen of the **BasicPub**^{hy} scheme. All responses to $\mathcal{A}$'s requests are valid, if $\mathcal{B}$ does not abort. Furthermore, the challenge ciphertext $C_{ch} = (U', V', W')$ is a valid encryption in SK-IBE for m_b where $b \in \{0, 1\}$ is random.

The remaining problem is to calculate the probability that $\mathcal{B}$ does not abort during simulation. Algorithm $\mathcal{B}$ could abort when one of the following events happens: (1) **Event 1**, denoted as $\mathcal{H}_1$: $\mathcal{A}$ queried a private key which is represented by $\perp$ at some point. Recall that only one private key is represented by $\perp$ in the whole simulation which could be inserted in an H_1 query (as the private key of ID_I) in Phase 1 or in the challenge phase (as the private key of ID_{ch}). Because of the rules of the game, the adversary will not query the private key of ID_{ch}. Hence, this event only happens when the adversary extracted the private key of $ID_I \neq ID_{ch}$, meanwhile $d_I = \perp$, i.e., $ID_I \neq ID_{ch}$ and $H_1(ID_I)$ was queried in Phase 1; (2) **Event 2**, denoted as $\mathcal{H}_2$: the adversary wants to be challenged on an identity $ID_{ch} \neq ID_I$ and $H_1(ID_I)$ was queried in Phase 1; (3) **Event 3**, denoted as $\mathcal{H}_3$: the adversary wants to be challenged on an identity $ID_{ch} \neq ID_I$ and $H_1(ID_I)$ was queried in Phase 2.

Notice that all the three events imply **Event 4**, denoted by $\mathcal{H}_4$, that the adversary did not choose ID_I as the challenge identity. Hence we have

$$Pr[\mathcal{B} \text{ does not abort}] = Pr[\neg\mathcal{H}_1 \wedge \neg\mathcal{H}_2 \wedge \neg\mathcal{H}_3] \geq Pr[\neg\mathcal{H}_4] \geq 1/q_1.$$

So, the lemma follows. $\qquad\square$

Remark 3. If an adversary only engages in the selective-identity adaptive chosen ciphertext attack game, the reduction could be tighter ($\mathcal{B}$ has the advantage $\epsilon(k)$ as $\mathcal{A}$), because $\mathcal{B}$ now knows exactly which identity should be hashed to h_0, so the game will never abort. Note that, in such game, $\mathcal{B}$ can pass the SK-IBE system parameters (the master public key) to $\mathcal{A}$ first, then $\mathcal{A}$ commits an identity ID_{ch} before issuing any oracle query. Hence the reduction could still be tightened to a stronger formulation than the one in [8] (see the separation in [20]).

Lemma 2. *Let H_3, H_4 be random oracles. Let $\mathcal{A}$ be an IND-CCA adversary against* **BasicPub**^{hy} *defined in Lemma 1 with advantage $\epsilon(k)$. Suppose $\mathcal{A}$ has running time $t(k)$, makes at most q_D decryption queries, and makes q_3 and q_4 queries to H_3 and H_4 respectively. Then there exists an IND-CPA adversary $\mathcal{B}$ against the following* **BasicPub** *scheme, which is specified by three algorithms:* **keygen, encrypt** *and* **decrypt.**
keygen: *Given a security parameter k, the parameter generator follows the steps.*

1. *Identical with step 1 in algorithm keygen of* **BasicPub**^{hy}.
2. *Identical with step 2 in algorithm keygen of* **BasicPub**^{hy}.
3. *Pick a cryptographic hash function $H_2 : \mathbb{G}_T \rightarrow \{0,1\}^n$ for some integer $n > 0$.*

The message space is $\mathbb{M} = \{0,1\}^n$. The ciphertext space is $\mathbb{C} = \mathbb{G}_1^* \times \{0,1\}^n$. The public key is $K_{pub} = (q, \mathbb{G}_1, \mathbb{G}_2, \mathbb{G}_T, \psi, \hat{e}, n, P_1, P_2, P_{pub}, h_0, (h_1, \frac{1}{h_1+s}P_2),$ $\ldots, (h_i, \frac{1}{h_i+s}P_2), \ldots, (h_{q_1-1}, \frac{1}{h_{q_1-1}+s}P_2), H_2)$ and the private key is $d_A = \frac{1}{h_0+s}P_2$. Again it has $\hat{e}(h_0 P_1 + P_{pub}, d_A) = \hat{e}(P_1, P_2)$.

encrypt: *Given a plaintext $m \in \mathbb{M}$ and the public key K_{pub}, choose a random $r \in \mathbb{Z}_q^*$ and compute ciphertext $C = (r(h_0 P_1 + P_{pub}), m \oplus H_2(g^r))$ where $g^r = \hat{e}(P_1, P_2)^r$.*

decrypt: *Given a ciphertext $C = (U, V)$, K_{pub}, and the private key d_A, compute $g' = \hat{e}(U, d_A)$ and plaintext $m = V \oplus H_2(g')$.*

with advantage $\epsilon_1(k)$ and running time $t_1(k)$ where

$$\epsilon_1(k) \geq \frac{1}{2(q_3+q_4)}[(\epsilon(k)+1)(1-\frac{2}{q})^{q_D} - 1]$$
$$t_1(k) \leq t(k) + O((q_3+q_4)\cdot(n+\log q)).$$

Proof: This lemma follows from the result of the Fujisaki-Okamoto transformation [17] and BF-IBE has a similar result (Theorem 4.5 [4]). We note that it is assumed that n and $\log q$ are of similar size in [4].   □

Lemma 3. *Let H_2 be a random oracle. Suppose there exists an IND-CPA adversary $\mathcal{A}$ against the **BasicPub** defined in Lemma 2 which has advantage $\epsilon(k)$ and queries H_2 at most q_2 times. Then there exists an algorithm $\mathcal{B}$ to solve the q_1-BDHI problem with advantage at least $2\epsilon(k)/q_2$ and running time $O(time(\mathcal{A}) + q_1^2 \cdot T_2)$ where T_2 is the time of a multiplication operation in $\mathbb{G}_2$.*

Proof: Algorithm $\mathcal{B}$ is given as input a random q_1-BDHI instance $(q, \mathbb{G}_1, \mathbb{G}_2, \mathbb{G}_T, \psi, \hat{e}, P_1, P_2, xP_2, x^2 P_2, \ldots x^{q_1} P_2)$ where x is a random element from $\mathbb{Z}_q^*$. Algorithm $\mathcal{B}$ finds $\hat{e}(P_1, P_2)^{1/x}$ by interacting with $\mathcal{A}$ as follows:

Algorithm $\mathcal{B}$ first simulates algorithm keygen of **BasicPub**, which was defined in Lemma 2, to create the public key as below. A similar approach is used in [1,2].

1. Randomly choose different $h_0, \ldots, h_{q_1-1} \in \mathbb{Z}_q^*$ and let $f(z)$ be the polynomial $f(z) = \prod_{i=1}^{q_1-1}(z+h_i)$. Reformulate f to get $f(z) = \sum_{i=0}^{q_1-1} c_i z^i$. The constant term c_0 is non-zero because $h_i \neq 0$ and c_i are computable from h_i.
2. Compute $Q_2 = \sum_{i=0}^{q_1-1} c_i x^i P_2 = f(x)P_2$ and $xQ_2 = \sum_{i=0}^{q_1-1} c_i x^{i+1} P_2 = xf(x)P_2$.
3. Check that $Q_2 \in \mathbb{G}_2^*$. If $Q_2 = 1_{\mathbb{G}_2}$, then there must exist an $h_i = -x$ which can be easily identified, and so, $\mathcal{B}$ solves the q_1-BDHI problem directly. Otherwise, $\mathcal{B}$ computes $Q_1 = \psi(Q_2)$ and continues.
4. Compute $f_i(z) = f(z)/(z+h_i) = \sum_{j=0}^{q_1-2} d_j z^j$ and $\frac{1}{x+h_i}Q_2 = f_i(x)P_2 = \sum_{j=0}^{q_1-2} d_j x^j P_2$ for $1 \leq i < q_1$.
5. Set $T' = \sum_{i=1}^{q_1-1} c_i x^{i-1} P_2$ and compute $T_0 = \hat{e}(\psi(T'), Q_2 + c_0 P_2)$.
6. Now, $\mathcal{B}$ passes $\mathcal{A}$ the public key $K_{pub} = (q, \mathbb{G}_1, \mathbb{G}_2, \mathbb{G}_T, \psi, \hat{e}, n, Q_1, Q_2, xQ_1 - h_0 Q_1, h_0, (h_1 + h_0, \frac{1}{h_1+x}Q_2), \ldots, (h_i + h_0, \frac{1}{h_i+x}Q_2), \ldots, (h_{q_1-1} + h_0, \frac{1}{h_{q_1-1}+x}Q_2), H_2)$ (i.e., setting $P_{pub} = xQ_1 - h_0 Q_1$), and the private key is $d_A = \frac{1}{x}Q_2$ which $\mathcal{B}$ does not know. H_2 is a random oracle controlled by $\mathcal{B}$. Note that $\hat{e}((h_i + h_0)Q_1 + P_{pub}, \frac{1}{h_i+x}Q_2) = \hat{e}(Q_1, Q_2)$ for $i = 1, \ldots, q_1 - 1$ and $\hat{e}(h_0 Q_1 + P_{pub}, d_A) = \hat{e}(Q_1, Q_2)$. Hence K_{pub} is a valid public key of **BasicPub**.

Now $\mathcal{B}$ starts to respond to queries as follows.

H_2-**query** (X_i): At any time algorithm $\mathcal{A}$ can issue queries to the random oracle H_2. To respond to these queries $\mathcal{B}$ maintains a list of tuples called H_2^{list}. Each entry in the list is a tuple of the form (X_i, ζ_i) indexed by X_i. To respond to a query on X_i, $\mathcal{B}$ does the following operations:

1. If on the list there is a tuple indexed by X_i, then $\mathcal{B}$ responds with ζ_i.
2. Otherwise, $\mathcal{B}$ randomly chooses a string $\zeta_i \in \{0,1\}^n$ and inserts a new tuple (X_i, ζ_i) to the list. It responds to $\mathcal{A}$ with ζ_i.

Challenge: Algorithm $\mathcal{A}$ outputs two messages (m_0, m_1) of equal length on which it wants to be challenged. $\mathcal{B}$ chooses a random string $R \in \{0,1\}^n$ and a random element $r \in \mathbb{Z}_q^*$, and defines $C_{ch} = (U, V) = (rQ_1, R)$. $\mathcal{B}$ gives C_{ch} as the challenge to $\mathcal{A}$. Observe that the decryption of C_{ch} is

$$V \oplus H_2(\hat{e}(U, d_A)) = R \oplus H_2(\hat{e}(rQ_1, \frac{1}{x}Q_2)).$$

Guess: After algorithm $\mathcal{A}$ outputs its guess, $\mathcal{B}$ picks a random tuple (X_i, ζ_i) from H_2^{list}. $\mathcal{B}$ first computes $T = X_i^{1/r}$, and then returns $(T/T_0)^{1/c_0^2}$. Note that $\hat{e}(P_1, P_2)^{1/x} = (T/T_0)^{1/c_0^2}$ if $T = \hat{e}(Q_1, Q_2)^{1/x}$.

Let $\mathcal{H}$ be the event that algorithm $\mathcal{A}$ issues a query for $H_2(\hat{e}(rQ_1, \frac{1}{x}Q_2))$ at some point during the simulation above. Using the same methods in [4], we can prove the following two claims:

Claim 1: $\Pr[\mathcal{H}]$ in the simulation above is equal to $\Pr[\mathcal{H}]$ in the real attack.

Claim 2: In the real attack we have $Pr[\mathcal{H}] \geq 2\epsilon(k)$.

Following from the above two claims, we have that $\mathcal{B}$ produces the correct answer with probability at least $2\epsilon(k)/q_2$.   □

Remark 4. In the proof, $\mathcal{B}$'s simulation of algorithm keygen of **BasicPub** is similar to the preparation step in Theorem 5.1 [1] (both follow the method in [28]. Note that in [1] ψ is an identity map, so $Q = Q_1 = Q_2$). However, the calculation of T_0 in [1] is incorrect, and should be computed as $T_0 = \prod_{i=0}^{q-1} \prod_{j=0}^{q-2} \hat{e}(g^{(\alpha^i)}, g^{(\alpha^j)})^{c_i c_{j+1}} \cdot \prod_{j=0}^{q-2} \hat{e}(g, g^{(\alpha^j)})^{c_0 c_{j+1}}$.

This completes the proof of Theorem 8.

4   Possible Improvements of SK-IBE

SK-IBE can be improved both on computation performance and security reduction. The only two known bilinear pairing instances so far are the Weil pairing and Tate pairing on elliptic curves (and hyperelliptic curves) [35]. When implementing these pairings, some special structures of these pairings can be exploited to improve the performance. As noticed by Scott and Barreto [33], the Tate pairing can be compressed when the curve has the characteristic 3 or greater than 3. The compressing technique not only can reduce the size of pairing, but also can speed up the computation of pairing and the exponentiation in $\mathbb{G}_T$. Pointed by

Galindo [19], an improved Fujisaki-Okamoto's transformation [18] has a tighter security reduction. Using the trick played in [25], the reduction can be further tightened by including the point rQ_A in H_2 (this also removes the potential ambiguity introduced by the compressed pairing). So, combined with these two improvements, a faster scheme (SK-IBE2) with better security reduction can be specified as follow.

Setup. Identical with SK-IBE, except that H_4 is not required and $H_2 : \mathbb{G}_1 \times \mathbb{F} \to \{0,1\}^{2n}$, where $\mathbb{F}$ depends on the used compressed pairing (see [33] for details).

Extract. Identical with SK-IBE.

Encrypt. Given a plaintext $m \in \mathbb{M}(\{0,1\}^n)$, the identity ID_A of entity A and the master public key M_{pk}, the following steps are performed.

1. Pick a random $\sigma \in \{0,1\}^n$ and compute $r = H_3(\sigma, m)$.
2. Compute $Q_A = H_1(ID_A)P_1 + P_{pub}$, $\varphi(g^r) = \varphi(\hat{e}(P_1, P_2)^r)$, where φ is the pairing compressing algorithm as specified in [33]. Note that φ and $\hat{e}$ can be computed by a single algorithm, so to improve the computation performance [33].
3. Set the ciphertext to $C = (rQ_A, (m\|\sigma) \oplus H_2(rQ_A, \varphi(g^r)))$.

Decrypt. Given a ciphertext $(U, V) \in C$, the identity ID_A, the private key d_A and M_{pk}, follow the steps:

1. Compute $\varphi(g') = \varphi(\hat{e}(U, d_A))$ and $m'\|\sigma' = V \oplus H_2(U, \varphi(g'))$.
2. Compute $r' = H_3(\sigma', m')$. If $U \neq r'(H_1(ID_A)P_1 + P_{pub})$, output $\perp$, else return m' as the plaintext.

Using the similar approach employed in the proof of Theorem 8 and the result of Theorem 5.4 in [18], we can reduce the security of SK-IBE2 to the k-BDHI assumption. We leave the details to the readers.

5   Comparison Between SK-IBE and BF-IBE

From the reduction described in Section 3.2, we have proved that SK-IBE is a secure IBE scheme based on the k-BDHI problem. The complexity analysis of k-DHI, k-sDH and k-BDHI in [1,2,41] has built confidence on these assumptions.

The security of BF-IBE is based on the BDH problem [4]. As shown in Theorem 6, BDH and 1-BDHI are polynomial time equivalent. It is obvious that the k-BDHI problem (when $k > 1$) is easier that the 1-BDHI problem, and therefore, is easier than the BDH problem as well. This certainly shows the disadvantage of current reduction for SK-IBE as compared with one for BF-IBE [4,19]. We leave it an open problem to find a tight reduction for SK-IBE based on a harder problem than k-BDHI.

However, the advantage of SK-IBE is that it has better performance than BF-IBE, particularly in encryption. We show a comparison of their performances in Table 1.

Table 1. Performance comparison between SK-IBE and BF-IBE

Scheme	pairings		multiplications		exponentiations		hashes	
	Encrypt	Decrypt	Encrypt	Decrypt	Encrypt	Decrypt	Encrypt	Decrypt
SK-IBE	0	1	2^{*1}	1	1	0	4	3
BF-IBE	1	1	1	1	1	0	4^{*2}	3

*1 An extra multiplication required than BF-IBE is used to map an identifier to an element in $\mathbb{G}_1$.

*2 BF-IBE requires the *maptopoint* operation to map an identifier to an element in $\mathbb{G}_1$ (or $\mathbb{G}_2$) which is slower than the hash function used in SK-IBE which maps an identifier to an element in $\mathbb{Z}_q^*$.

If taking a closer look between SK-IBE and BF-IBE, SK-IBE is faster than BF-IBE in two aspects. First, in the Encrypt algorithm of SK-IBE, no pairing computation is required because $\hat{e}(P_1, P_2)$ can be pre-computed. Second, in operation of mapping an identity to an element in $\mathbb{G}_1$ or $\mathbb{G}_2$, the *maptopoint* algorithm used by BF-IBE is not required. Instead of that, SK-IBE makes use of an ordinary hash-function.

6   Conclusion

In this paper, an identity-based encryption scheme, SK-IBE, is investigated. SK-IBE provides an attractive performance. We prove that SK-IBE is secure against adaptive chosen ciphertext attacks in the random oracle model based on the k-BDHI assumption.

Acknowledgements

We thank Keith Harrison, John Malone-Lee and Nigel Smart for helpful discussions and comments on this work, David Galindo for his valuable comments on the manuscript and for sharing with us the latest work on the security notions of IBE [20], Alex Dent for his detailed and valuable comments on the manuscript, and anonymous referees for useful feedback.

References

1. D. Boneh and X. Boyen. Efficient selective-ID secure identity-based encryption without random oracles. In *Proceedings of Advances in Cryptology - Eurocrypt 2004*, LNCS 3027, pp. 223–238, Springer-Verlag, 2004.

2. D. Boneh and X. Boyen. Short signatures without random oracles. In *Proceedings of Advances in Cryptology - Eurocrypt 2004*, LNCS 3027, pp. 56–73, Springer-Verlag, 2004.

3. D. Boneh and X. Boyen. Secure identity-based encryption without random oracles. In *Proceedings of Advances in Cryptology - Crypto 2004*, Springer-Verlag, 2004.

4. D. Boneh and M. Franklin. Identity based encryption from the Weil pairing. In *Proceedings of Advances in Cryptology - Crypto 2001*, LNCS 2139, pp.213–229, Springer-Verlag, 2001.

5. D. Boneh, B. Lynn and H. Shacham. Short signatures from the Weil pairing. *Advances in Cryptology – Asiacrypt 2001*, Springer-Verlag LNCS 2248, 514–532, 2001.

6. M. Bellare and P. Rogaway. Random oracles are practical: a paradigm for designing efficient protocols. In *Proceedings of the First Annual Conference on Computer and Communications Security*, ACM, 1993.

7. X. Boyen. Multipurpose identity-based signcryption: a swiss army knife for identity-based cryptography. In *Proceedings of Advances in Cryptology - CRYPTO 2003*, LNCS 2729, pp. 382–398, Springer-Verlag, 2003.

8. R. Canetti, S. Halevi and J. Katz. A forward-secure public-key encryption scheme. In *Proceedings of Advances in Cryptology - Eurocrypt 2003*, LNCS 2656, pp. 255-271, Springer-Verlag, 2003.

9. R. Canetti, S. Halevi and J. Katz. Chosen-ciphertext security from identity-based encryption. In *Proceedings of Advances in Cryptology - Eurocrypt 2004*. Springer-Verlag, 2004. See also Cryptology ePrint Archive, Report 2003/182.

10. J. C. Cha and J. H. Cheon. An identity-based signature from gap Diffie-Hellman groups. In *Proceedings of Practice and Theory in Public Key Cryptography - PKC 2003*, LNCS 2567, pp. 18–30, Springer-Verlag, 2003. See also Cryptology ePrint Archive, Report 2002/018.

11. L. Chen and Z. Cheng. Security proof of Sakai-Kasahara's identity-based encryption scheme. Cryptology ePrint Archive, Report 2005/226.

12. L. Chen and C. Kudla. Identity-based authenticated key agreement from pairings. In *Proceedings of the 16th IEEE Computer Security Foundations Workshop*, pp. 219-233, IEEE, 2003. See also Cryptology ePrint Archive, Report 2002/184.

13. L. Chen and J. Malone-Lee. Improved identity-based signcryption. In *Proceedings of Public Key Cryptography - PKC 2005*, LNCS 3386, pages 362–379, Springer-Verlag, 2005. See also Cryptology ePrint Archive, Report 2004/114.

14. Z. Cheng and L. Chen. On security proof of McCullagh-Barreto's key agreement protocol and its variants. Cryptology ePrint Archive, Report 2005/201.

15. Z. Cheng, M. Nistazakis, R. Comley and L. Vasiu. On the indistinguishability-based security model of key agreement protocols-simple cases. In *Proceedings of ACNS 2004*. Full version available on Cryptology ePrint Archive, Report 2005/129.

16. C. Cocks. An identity-based encryption scheme based on quadratic residues. In *Proceedings of Cryptography and Coding*, LNCS 2260, pp. 360–363, Springer-Verlag, 2001.

17. E. Fujisaki and T. Okamoto. Secure integration of asymmetric and symmetric encryption schemes. In *Proceedings of Advances in Cryptology - CRYPTO '99*, LNCS 1666, pp. 535-554, Springer-Verlag, 1999.

18. E. Fujisaki and T. Okamoto. How to enhance the security of public-key encryption at minimum cost. *IEICE Trans. Fundamentals*, E83-9(1):24-32, 2000.

19. D. Galindo. Boneh-Franklin identity based encryption revisited. In *Proceedings of the 32nd International Colloquium on Automata, Languages and Programming, ICALP 2005*. Also available on Cryptology ePrint Archive, Report 2005/117.

20. D. Galindo and I. Hasuo. Security Notions for Identity Based Encryption. Manuscript, 2005.

21. S. Goldwasser, S. Micali and R. Rivest. A digital signature scheme secure against adaptive chosen-message attacks. *SIAM Journal on Computing*, 17(2):281–308, 1988.

22. F. Hess. Efficient identity based signature schemes based on pairings. In *Proceedings of Selected Areas in Cryptography – SAC 2002*, LNCS 2595, pp. 310–324, Springer-Verlag, 2002.
23. ISO/IEC 11770-3:1999. Information technology - Security techniques - Key management - Part 3: Mechanisms using asymmetric techniques.
24. ISO/IEC 14888-2:1998. Information technology - Security techniques - Digital signatures with appendix - Part 2: Identity-based mechanisms.
25. ISO/IEC 2nd FCD 18033-2:2004-12-06. Information technology - Security techniques - Encryption algorithms - Part 2: Asymmetric ciphers.
26. N. McCullagh and P. S. L. M. Barreto. Efficient and forward-secure identity-based signcryption. Available on Cryptology ePrint Archive, Report 2004/117.
27. N. McCullagh and P. S. L. M. Barreto. A new two-party identity-based authenticated key agreement. In *Proceedings of CT-RSA 2005*. See also Cryptology ePrint Archive, Report 2004/122.
28. S. Mitsunari, R. Sakai and M. Kasahara. A new traitor tracing. *IEICE Trans. Fundamentals*, E85-A(2):481–484, 2002.
29. R. Sakai and M. Kasahara. ID based cryptosystems with pairing on elliptic curve. Cryptology ePrint Archive, Report 2003/054.
30. R. Sakai, K. Ohgishi and M. Kasahara. Cryptosystems based on pairing. *The 2000 Symposium on Cryptography and Information Security*, Okinawa, Japan, January 2000.
31. R. Sakai, K. Ohgishi and M. Kasahara. Cryptosystems based on pairing over elliptic curve (in Japanese). *The 2001 Symposium on Cryptography and Information Security*, Oiso, Japan, January 2001.
32. M.Scott. Computing the Tate pairing. In *Proceedings of CT-RSA 2005*, LNCS 3376, pp. 293–304, Springer-Verlag, 2005.
33. M. Scott and P. S. L. M. Barreto. Compressed pairings. In *Proceedings of Advances in Cryptology - Crypto 2004*, LNCS 3152, 2004, Springer-Verlag, 2004. See also Cryptology ePrint Archive, Report 2004/032.
34. A. Shamir. Identity-based cryptosystems and signature schemes. In *Proceedings of Advances in Cryptology - Crypto '84*, LNCS 196, pp.47–53, Springer-Verlag, 1985.
35. J. Silverman. *The arithmetic of elliptic curve*. Springer-Verlag, 1986.
36. N. P. Smart. An identity based authenticated key agreement protocol based on the Weil pairing. *Electronics Letters*, 38(13):630–632, 2002. See also Cryptology ePrint Archive, Report 2001/111.
37. N. Smart and F. Vercauteren. On computable isomorphisms in efficient pairing based systems. Cryptology ePrint Archive, Report 2005/116.
38. V.D. Tô, R. Safavi-Naini and F. Zhang. New traitor tracing schemes using bilinear map. In *Proceedings of 2003 DRM Workshop*, 2003.
39. B. R. Waters. Efficient identity-based encryption without random oracles. In *Proceedings of Advances in Cryptology - Eurocrypt 2005*, Springer-Verlag, 2005. See also Cryptology ePrint Archive, Report 2004/180.
40. V. Wei. Tight Reductions among Strong Diffie-Hellman Assumptions. Cryptology ePrint Archive, Report 2005/057.
41. F. Zhang, R. Safavi-Naini and W. Susilo. An efficient signature scheme from bilinear pairings and its applications. In *Proceedings of International Workshop on Practice and Theory in Public Key Cryptography - PKC 2004*, 2004.

Author Index

Lecture Notes in Computer Science

For information about Vols. 1–3706

please contact your bookseller or Springer

Vol. 3758: Y. Pan, D.-x. Chen, M. Guo, J. Cao, J.J. Dongarra (Eds.), Parallel and Distributed Processing and Applications. XXIII, 1162 pages. 2005.

Vol. 3757: A. Rangarajan, B. Vemuri, A.L. Yuille (Eds.), Energy Minimization Methods in Computer Vision and Pattern Recognition. XII, 666 pages. 2005.

Vol. 3756: J. Cao, W. Nejdl, M. Xu (Eds.), Advanced Parallel Processing Technologies. XIV, 526 pages. 2005.

Vol. 3754: J. Dalmau Royo, G. Hasegawa (Eds.), Management of Multimedia Networks and Services. XII, 384 pages. 2005.

Vol. 3753: O.F. Olsen, L.M.J. Florack, A. Kuijper (Eds.), Deep Structure, Singularities, and Computer Vision. X, 259 pages. 2005.

Vol. 3752: N. Paragios, O. Faugeras, T. Chan, C. Schnörr (Eds.), Variational, Geometric, and Level Set Methods in Computer Vision. XI, 369 pages. 2005.

Vol. 3751: T. Magedanz, E.R. M. Madeira, P. Dini (Eds.), Operations and Management in IP-Based Networks. X, 213 pages. 2005.

Vol. 3750: J.S. Duncan, G. Gerig (Eds.), Medical Image Computing and Computer-Assisted Intervention – MICCAI 2005, Part II. XL, 1018 pages. 2005.

Vol. 3749: J.S. Duncan, G. Gerig (Eds.), Medical Image Computing and Computer-Assisted Intervention – MICCAI 2005, Part I. XXXIX, 942 pages. 2005.

Vol. 3748: A. Hartman, D. Kreische (Eds.), Model Driven Architecture – Foundations and Applications. IX, 349 pages. 2005.

Vol. 3747: C.A. Maziero, J.G. Silva, A.M.S. Andrade, F.M.d. Assis Silva (Eds.), Dependable Computing. XV, 267 pages. 2005.

Vol. 3746: P. Bozanis, E.N. Houstis (Eds.), Advances in Informatics. XIX, 879 pages. 2005.

Vol. 3745: J.L. Oliveira, V. Maojo, F. Martín-Sánchez, A.S. Pereira (Eds.), Biological and Medical Data Analysis. XII, 422 pages. 2005. (Subseries LNBI).

Vol. 3744: T. Magedanz, A. Karmouch, S. Pierre, I. Venieris (Eds.), Mobility Aware Technologies and Applications. XIV, 418 pages. 2005.

Vol. 3740: T. Srikanthan, J. Xue, C.-H. Chang (Eds.), Advances in Computer Systems Architecture. XVII, 833 pages. 2005.

Vol. 3739: W. Fan, Z.-h. Wu, J. Yang (Eds.), Advances in Web-Age Information Management. XXIV, 930 pages. 2005.

Vol. 3738: V.R. Syrotiuk, E. Chávez (Eds.), Ad-Hoc, Mobile, and Wireless Networks. XI, 360 pages. 2005.

Vol. 3735: A. Hoffmann, H. Motoda, T. Scheffer (Eds.), Discovery Science. XVI, 400 pages. 2005. (Subseries LNAI).

Vol. 3734: S. Jain, H.U. Simon, E. Tomita (Eds.), Algorithmic Learning Theory. XII, 490 pages. 2005. (Subseries LNAI).

Vol. 3733: P. Yolum, T. Güngör, F. Gürgen, C. Özturan (Eds.), Computer and Information Sciences - ISCIS 2005. XXI, 973 pages. 2005.

Vol. 3731: F. Wang (Ed.), Formal Techniques for Networked and Distributed Systems - FORTE 2005. XII, 558 pages. 2005.

Vol. 3729: Y. Gil, E. Motta, V. R. Benjamins, M.A. Musen (Eds.), The Semantic Web – ISWC 2005. XXIII, 1073 pages. 2005.

Vol. 3728: V. Paliouras, J. Vounckx, D. Verkest (Eds.), Integrated Circuit and System Design. XV, 753 pages. 2005.

Vol. 3726: L.T. Yang, O.F. Rana, B. Di Martino, J.J. Dongarra (Eds.), High Performance Computing and Communications. XXVI, 1116 pages. 2005.

Vol. 3725: D. Borrione, W. Paul (Eds.), Correct Hardware Design and Verification Methods. XII, 412 pages. 2005.

Vol. 3724: P. Fraigniaud (Ed.), Distributed Computing. XIV, 520 pages. 2005.

Vol. 3723: W. Zhao, S. Gong, X. Tang (Eds.), Analysis and Modelling of Faces and Gestures. XI, 4234 pages. 2005.

Vol. 3722: D. Van Hung, M. Wirsing (Eds.), Theoretical Aspects of Computing – ICTAC 2005. XIV, 614 pages. 2005.

Vol. 3721: A.M. Jorge, L. Torgo, P.B. Brazdil, R. Camacho, J. Gama (Eds.), Knowledge Discovery in Databases: PKDD 2005. XXIII, 719 pages. 2005. (Subseries LNAI).

Vol. 3720: J. Gama, R. Camacho, P.B. Brazdil, A.M. Jorge, L. Torgo (Eds.), Machine Learning: ECML 2005. XXIII, 769 pages. 2005. (Subseries LNAI).

Vol. 3719: M. Hobbs, A.M. Goscinski, W. Zhou (Eds.), Distributed and Parallel Computing. XI, 448 pages. 2005.

Vol. 3718: V.G. Ganzha, E.W. Mayr, E.V. Vorozhtsov (Eds.), Computer Algebra in Scientific Computing. XII, 502 pages. 2005.

Vol. 3717: B. Gramlich (Ed.), Frontiers of Combining Systems. X, 321 pages. 2005. (Subseries LNAI).

Vol. 3716: L. Delcambre, C. Kop, H.C. Mayr, J. Mylopoulos, Ó. Pastor (Eds.), Conceptual Modeling – ER 2005. XVI, 498 pages. 2005.

Vol. 3715: E. Dawson, S. Vaudenay (Eds.), Progress in Cryptology – Mycrypt 2005. XI, 329 pages. 2005.

Vol. 3714: H. Obbink, K. Pohl (Eds.), Software Product Lines. XIII, 235 pages. 2005.

Vol. 3713: L.C. Briand, C. Williams (Eds.), Model Driven Engineering Languages and Systems. XV, 722 pages. 2005.

Vol. 3712: R. Reussner, J. Mayer, J.A. Stafford, S. Overhage, S. Becker, P.J. Schroeder (Eds.), Quality of Software Architectures and Software Quality. XIII, 289 pages. 2005.

Vol. 3711: F. Kishino, Y. Kitamura, H. Kato, N. Nagata (Eds.), Entertainment Computing - ICEC 2005. XXIV, 540 pages. 2005.

Vol. 3710: M. Barni, I. Cox, T. Kalker, H.J. Kim (Eds.), Digital Watermarking. XII, 485 pages. 2005.

Vol. 3709: P. van Beek (Ed.), Principles and Practice of Constraint Programming - CP 2005. XX, 887 pages. 2005.

Vol. 3708: J. Blanc-Talon, W. Philips, D.C. Popescu, P. Scheunders (Eds.), Advanced Concepts for Intelligent Vision Systems. XXII, 725 pages. 2005.

Vol. 3707: D.A. Peled, Y.-K. Tsay (Eds.), Automated Technology for Verification and Analysis. XII, 506 pages. 2005.